Chemistry for Biologists

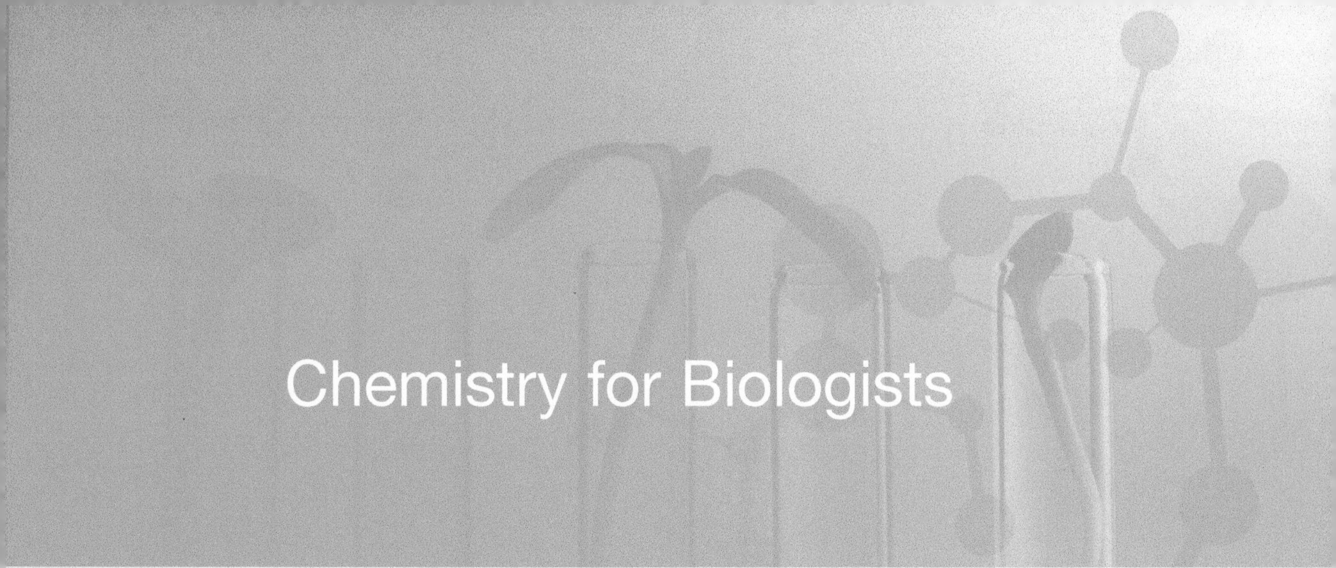

Chemistry for Biologists

David Reed

PEARSON

Harlow, England • London • New York • Boston • San Francisco • Toronto • Sydney
Auckland • Singapore • Hong Kong • Tokyo • Seoul • Taipei • New Delhi
Cape Town • São Paulo • Mexico City • Madrid • Amsterdam • Munich • Paris • Milan

Pearson Education Limited
Edinburgh Gate
Harlow CM20 2JE
Tel: +44 (0)1279 623623
Fax: +44 (0)1279 431059
Website: www.pearson.com/uk

First published 2013 (print and electronic)

ISBN: 978-1-4082-8082-9 (print)
        978-1-4082-8084-3 (PDF)
        978-0-273-77961-2 (eText)

British Library Cataloguing-in-Publication Data
A catalogue record for the print edition is available from the British Library

Library of Congress Cataloging-in-Publication Data
A catalog record for the print edition is available from the Library of Congress

ARP Impression 98

Print edition typeset in 9.25/12.5 ITC Stone Serif by 73
Printed by Ashford Colour Press Ltd

NOTE THAT ANY PAGE CROSS REFERENCES REFER TO THE PRINT EDITION

Brief Contents

Contents

8　Reaction mechanisms　236

Lecturer Resources

For password-protected online resources tailored to support the use of this textbook in teaching, please visit www.pearsoned.co.uk/reed

ON THE WEBSITE

Preface

Biology is a very diverse topic, spanning biochemistry, molecular biology, genetics, taxonomy and zoology to name but a few of the sub-disciplines. Each of these requires students to gain a fundamental grounding in the basics of biology, and an understanding of chemistry greatly enhances the student's opportunity to grasp those basic principles of biology. To allow students to develop these skills there is a need for an introductory university level text which is *focused*, yet *chemically and mathematically rigorous*, and which covers the chemistry needed for biology students, in order to allow them to pursue their further studies, and giving them the best possible opportunity to understand the basic biological processes at a fundamental level.

Every effort has been made to ensure that the content and style of the book are relevant to the needs of students of biology who are taking just a year of chemistry. Therefore during the process of introducing and developing the basic principles of chemistry in the first half of the book, the relevance of this material to biology is always emphasised. In the second half of the book the biological emphasis becomes more apparent, covering, for example,

- the structural features of biological molecules such as peptides and proteins, nucleic acids, carbohydrates and lipids;
- the background to enzyme kinetics and the impact of enzyme inhibitors;
- the relevance of principles of electrochemistry to the transportation of ions across membranes;
- the principles of organic reaction mechanisms, and their role in understanding the individual stages of metabolic pathways.

This is, however, a chemistry book and not a biology one, and biology is referred to principally in order to highlight the relevance of the chemistry.

The content of the book derives from the author's experience, over a period of about 8 years, of being the course organiser for the first year 'Biological Chemistry' courses at the University of Edinburgh. During this time the author worked with colleagues and students, within both the School of Chemistry and the School of Biology, to develop courses which fulfilled the needs of the students.

The approach of this book is to introduce the topics in an order which will allow students to develop their knowledge and skills as they progress through the book. In addition to the main text some topics have been highlighted in text boxes. These serve a range of functions, including:

- emphasising some biological relevance to specific topics;
- looking in more detail at key concepts;
- illustrating key concepts.

There is an appendix dedicated to supporting the mathematical needs of students. The approach in this has been to present the key mathematical concepts, but to illustrate how these are applied by using real chemical examples.

Acknowledgements

Firstly I would like to thank the University of Edinburgh for granting me an Honorary Fellowship after my retirement. This provided me with access to library facilities which was invaluable in supporting background research for this book.

I would like to thank a number of ex-colleagues, whose help and advice have contributed significantly to the writing of this book. Over the years many staff members contributed to the first year 'Biological Chemistry' courses, as lecturers, tutors, etc., and their contributions to, and feedback about, the courses all helped me to formulate the structure of this book. Specifically, however, there are some people whose input I particularly wish to recognise: Dr Ian Sadler, for allowing me to adapt some of his unpublished material on organic chemistry; Dr Andrew Mount for developing the bioenergetics lecture course, the content of which formed the basis of the chapter on that subject within this book; and Professor Andrew Harrison, whose lectures on thermodynamics and equilibria showed how challenging topics could be handled sympathetically. I would also like to thank those members of staff who spent time either reading through my early drafts, or discussing the content and layout with me. In particular Drs Steve Henderson, Peter Kirsop, Murray Low and Gordon McDougall have all provided helpful insights during the writing of this book.

Finally I would like to thank the members of my family, Eunice, Joanna and Caitlin, for their encouragement, support and patience during this time. I would particularly like to thank my older daughter Joanna for reading through many of the earlier chapters and providing a 'students' eye view' of the content and presentation.

David Reed, May 2012

Publisher's acknowledgements

The publisher would like to thank the following for their kind permission to reproduce their photographs:
Science Photo Library Ltd: Pasieka 436, Charles D. Winters 158.

Cover images: iStockphoto.
In some instances we have been unable to trace the owners of copyright material, and we would appreciate any information that would enable us to do so.

1 Basic concepts

By the end of this chapter students should be able to:

- know of the different states of matter;
- understand the difference between elements, compounds and mixtures;
- know about the different units of measurement, and have a sense of the scales involved;
- describe the different components of atomic structure;
- understand what is meant by the term isotope;
- understand the concept of the mole;
- know how to perform simple mole calculations;
- describe how to make solutions of different concentrations.

1.1 Introduction

Biology is a very diverse topic, and an understanding of chemistry forms a vital role in underpinning the basic principles of biology. At the most basic level biological processes are chemical reactions. To take an example, let us consider how an animal gets the energy it needs to carry out its daily routine, whether it be a rabbit hopping around a field or a human doing their day-to-day tasks. In order to do this we will briefly review the processes known as photosynthesis and cellular respiration.

The primary source of energy is the Sun, and plants use photosynthesis, a process that makes use of the energy provided by the Sun to convert carbon dioxide (CO_2) and water into glucose in plants, thereby providing a foodstock for the animal kingdom. The glucose in plants is largely stored in polymeric forms (called polysaccharides), as starch or cellulose.

When animals consume plants, their digestive systems break down the polysaccharides back into glucose, and they use this glucose as a source of energy. This is achieved by a sequence of metabolic pathways, referred to as cellular respiration. Glucose is broken down into pyruvic acid during a process called glycolysis. The pyruvic acid is then converted into acetyl coenzyme A, and this then enters the tricarboxylic acid (TCA; also called the Krebs or citric acid) cycle, which is responsible for the production of energy (in the form of ATP, adenosine triphosphate) and the 'waste' products CO_2 and water. Figure 1.1 gives a schematic representation of this whole process.

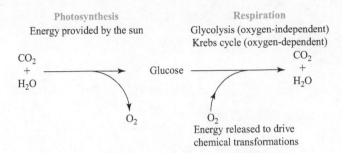

Figure 1.1 **Overview of photosynthesis and cellular respiration.**

Both glycolysis (which converts glucose to pyruvate in ten steps) and the Krebs cycle (made up of eight steps) might appear very complicated, involving in each case a variety of different chemical transformations, but, in fact, each of the steps involves relatively simple chemical processes. During the course of this book we will provide the tools necessary to analyse such processes and to understand not just what is happening, but how and why reactions happen the way they do. Later in the book (in Chapter 12) the specifics of the chemistry involved in the cellular respiration processes will be looked at in more detail.

1.1.1 What is chemistry?

The link between biology and chemistry makes a knowledge of the basics of chemistry a fundamental component of the study of biology. Chemistry itself deals with the study of the composition of substances (elements and molecules), how they react with each other, and the factors that drive these reactions. So before exploring those details of chemistry relevant to intending biologists, this chapter will be used to introduce certain key concepts:

- What are solids, liquids, gases, elements, compounds and mixtures?
- Ideas of measurement and units.
- What is an atom? What are its components?
- What are molecules?
- How much of one material will interact with a different one?

1.1.2 The states of matter

As chemistry is the study of the composition and behaviour of matter, then it is important to know the various forms in which matter exists. Matter comes in three principal forms (physical states), these being solids, liquids and gases. Table 1.1 summarises some examples of each.

For a given substance the states of matter can be interconverted by heating or cooling. For example, water exists in three forms, as indicated below:

$$\text{Ice} \underset{\text{cool}}{\overset{\text{heat}}{\rightleftharpoons}} \text{Water} \underset{\text{cool}}{\overset{\text{heat}}{\rightleftharpoons}} \text{Steam}$$

Table 1.1 **The states of matter***

Physical state	Examples	Characteristic features
Solid	Iron, graphite, concrete	Rigid, constant volume, need force to alter shape
Liquid	Water, oil	Mobile, can flow but adopt shape of container, constant volume
Gases	Nitrogen, oxygen, carbon dioxide	Mobile, do not have constant volume, will fill any container

*Under standard conditions of temperature and pressure, where the temperature is usually taken as 298 K and the pressure is 1 atmosphere ($101\,325\,N\,m^{-2}$).

1.1.3 Elements, compounds and mixtures

Each of the physical states of matter described above can be further divided into three types, namely *elements, compounds and mixtures.*

Elements

Elements are substances which cannot be split up into anything simpler by any chemical process. Elements can be solids, liquids or gases in their naturally occurring states. For example, copper, carbon or lead are examples of solid elements, mercury and bromine are examples of liquid elements and nitrogen, oxygen and helium are examples of gaseous elements.

Elements are often regarded as the building blocks of matter, as all matter is made up from elements.

Compounds

When two or more different elements combine chemically in a fixed proportion by weight, compounds are formed. For example, sodium and chlorine can combine to form sodium chloride (common household salt). Compounds usually bear no resemblance, physically or chemically, to their parent elements. Thus, sodium chloride is a pleasant white solid, soluble in water, whereas sodium is a silvery coloured metallic element which will react violently with water, and chlorine is a toxic green gas.

Mixtures

When two or more elements or compounds are physically mixed, a mixture is formed. Any quantity of each material may be used (not fixed ratios) and the components of a mixture may be separated by a physical process.

Thus, a mixture of the two elements, iron filings and sulfur powder, can be separated by using a magnet to extract the iron from the mixture. Similarly, a mixture of two liquids can be separated by making use of their different boiling points (distillation). Even the very air that we breathe is a mixture of gases, some elements, some compounds, the four main components being nitrogen, N_2, 78.084%; oxygen, O_2, 20.9476%; argon, Ar, 0.934%; and carbon dioxide, CO_2, 0.0314%.

1.2   Measurement and units

In any scientific discipline it is important to be able to gather quantitative data, to have a scale by which the data can be gauged, and to have a sense of the reliability of the data. The gathering of such data is the act of measurement, and in order to compare data it is necessary to have a sense of scale. In other words units are needed. This section will introduce some of the key measurements that are used in chemistry and biology, and the units that correspond to these measurements. Other types of data that are encountered throughout the book will be addressed as necessary.

The *Système International (SI)* is the currently accepted system of units used by scientists worldwide. In this system there are seven 'base units', from which others can be derived. Table 1.2 shows these base units, and Table 1.3 shows some of the key derived units.

1.2.1   Scales of units

In the various scientific disciplines quantities can be either:

- very large, for example in astronomy, where even the nearest star is approximately 39 700 000 000 000 000 m away from our own Sun; or
- very small, for example the lengths of C—H bonds in many organic compounds is approximately 0.000 000 000 110 m.

Numbers presented in such a manner are clearly very difficult to deal with, and there is great potential for errors to occur when attempting to reproduce such numbers. These numbers could, of course, be simplified to 3.97×10^{16} m and to 1.10×10^{-10} m respectively, but people prefer to work with simple digits wherever possible.

In the case of astronomy the issue was resolved by introducing a non-SI unit called the light year, equivalent to 9 460 730 472 581 km, which results in the distance

Table 1.2   **Base SI units**

Quantity	Unit (Abbreviation)
Mass	kilogram (kg)
Length	metre (m)
Temperature	kelvin (K)
Time	second (s)
Quantity of substance	mole (mol)*
Electrical current	ampere (A)
Luminous intensity	candela (cd)

*This will be examined in detail in Section 1.4.

Table 1.3   Some common derived SI units

Quantity	Dimensions (M = mass, L = length, T = time)	Unit (common name, abbreviation)
Area	length × length (L^2)	m^2 (square metre)
Volume	length × length × length (L^3)	m^3 (cubic metre)
Density	mass/volume ($M \times L^{-3}$)	$kg\,m^{-3}$
Velocity	length/time ($L \times T^{-1}$)	$m\,s^{-1}$
Acceleration	velocity/time ($L \times T^{-2}$)	$m\,s^{-2}$
Frequency	cycles/time (T^{-1})	s^{-1} (hertz, Hz)
Force	mass × acceleration ($M \times L \times T^{-2}$)	$kg\,m\,s^{-2}$ (newton, N)
Pressure	force/area ($M \times L^{-1} \times T^{-2}$)	$kg\,m^{-1}s^{-2}$ (pascal, Pa)
Energy (or heat, or work)	force × distance ($M \times L^2 \times T^{-2}$)	$kg\,m^2s^{-2}$/(joule, J)

to the nearest star being recorded as 4.22 light years. In an ideal case, however, it would be preferable to retain some integrity with the SI units. This has been achieved with bond lengths by using recognised prefixes to the SI units. Thus, using the C—H example, 1.10×10^{-10} m can be rewritten as either 110×10^{-12} m or as 0.110×10^{-9} m.

Table 1.4 identifies some accepted prefixes making it possible to reproduce these lengths as either 110 pm or as 0.110 nm. In this text in discussing bond lengths the picometre, pm, will be used. It is worth mentioning, however, that there is also a historic unit of length, called the angstrom (Å). One angstrom equals 1.00×10^{-10} m, so for the C—H bond length described above this would give 1.10 Å.

1.2.2   A review of some commonly used measurements

Mass

The mass of an object is a measure of the quantity of material that it contains, and it is the property that gives rise to objects having weight. The SI unit of mass is the kilogram, though in laboratories this is generally too large a unit for practical purposes, and the gram (g = 10^{-3} kg) is more common, and where materials are very expensive even the milligram (mg = 10^{-6} kg or 10^{-3} g) or the microgram (μg = 10^{-9} kg or 10^{-6} g) are likely to be used.

Length

This is the distance between two points, and the SI unit of length is the metre. In the laboratory environment this may again be too large a unit for many practical purposes, and the more common unit is the centimetre (cm = 10^{-2} m).

Table 1.4 **Prefixes for metric numbers**

Prefix	Symbol	Magnitude	Meaning (multiply by)
giga-	G	10^9	1 000 000 000
mega-	M	10^6	1 000 000
kilo-	k	10^3	1000
hecto-	h	10^2	100
deka-	da	10	10
deci-	d	10^{-1}	0.1
centi-	c	10^{-2}	0.01
milli-	m	10^{-3}	0.001
micro-	μ*	10^{-6}	0.000 001
nano-	n	10^{-9}	0.000 000 001
pico-	p	10^{-12}	0.000 000 000 001

*This is the Greek letter mu.

Volume

The volume is a measure of the amount of space something occupies. Although the base SI unit of volume is the cubic metre, m^3, this is not a very practical unit for laboratory purposes, and a more commonly used unit is the litre (L), or even the millilitre (mL). The litre is a metric unit and is $(10\,cm \times 10\,cm \times 10\,cm) = 1000\,cm^3$.

As a millilitre (mL) is 0.001 L, then clearly $1\,mL = 1\,cm^3$. Furthermore, as can be deduced from Table 1.4,

$$10 \text{ centimetres} = 1 \text{ decimetre (dm)}$$

Therefore, given that

$$1\,L = 10\,cm \times 10\,cm \times 10\,cm$$

and that

$$1\,dm = 10\,cm$$

then

$$10\,cm \times 10\,cm \times 10\,cm = 1\,dm \times 1\,dm \times 1\,dm$$

or

$$1000\,cm^3 = 1\,dm^3 = 1\,L$$

In words, one litre equals one cubic decimetre. This is important because many scientific books and journals refer to volumes in terms of $1\,dm^3$ instead of $1\,L$.

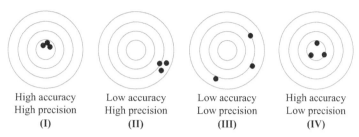

High accuracy   Low accuracy   Low accuracy   High accuracy
High precision   High precision   Low precision   Low precision
(I)   (II)   (III)   (IV)

Figure 1.2 **Representations of the concepts of accuracy and precision.**

Temperature

Temperature is the measure of heat content. The SI unit of temperature is the kelvin (K), and this is the temperature scale that is used in science. Most thermometers, however, measure temperature in Celsius, and any readings that are taken have to be corrected. This is a straightforward process, because in order to change from the Celsius scale to the kelvin scale it is only necessary to add 273.15. Thus, the freezing point of water on the Celsius scale is 0.00°C, but on the kelvin scale it is 273.15 K.

The boiling point of water on the Celsius scale is 100.00°C, but on the kelvin scale it is 373.15 K. Note that a degree symbol is used with the Celsius scale, but not with the kelvin scale. 0.00 K is referred to as the absolute zero of temperature, which is the lowest temperature possible. In general, *298.15 K* or 25°C, is referred to as the *standard temperature*.

1.2.3 Accuracy and precision

In common usage people often confuse the terms accuracy and precision, but in science there is a clear distinction between the two terms:

- The accuracy of a measurement (or group of measurements) describes how close to the expected/true value it is (or they are).
- Precision describes the reproducibility of measurements.

To illustrate this, look at Figure 1.2, in which the terms accuracy and precision are described in reference to a series of arrow strikes at an archery target.

- Compare I and IV. Both sets deliver three bullseye strikes, but the grouping in I is much tighter; they have a greater level of precision.
- Looking at II and III, clearly the results in II are reproducible, but are consistently of low accuracy, whereas in III the results have a similar level of inaccuracy as those in II, but are much more scattered.

1.3 Atoms

Atoms are the fundamental building blocks of matter. Different elements, which are the simplest forms of matter that are considered by chemists, have different atomic compositions.

Table 1.5 **The components of an atom**

Name	Charge	Mass (kg)	Mass (atomic mass units, amu)
Proton	+1	1.6725×10^{-27}	1
Neutron	0	1.6750×10^{-27}	1
Electron	−1	9.1095×10^{-31}	5.4×10^{-4}

An atom is electrically neutral, and is made up of:

- A nucleus, which is itself composed of protons and neutrons (except for the hydrogen atom which has no neutrons; see later). Protons are positively charged and neutrons are neutral, so overall a nucleus is positively charged, and it occupies a very small volume.

- Electrons, which surround the nucleus in designated orbitals, are negatively charged, occupying a large volume, but with comparatively little mass.

The number of negatively charged electrons must equal the number of positively charged protons in order to make the atom electrically neutral. The nucleus is the heaviest part of an atom, with the electrons having a very small mass. The key facts are summarised in Table 1.5.

These points can be well illustrated by considering two of the most important components of life, namely the *elements* hydrogen and carbon. Hydrogen in its most common form (see isotopes later to elaborate on this) has a nucleus made up of a single proton. The hydrogen nucleus does not possess a neutron. In order to maintain electrical neutrality the atom possesses a single electron. Carbon, in its most common form, has a nucleus which contains six protons and six neutrons. This means that, in order to maintain electrical neutrality, the nucleus will be surrounded by six electrons.

1.3.1 Representation of atoms

It is important to have a method of representing atoms to summarise all of these key points as clearly and concisely as possible. The notation that allows this to be achieved, for any element E, is:

$$_A^M E$$

In this notation:

- M is the mass number (in atomic mass units, amu; see Section 1.4.1) and equals the number of protons plus the number of neutrons.
- A is the number of protons (and hence also the number of electrons).
- The number of neutrons is therefore (M − A).

Thus the most common form of the hydrogen atom, H-1, as referred to above, is represented by:

$$_1^1 H$$

The most common carbon atom, carbon-12, is represented by:

$$^{12}_{6}\text{C}$$

1.3.2 Isotopes

Elements can exist in different forms, known as isotopes, where the numbers of protons/electrons remain the same, but the number of neutrons has changed.

Isotopes of hydrogen

For hydrogen there are three isotopes, these being:

$$^{1}_{1}\text{H} \qquad ^{2}_{1}\text{H} \qquad ^{3}_{1}\text{H}$$

These are referred to as hydrogen-1, hydrogen-2 and hydrogen-3 respectively, though they are also often referred to by the names protium, deuterium and tritium. Because each of them has only one proton, the symbolic representation is often reduced to ^{1}H, ^{2}H and ^{3}H respectively. In each case the atoms contain one proton and one electron, but for deuterium and tritium respectively there are one and two neutrons.

In nature the most commonly found isotope of hydrogen (by far) is hydrogen-1, which forms 99.98% of all naturally occurring hydrogen found. The two stable isotopes of hydrogen are represented in Figure 1.3.

Isotopes of carbon

It has been internationally agreed that the mass of carbon-12 is 12 amu. The mass of carbon generally quoted for carbon in nature is 12.01 amu, and this is because naturally occurring carbon is composed of carbon-12, ^{12}C (natural abundance close to 98.9%), and carbon-13, ^{13}C (natural abundance approximately 1.1%), and carbon-14, ^{14}C (natural abundance very very small).

Not all isotopes are stable, and this instability gives rise to the phenomenon of radioactivity (carbon-14 is an example of such an isotope), where the isotope decays into a more stable form. Although an in-depth discussion of the principles of isotope stability and the different types of radioactivity is outside the remit of this book, it is important to acknowledge that the subject is important medically, for example in the application of radiotherapy treatments for various cancers. The subject of radioactivity will now be reviewed briefly.

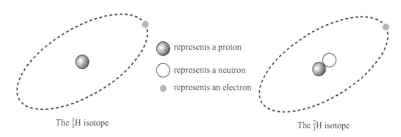

represents a proton

represents a neutron

represents an electron

The $^{1}_{1}\text{H}$ isotope

The $^{2}_{1}\text{H}$ isotope

Figure 1.3 **The two stable isotopes of hydrogen.**

1.3.3 Isotopes, radioactivity and the types of radiation

Beta decay

Carbon-14, which has six protons and eight neutrons, is found to be an unstable isotope of carbon. It breaks down as indicated in the equation below:

$$^{14}_{6}C \rightarrow \,^{14}_{7}N + \,^{0}_{-1}e + \,^{0}_{0}\bar{\nu}$$

What this shows is that a neutron in the carbon breaks down to form a proton plus an electron (the electron produced this way is represented by the symbol $^{0}_{-1}e$) along with another type of particle called an antineutrino (the symbol for which is $^{0}_{0}\bar{\nu}$).

Note how this type of radiation, which is called *beta decay*, results in a change in the element.

Another example, which will be returned to in the section on isotopes in medicine, is the decay of iodine-131:

$$^{131}_{53}I \rightarrow \,^{131}_{54}Xe + \,^{0}_{-1}e + \,^{0}_{0}\bar{\nu}$$

A pictorial representation of a nucleus undergoing beta decay is given in Figure 1.4.

Alpha decay

The element radium has many isotopes, none of which is stable. Below is the equation representing the radioactive decay of one of them, radium-222:

$$^{222}_{88}Ra \rightarrow \,^{218}_{86}Rn + \,^{4}_{2}He$$

In this case the radium nucleus has split into two parts, yielding two nuclei, namely radon-218 and helium-4, with the helium-4 nucleus being at relatively high energy.

The radioactive isotopes of many elements decay with the loss of helium-4, and this is known as alpha decay, with the helium-4 nucleus itself being referred to as an alpha particle. Another example of alpha decay is the decomposition of polonium-208:

$$^{208}_{84}Po \rightarrow \,^{204}_{82}Pb + \,^{4}_{2}He$$

Polonium is named after Poland, the birthplace of the famous scientist Marie Curie, who was one of the foremost researchers into the development of the theories behind radioactivity. Figure 1.5 gives a pictorial representation of a nucleus undergoing alpha decay.

Gamma radiation

Gamma radiation is a form of ionising radiation that results from a range of different sources, but usually involves the interaction of subatomic particles (the particles

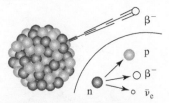

The grey balls represent neutrons.
The orange balls represent protons.
The small ball represents an electron.

Figure 1.4 Pictorial representation of beta decay.

(Adapted from http://en.wikipedia.org/wiki/File:Beta-minus_Decay.svg)

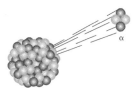

The grey balls represent neutrons.
The orange balls represent protons.
The alpha particle expelled is made up from
2 protons + 2 neutrons.

Figure 1.5 **Pictorial representation of alpha decay.**

(Adapted from http://en.wikipedia.org/wiki/File:Alpha_Decay.svg)

that make up an atom). It is part of the electromagnetic spectrum of radiation and is at the high energy end of that spectrum. As a result gamma radiation is the most dangerous of the three types of radioactive behaviour listed here.

Gamma rays from radioactive gamma decay are produced with the other forms of radiation discussed above, namely alpha or beta, and they are produced after the other types of decay have happened. Thus when a nucleus emits an α or β particle, the resultant nucleus (often called the daughter nucleus) is left in an excited state, and gets to a lower energy state by emitting a gamma ray.

Table 1.6 summarises the properties of the different types of radiation.

Half-life and radioactive decay

A radioactive nuclide (nuclide is the term for a nucleus with one or more orbiting electrons) emits radioactivity at a characteristic rate, different from that of other nuclides, and the rate of decay of a radioactive nuclide is measured by its *half-life*. This is the time required for one half of the atoms in any starting sample of a radio-isotope to decay. If the half-life, $t_{1/2}$, of a radioactive nuclide is known, its decay constant can be calculated by:

$$\lambda = 0.693/t_{1/2}$$

and

$$N_t = N_0 e^{(-\lambda t)}$$

Table 1.6 **Summary of properties of different types of radiation**

Name	Form	Charge	Potency
Alpha	Helium nuclei	+ 2	Low energy, stopped by paper sheet. Harmful if ingested.
Beta	Electrons	− 1	Passes through paper and thin aluminium sheet. Stopped by thin lead sheet. Can cause burns to skin and harmful if ingested.
Gamma	Electromagnetic radiation	0	Highly penetrating, can go through several cm of lead/concrete. Very dangerous to living tissue.

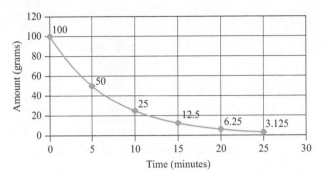

Figure 1.6 **Graphical illustration of concept of half-life: amount of radioactive material vs time**

In these equations:

- N_0 is the starting number of nuclei;
- N_t is the number of nuclei remaining after time t;
- λ is the decay constant;
- e = 2.718.

The units for the decay constant would be s^{-1} (or disintegrations per second) if the half-life is expressed in seconds. This is based on statistics and probability, from an examination of the behaviour of a large number of individual situations. It does not give any indication when a particular nucleus will undergo decay, but does show the amount of time needed for a certain proportion of the nuclei in the sample to decay. The idea of half-life is easily demonstrated graphically, as shown in Figure 1.6, in which an example of a radioactive decay with a half-life of 5 minutes is shown. It is clear that after every half-life (5 minutes) the amount of radioactive material is half of its preceding value.

Box 1.1 shows an example of the use of the half-life equation in the topic of radioactive carbon dating.

Isotopes in medicine

Radioactive isotopes have many uses in medicine; for example:

- Hyperthyroidism is a condition where the thyroid gland is malfunctioning, resulting in overproduction of thyroid hormones. A treatment for this involves the use of iodine-131 which, as was seen earlier, decays via beta emission (as well as some gamma emission). In sufficient doses, the iodine is absorbed by the thyroid gland and the radiation kills the problematic cells.
- Cobalt-60 is used as a gamma source, and the gamma rays can be focused on a tumour site, leaving healthy cells unaffected.

1.3.4 Electrons

Chemistry is concerned primarily with electrons and how they behave. Below are a few key ideas which summarise the current understanding:

- Electrons in atoms are said to be found in orbitals, with each orbital containing a maximum of two electrons.

Box 1.1

Radiocarbon dating - a numerical example

Living matter has a radioactivity of 750 counts per hour (cph) per gram of carbon resulting from the decomposition of ^{14}C in the atmosphere. Cypress wood obtained from the tomb of Sneferu (the founder of the fourth Egyptian dynasty, and father of Khufu, who built the Great Pyramid at Giza) was measured as having a radioactivity of 413 cph per gram of carbon. The half-life of ^{14}C is 5730 years. We will use this to estimate the age of the wood found in the tomb.

Recalling the relationships

$$\lambda = 0.693/t_{1/2} \qquad [1]$$

and

$$N_t = N_0 e^{(-\lambda t)} \qquad [2]$$

We will use the half-life equation, [1], to work out the value of the decay constant, and then use this value in equation [2] to allow us to calculate the age of the wood, given by t:

$$\lambda = 0.693/t_{1/2}$$

$$= 0.693/5730$$

$$= 1.21 \times 10^{-4}$$

Using equation [2], and recognising that the values of N_0 and N_t will be directly proportional to the radioactivity of the wood at the two times, we can therefore write:

$$N_t = N_0 e^{(-\lambda t)}$$

and rearrange this to give:

$$N_t/N_0 = e^{(-\lambda t)}$$

In taking logarithms to the base e (see Appendix 1 on mathematical techniques), this then gives:

$$\ln (N_t/N_0) = -\lambda t$$

and rearranging this to get t on its own results in:

$$t = -1/\lambda \ln (N_t/N_0)$$

Using the rules of logarithms (again see Appendix 1), which state that $\ln (x/y) = - \ln (y/x)$, we then get:

$$t = 1/\lambda \ln (N_0/N_t)$$

In this case, $1/\lambda = 8268$, $N_0 = 750$ and $N_t = 413$, so

$$t = 8268 \ln (750/413)$$

$$= 8268 \times 0.597$$

$$= 4933 \text{ years}$$

This is a bit less than one half-life, which can be inferred by the wood's activity value being > 375 cph.

- Electrons possess a property known as spin, and this is clockwise and anticlockwise. If a pair of electrons are in the same orbital then they *must* have opposing spins, which is often represented thus:

$$\uparrow\downarrow \quad \text{or} \quad \boxed{\uparrow\downarrow}$$

This is known as *Pauli's exclusion principle*.

- Orbitals are arranged in energy levels (or shells). Larger values of the level numbers indicate that the electrons in these levels are further from the nucleus. The first level is closest to the nucleus and is said to have a *principal quantum number $n = 1$*. The second level, with a principal quantum number $n = 2$, is further from the nucleus, etc.

In the next chapter these concepts and ideas will be explored further, highlighting how electrons are arranged within shells, and then to show how the chemical behaviour of elements can be rationalised by reference to the resulting electron configurations.

1.3.5 Molecules

Molecules are, simply, two or more atoms joined tightly together. When a molecule involves just two atoms it is said to be diatomic, while if it involves three atoms it is said to be triatomic. Molecules can be made up of atoms of the same element (in which case they are called homonuclear) or atoms of different elements (in which case they are called heteronuclear). When molecules are composed of several atoms they tend to be referred to as polyatomic.

Some examples are:

- Homonuclear diatomic: O_2, dioxygen; N_2, dinitrogen; Br_2, bromine
- Homonuclear triatomic: O_3, ozone
- Heteronuclear diatomic: HBr; HCl; CO (carbon monoxide)
- Heteronuclear triatomic: H_2O (water); CO_2 (carbon dioxide)
- More complex molecules: CH_3OH (methanol); $CH_3CH(NH_2)(COOH)$ (alanine).

The bonding between atoms will be looked at more extensively in the next chapter.

1.4 The concepts of stoichiometry: calculations of quantity in chemistry

1.4.1 Introduction

There are less than one hundred naturally occurring elements, each of which has a characteristic relative atomic mass (often referred to as its atomic weight). The units in which they are quoted are atomic mass units, amu.

Most periodic tables include the relative atomic mass of an element in the box with the element. The relative atomic mass is usually not an integer because it is dependent on the number of protons plus the average number of neutrons of an element, and elements with greater numbers of protons + neutrons will have greater relative atomic masses. Because of this, 1 gram of an element of a low relative atomic mass will contain more atoms than 1 gram of an element with a higher relative atomic mass.

In science it is necessary to be able to assess the amount of material (whether as an element or as a compound) in the context of the number of atoms or molecules that are present. Consider a series of elements; for example, helium, magnesium and carbon:

- The relative atomic mass of helium is 4.003 amu.
- The relative atomic mass of carbon is 12.011 amu.
- The relative atomic mass of magnesium is 24.31 amu.

1.4.2 Avogadro's number and the concept of the mole

The modern system of atomic masses is based on ^{12}C (carbon-12) as the standard. Thus in this system ^{12}C is assigned a mass of exactly 12 amu. The masses of all other atoms are given relative to this standard.

In tables of the elements, the atomic mass of carbon is shown as 12.01. This is because natural carbon is a mixture of two main isotopes: ^{12}C (98.89%) and ^{13}C (1.11%). Thus the average atomic mass of C is 12.01.

The idea of the mole

The mole is defined as the number of atoms in 12 g (sometimes quoted as 0.012 kg) of pure ^{12}C. This has been determined to be $6.022\,137 \times 10^{23}$, which is known as Avogadro's number (symbol N_A, sometimes given the symbol L). In practice, 6.022×10^{23} is accurate enough for most purposes.

Using this definition, a mole of any pure substance has a mass in grams exactly equal to the molecular or atomic mass of that substance.

Avogadro's number is very large and, to give an example of its size, it is estimated that a mole of typical marbles would cover the entire surface of the Earth to a depth of 80 km (50 miles). For example:

- A mole of aluminium is 27.0 grams of aluminium atoms. Aluminium is an element, so the fundamental particles of aluminium are atoms. There are Avogadro's number of aluminium atoms in 27.0 grams of it.
- But 1.008 grams of hydrogen is *not* a mole of hydrogen! The reason for this is that hydrogen is one of the diatomic gases. There are, in nature, no such things as loose hydrogen atoms. The total mass of a single hydrogen diatomic molecule (H_2) is 2.016 amu. Therefore, because hydrogen exists as a diatomic molecule, a mole of hydrogen gas, which contains 6.022×10^{23} molecules, has a mass of 2.016 grams.

The masses of atoms are known as the relative atomic masses.

1.4.3 Formulae and molecular mass

Chemical compounds have formulae and the sum of the atomic masses of all the atoms in the formula is called the relative formula mass (M_r). For example, the ionic compound NaCl (common salt) has the formula mass 22.99

(for Na) + 35.45 (for Cl) = 58.44 (total for NaCl). A mole of NaCl contains 6.022×10^{23} NaCl units.

Similarly, a mole of water is 18.015 g because each water molecule has two hydrogen atoms, each with mass 1.008, and one oxygen atom, which has a mass of 15.999. A mole of water has 6.022×10^{23} (Avogadro's number) of water molecules in it.

Another way to view the same thing is that a formula mass is the total mass of a formula in amu expressed with units of grams per mol (g mol^{-1}). When a substance is known to consist of only molecules then the term molecular mass is often used. This is the same as formula mass and is used specifically for molecular substances.

Summary

The name for Avogadro's constant of ANYTHING (atoms or molecules or ions) is a mole (or mol).

Example 1.1 What is the molecular mass of methanol, CH_3OH?

List the number of each type of atom, along with its atomic mass. There is one C atom, four H atoms and one O atom. Therefore:

12.011 (C) + 4 × 1.008 (H) + 15.999 (O) = 32.042 (= molecular mass of CH_3OH)

So, 1 mol of methanol has a mass of 32.042 g

Example 1.2 What is the formula mass of calcium sulfate, $Ca(SO_4)$? (To 2 decimal places only)

$Ca(SO_4)$ contains $\qquad$ Ca + S + (4 × O)

Thus, the formula mass of calcium sulfate = (40.08) + (32.07) + (4 × 16.00)

= 40.08 + 32.07 + 64.00

= 136.15

So one mol of $Ca(SO_4)$ has a mass of 136.15 g.

Example 1.3 How many molecules of methanol, CH_3OH, are present in 0.5 moles? How many atoms are present?

In 0.5 moles there will be $0.5 \times 6.022 \times 10^{23}$ molecules = 3.011×10^{23} molecules. Each molecule contains [(1 × C) + (1 × O) + (4 × H)] = 6 atoms. Therefore there will be $6 \times 3.011 \times 10^{23}$ atoms in 0.5 moles of methanol = $18.066 \times 10^{23} = 1.8066 \times 10^{24}$ atoms.

Example 1.4 State the number of moles of sodium ions, anions and the total number of moles of ions that are present in aqueous solutions of each of the following:

(a) 1.00 mole of Na_2CO_3 (sodium carbonate)

(b) 4.00 moles of NaOH (sodium hydroxide)

(c) 0.75 moles of Na_3PO_4 (sodium phosphate)

All of the compounds above are ionic salts, and will dissolve in water as shown:

(a) Na_2CO_3 (s) dissolves to give $2Na^+$ (aq) + CO_3^{2-} (aq)
  - 1 mole of the formula unit Na_2CO_3 dissolves to give 2 moles of sodium ions and 1 mole of carbonate ions in the solution.
  - This means a total of 3 moles of ions in solution.

(b) NaOH (s) dissolves to give Na^+ (aq) + OH^- (aq)
  - 1 mole of the formula unit NaOH dissolves to give 1 mole of sodium ions and 1 mole of hydroxide ions.
  - Therefore if 4 moles are dissolved there will be 4 moles of sodium ions and 4 moles of hydroxide ions in the solution.
  - This means a total of 8 moles of ions in the solution.

(c) Na_3PO_4 (s) dissolves to give $3Na^+$ (aq) + PO_4^{3-} (aq)
  - 1 mole of the formula unit Na_3PO_4 dissolves to give 3 moles of sodium ions and 1 mole of phosphate ions.
  - Therefore if 0.75 moles are dissolved this means there will be (3×0.75) moles of sodium ions, which is 2.25 moles of sodium ions, and 0.75 moles of phosphate ions in solution.
  - This means that there will be a total of $(2.25 + 0.75) = 3$ moles of ions in solution.

1.4.4 Mass percent composition

This is often useful in determining the composition of a compound. The mass percent of an element in a compound can be calculated by comparing the mass of the element in 1 mole of the compound with the total mass of 1 mole of the compound and multiplying by 100.

Example 1.5 Calculate the percentage of oxygen in methanol.

Oxygen has a relative atomic mass of 15.999 g mol^{-1}, and methanol has a formula mass of 32.042 g mol^{-1} (see earlier).

Therefore, the percentage of oxygen in the molecule is $(15.999/32.042) \times 100 = 49.931\%$.

Example 1.6 What is the percentage of chloride in potassium chloride, KCl? (To 2 decimal places.)
The relative atomic mass of potassium is 39.10 g mol^{-1}. That of chlorine is 35.45 g mol^{-1}.
So the formula mass of potassium chloride is $(39.10 + 35.45) = 74.55$ g mol^{-1}.

Therefore the percentage of chloride is $(35.45/74.55) \times 100 = 47.55\%$.

Example 1.7 C_2H_4 is the molecular formula for ethene.

(a) How many atoms are in one molecule?

2 carbons + 4 hydrogens = 6 atoms.

(b) Which atoms make up ethene?

Carbon and hydrogen.

(c) How many molecules are in 5.3 moles of ethene?

In 1 mole there are 6.022×10^{23} molecules. Therefore in 5.3 moles there are $6.022 \times 5.3 \times 10^{23}$ molecules. This is $31.92 \times 10^{23} = 3.192 \times 10^{24}$.

(d) How many atoms are in a mole of ethene?

In a molecule of ethene there are 6 atoms, so in a mole there are $6 \times 6.022 \times 10^{23}$ $= 36.13 \times 10^{23}$ or 3.613×10^{24} atoms.

(e) What is the molecular mass of ethene?

There are 2 C atoms and 4 H atoms. Therefore
$$\text{molecular mass} = (2 \times 12.011) + (4 \times 1.008)$$
$$= 24.022 + 4.032 = 28.054 \, \text{g mol}^{-1}.$$

(f) What is the percentage of carbon in ethene?

The percentage, using the details from (e) above, is $(24.022/28.054) \times 100 = 85.63\%$.

1.4.5 Empirical and molecular formulae

- An empirical formula is one in which the subscripts are the *smallest* whole numbers that describe the ratios of the atoms in the substance.
- A molecular formula specifies the actual composition of one molecule.

Thus P_4O_{10} is a molecular formula but P_2O_5 is the empirical formula for the same compound. Materials can be readily analysed to show how much (in g) of each element is present, but this information is only of limited value. If, however, we could work out the number of moles of each element that is present, we could then get an insight into the formulae of the materials. To do this we divide the mass of the element in the sample (in g) by its atomic mass (in g mol^{-1}).

The following examples illustrate ways in which this can be done.

Example 1.8 A sample of an unknown compound with mass 3.038 g was found to contain 1.102 g of carbon and 2.936 g of oxygen. What is the empirical formula?

Step 1: Work out the number of moles present:

1 mol of C is 12.01 g so 1.102 g corresponds to $1.102/12.01 = 0.092$ mol
1 mol of O is 16.00 g so 2.936 g corresponds to $2.936/16.00 = 0.184$ mol

Step 2: Divide through by the smallest number of moles

For C get $0.092/0.092 = 1.00$
For O get $0.184/0.092 = 2.00$

Step 3: Determine the empirical formula

So the ratio of C to O is 1:2, giving an empirical formula of CO_2 (i.e. carbon dioxide).

This approach can be presented differently in a tabulated form, for example:

	C	O
Mass of each element	1.102	2.936
Number of moles of each element (mass/atomic mass)	$1.102/12.01 = 0.092$	$2.936/16.00 = 0.184$
Divide through by smallest number of moles	$0.092/0.092 = 1.0$	$0.184/0.092 = 2.0$

So, as we saw earlier, the ratio of C to O is 1:2, giving an empirical formula of CO_2 (i.e. carbon dioxide).

Example 1.9 A white powder contains 43.64% P and 56.36% O by mass. The compound has a molar mass of 283.88 g. What are the compound's empirical and molecular formulae?

In this case the total mass used is unknown, but because the percentages of the elements present are known, then it is possible to say that in 100 g of this compound there would be 43.64 g of P and 56.36 g of O.

Summarising the information in a table gives:

	P	O
%	43.64	56.36
Divide % by atomic mass of element	43.64/30.97 = 1.41	56.36/16.00 = 3.52
Divide by the smallest ratio	1.41/1.41 = 1.0	3.52/1.41 = 2.5
Multiply through to give whole numbers	2.0	5.0

This means the empirical formula is P_2O_5.

Based on the empirical formula, the mass would be $(2 \times 30.97) + (5 \times 16.00) = 141.94$ g, but the molar mass is known to be 283.88, and as $283.88/141.94 = 2$ then it is clear that the molecular formula must be P_4O_{10}.

1.4.6 Writing and balancing chemical equations

A chemical reaction is the *process* of chemical change, with a chemical equation being a shorthand quantitative description of a chemical reaction.

The starting materials are called *reactants* and the substances formed are called *products*. An arrow is used to symbolise a reaction:

$$\text{Reactants} \rightarrow \text{Products}$$

To illustrate this consider the reaction undergone between potassium, K, and water, H_2O. Initially write a *skeletal equation* listing the reactants and products of the reaction, using their molecular formulae thus:

$$K + H_2O \rightarrow KOH + H_2 \qquad [1]$$

This has given a qualitative summary of the reaction, but chemists are much more interested in describing reactions *quantitatively* (for example, in determining how much product is obtained from a given amount of reactant). This employs the *Law of conservation of mass,* which states that atoms are neither created nor destroyed in a chemical reaction.

The skeletal equation [1] does not obey the law of conservation of mass. The reactant side contains two hydrogen atoms and the product side contains three hydrogen atoms.

The skeletal equation is balanced to ensure that the same numbers of atoms of each element are shown on each side of the arrow. Thus the reaction between potassium and water can be rewritten as [2] below:

$$2K + 2H_2O \rightarrow 2KOH + H_2 \qquad [2]$$

19

The numbers multiplying in front of the reactants and the products leading to this balanced equation are called *stoichiometric coefficients*.This simple, balanced, equation tells us that for every two atoms of potassium and two molecules of water that react one molecule of hydrogen is produced.

Stoichiometric coefficients in any balanced chemical equation show the relative numbers of moles of each substance consumed or produced in the reaction.

State symbols, (s), (l), (g) or (aq), given as either subscripts, or in brackets, after the individual reactants/products, denote the physical state of each reactant or product, solid, liquid, gas or aqueous solution.

Method 1: $2K(s) + 2H_2O(1) \rightarrow 2KOH(aq) + H_2(g)$

Method 2: $2K(s) + 2H_2O(l) \rightarrow 2KOH(aq) + H_2(g)$

In this text, the representation of state symbols will follow that shown in method (2).

1.4.7 Balancing equations: a systematic approach

Consider one element at a time to ensure that the same number of atoms occurs in the products as in the reactants.

Step 1: Determine what reaction is occurring. Establish what are the reactants and what are the products and determine in what physical states (e.g. gaseous, liquid, etc.) they are operating.

Step 2: Write the unbalanced simplest equation that summarises all of the above information.

Step 3: Balance the equation by inspection. A good approach is to start with the most complicated molecule. Work out what coefficients are necessary to ensure that the same number of each type of atom appears on both reactant and product sides and *do not change* the identities (formulas) of any of the reactants or products.

Example 1.10 Write a balanced equation for the reaction when methane is burned in air to produce carbon dioxide and water.

Step 1: The skeletal equation for this needs to incorporate methane, carbon dioxide and water and the reactive component in air, which is oxygen (or more precisely dioxygen, O_2, as this is the form oxygen exists in naturally).

Step 2: Thus the skeletal equation is:

$$CH_4 + O_2 \rightarrow CO_2 + H_2O$$

Step 3: We need to balance the numbers of atoms. First consider the element(s) occurring in the fewest formulae. In this case this is C and H which occur in two formulae (O is in three formulae).
- C is already balanced – one atom on each side
- H has four atoms on the left hand side of the equation but only two on the right. To balance this we apply a stoichiometry coefficient of 2 to the water. This gives:

$$CH_4 + O_2 \rightarrow CO_2 + 2H_2O$$

- O now has two atoms on the left hand side of the equation and four on the right hand side. To balance this we apply a stoichiometry coefficient of 2 to the dioxygen to give:

$$CH_4 + 2O_2 \rightarrow CO_2 + 2H_2O$$

Checking through the atom types, it is clear that we have a balanced equation.

It is important to remember to define the states of the species. At the high temperatures in a flame the water is produced as vapour (steam):

$$CH_4\,(g) + 2O_2\,(g) \rightarrow CO_2\,(g) + 2H_2O\,(g)$$

In many reactions polyatomic ions such as NH_4^+ or SO_4^{2-}, called the ammonium ion and the sulfate ion respectively, remain intact throughout the reaction. They can then be balanced as *single entities* in much the same way as, for example, K^+ or Cl^- would be.

1.4.8 Moles and masses

It is possible to compare the amounts of any materials in the same chemical equation using the formula masses and the coefficients of the materials in the equation.

Consider the equation for the Haber process, where nitrogen gas and hydrogen gas are combined to make ammonia. The formula for nitrogen is N_2 and the formula for hydrogen is H_2, and the formula for ammonia is NH_3.

The balanced equation requires one nitrogen molecule and three hydrogen molecules reacting to make two ammonia molecules, meaning that one nitrogen molecule reacts with three hydrogen molecules to make two ammonia molecules or one *mole* of nitrogen and three *moles* of hydrogen make two *moles* of ammonia. This is summarised by the equation:

$$N_2\,(g) + 3H_2\,(g) \rightarrow 2NH_3\,(g)$$

As scientific balances read in grams and not moles, in order to work out how much material to use, numbers of moles will have to be converted to mass. So for the Haber process:

- 28 grams ($14\,g\,mol^{-1} \times 2$ atoms of nitrogen per molecule) of nitrogen and
- 6 grams of hydrogen ($1\,g\,mol^{-1} \times 2$ atoms of hydrogen per molecule times 3 moles) make
- 34 grams of ammonia (2 moles of a molecule containing $1 \times N$ and $3 \times H$).

Notice that no mass is lost or gained, since the formula mass for ammonia is 17 (one nitrogen at $14\,g\,mol^{-1}$ and three hydrogens at $1\,g\,mol^{-1}$) and there are two moles of ammonia made.

1.4.9 Concentration of solutions

The relationship between concentrations, moles and volumes is shown in Box 1.2. The key points are:

- a solution is a mixture of a fluid (often water, but not always) and another material;
- the fluid is referred to as the solvent;

- the material mixed in with the solvent is called the solute;
- the volume of a solution, V, is measured the same way the volume of a pure liquid is measured;
- the concentration of a solute in a solution is the amount of material per unit volume. Therefore, if concentration is called C, and the amount of material is called n (usually in moles), with the volume being V, then $C = n/V$.

Concentration can be expressed in a number of ways, but the most common in chemistry/biology is to use molarity. A 1-molar solution (shown either as 1 M or 1 $mol\,dm^{-3}$ or 1 $mol\,L^{-1}$) is one mole of solute (i.e. the amount) in 1 litre of solvent. It is important to notice that the fluid is usually nothing more than a diluting agent and, generally, does not participate in any reaction. (In biology the concentrations worked with are very low, and biological solutions are often referred to in *millimolar* concentrations, represented by mM, where a 1-mM solution is 0.001 M, or 1/1000 M.)

Box 1.2

Concentrations, moles and volumes

Concentration is, by definition, amount of material per unit volume. The amount of material is usually in moles (mol) and the volume is usually in litres (L). Therefore concentration is in moles per litre $(mol\,L^{-1})$, sometimes called molar (M).

Mathematically this is given by

$$C = \frac{n}{V}$$

If we know the concentration, C, that we need to make, and the amount of material, n (in moles), that we possess, then we can determine the volume of solution we need to prepare by rearranging the above to

$$V = \frac{n}{C}$$

Similarly, if we know the intended concentration, along with the volume we need to prepare, then we can determine the amount of material (in moles) that we need to use from the alternative rearrangement:

$$n = C \times V$$

Example 1.11 How would you make 100 mL (=0.1 L) of a 0.1M solution of ethanoic (also known as acetic) acid in water?

Method 1

- The relative molecular mass (M_r) of ethanoic acid (molecular formula CH_3COOH, or $C_2H_4O_2$) is:
  $$M_r = (2 \times 12) + (4 \times 1) + (2 \times 16) = 60\ amu$$
- A 1 M solution of ethanoic acid in water would equate with 60 g of ethanoic acid dissolved in water and made up to 1 L.
- To make 0.1 L (i.e. 100 mL) of a 1 M solution, we would need 0.1 × the amount of ethanoic acid.

- Therefore 6 g of ethanoic acid dissolved in water and made up to 100 mL is also a 1 M solution.
- Therefore, to make 100 mL of a 0.1 M solution of ethanoic acid in water, we would need to use less (1/10) of the solute.

Thus, 0.6 g of ethanoic acid dissolved in water and made up to 100 mL would give a 0.1 M solution.

(Note, when learning about moles and molarities, it is good practice to work through calculations of this type in the stepwise manner laid out above. Any errors made are easier to identify!)

Method 2

- Using $n = C \times V$ to determine the number of moles of ethanoic acid needed,

$$n = 0.1 \times 0.1 = 0.01 \text{ mol} \quad (100 \text{ mL} = 0.1 \text{ L})$$

- 1 mol ethanoic acid $= 60$ g, so 0.01 mol $= (0.01 \times 60) = 0.60$ g

Thus, 0.6 g of ethanoic acid dissolved in water and made up to 100 mL would give a 0.1 M solution.

Questions

1.1 Which of the following endings (a to d) is correct for this partial statement? Isotopes of the same element have identical ...

(a) nuclei
(b) mass numbers
(c) numbers of neutrons
(d) numbers of protons.

1.2 Brine can be made by dissolving common salt in water. Which of the terms listed below describes the water used when making the brine?

(a) Solute
(b) Saturated
(c) Solvent
(d) Solution

1.3 The element boron is made up of two naturally occurring types of atom shown in the table below. Complete the table and state the term used to describe the different types of boron atom.

	Number of protons	Number of neutrons	Number of electrons
$^{10}_{5}\text{B}$			
$^{11}_{5}\text{B}$			

1.4 Phosphorus-32 is an isotope of phosphorus, and is a beta emitter with a half-life of 14 days.

(a) Show the equation representing the decay process, and identify the element that results from this process.
(b) Calculate the remaining mass of the sample after 112 days.

1.5 What is the percentage of sulfate in beryllium sulfate tetrahydrate, $Be(SO_4).4H_2O$?

1.6 What is the percentage composition of oxygen in each of the following materials:

(a) CO
(b) CO_2
(c) $(PO_4)^{3-}$
(d) ethyl alcohol, C_2H_6O
(e) calcium phosphate
(f) $Cu(SO_4).5H_2O$?

1.7 (a) How many oxygen atoms are in a mole of carbon dioxide, CO_2?
(b) How many molecules are there in 6.2 moles of carbon dioxide?

1.8 Balance the equations below:

(a) ____ $KClO_3 \rightarrow$ ____ $KCl +$ ____ O_2
(b) ____ $Pb(OH)_2 +$ ____ $HCl \rightarrow$ ____ $H_2O +$ ____ $PbCl_2$
(c) ____ $AlBr_3 +$ ____ $K_2SO_4 \rightarrow$ ____ $KBr +$ ____ $Al_2(SO_4)_3$
(d) ____ $C_6H_{14} +$ ____ $O_2 \rightarrow$ ____ $CO_2 +$ ____ H_2O

1.9 For the Haber process, involving the production of ammonia from nitrogen and hydrogen:

(a) Write the balanced equation for this reaction. Include state symbols.
(b) How many moles are there in 37.5 g of H_2?
(c) How much N_2 (quoted both in moles and grams) would react with 37.5 g of H_2 to produce ammonia?
(d) What amount of ammonia would you make if you started with 37.5 g of H_2?

1.10 Explain how you would make up 500 mL of a 1 M solution of sodium hydroxide.

1.11 You are asked to take 10 mL of the solution from question 1.10 and dilute it to 100 mL. What is its concentration? How many moles are in the new solution?

1.12 What is the concentration (in $mol\,L^{-1}$) of a sugar (sucrose, $C_{12}H_{22}O_{11}$) solution if 35 g are dissolved in enough water to make 3 L?

1.13 Ethanol (C_2H_5OH) has a density of $0.789\,kg\,L^{-1}$. What volume of it is needed to add to water to make 3 L of a 0.15 M solution?

2 Atoms, periodicity and chemical bonding

Learning outcomes

By the end of this chapter students should be able to:

- understand the concepts behind the de Broglie relationship, the uncertainty principle and the Schrödinger equation;
- understand the differences between s, p and d atomic orbitals;
- use the aufbau principle, Hund's rule and the Pauli exclusion principle to predict the electron configuration of an element and establish a set of four quantum numbers to any electron;
- predict the expected chemical properties of an element from its position in the periodic table and to have an understanding of periodic trends;
- predict the nature of the bonding in a compound from a knowledge of relative electronegativity values, and draw Lewis structural representations of ionic and covalent structures;
- predict the shape of covalent molecules from valence shell electron pair repulsion (VSEPR) theory and draw 3D representations of the molecules;
- understand the concept of hybridised atomic orbitals, and the formation of molecular orbitals from the combination of atomic orbitals in describing covalent bonds, and extend the approach to understand the nature of multiple bonds;
- predict the nature of intermolecular forces that can occur between molecules and rationalise their effect on physical properties.

2.1 Introduction

This chapter will focus on:

- developing the ideas about atomic structure introduced in Chapter 1, specifically looking at both the nature of the electron itself, and the detailed electron structures (configurations) in atoms;
- showing how elements with similar electron arrangements can have similar chemical properties (periodicity) which will lead to an understanding of how the periodic table was developed;

25

- exploring the different ways in which atoms can interact, and introducing the ideas behind the different types of chemical bonding.

In the context of biology this may all seem esoteric, but an understanding of the structure of atoms is very important in developing an understanding of:

- molecular shape and reactivity;
- the workings of the different analytical methods available to chemical and biological scientists.

2.2 Electronic structure

In order to fully appreciate the nature of the electron, and how electrons are arranged in atoms, it is helpful to briefly overview how the principles were developed in the early part of the twentieth century. It was during this period that experiments were carried out that resulted in the basic model of the atom being revealed, with the negative electrons being arranged around a central, positive nucleus. The intuitive view of this is to picture electrons as small particles orbiting around the central nucleus (represented in Figure 2.1 for a helium-4, $^4_2$He, atom); however, the reality is somewhat more complex.

It is the electrons that are the active participants in chemical processes and, as such, it is important to develop an understanding of the electronic structure of atoms in order to explain chemical behaviour. Key to the understanding of electronic structure and the behaviour of electrons is the study of electromagnetic radiation.

2.3 Electromagnetic radiation

Electromagnetic radiation is a wave-like form of energy that has both electrical and magnetic components. It is generally described in terms of the frequency of its wave, though the wavelength would be equally valid. The waves travel in a vacuum at the velocity of light (approximately $3 \times 10^8 \, \text{m s}^{-1}$). The range of wavelengths found (often referred to as the electromagnetic spectrum) is between 10^{-14} m and 10^4 m, as shown in Figure 2.2.

From the figure it is clear that visible light is one (relatively small) part of the spectrum shown, and it ranges between wavelengths of 4×10^{-7} m (the violet end of the visible spectrum) and 7×10^{-7} m (the red end of the visible spectrum).

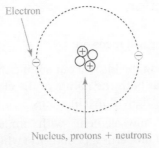

Figure 2.1 Schematic representation of a helium atom.

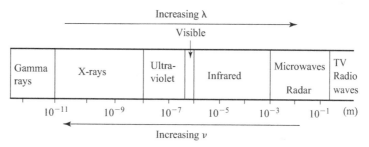

Figure 2.2 **The electromagnetic spectrum.**

There is a relationship between the velocity of a wave and its frequency and wavelength, and this is:

$$\text{velocity (ms}^{-1}) = \text{wavelength (m)} \times \text{frequency (s}^{-1})$$

In mathematical symbolism this would be, for electromagnetic radiation:

$$c = \lambda\, v$$

where

- c = the velocity of light;
- the wavelength is represented by λ;
- the frequency is represented by v.

(See Box 2.1 for a look at some definitions of the terms used here.)

When electromagnetic radiation interacts with matter there is a transfer of energy from the radiation to the matter. For example, microwaves fall within the electromagnetic spectrum, and can be used to heat food. This leads to the question of how this energy is transferred.

In the late nineteenth century and the early part of the twentieth century, a variety of experiments were carried out involving light. As a result of looking at how heated solids emitted light, and how different temperatures produced different coloured radiance (white hot is hotter than red hot), Max Planck proposed that the energy put into the solids could only be taken in as discrete amounts, or *quanta*. He proposed that the energy associated with the radiation is related to its wavelength or frequency, as shown in the equation:

$$\text{Energy for a quantum} = E = h\,v\ (E \text{ is in joules, J})$$

where h is a constant, known as Planck's constant $= 6.63 \times 10^{-34} \text{J s}^{-1}$. Hence, as $c = \lambda\, v$, then it follows that $v = c/\lambda$, and so:

$$E = h\,c/\lambda$$

Einstein developed these ideas by considering the results of experiments involving shining light at the surfaces of certain metals. This resulted in a phenomenon known as the photoelectron effect, where electrons were expelled from the metal as a result of the light hitting it. His observations suggested that an electron is 'knocked out' of the metal surface through a collision with one particle of light. These particles were

27

Box 2.1

Waves: description

There are three main features that characterise waveforms, these being wavelength, frequency and amplitude. In the context of this book we are most interested in frequency and wavelength. The wavelength is illustrated below:

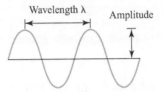

The unit of wavelength is metres (m). The frequency (the symbol for which is v, which is the Greek letter nu), is the number of complete waves passing a point in a given time (usually per second), and its SI unit is s^{-1} or Hz. To illustrate frequency, consider the figures below:

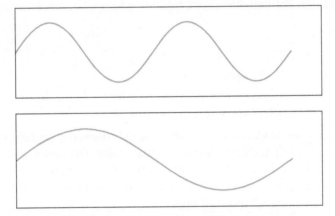

Note the greater number of peak maxima and minima in the upper trace compared with the lower one; it has a greater frequency (it has a frequency of $2 \times$ the lower trace).

---

called *photons*, and they required a minimum amount of energy (called the work function) in order to eject electrons.

The energy associated with the radiation is related to its wavelength or frequency, as shown in the equation:

$$\text{Energy for a photon} = E = h\,v$$

This relationship can be easily extended to one mole of photons by redrafting the equation to incorporate Avogadro's number:

$$\text{Energy of 1 mole of photons} = E = N_A\,h\,v$$

(where E is now in $J\,mol^{-1}$). This behaviour is known as *wave–particle duality*.

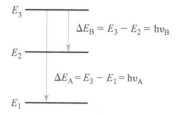

Figure 2.3 **Transitions between defined electron energy levels.**

2.4   The Bohr model of the atom

After the work outlined above, Niels Bohr looked at the phenomenon of line spectra, and as a result he made a number of key assumptions about the structures of atoms in general:

- Electrons adopt a circular orbit around the central nucleus.
- These electrons could only adopt defined orbits, and these orbits would have allowed energy states.
- For electrons to move between energy states a photon of light would need to be absorbed or emitted.
- The photon of light would need to equal the energy difference between the two energy levels, where the value of the energy difference, ΔE, is related to the frequency of the photon by $\Delta E = h\nu$.

The movement of electrons between defined energy levels is illustrated diagrammatically in Figure 2.3. The energy levels are said to be *quantised*.

Bohr's theory was a major step forward, making, as it did, the assumption that electrons could be quantised, and that defined quanta of energy were required to move electrons between energy levels. Indeed, Bohr's theory was very good (though not perfect) in rationalising the emission spectrum of hydrogen.

Bohr's theory was further developed by the French scientist Louis de Broglie, who proposed that electrons, which had been considered as particles, can behave like waves. He suggested that the wavelength of the wave would be related to the mass and the velocity of the particle by the relationship:

$$\lambda = \frac{h}{mv}$$

where λ is the wavelength, m is the mass, v is the velocity and h is Planck's constant. The quantity mv is the momentum of the object.

2.5   An introduction to atomic orbitals

Summarising the key points made so far:

- Electrons can be found in orbitals around the central nucleus of atoms.
- Electrons possess both particle and wave properties.

29

- The idea of electrons in circular orbits around the nucleus is too simplistic. There were features of emission spectra that were not explained by his model.

In Bohr's theory of the electronic structure in atoms, the electron was thought to be orbiting the nucleus, rather like the Earth orbits the Sun. This would, however, mean that it was possible to know the position of the electron at all times. In 1927, Heisenberg demonstrated that it is not possible to know simultaneously with absolute precision both the position and momentum of an electron. This is the Heisenberg uncertainty principle.

The mathematical form of this is:

$$\Delta x \cdot \Delta(mv) \geq \frac{h}{4\pi}$$

where:

- Δx = the uncertainty in the position;
- $\Delta(mv)$ = the uncertainty in the momentum;
- h = Planck's constant.

(Note that as the mass of the electron is well known, any uncertainty in momentum will arise from uncertainty regarding velocity, so $\Delta(mv)$ is often thought of in terms of the error in the velocity of the electron, Δv.)

In 1926, Erwin Schrödinger developed a theory for the wave properties of electrons in atoms, which is now known as the Schrödinger wave equation. Using this, it is possible to devise a picture to establish where there is a 95% probability of finding an electron. For a hydrogen atom it is spherical, as shown in Figure 2.4.

Key outcomes:

- Figure 2.4 represents a diagrammatic representation of a solution of the Schrödinger equation and this is called an orbital.

- The orbital represented in Figure 2.4 is called an s orbital.

The solutions to the Schrödinger wave equation for the hydrogen atom showed that electrons in an atom can be described by four quantum numbers, three of which define the orbitals in which electrons can be found, and one describes the electron spin. It is unnecessary to go into the mathematics surrounding this, but it is useful to summarise the main outcomes, discussing their significance.

2.5.1 Quantum numbers

Principal quantum number, n

When describing the Bohr representation of the atom, the theory suggested that electrons could be found in different energy levels, often referred to as shells, and

Figure 2.4 Representation of electron distribution in a hydrogen atom.

the principal quantum number, n, is the one that represents these levels. For the first level, or shell, $n = 1$; for the second shell $n = 2$ and so on.

Higher values of n represent higher potential energies associated with the shell, meaning that electrons in higher shells are likely to be further from the nucleus.

Angular momentum quantum number, l

This quantum number, also referred to as the azimuthal quantum number, defines the shape of the orbital. It can have integral values that range from 0 to $n - 1$. Thus,

- for a principal quantum number $n = 1$, then there can only be one value of angular quantum number l, which is 0;
- for a principal quantum number $n = 2$, there can be two values of angular quantum number l, namely 0 and 1;
- for a principal quantum number $n = 3$, there can be three values of angular quantum number l, namely 0, 1 and 2;
- for a principal quantum number $n = 4$, there can be four values of angular quantum number l, namely 0, 1, 2 and 3;

and so on.

There are symbols assigned to the different values of l, these being:

- $l = 0$ has the symbol s;
- $l = 1$ has the symbol p;
- $l = 2$ has the symbol d;
- $l = 3$ is given the symbol f.

For any given set of angular momentum quantum numbers, the s orbitals have the lowest energy, followed by the p, then the d and then the f (as an example, in order of increasing energy, 3s < 3p < 3d). These s, p, d, f ..., etc. orbitals are often called subshells.

Magnetic quantum number, m_l

This quantum number defines the spatial orientation of the orbitals. It can have values that range from $-l$, through 0, to $+l$. Thus,

- for an s orbital there is one value of m_l, namely 0;
- for a p orbital there are three values of m_l, namely -1, 0 and $+1$;
- for a d orbital there are five values of m_l, namely -2, -1, 0, $+1$ and $+2$
- for an f orbital there are seven values of m_l, namely -3, -2, -1, 0, $+1$, $+2$ and $+3$;

and so on. Table 2.1 summarises the results described above.

Earlier it was shown that, for the hydrogen atom in its ground state, the electron is found in an s orbital. This is, in fact, the 1s *atomic orbital* and, as the orbital is spherical, there is an equal chance of the electron being found at any point about the nucleus. The other possible orbitals that were identified by solving the Schrödinger equation are said to be unoccupied. One important consequence of 'doing the maths' and solving the Schrödinger equation is that other orbitals will have different spatial orientations. The p orbitals, for example, are usually described as being a dumbbell shape, as indicated in Figure 2.5.

Table 2.1 **Relating *n*, *l* and *m*ₗ**

Value of n	Possible values of l	Subshell name	Possible values of mₗ
1	0	1s	0
2	0	2s	0
2	1	2p	$-1, 0, +1$
3	0	3s	0
3	1	3p	$-1, 0, +1$
3	2	3d	$-2, -1, 0, +1, +2$
4	0	4s	0
4	1	4p	$-1, 0, +1$
4	2	4d	$-2, -1, 0, +1, +2$
4	3	4f	$-3, -2, -1, 0, +1, +2, +3$

Figure 2.5 **A representation of a p orbital.**

Since for the p orbitals $l = 1$, there are three possible values of m_l, namely -1, 0 and $+1$, then there are three p atomic orbitals *of equal energy*. When orbitals have equal energy they are said to be *degenerate*, but as they have three different values of m_l, they will have different spatial orientations. They are, in fact, arranged along

Box 2.2

```
Some key points about atomic orbitals
```

- An atomic orbital defines the region of space around an atomic nucleus where there is a 95% probability of an electron residing.

- s orbitals are spherical about the nucleus.

- p, d, f, etc. orbitals are not spherical. They have directional components.

- For increasing values of principal quantum number, *n*, the electrons are most probably found further away from the nucleus than those of smaller values of *n*. Thus a 3s orbital covers more volume than a 2s, which is in turn bigger than a 1s orbital, as illustrated below:

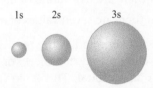

- Application of energy can result in electrons being promoted from lower energy orbitals to higher energy ones. These electrons then return to their lower energy states with the release of energy (hence the lines observed in atomic spectra).

- An atomic orbital can accommodate a maximum of two electrons.

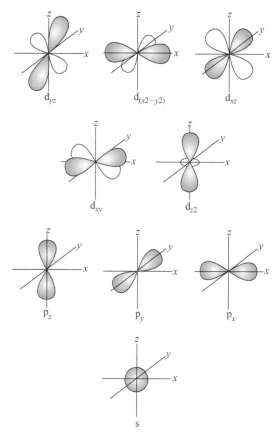

Figure 2.6 Shapes of atomic orbitals.

three mutually perpendicular axes (the orbitals, therefore, are called p_x, p_y and p_z). Similarly for the d orbitals, as there are five values of m_l, there are five degenerate orbitals.

Figure 2.6 illustrates the range of different spatial arrangements found for the various s, p and d orbitals. Note that the s orbital is spherical, the p orbitals are all at 90° to each other, and for the d_{xz}, d_{yz} and d_{xy} orbitals, the lobes all lie in between the axes mentioned – the lobes of the d_{xz} orbital lie between the x and z axes, etc.

The fourth quantum number, m_s, the spin quantum number

This relates to a property of the electron known as spin. The electron can be thought of as having spin in two opposite directions, and therefore the spin quantum number can have either of two values, $+\frac{1}{2}$ or $-\frac{1}{2}$.

Quantum numbers – a summary

There are, therefore, a total of four quantum numbers that will describe the energy and likely location of a single electron in a many electron system, these being n, l, m_l and m_s. These, and their main properties, are summarised in Table 2.2.

Table 2.2 **The quantum numbers and their properties**

Type of quantum number	Symbol	Value	Principal property of orbital/electron
Principal	n	1, 2, 3, ...	Orbital size/energy
Angular (azimuthal)	l	$(n-1)$, i.e. 0, 1, 2, ...	Orbital shape
Magnetic	m_l	$-1, ... 0, ... +1$	Orbital orientation and multiplicity (how many)
Spin	m_s	$+\frac{1}{2}$ or $-\frac{1}{2}$	Electron spin

2.5.2 The Pauli exclusion principle

Wolfgang Pauli, in 1925, stated that 'No two electrons in a single atom can have the same set of four quantum numbers'. This statement leads to two very important conclusions:

- There can be no more than two electrons in any atomic orbital.
- If there are two electrons in an orbital then they must have opposite spin (the terminology often used for this is that they must be antiparallel).

Some key points about atomic orbitals are summarised in Box 2.2 (see p. 32).

2.6 Electron configurations in atoms

To summarise the above discussions, it is relatively easy to establish that the maximum number of electrons that can be held in the principal energy levels is $2n^2$, where n is the principal quantum number. The outcome for the first four shells is:

- The energy level with $n = 1$ can contain a maximum of $2 \times 1^2 = 2$ electrons.
- The energy level with $n = 2$ can contain a maximum of $2 \times 2^2 = 8$ electrons.
- The energy level with $n = 3$ can contain a maximum of $2 \times 3^2 = 18$ electrons.
- The energy level with $n = 4$ can contain a maximum of $2 \times 4^2 = 32$ electrons.

2.6.1 How electrons are arranged in shells and subshells

Consider the hydrogen atom. Hydrogen-1, ^{1}H, has a nucleus consisting of just one proton, and it has one electron. In the most stable form of the atom (known as its *ground state*), the electron is found in the 1s orbital, and this situation can be represented as $1s^1$, where the superscript represents the number of electrons in the orbital.

Now consider the second smallest atom, helium. Helium-4, ^{4}He, has a nucleus comprising two protons and two neutrons, and there are two electrons. The ground state form of this locates both of the electrons in the 1s orbital also, and this is therefore written as $1s^2$. Sometimes, however, it is useful to be explicit about the antiparallel nature of electrons in orbitals. This can be achieved by having a box representing an orbital, and arrows representing the electrons.

Using this type of representation for hydrogen and for helium, the configurations can be shown as:

Hydrogen   Helium

1s $\uparrow$ 1s $\uparrow\downarrow$

The situation for helium is consistent with the Pauli exclusion principle. Whilst the electrons both have three quantum numbers that are identical, namely $n = 1$, $l = 0$, $m_l = 0$, they differ in that one has $m_s = +\frac{1}{2}$ and the other has $m_s = -\frac{1}{2}$. The question that remains to be answered is how are orbitals filled in multi-electron atoms?

This question is answered by drawing on two further principles, namely the *aufbau principle* and *Hund's rule of maximum multiplicity*.

2.6.2 The aufbau principle

The aufbau principle states that electrons will occupy orbitals of the lowest energy first. This relies on possessing a knowledge of the relative energies of orbitals, and this information was provided by analysis of spectroscopic data. The order of relative energy (from low to high) is:

1s 2s 2p 3s 3p 4s 3d 4p 5s 4d 5p 6s 4f 5d 6p 7s 5f 6d 7p

This does not seem like a very intuitive order, but there are methods by which it can be memorised. For example, there is the $(n + l)$ method, in which it is recognised that orbitals are filled in order of the increasing value of the sum of the two quantum numbers n and l. If the value of $(n + l)$ is identical, then the lower value of n has the lower energy.

Redrafting the order shown above in a table, with the $(n + l)$ values incorporated produces the results shown in Table 2.3, which is consistent with the $(n + l)$ method as outlined.

2.6.3 Hund's rule of maximum multiplicity

The aufbau principle describes how orbitals are filled in terms of their relative energy, but it does not address what happens in the case of degenerate orbitals, which possess the same energy. For example, for any given value of n (except 1), there will be three p orbitals of equal energy. In these cases Hund's rule of maximum multiplicity is used, and it states that:

When electrons occupy degenerate orbitals, they fill each singly with parallel spins before spin pairing occurs.

The application of these rules is most readily illustrated with a few examples. Figure 2.7, for example, shows the electron configurations that exist for the atoms

Table 2.3 **Ordering for filling atomic orbitals**

Orbital	1s	2s	2p	3s	3p	4s	3d	4p	5s	4d	5p	6s	4f	5d	6p	7s	5f	6d	7p
$(n + l)$	1	2	3	3	4	4	5	5	5	6	6	6	7	7	7	7	8	8	8

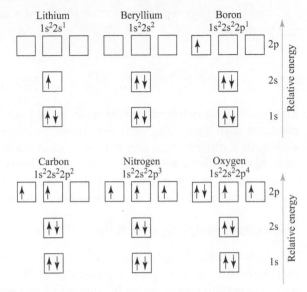

Figure 2.7 Electron configurations for the elements with atomic numbers 3–8, lithium to oxygen.

from lithium to oxygen. It can be seen that the 2p orbitals are filled up sequentially in going from boron to nitrogen, and only with oxygen does pairing of electrons in the 2p orbitals occur.

In each of the examples in Figure 2.7, the outermost electrons are found in the shell with principal quantum number $n = 2$. Because these electrons are in the outermost shell of the atom, these are the electrons that will participate in chemical bonding (to be discussed further later in this chapter), and these electrons are referred to as *valence electrons*.

The way of drawing these configurations as represented in Figure 2.7, whilst demonstrating in an intuitive sense the relative energies (and in particular the degeneracy where appropriate) of the respective orbitals, is wasteful in terms of both time and space. Figure 2.8 shows an alternative presentation, where the same information is represented more succinctly by redrafting the energy levels horizontally. Although

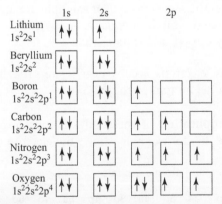

Figure 2.8 Electron configurations for the elements with atomic numbers 3–8, lithium to oxygen; alternative representation to Figure 2.7.

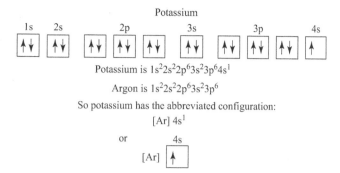

Potassium is $1s^2 2s^2 2p^6 3s^2 3p^6 4s^1$

Argon is $1s^2 2s^2 2p^6 3s^2 3p^6$

So potassium has the abbreviated configuration:

$[Ar]\,4s^1$

Figure 2.9 Full and abbreviated electron configuration of potassium.

each individual orbital can only hold two electrons, for nitrogen and oxygen there are three and four electrons respectively in the 2p orbitals as a whole, hence the representations as $1s^2$, $2s^2$, $2p^3$ for nitrogen and $1s^2$, $2s^2$, $2p^4$ for oxygen.

2.6.4 The noble gases and abbreviated configurations

The elements helium, He; neon, Ne; argon, Ar; krypton, Kr; xenon, Xe; and radon, Rn, which are collectively known as the noble gases, are unique in that they have full valence electron shells. Helium, as we have seen, has the configuration $1s^2$; neon has the configuration $1s^2$, $2s^2$, $2p^6$ and argon has the configuration $1s^2$, $2s^2$, $2p^6$, $3s^2$, $3p^6$. Because these elements have full valence shells, they are chemically extremely inert, and they are often used as reference points for abbreviating electron configurations. To explain this further, consider the element potassium, the main isotope of which is ^{39}K.

This isotope has 19 protons and 20 neutrons forming the nucleus, and it has, therefore, 19 electrons. The different ways of representing the configuration are shown in Figure 2.9.

2.7 The periodic table

The Russian scientist Dimitri Mendeleev, in 1869, was credited with listing the chemical elements in a tabulated form, based on the chemical properties displayed by those elements. Since that time, the understanding of chemical reactivity has developed, and the rationale for the table has been updated, but in essence the modern version is only marginally changed from the table proposed by Mendeleev. Figure 2.10 shows a modern form of the table.

The current rationalisation for the layout of the table is based on the electron configurations of the elements. In order to illustrate this, look at the column labelled 15, containing the elements N, P, As, Sb and Bi. The electron configurations for these elements are, in abbreviated form:

N is $[He]\,2s^2\,2p^3$
P is $[Ne]\,3s^2\,3p^3$
As is $[Ar]\,4s^2\,3d^{10}\,4p^3$
Sb is $[Kr]\,5s^2\,4d^{10}\,5p^3$
Bi is $[Xe]\,6s^2 4f^{14}\,5d^{10}\,6p^3$

1	2	3	4	5	6	7	8	9	10	11	12	13	14	15	16	17	18
1 H 1.008																	2 He 4.003
3 Li 6.941	4 Be 9.012											5 B 10.811	6 C 12.011	7 N 14.007	8 O 15.999	9 F 18.998	10 Ne 20.180
11 Na 22.99	12 Mg 24.31											13 Al 26.98	14 Si 28.09	15 P 30.97	16 S 32.07	17 Cl 35.45	18 Ar 39.95
19 K 39.10	20 Ca 40.08	21 Sc 44.96	22 Ti 47.88	23 V 50.94	24 Cr 52.00	25 Mn 54.93	26 Fe 55.85	27 Co 58.93	28 Ni 58.69	29 Cu 63.55	30 Zn 65.39	31 Ga 69.72	32 Ge 72.61	33 As 74.92	34 Se 78.96	35 Br 79.90	36 Kr 83.80
37 Rb 85.47	38 Sr 87.62	39 Y 88.91	40 Zr 91.22	41 Nb 92.91	42 Mo 95.94	43 Tc (98)	44 Ru 101.07	45 Rh 102.91	46 Pd 106.42	47 Ag 107.87	48 Cd 112.41	49 In 114.82	50 Sn 118.71	51 Sb 121.75	52 Te 127.60	53 I 126.90	54 Xe 131.29
55 Cs 132.91	56 Ba 137.33	57 La* 138.91	72 Hf 178.49	73 Ta 180.95	74 W 183.85	75 Re 186.21	76 Os 190.23	77 Ir 192.22	78 Pt 195.08	79 Au 196.97	80 Hg 200.59	81 Tl 204.38	82 Pb 207.2	83 Bi 208.98	84 Po (209)	85 At (210)	86 Rn (222)
87 Fr (223)	88 Ra 226.03	89 Ac* 227.03															

| | | 58
Ce
140.12 | 59
Pr
140.91 | 60
Nd
144.24 | 61
Pm
(145) | 62
Sm
150.36 | 63
Eu
151.97 | 64
Gd
157.25 | 65
Tb
158.93 | 66
Dy
162.50 | 67
Ho
164.93 | 68
Er
167.26 | 69
Tm
168.93 | 70
Yb
173.04 | 71
Lu
174.97 | |
| | | 90
Th
232.04 | 91
Pa
231.04 | 92
U
238.03 | 93
Np
237.05 | 94
Pu
(244) | 95
Am
(243) | 96
Cm
(247) | 97
Bk
(247) | 98
Cf
(251) | 99
Es
(252) | 100
Fm
(257) | 101
Md
(258) | 102
No
(259) | 103
Lr
(260) | |

Figure 2.10 Periodic table of the elements.

For radioactive elements that do not occur in nature, the mass number of the most stable isotope is given in parentheses.
*The elements with atomic numbers 58–71 and 90–103, in the bottom two rows, are formally often regarded as part of the f-block groups known as the lanthanides and actinides.

There is a constant theme, in that the outermost electrons are those found in the p orbitals. The elements in this column are said to be in a group, and it is called group 15. Similarly, for the elements in groups 13, 14, 16, 17 and 18 (except helium), the outermost electrons are in the p orbitals. The elements in these groups (except helium) are referred to as p-block elements.

There are patterns found in other areas of the table also. The elements in groups 1 and 2 have their outermost electrons in s orbitals and are, therefore, often referred to as s-block elements, while those elements in groups 3–12 are called d-block elements, as their outermost electrons are in d orbitals.

2.7.1 Periodic properties

The periodic table is a listing of elements arranged by their atomic number. It illustrates the way in which their physical and chemical properties vary periodically.

Atomic radius

The atomic radius refers to the distance between an atom's nucleus and its outermost (valence) electrons. Electrons that are closer to the nucleus have lower energy and are more tightly held. Consider the options:

- Moving from left to right in a line across the table is referred to as moving across a period and, in doing so, the atomic radius decreases. This is because in moving from left to right the nucleus of the atom gains protons, increasing the positive charge of the nucleus and increasing the attractive force of the nucleus upon the negatively charged electrons. In order for the atoms to maintain neutrality, electrons are also added as the elements move from left to right across a period, but these electrons reside in the same energy shell. The consequence of this is that the newly added electrons do not receive effective shielding from the increased positive charge of the nucleus, and so are held more tightly, thereby resulting in the atomic radius decreasing.

- Moving down a group results in an increase in the atomic radius. In this case protons are added on moving down a group, but so are new energy levels (shells) of electrons. The new energy shells provide effective shielding, allowing the valence electrons to experience only a minimal amount of the increased positive charge of the protons.

Below is a pictorial representation of the atomic radii (in picometres, pm) along the second period, and down group 17.

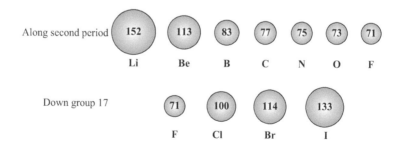

Ionisation and ionisation energy

Ionisation occurs when an atom gains or loses electrons, leading to either negatively charged or positively charged species, known as ions. There are energy changes associated with such processes and there are two common ways to measure these energy changes, namely as:

- *ionisation energies* for formation of positive ions, called *cations* (loss of electrons);
- *electron affinities* for formation of negative ions, called *anions* (gain of electrons).

The ionisation energy is the energy change needed to fully remove an electron from the atom. Because energy needs to be added to the atom, the energy change is positive. When multiple electrons are removed from an atom, the energy required to remove the first electron is called the first ionisation energy, the energy required to remove the second electron is the second ionisation energy, and so on (see Box 2.3).

In general, the second ionisation energy is greater than the first ionisation energy. This is because the first electron removed feels the effect of shielding by the second electron and is therefore less strongly attracted to the nucleus. The second electron, however, does not have this protection from the nucleus, and so is more tightly held, therefore requiring more energy to remove it. If the removal of an electron empties a subshell, the next ionisation energy is substantially larger, rather than following its normal gently increasing trend. This fact helps to show that just as atoms are more stable when they have a full valence shell, they are also relatively more stable when they at least have a full subshell.

The variation of ionisation energy for the s and p group elements of the first three periods is illustrated very clearly in Table 2.4 and in the graph shown in Figure 2.11.

Box 2.3

The ionisation of calcium

Only valence electrons are removed during a chemical reaction. For calcium the values of the first three ionisation energies (IE) are shown.

$$1\text{st IE Ca (g)} \rightarrow Ca^+(g) + e^- \quad 590 \text{ kJ mol}^{-1}$$

$$2\text{nd IE Ca}^+(g) \rightarrow Ca^{2+}(g) + e^- \quad 1146 \text{ kJ mol}^{-1}$$

$$3\text{rd IE Ca}^{2+}(g) \rightarrow Ca^{3+}(g) + e^- \quad 4912 \text{ kJ mol}^{-1}$$

The third ionisation energy is much greater than the first two because the third electron is in a lower shell – it is not a valence electron, and as such chemical processes only make the Ca^{2+} ion.

The variation of ionisation energy across a period

Ionisation energy generally increases moving across the periodic table from left to right. When looking at the variation of atomic radius with position in the periodic table, it was seen that, on moving from left to right, the number of protons increased, but although the number of electrons also increased, there was a very limited increase in shielding. From left to right, the electrons therefore become more tightly bound, meaning more energy is needed to remove them. There are, however, some deviations from this trend. Looking at the period starting with lithium, Li, there are dips at boron, B, and at oxygen, O. In the case of boron, this is thought to occur because the outermost electron is on its own in a p orbital, which is a higher energy orbital, thereby making this electron easier to remove. The rationale

Table 2.4 **Ionisation data for selected elements**

Element	First ionisation energy	Second ionisation energy	Electron affinity
H	131	—	−71
He	237	525	—
Li	520	729	−60
Be	899	175	—
B	801	242	−27
C	108	235	−122
N	140	285	0
O	131	338	−141*
F	168	337	−328
Ne	208	395	—
Na	496	456	−53
Mg	738	145	—
Al	577	181	−44
Si	786	157	−134
P	101	190	−72
S	100	225	—
Cl	125	229	−349
Ar	152	266	—
K	419	306	−48
Ca	590	114	—
Br	114	208	−325
Kr	135	236	—
Rb	403	265	−47
Sr	549	106	—
I	100	184	−295
Xe	117	204	—
Cs	376	242	−45
Ba	503	965	—

*Estimated second electron affinities for O and S are +791 and +565 kJ mol^{-1} respectively.
Ionisation energies are the energies required to remove a mole of electrons from a mole of gaseous atoms (in kJ mol^{-1}). They are always positive for neutral atoms or cations.
Electron affinities are the energies required to add a mole of electrons to a mole of gaseous atoms. Second electron affinities are always positive.

for oxygen is that the presence of a fourth electron in the p subshell can only be achieved by pairing electrons within an orbital, which causes the electrons in this situation to be of slightly higher energy, so causing the first ionisation energy to drop a little compared with nitrogen.

The variation of ionisation energy going down a group

Ionisation energy decreases moving down a group for the same reason that atomic size increases: electrons filling more levels create extra shielding, protecting the outermost

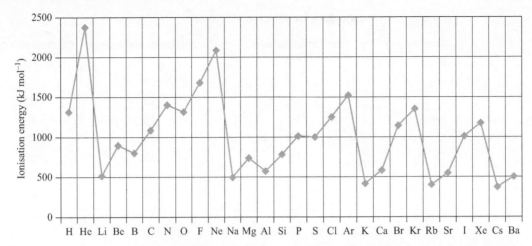

Figure 2.11 **The variation of ionisation energies with atomic number for the s and p block elements.**

ones from the additional protons. The atomic radius increases as we go down a group, and so does the energy of the valence electrons. Therefore less energy is needed to re-move an electron, resulting in decreased ionisation energy measurements. This is very apparent when looking at how the ionisation energies of the noble gases change with atomic mass; the larger the atom the lower the ionisation energy.

Electron affinity

The electron affinity of an atom is the energy change that results from the atom gaining an electron. When the resultant ion is more stable than the atom, then it has a lower potential energy associated with it and so the energy change is negative.

The variation of electron affinities across a period

Electron affinities show a generally increasingly negative trend on going from left to right across a period, though there is a break in this pattern at group 15.

The variation of electron affinities moving down a group

Electron affinities do not vary much on moving down a group, though they do generally become slightly more positive. The exceptions to this trend are the third period elements, which often have more negative electron affinities than the cor-responding elements in the second period. For this reason, chlorine, Cl (group 17, period 3), has the most negative electron affinity.

The octet rule and ionisation

The noble or inert gases, those elements in column 18 of the periodic table, have either two valence electrons (He) or eight valence electrons (Ne, Ar, Kr, Xe, Rn). They are said to have a full outer electron shell: this is a particularly stable electron configuration, hence the name noble or inert gases. Other elements are more reac-tive, and are regarded as trying to achieve a stable, noble gas, configuration, which forms the basis for the octet (8) rule. This states that elements will react to attain the stable octet of electrons.

Consider, for example, sodium, Na. Sodium has a total of 11 electrons with the configuration $1s^22s^22p^63s^1 = [Ne]3s^1$. To get to the octet, the nearest stable noble gas configuration, it needs to lose one electron (its valence electron, the $3s^1$) to give the positively charged ion Na^+, which has the same electron configuration as neon, that is $1s^22s^22p^6$.

Compare this with aluminium, Al. Aluminium has 13 electrons with the configuration $1s^22s^22p^63s^23p^1 = [Ne]3s^23p^1$. To achieve an octet it needs to lose three electrons (its valence electrons, the $3s^23p^1$) to give Al^{3+} which has the same configuration as neon.

In other words,

- Na^+, Al^{3+} and Ne all have the same electron configuration: $1s^22s^22p^6$.
- They are said to be *isoelectronic* (they have the same number of electrons).
- Both Na^+ and Al^{3+} are *cations*.

Elements on the right hand side of the periodic table tend to gain (add) electrons in order to get to a noble gas configuration (they would have to lose too many electrons to go the other way). For example, take fluorine, F, which has a total of nine electrons with the configuration $1s^22s^22p^5$. In order to get to the octet it is necessary to add one electron, to give F^-, which has the configuration $1s^22s^22p^6$. This is the fluoride ion which has a stable octet electronic configuration.

Compare this with nitrogen, N, the atom of which has the electron configuration $1s^22s^22p^3$. In order to achieve the nearest octet it is necessary to add three electrons to give N^{3-}, with an electronic configuration of $1s^22s^22p^6$. This is the nitride ion which is stable.

- Both F^- and N^{3-} are *anions*.
- The species Na^+, Al^{3+}, F^-, N^{3-} and Ne are all isoelectronic with a total of ten electrons.

The octet rule is particularly helpful for the elements between boron and magnesium. The elements with lower atomic numbers are governed principally by the helium configuration, whereas d orbitals can be implicated for those elements with higher atomic number.

Transition metals also lose electrons to form positively charged ions (cations) but they can form various stable electron configurations, e.g. Fe^{2+} or Fe^{3+}, and so the charge on a transition metal ion cannot always be readily predicted.

Ion size

- Positive ions (cations) are smaller than the parent atom for two reasons:
  - Cations have more protons than electrons and increased nuclear charge pulls electrons in towards the nucleus.
  - When the cation has a noble gas configuration then it has lost its outermost electrons to give a lower n value noble gas configuration, therefore it is smaller.
- Negative ions (anions) are larger than the parent atom as anions have more electrons than protons and the nucleus has less pull.
- Multiply charged species are larger (for anions) or smaller (for cations) than the corresponding singly charged species. Thus, for example, $O^{2-} > O^- > O$ or $Cu^{2+} < Cu^+ < Cu$.

2.8 An introduction to bonding: how atoms become molecules

2.8.1 Introduction

Atoms are held together to form molecules by chemical bonds, and an understanding of chemical bonding is core to an understanding of chemical (and hence biological) reactions. The breaking and making of chemical bonds is essentially what is happening in chemical reactions. A chemical bond is formed when atoms or molecules rearrange their electrons to bring about a lower energy situation than existed prior to the bond formation, and the lower the energy the more stable the outcome. In the previous section it was shown that some elements lose electrons to form cations, whereas some gain electrons, resulting in anion formation. Thus there is a separation of the periodic table into metals and non-metals, in which metals generally lose electrons to form cations whereas non-metals tend to gain electrons to form anions.

The types of chemical bond that will be described here are:

- ionic bonds, usually (not always) result from the interaction of a metal with a non-metal;
- covalent bonds, which usually result from interaction between non-metal and non-metal;
- metallic: metal with metal.

Before discussing bonding in more detail, however, we need to look at a method of showing valence electrons in atoms in order to help with bonding. The approach that will be used is known as the Lewis electron-dot method. As an example, consider the atom of carbon, C, for which the full electron configuration is $1s^2 2s^2 2p^2$. This has four valence electrons (those in the principal shell 2, namely the $2s^2 2p^2$ electrons). Using the Lewis electron-dot method this is shown such that all core electrons are ignored and the valence electrons are put at the N-S-E-W points of compass only:

This approach is now extended to atomic sulfur, S, for which the electron configuration is $1s^2 2s^2 2p^6 3s^2 3p^4$, for which there are six valence electrons, $3s^2 3p^4$, depicted as shown:

These representations can be used to help illustrate different bonding types.

The elements, with the exceptions of the noble gases, do not exist naturally in atomic form, but rather as molecules or as extended arrays. This is because the situation where unpaired electrons exist is relatively unstable, and atoms in elements will find ways of bonding with each other to create a more stable system. Likewise, unpaired electrons in molecular systems are inherently unstable.

2.8.2 Ionic bonding

This type of bonding generally occurs between a metal and a non-metal – between ions. The metal loses electrons to attain a stable octet configuration, whereas the non-metal gains electrons to achieve the stable octet configuration. In order for this to happen there is an electron transfer from the metal to the non-metal, and then an electrostatic attraction between the resulting positive cations and negative anions, rather like the N–S attraction of a magnet. Ionic bonding leads to three-dimensional solids. Consider the example of lithium fluoride, LiF. Representing the situation in terms of electron configurations:

move electron

$$Li\ 1s^22s^1\ +\ F\ 1s^22s^22p^5\ \longrightarrow\ Li^+\ 1s^2\ +\ F^-\ 1s^22s^22p^6$$

stable [He]   stable [Ne]
configuration   configuration

In short, Li transfers an electron to F, resulting in Li^+F^-.

It is possible to represent this in a Lewis dot diagram also, as shown:

$$Li\bullet\ +\ \bullet\ddot{\underset{\bullet\bullet}{F}}\colon\ \longrightarrow\ Li^+\ +\ \colon\ddot{\underset{\bullet\bullet}{F}}\colon^-$$

The same approach can be extended even where the cation : anion ratio is not 1:1. Consider, for example, the interaction between calcium and fluorine to form calcium fluoride:

Step 1

$$Ca\colon\ +\ \bullet\ddot{\underset{\bullet\bullet}{F}}\colon\ \longrightarrow\ Ca^+\bullet\ +\ \colon\ddot{\underset{\bullet\bullet}{F}}\colon^-$$

Step 2

$$Ca^+\bullet\ +\ \bullet\ddot{\underset{\bullet\bullet}{F}}\colon\ \longrightarrow\ Ca^{2+}\ +\ \colon\ddot{\underset{\bullet\bullet}{F}}\colon^-$$

$$\left(+\ \colon\ddot{\underset{\bullet\bullet}{F}}\colon^-\right)$$
from step 1

Overall

$$Ca\colon\ +\ 2\bullet\ddot{\underset{\bullet\bullet}{F}}\colon\ \longrightarrow\ Ca^{2+}\ +\ 2\colon\ddot{\underset{\bullet\bullet}{F}}\colon^-$$

Properties of ionic compounds

Using the example of sodium chloride, NaCl, the so-called 'common salt', ionic solids exhibit an extended 3D structure. Two representations of NaCl are shown in Figure 2.12. (A more detailed description of the different categories of solid state structures will be provided in Chapter 11.)

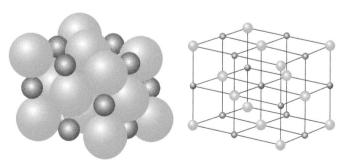

Figure 2.12 **Representations of the structure of sodium chloride.**

In general, ionic compounds share a range of common physical properties. They tend to be hard, rigid, brittle with high melting and boiling points; for example, NaCl has a melting point of 801°C, and a boiling point of 1413°C. This is because the electrostatic forces between ions are strong. Also, although they do not conduct electricity in the solid state, they do conduct it in solution and in the molten phase. In the solid phase the ions are held together very strongly and so are not free to move, but in solution or in the molten phase the ions are free to move.

2.8.3 Covalent bonding

This type of bonding generally occurs between non-metal and non-metal and arises through electron sharing. A bond between two non-metals involves two electrons in a single bond with one electron contributed by each non-metal. The shared electron pair is considered to be localised between the two non-metals. Covalent bonding usually results in formation of discrete molecules (except for some covalent network compounds such as diamond). Therefore the chemical formula reflects the actual number of atoms in the molecule, e.g. CO_2 consists of one C atom and two O atoms. Covalent bonds are largely responsible for the bonding found in simple systems such as H_2 to large biological molecules such as proteins and DNA itself.

Representations of covalent bonding

Atoms still work towards a full valence electron shell (noble gas configuration). In covalent systems, however, each atom in a covalent bond 'counts' the shared electrons as belonging entirely to itself. Taking, for example, the simplest example of the hydrogen molecule, H_2, consider the way in which these atoms can link. Atomic hydrogen has only one electron, whereas helium, the most stable noble gas configuration near to this, has two electrons. By sharing a pair of electrons, both hydrogen atoms now have two electrons. This can be depicted as:

$$H \bullet \ \bullet H \quad \text{join to give} \quad H \overset{\bullet}{\underset{\bullet}{}} H$$

A more straightforward way of representing the bonding of two hydrogen atoms is H–H. The pair of electrons linking the two atoms is referred to as a *bonding pair* of electrons.

Consider now the example of hydrogen fluoride, HF. This can be depicted as:

$$H \bullet \ \bullet \overset{\bullet \bullet}{\underset{\bullet \bullet}{F}} \quad \text{join to give} \quad H \overset{\bullet \bullet}{\underset{\bullet \bullet}{F}}$$

Again, the H has now attained the configuration of helium, whereas the fluorine, F, has achieved the configuration corresponding to neon, Ne.

Note with HF that the fluorine is bonded to the hydrogen by sharing two electrons, but around the fluorine there are also three additional pairs of electrons. These are referred to as *lone pairs* of electrons. These play an important role in determining molecular shape (as will be demonstrated later in this chapter).

Multiple bonds and the idea of bond order

In the examples above, in both HF and H_2 the atoms are held together by a single bonding pair of electrons, referred to as a single bond. They are said to have a bond order of one. As well as single bonds, however, multiple bonds are also found.

Consider carbon dioxide, CO_2, a product of our breathing and also of the combustion of fossil fuels. The constituents of this are carbon, C, which has four valence electrons, and oxygen, O, which has six valence electrons. The sharing of electrons between the C and O should be done in such a way that no unpaired electrons are left.

Atomic carbon and oxygen have the valence electron representations:

$$\cdot \overset{\cdot \cdot}{\underset{\cdot}{C}} \cdot \qquad \overset{\cdot \cdot}{\underset{\cdot \cdot}{\cdot \overset{\cdot}{O} \cdot}}$$

In order to generate the molecule CO_2 and have no unpaired electrons, the C and O atoms will need to be linked by double bonds:

$$\overset{\cdot \cdot}{\underset{\cdot \cdot}{O}} :: C :: \overset{\cdot \cdot}{\underset{\cdot \cdot}{O}}$$

which can also be shown as:

$$\overset{\cdot \cdot}{\underset{\cdot \cdot}{O}} = C = \overset{\cdot \cdot}{\underset{\cdot \cdot}{O}}$$

Therefore the carbon and oxygen atoms are linked by double bonds, and they are said to have a bond order = 2. The oxygen atoms each have two lone pairs of electrons associated with them.

The same approach can be used to explain the presence of a triple bond, for example in dinitrogen, N_2. Nitrogen, N, has five valence electrons and two atoms combining in the following way.

The two atoms of nitrogen have the valence electron representations:

$$\cdot \overset{\cdot \cdot}{N} \cdot \qquad \cdot \overset{\cdot \cdot}{N} \cdot$$

In order to generate the molecule N_2 and have no unpaired electrons, the two N atoms will need to be linked by triple bonds:

$$: N \vdots N :$$

which can also be shown as:

$$: N \equiv N :$$

Each nitrogen atom in the molecule has eight electrons associated with it, and the bond order is 3.

Electronegativity

The term electronegativity of an atom refers to its ability to attract the electrons of another atom to it when those two atoms are associated through a bond. Electronegativity is based on an atom's ionisation energy and electron affinity. For that reason, electronegativity follows similar trends as its two constituent measures. The most widely accepted (but by no means the only) measure of electronegativity is the method proposed by Linus Pauling, and the values that derive from his method are shown in the periodic table of Figure 2.13.

1	2	3	4	5	6	7	8	9	10	11	12	13	14	15	16	17	18
1 H 2.1																	2 He –
3 Li 1.0	4 Be 1.5											5 B 2.0	6 C 2.5	7 N 3.0	8 O 3.5	9 F 4.0	10 Ne –
11 Na 0.9	12 Mg 1.2											13 Al 1.5	14 Si 1.8	15 P 2.1	16 S 2.5	17 Cl 3.0	18 Ar –
19 K 0.8	20 Ca 1.0	21 Sc 1.3	22 Ti 1.5	23 V 1.6	24 Cr 1.6	25 Mn 1.5	26 Fe 1.8	27 Co 1.8	28 Ni 1.8	29 Cu 1.9	30 Zn 1.6	31 Ga 1.6	32 Ge 1.8	33 As 2.0	34 Se 2.4	35 Br 2.8	36 Kr 3.0
37 Rb 0.8	38 Sr 1.0	39 Y 1.2	40 Zr 1.4	41 Nb 1.6	42 Mo 1.8	43 Tc 1.9	44 Ru 2.2	45 Rh 2.2	46 Pd 2.2	47 Ag 1.9	48 Cd 1.7	49 In 1.7	50 Sn 1.8	51 Sb 1.9	52 Te 2.1	53 I 2.5	54 Xe 2.6
55 Cs 0.7	56 Ba 0.9	57 La 1.1	72 Hf 1.3	73 Ta 1.5	74 W 1.7	75 Re 1.9	76 Os 2.2	77 Ir 2.2	78 Pt 2.2	79 Au 2.4	80 Hg 1.9	81 Tl 1.8	82 Pb 1.9	83 Bi 1.9	84 Po 2.0	85 At 2.2	86 Rn –
87 Fr 0.7	88 Ra 0.9	89 Ac 1.1															

| | | | 58
Ce
1.1 | 59
Pr
1.1 | 60
Nd
1.1 | 61
Pm
1.2 | 62
Sm
1.2 | 63
Eu
1.1 | 64
Gd
1.2 | 65
Tb
1.2 | 66
Dy
1.2 | 67
Ho
1.2 | 68
Er
1.2 | 69
Tm
1.2 | 70
Yb
1.2 | 71
Lu
1.3 | |
| | | | 90
Th
1.3 | 91
Pa
1.5 | 92
U
1.7 | 93
Np
1.3 | 94
Pu
1.3 | 95
Am
1.3 | 96
Cm
1.3 | 97
Bk
1.3 | 98
Cf
1.3 | 99
Es
1.3 | 100
Fm
1.3 | 101
Md
1.3 | 102
No
1.5 | 103
Lr
– | |

Figure 2.13 Periodic table of the elements: each entry shows atomic number (top), element symbol (middle) and Pauling electronegativity value (bottom). Group numbers (bold) run along the top of the table and the arrows indicate increasing electronegativity values.

Electronegativity generally increases moving left to right across a period and decreases moving down a group. Fluorine (F), in group 17 and period 2, is the most powerfully electronegative of the elements, with a Pauling value of 4.0; oxygen (O), in group 16 is the next most electronegative at 3.5, with nitrogen and chlorine (N and Cl) both having values of 3.0. The least electronegative elements are caesium, Cs, and francium, Fr, both of which have Pauling values of 0.7.

2.8.4 Formal oxidation states

The concept of formal oxidation states provides a way of keeping track of electrons. There is a set of simple rules for assigning oxidation states as follows:

1. The oxidation state of an atom in an element is 0. For example, in copper metal the oxidation state of Cu is 0 – written as Cu(0).

2. The oxidation state of a monoatomic ion is the same as its charge. For example, Na^+ (where Na has lost an electron) is Na(I).

3. Oxygen is assigned an oxidation state of −2 in its covalent compounds; for example, in CO, CO_2, SO_2.

4. Hydrogen is assigned an oxidation state of +1 in its normal covalent compounds; for example, HCl, H_2O.

5. In binary compounds the assignment follows electronegativities. Therefore in compounds which contain fluorine, F, the fluorine is always −1 because it is the most electronegative element. In ammonia, NH_3, N is −3 and H is +1, because N is the more electronegative.

6. The sum of the oxidation states must be zero for an electrically neutral compound.

Box 2.4 gives examples of the application of these rules.

Box 2.4

Some examples of formal oxidation states

Assign oxidation states to all the atoms in the following:

(a) CO_2
(b) SF_6
(c) NO_3^-

Answers

(a) O is assigned the oxidation state −2, and there are two oxygens. Therefore, in order for the compound to be neutral, C must be assigned an oxidation state of +4.
(b) F is assigned the oxidation state −1, and there are six of them. Therefore, in order to account for the neutral molecule, the oxidation state of S must be +6.
(c) O is assigned −2, and there are three oxygen atoms. But this is a nitrate ion, and it has an overall 1− charge associated with it. This means that the formal oxidation state of N can only be +5.

2.8.5 Polarisation: covalent or ionic bonding?

The idea of covalency has been described as the sharing of electrons between two atoms, but this is not necessarily a comprehensive picture of what is occurring. For example, although it is entirely reasonable to expect the two H atoms in dihydrogen, H_2, to share the electrons linking them, the situation for hydrogen fluoride, HF, is not so balanced. There is a significant difference in the electronegativity values between hydrogen (at 2.1) and fluorine (at 4.0), and this results in an unequal sharing of the electrons between the atoms, with the electrons located closer to the fluorine than the hydrogen.

There are different ways this can be represented:

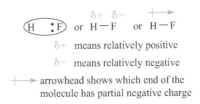

Table 2.5 **Electronegativity difference as a function of bond polarity**

ΔEN	Ionic character
>1.7	Mostly ionic
0.5–1.7	Polar covalent
≤0.4	Mostly covalent
0	Non-polar covalent

HF has a polar covalent bond, and it possesses a dipole moment as a result of the difference in charges at either end of the molecule.

Electronegativity values and ionic character

If the electronegativity difference between two bonded atoms = ΔEN, then it is fair to say that the value ΔEN is a measure of the extent of the ionic character of the bond. This is summarised in Table 2.5.

As the value of ΔEN decreases, the bond becomes more covalent. This means that C and H have almost the same electronegativities (2.5 and 2.1 respectively). Therefore C—H bonds, which form a large part of all organic and biological compounds, are considered as effectively non-polar.

Covalent bond strengths and lengths

Energy is needed to break bonds. The bond energy (BE) (also known as bond enthalpy or bond strength) is usually regarded as the energy required to break covalent bonds in 1 mole of gaseous molecules, and a positive quantity of energy is needed to be put into a system in order to break bonds.

Some typical examples of bond energies are listed in Table 2.6, alongside the corresponding bond lengths. From this it is clear that there is a correlation found between bond energies and in bond lengths, such that as bond energy increases,

Table 2.6 **Selected bond strengths and lengths**

Bond type	Bond energy/kJ mol^{-1}	Bond length*/pm
H—H	432	74
H—F	565	92
C—O	358	143
C=O	745	123
C—Cl	339	177
C—Br	276	194
C—I	216	213
N—N	160	146
N=N	418	122
N≡N	945	110

*Bond length = distance between two nuclei.

so bond length decreases. Therefore, as is clear from the table, a couple of factors emerge:

- Triple bonds are stronger and shorter than double bonds which are, in turn, stronger and shorter than single bonds.
- Taking the examples of carbon–halogen bonds, as covalent radius increases, so bond strength decreases.

2.8.6 Metallic bonding

This is not very relevant in the context of a book on chemistry for biologists, but for completeness, and also in order to get the best possible grasp of the range of bonding types, it will be referred to briefly. Generally metallic elements have low ionisation energy values, and this means that it is relatively easy for them to lose electrons. In metallic bonding all the metal atoms give up their valence electrons to a 'sea' of electrons that 'flow' between and around the metal-ion cores. The metal-ion core is arranged in an orderly 3D array. Electrons in metallic bonding are delocalised and they move freely throughout the piece of metal. Metals can exist in pure element form, but can also exist as mixtures of metals, known as alloys. Examples of alloys are found in many everyday materials, such as coinage, steel and so on.

2.8.7 Shapes of molecules: the VSEPR approach

Molecular shape is very important in chemistry and biology, and one relatively straightforward and intuitive method of predicting and rationalising molecular shape uses the valence shell electron pair repulsion (VSEPR) theory. The underlying principle of the VSEPR theory is that each group of valence electrons around a central atom is located as far away as possible from the other electron groups so as to minimise repulsions (electrons are negatively charged, so there will be electrostatic repulsion).

An electron group may consist of a single bond, a double bond, a triple bond, linking the central atom to an atom or group, or it may be a lone pair of electrons. The shape that will allow the electron groups to be as far away from each other as possible depends on the number of such groups, and among the more common ones are:

6 electron groups:	octahedral shape;
5 electron groups:	trigonal bipyramidal shape;
4 electron groups:	tetrahedral shape.
3 electron groups:	trigonal planar shape;
2 electron groups:	linear shape.

Pictorially these are shown in Figure 2.14; the central atom can be thought of as being in the centre of a sphere, with the electron groups being represented about the surface of the sphere. In addition, factors that can affect the precise shape adopted are:

- Lone pairs occupy more space than bonding electron pairs.
- Double bonds occupy more space than single bonds.

51

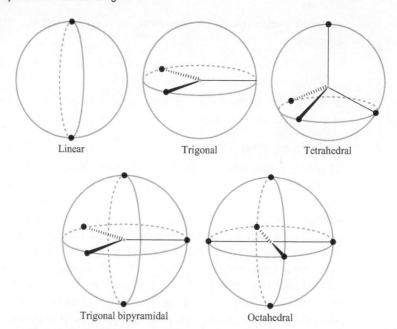

Linear   Trigonal   Tetrahedral

Trigonal bipyramidal   Octahedral

Figure 2.14 Shapes adopted as a consequence of electron repulsion in the VSEPR model.

Some examples of molecular shape

Linear

The angle at the central atom is 180°. A good example of this is CO_2, for which the Lewis structure has already been discussed (Section 2.8.3), and is shown again below:

$$\ddot{\text{:}}\text{O}=\text{C}=\text{O}\ddot{\text{:}}$$

Each double bond is considered as a separate electron group, therefore there are two electron groups around the central C atom. The C has no lone pairs of electrons and therefore the molecule has a linear shape.

Trigonal planar

The angle at the central atom = 120°. An example of this type of molecule is boron trifluoride, BF_3.

$$\text{:}\ddot{\text{F}}\text{,,,}_{B}-\ddot{\text{F}}\text{:}$$
$$\text{:}\ddot{\text{F}}$$

- A B atom has three valence electrons.
- One is contributed from each of the fluorine atoms (three in total).
- Therefore three electron pairs surround the B atom.
- Hence trigonal planar.

Note with this example that the boron does not have an octet; there are only six electrons about the B atom, and it is said to be electron-deficient. The fluorine atoms, by contrast, all have complete octets.

Tetrahedral

The angle at the central atom in a tetrahedral environment is 109.5°, and there is a clear 3D shape. The way in which it can be represented on paper is with straight lines in the plane of the paper, a solid wedge coming out of the paper, and a dashed line going into the paper (away from the reader). An example of a tetrahedrally shaped molecule is methane, CH_4.

$$\cdot \overset{\displaystyle \cdot}{\underset{\displaystyle \cdot}{C}} \cdot \ + \ 4 \times \cdot H \qquad H \diagup \overset{\displaystyle \overset{\textstyle H}{|}}{\underset{\textstyle H}{C}} \diagdown H$$

- In the Lewis structure C has four valence electrons.
- Each H has one valence electron.
- Eight electrons, so four electron groups, and hence tetrahedral about C.

A lone pair of electrons also counts as an electron group. Therefore ammonia, NH_3, has a shape that is based on a tetrahedron and not on a trigonal plane, because of the presence of a lone pair of electrons on N.

$$\cdot \overset{\displaystyle \cdot \cdot}{\underset{\displaystyle \cdot}{N}} \cdot \ + \ 3 \times \cdot H \qquad H \diagup \overset{\displaystyle \overset{\textstyle \cdot \cdot}{N}}{\underset{\textstyle H}{}} \diagdown H$$

- N has five valence electrons, H has one valence electron.
- There are three N—H bonds + a lone pair of electrons.
- This is four electron groups.
- Shape must be based on a tetrahedron.
- Including lone pair shape is tetrahedral.
- Ignoring lone pair shape is called trigonal pyramidal.
- Lone pair of electrons takes up more space than single bond so that H—N—H angle is less than tetrahedral angle of 109.5°.
- In ammonia the angle is 107.3°.

Self-test

Water, H_2O, also has a shape based on the tetrahedron. Why? How do we describe the shape based on atoms only?

Trigonal bipyramidal

Considering the general structure MX_5, the trigonal bipyramidal shape is:

$$\begin{array}{c} X_a \\ | \\ X_{e} \cdots M — X_e \\ X_e \\ | \\ X_a \end{array}$$

This shape has two different types of X groups. Two are referred to as axial, denoted X_a, and three are referred to as equatorial, denoted X_e, where the axial groups lie along an axis about which the structure can be rotated, and the equatorial groups lie about a 'belt' around the middle, hence the name equatorial from equator. In a perfect trigonal bipyramidal shape, the angle between axial and equatorial groups is 90° and between the equatorial groups is 120°.

All molecules with five or six electron groups have a central atom located in period 3 or higher in the periodic table. Because these atoms do not follow the octet rule, they can have more than eight electrons in a valence shell because they can use empty available d orbitals. It is still necessary to check for lone pairs at the central atom. These ideas can be illustrated by a couple of examples.

- PF_5. In this molecule, the P has five valence electrons and the F has seven valence electrons. The molecule has five P—F bonds, indicating that the P uses all its valence electrons in forming P—F bonds, with no lone pairs on the P. Each F contributes one electron to the P—F bonds, thereby ensuring that each F has eight electrons. Therefore, at P, there are five electron groups and the shape is trigonal pyramidal.

- SF_4. In this case the central atom is sulfur, S, which is in group 16 and has six valence electrons. Fluorine, F, has seven valence electrons, and SF_4 has four S—F bonds which leaves two non-bonding electrons on S, which will leave one lone pair. Each F has an octet as a result of the bonding with S. Therefore at S there are five electron groups, comprising four S—F bonds and one lone pair. The shape is, therefore, based on a trigonal bipyramid. There are two choices for the location of the lone pair, either axial or equatorial. Because lone pairs do take up more room than single bonds, the lone pair is found in an equatorial position.

In addition to the lone pair being in an equatorial position, the extra room it requires also has an effect on the equatorial F—S—F bond angle, which is lower than the ideal 120° expected for a trigonal bipyramidal shape, and it is, in fact, 101.5°.

Octahedral

An octahedral structure has the shape:

where all of the X–M–X angles are either 90° or 180°.

A classic example of such a compound is another example of a sulfur halide, this time the hexafluoride SF_6. Recalling that S has six valence electrons and F has seven valence electrons, then SF_6 possesses six S—F bonds which uses all of the valence electrons associated with the S, meaning that there are no lone pairs on the sulfur, S, and therefore the shape is octahedral, as shown:

It is also possible to prepare compounds between halogens, and BrF_5 is one such compound. Bromine, Br, has seven valence electrons and fluorine, F, likewise, has seven valence electrons. There are five Br—F single bonds formed, so that leaves a pair of electrons as a lone pair. The resulting structure is:

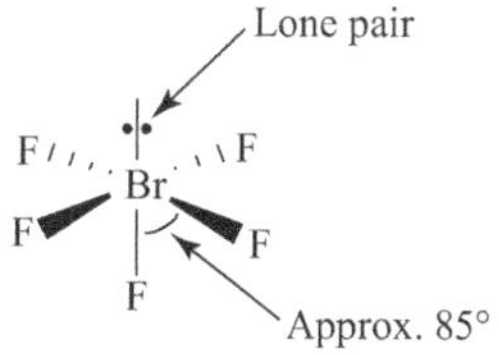

The fluorine situated opposite from the lone pair is now not at 90° to the other four F atoms, but at approximately 85°. This is because of extra room taken up by the lone pair. In terms of just the atoms present, the structure is square pyramidal.

Self-test

Despite being a noble gas, xenon does exhibit some chemical behaviour. One compound of xenon, xenon tetrafluoride, XeF_4, has a shape based on the octahedron. Why?

How do we describe the shape based on atoms only?

2.8.8 Resonance

The carbonate ion, CO_3^{2-}, can readily be shown to adopt a trigonal planar structure, as shown:

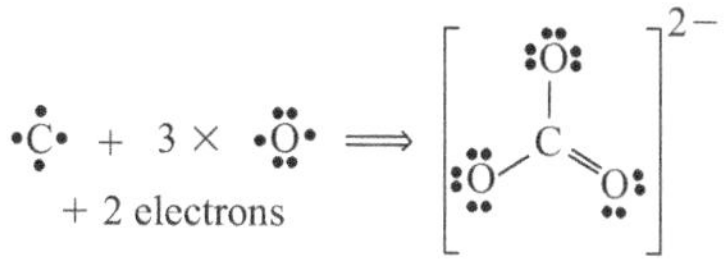

- C has four valence electrons, O has six valence electrons.
- Two 'extra' electrons because it is a dianion.
- Eight electrons about C, only three O groups, so need a C=O double bond.

If this was the structure, however, then the O—C—O angles would not all be the same since the C=O would take up more space, and this would then make the angle between the single-bonded O—C—O atoms less than 120°. In reality, however, all of the O—C—O angles are the same, as are all three of the C—O bond lengths, and this is due to resonance.

The process of resonance is represented as shown below:

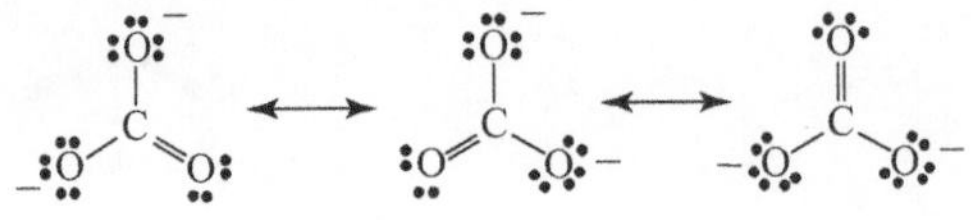

The double headed arrow means the structures on either side are in resonance with each other. The key point, however, is that the molecule does not change back and forwards between three forms – it is an average of the three forms.

All of the C—O bonds are equivalent and each bond is 4/3 of a bond (from the average of one double and two single bonds), as indicated below:

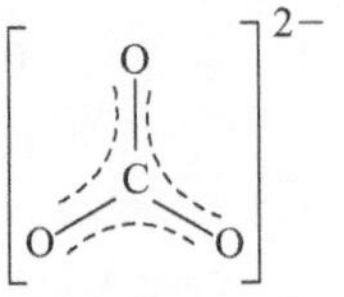

In this representation the dotted lines indicate electron delocalisation, which spreads the electron density over a greater volume which reduces electron–electron repulsions and hence lowers the energy of the molecule (in other words, it stabilises the molecule).

The classic example for illustrating resonance as a concept is benzene. Benzene has the formula C_6H_6, and it is a six-membered ring, with the bonding for this being explained with alternating double and single bonds, as shown:

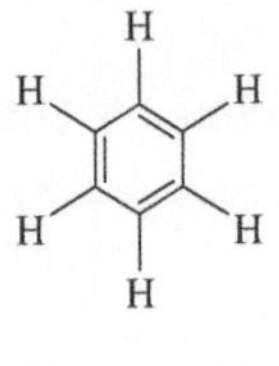

This structure, however, cannot be correct as it stands, because all of the C—C bond lengths are found experimentally to be identical. Resonance structures can be drawn to rationalise this observation,

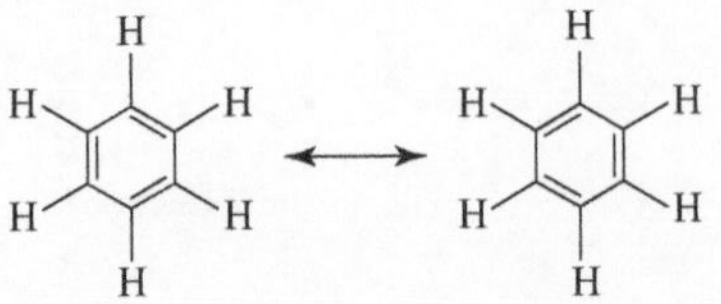

Neither of the two above structures have any physical reality, and the next structure more accurately represents the true state:

This is often shown without the H atoms present, thus:

2.9 Covalent bonding: atomic and molecular orbitals

So far the main focus has been on developing the concepts of bonding involving atoms largely in terms of how many electron groups there were arranged about them, and have considered molecular shape in terms of the VSEPR approach, which is a simplistic approach based on electrostatic repulsion. This is fine as far as it goes, but there is a more rigorous way of looking at bonding, and at the shapes that arise as a consequence, involving *atomic orbitals* (AOs) and *molecular orbitals* (MOs). These approaches are also useful in later chapters when we consider both reaction mechanisms and structural determination.

The most obvious place to start is with a look at the bonding that is found in the simplest molecule, H_2.

2.9.1 The hydrogen molecule

Hydrogen atoms each possess one electron and this occupies the (lowest energy) 1s AO. The possible higher energy (2s, 2p, etc.) AOs are empty. Two 1s AOs, one from each H atom, *overlap* and interact to form two new molecular orbitals (MOs) of different energies.

Note: The interaction of *two* AOs results in the creation of *two* new MOs – *you get no more or less than you put in.*

This is represented in Figure 2.15. Two 1s AOs combine to give two MO, labelled σ and σ^*.

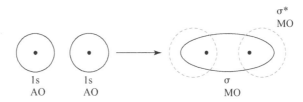

Figure 2.15 Schematic representation of combination of atomic orbitals to form molecular orbitals in H_2.

In terms of relative energies, these are represented below (lower energy more stable):

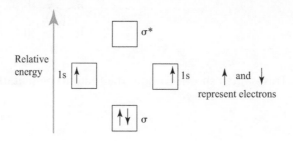

The MOs can each take up to two electrons (the same as the AOs). Since there are only two electrons associated with the hydrogen molecule, these both go into the lower energy (σ) MO. This orbital has a lower energy than the starting 1s AOs; the electrons spend most of their time between the atoms and the σ MO is known as a *bonding orbital*. The higher energy (σ^*) MO (which covers two regions of space) is empty and we can *usually* (though not always) ignore it – this MO is known as an *antibonding orbital*.

These ideas will now be extended to more complex systems, focusing on organic molecules.

2.9.2 Bonding in alkane hydrocarbons

Hybridisation at carbon

Each carbon atom possesses six electrons occupying $1s^2$, $2s^2$, $2p_x^1$, $2p_y^1$, $2p_z^0$ AOs. There are only two electrons not paired so there might be an expectation that it should only form two bonds. But it is known that carbon forms four bonds, as evidenced in methane which has four C—H bonds. This is rationalised by recognising that the 2s AO energy is quite close in energy to the 2p AOs and if one 2s electron were moved to the $2p_z$ orbital there would be four unpaired electrons which could each form a bond to hydrogen.

However, this would not give the correct bond angles as the 2p AOs are at 90° to each other. It is known that alkanes are tetrahedral about carbon, so *four equivalent orbitals* on each carbon with directions pointing towards the corners of a tetrahedron are needed. This can be obtained if the 2s and the three 2p AOs interact and 'reorganise' themselves into four new, equivalent, AOs of the same energy, as illustrated in Figure 2.16. (*This 'reorganisation' is often called 'hybridisation'.*)

Each new orbital is 3/4 p and 1/4 s in character and is known as an sp^3 *hybrid atomic orbital*. Each sp^3 hybrid AO will then take one electron. Methane will be used as an illustration.

Bonding within methane

Each sp^3 hybrid AO can overlap and interact with a hydrogen 1s AO to form a σ-bonding MO (and, of course, a σ^*-antibonding MO – which can be ignored). This bonding MO will take two electrons (one from C and one from H). Thus four two-electron bonds are formed, as shown in Figure 2.17.

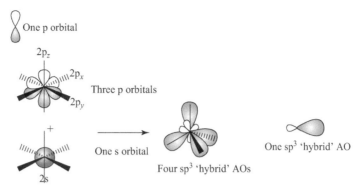

Figure 2.16 Hybridisation of s and p atomic orbitals to form sp^3 atomic orbitals.

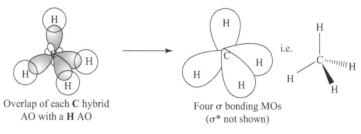

Figure 2.17 Bonding in methane.

Bonding within ethane

The carbon—carbon bond is formed by overlap of one sp^3 hybrid AO from each carbon, and the remaining three sp^3 hybrid AOs on each carbon form bonds to hydrogen as in methane (Figure 2.18). Again this ignores the antibonding MOs. Rotation is possible about the C—C bond so the hydrogen atoms at one end of the molecule can move relative to those at the other. (This will be explored more in Chapter 3 when looking at conformational isomerism.)

2.9.3 Bonding in alkene hydrocarbons

Hybridisation at carbon

In alkenes it is necessary to explain the presence of a C=C double bond, with a bond angle of 120°, and also to account for the C=C double bond being less than twice the bond strength of a C—C single bond. This is rationalised by having only

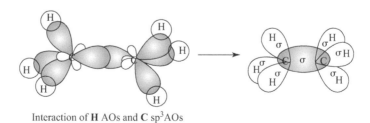

Interaction of **H** AOs and **C** sp^3 AOs

Figure 2.18 Bonding in ethane.

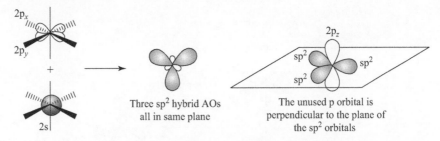

Figure 2.19 Hybridisation of s and p atomic orbitals to form sp² atomic orbitals.

the $2p_y$ and $2p_x$ AOs interacting with the 2s AO and reorganising themselves into three new equivalent 'hybrid' AOs and the $2p_z$ AO remains as it is. Each new orbital is 2/3 p and 1/3 s and is known as an sp² hybrid AO. These three lie in a plane at 120° to each other and the unaltered $2p_z$ AO lies at right angles to this plane, as illustrated in Figure 2.19.

Bonding in ethylene (ethene)

Consider the bonding in the simplest alkene hydrocarbon, ethene. The first carbon—carbon bond (σ) is formed by overlap of one sp² hybrid AO from each C atom and the remaining two sp² hybrid AOs from each C form bonds to hydrogen as in methane. *Provided all the atoms lie in the same plane*, the remaining $2p_z$ AO on each carbon can now overlap *sideways on* to form two new molecular orbitals known as the π (lower energy or bonding) MO and the π* (higher energy or antibonding) MO (Figure 2.20). The two electrons go into the lower energy π MO (the π* is empty and is ignored as before).

Important features about double bonds

- The p orbital overlap effectively locks the atoms in the plane and rotation about the C—C bond is not possible – this would break the new bond because the $2p_z$ AOs could not overlap properly. Thus we have the condition where double-bonded compounds can exhibit geometric isomerism, which is explored in Chapter 3.

- The $2p_z$ AOs on each carbon do not overlap as much as the sp² AOs that make the C—C σ MO, and so the additional bond (the π bond) is weaker than the σ bond. This is why a C=C double bond is not twice as strong as a C—C single bond.

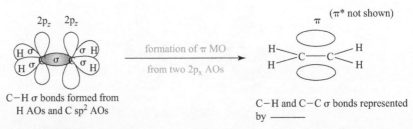

Figure 2.20 Bonding in ethene.

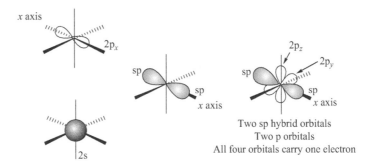

Figure 2.21 **Hybridisation of s and p atomic orbitals to form sp atomic orbitals.**

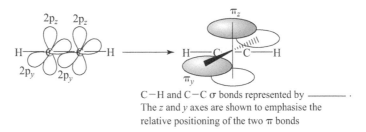

C—H and C—C σ bonds represented by ———— .
The z and y axes are shown to emphasise the
relative positioning of the two π bonds

Figure 2.22 **Bonding in ethyne.**

2.9.4 Bonding in alkyne hydrocarbons

Hybridisation at carbon

In alkynes the bond angle is 180° and the second and third bonds are weaker than the first. The explanation for this is an extension of the approach described above, for alkenes, but in this case only the $2p_x$ AO interacts with the 2s AO and these form themselves into two new equivalent 'hybrid' AOs, with the $2p_z$ and $2p_y$ AOs remaining unaltered. Each new orbital is $\frac{1}{2}$ p and $\frac{1}{2}$ s in character and is known as an *sp hybrid AO*. These two lie along a line, and point in opposite directions, with the unaltered $2p_z$ and $2p_y$ AOs lying at right angles to this line and to each other (see Figure 2.21).

Bonding in ethyne

As an alkyne, ethyne has a carbon–carbon triple bond. The first carbon–carbon (σ) bond is formed by overlap of one sp hybrid AO from each C and the remaining one from each carbon forms bonds to hydrogen as in methane. The remaining $2p_z$ AOs on each carbon can now overlap sideways on to form one π MO (also a π^* is formed, but it can be ignored) and similarly the $2p_y$ AOs on each carbon form another π MO. This is illustrated in Figure 2.22. As in the alkenes the π bonds are weaker than the σ bonds, so the triple bond is significantly less than three times stronger than a C—C single bond.

Conjugation

Two double bonds in a molecule give a diene. If the double bonds are isolated, as in pent-1,4-diene shown below, then the double bonds are no different in character from those in ethene.

61

Penta-1, 4-diene

There arises a special case, however, when the double bonds are adjacent (separated by only one σ bond), and such a system is said to be *conjugated*. The two π bonds interact by sharing electrons. The consequence is that the central σ bond is shorter. For example, buta-1,3-diene:

In this case, if the picture is drawn indicating the presence of the molecular orbitals, as shown in Figure 2.23, then it is easier to see that the relative positions of the two π bonds allows them to interact as described above, thereby ensuring that the intervening single bond has a little double bond character, and is hence shorter. This has consequences in, for example, physical properties such as UV–visible spectra (see Chapter 13).

Benzene

Benzene was described earlier in terms of Lewis structures and of resonance forms. A more complete analysis of its structure can be made by considering the molecular orbital picture.

Benzene is a six-membered ring in which all of the carbon–carbon bond lengths are the same, and the bond angles are all 120°. This arises as a consequence of a development in the concept of conjugation.

Considering the basics first, the bond angles will be correct if we use carbon atoms with sp^2 hybridised AOs, with the carbon atoms lying at the corners of a regular hexagon. Each carbon uses two sp^2 hybrid AOs to make σ bonds to the carbon atoms on either side and uses the third to make a σ bond to hydrogen. The remaining six unused $2p_z$ AOs, one from each C atom, lie perpendicular to this plane – each overlapping to the same extent with those from the C atom on each side of it.

- All six $2p_z$ AOs interact with each other to produce six new π MOs.
- All six MOs occupy the same ring-doughnut regions of space above and below the plane of the atoms.

Three of these MOs are lower in energy than the original $2p_x$ AOs and three are higher. The six electrons, one from each $2p_x$ AO, occupy the three lower energy (bonding)

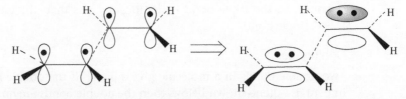

Figure 2.23 Conjugation in terms of orbital overlap.

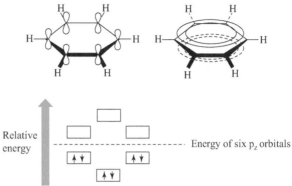

Relative energies of the six π molecular orbitals.
Three are bonding orbitals, three are antibonding.

Figure 2.24 **Bonding in benzene.**

orbitals and are all associated equally with all six carbon atoms (significance – electrons are not always associated with only two atoms). Figure 2.24 shows the bonding in benzene.

Six p_z orbitals overlap to form a π system delocalised across all six carbon atoms.

2.9.5 Hybridisation in atoms other than carbon

As well as carbon and hydrogen, other atoms occur commonly in biomolecules, and oxygen and nitrogen are particularly common. As with carbon, these too undergo hybridisation. In other words, they too form single, double and triple bonds, and the principles of hybridisation described previously explain these features. Water, H_2O, will be used as an example, and then the approach extended into organic compounds.

The molecule of water can be described as 'bent'. This is because:

- an atom of oxygen has six electrons;
- when covalently bonded with two hydrogen atoms, each of which contributes one electron, oxygen achieves its octet;
- each $O-H$ bond requires two electrons;
- there are two lone pairs;
- the electron pairs (bonding pairs *and* lone pairs) are arranged to maximise their distance from one another, and are arranged (approximately) tetrahedrally. Schematically:

H
|
$\cdot$;;;;;;;O
$\cdot\cdot$ ╲
$\cdot\cdot$ H

This, however, is a simplistic way of looking at matters. The electrons are in orbitals, so what are the orbitals associated with oxygen? Oxygen is in the same period as carbon in the periodic table so, as with carbon, there are only s and p orbitals involved in the valence shell. The arrangement of electron groups about the oxygen is based on a tetrahedral shape. Thus there are *four equivalent orbitals* on the O, with their directions pointing towards the corners of a tetrahedron.

This is explained in the same way as the arrangement of orbitals in methane above, if the 2s and the three 2p AOs interact and reorganise themselves into four new equivalent AOs of same energy. *In other words, the oxygen is sp³ hybridised.*

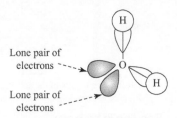

The same kind of argument can be applied to nitrogen, for example in ammonia, NH_3. This has three N—H bonds and one lone pair or, put another way, four electron groups. *The nitrogen atom is sp³ hybridised.* Note how one of the sp³ orbitals is accommodating the lone pair of electrons:

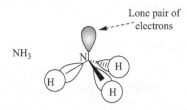

Extending this argument therefore, a whole range of oxygen- and nitrogen-containing compounds are explained this way, e.g. with the cases of carbonyl compounds, imines and nitriles.

Bonding in carbonyl, C=O, compounds

This is similar to that in ethene; carbon and oxygen are both sp². Looking at the simple aldehyde methanal, which is more trivially named acetaldehyde, the bonding analogy with ethene is clear, as demonstrated in Figure 2.25.

Box 2.5

Simple summary of bonding points

- If a carbon has four electron groups (it has a tetrahedral arrangement of groups about it, e.g. in alkanes, alkyl groups, etc.), then the hybridisation is sp³.

- If a carbon has three electron groups (it has a trigonal arrangement of groups about it, e.g. in alkenes, benzene rings, carbonyl groups, etc.), then the hybridisation is sp².

- If a carbon has two electron groups (it has a linear arrangement of groups about it, e.g. in alkynes, nitriles, etc.), then the hybridisation is sp.

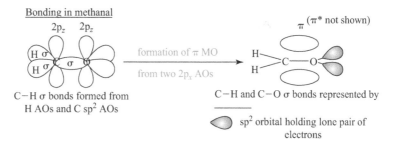

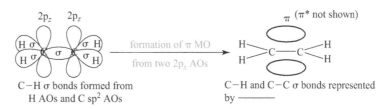

Figure 2.25 Comparison of the bonding in methanal and ethene.

2.10 ## Intermolecular forces

So far the focus has been on the factors holding atoms together within molecules, that is, the bonding within molecules. Covalent bonds are *intramolecular*: that is, they occur within a molecule. *Intermolecular* forces occur between molecules. They arise from electrostatic attractions between molecules as a result of partial charges. Usually intramolecular forces are stronger than intermolecular forces.

There are different types of intermolecular forces:

- dipole–dipole;
- dispersion (London);
- hydrogen bonding.

2.10.1 Dipole–dipole interactions

The positive end of one molecule attracts the negative end of another, for example $H—Cl$. The electronegativity of Cl is greater than that of H, and therefore there is a transfer of partial charge, with Cl taking more of the bonding pair of electrons. When two $H—Cl$ molecules are near to each other they will arrange themselves like this:

$$\delta+ \quad \delta- \quad \delta+ \quad \delta-$$
$$H—Cl\text{-------}H—Cl$$

The negative end (Cl) of one molecule will point at the positive (H) end of the neighbouring molecule. The greater the dipole moment the greater the dipole–dipole force between molecules. This attraction between molecules gives rise to increases in measurable quantities such as melting/boiling points, because energy needs to be put into a system in order to break the (relatively) strong dipole–dipole forces holding the molecules together in order for it to boil.

2.10.2 Dispersion (London) forces

Not all substances possess significant dipole moments, but they may still have melting/boiling points significantly above absolute zero. There must, therefore, be something responsible for attraction between molecules. In 1930 Fritz London (hence the name) suggested how to account for a weak attraction between molecules/atoms. For example, a noble gas is not molecular, and has no overall dipole. At any instant, however, more electrons may be on one side of the atomic nucleus than on the other, thereby creating an instantaneous dipole. Also, a number of atoms together might influence the motions of each other's electrons. These are referred to as dispersion forces, their formal definition being 'weak attractive forces between molecules resulting from the small, instantaneous dipoles that occur due to the varying positions of electrons during their motion about nuclei'.

Therefore dispersion forces mean that non-polar molecules, such as H_2, will condense and solidify at temperatures greater than absolute zero. Because dispersion forces are caused by momentary oscillations of electron charge, molecules need to be close together for this force to have an effect, which means properties such as boiling points will occur at low temperatures. Dispersion forces occur between all molecules and are the dominant intermolecular force for non-polar molecules. Intermolecular dispersion forces increase with increasing number of electrons because there is a greater chance of instantaneous dipoles, and therefore these forces tend to increase with molecular (atomic) weight, as larger molecules (atoms) have more electrons. Also, larger atoms are more polarisable, because their electrons are more easily perturbed, thereby producing instantaneous dipoles.

Consider some examples:

- For the halogens, increased boiling points are observed on going down the periodic group; the boiling point of F_2 occurs at 85 K; that of Cl_2 occurs at 239 K; for Br_2 it is at 333 K, and for I_2 the boiling point is 458 K.

- At normal room temperature (25 °C or 298 K), methane, CH_4, is gaseous whereas hexane, C_6H_{14}, is a liquid.

2.10.3 Hydrogen bonding

On the basis of the descriptions of intermolecular interactions arising from London dispersion forces and from dipole–dipole interactions, then comparing CH_3F and CH_3OH, which possess similar molecular masses (34 vs 32) and similar values of dipole moment (1.81 and 1.70 D), then it would be expected that their boiling points might be very similar. In fact, however, they are 195 K for CH_3F and 358 K for CH_3OH. This is a very large difference, so there is clearly a further, and very significant, attractive force involved with CH_3OH. The difference arises because of a phenomenon known as *hydrogen bonding*.

A hydrogen bond is the attractive interaction of a hydrogen atom with an electronegative atom, such as nitrogen, oxygen or fluorine, that comes from another

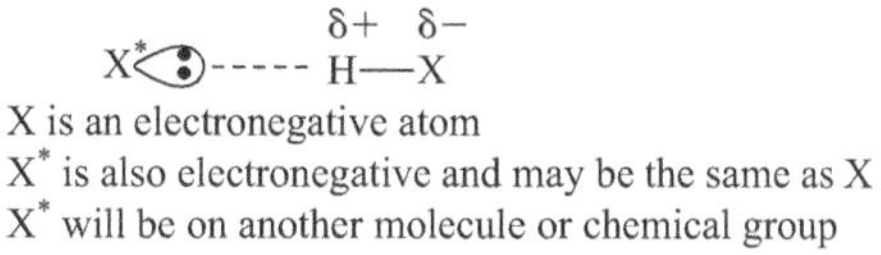

$$\overset{\delta+\quad\delta-}{X\overset{*}{\bullet}\text{-----}H\!-\!X}$$

X is an electronegative atom
X* is also electronegative and may be the same as X
X* will be on another molecule or chemical group

Figure 2.26 **Criteria for hydrogen bonding.**

molecule or chemical group. The hydrogen must also be covalently bonded to another electronegative atom to create the bond (see Figure 2.26).

To illustrate how this occurs, using a real case, consider the example of CH_3OH referred to above (Figure 2.27).

Because the $O\!-\!H$ bond is highly polarised, meaning that the hydrogen that is joined directly to the very electronegative oxygen atom is $\delta+$, then the hydrogen of this bond will be attracted to a lone pair of electrons situated on an oxygen atom of a neighbouring molecule of CH_3OH. As oxygen has two lone pairs of electrons, then each of them could be involved in an interaction with the $O\!-\!H$ hydrogen of a neighbouring molecule, as shown.

In the case of CH_3F there is no hydrogen bonding observed. This is because the very electronegative fluorine does not have any H atoms directly attached to it, and so there is no opportunity for a strong electrostatic interaction.

$$F\!-\!CH_3$$

Hydrogen bonding in water

Considering water, H_2O, which has a bent shape like that of CH_3OH, the same phenomenon is observed (Figure 2.28), and because both H atoms are linked to the

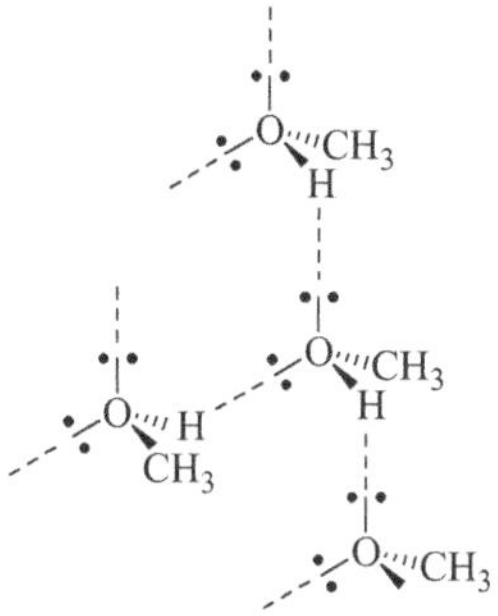

Figure 2.27 **Hydrogen bonding in methanol. The dotted line is the intermolecular hydrogen bond which is weaker and longer than the intramolecular polar $O\!-\!H$ bond.**

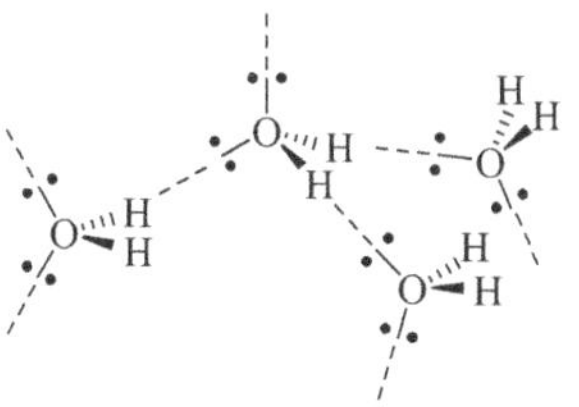

Figure 2.28 **Hydrogen bonding in water.**

Table 2.7 **Boiling points of molecular hydrides from groups 14 and 16**

Boiling points, group 14 hydrides (K)	Boiling points, group 16 hydrides (K)
$CH_4 = 113$	$H_2O = 373$
$SiH_4 = 163$	$H_2S = 213$
$GeH_4 = 183$	$H_2Se = 243$
$SnH_4 = 198$	$H_2Te = 263$

oxygen, then the intermolecular interactions are even more extensive, which will give rise to the much higher than expected boiling point of water (373 K).

As intimated earlier, hydrogen bonding has implications for physical properties. Thus, looking at the boiling points of water and other H_2X molecules in the same column of the periodic table, and comparing these values with a similar list from group 14 gives the results illustrated in Table 2.7.

For the group 14 hydrides, on descending the column, there is an increase in boiling points due to increases in London (dispersion) forces. For the group 16 hydrides, basing predictions on a consideration of London forces, the boiling point of H_2O would be expected to be about 183 K. Hydrogen bonding causes an almost 200-degree increase from the boiling point as predicted using those criteria.

Biological implications of hydrogen bonding

The presence of hydrogen bonding interactions in water are of paramount importance to life as it exists on Earth. In the very simplest terms, given the prediction described above, without their existence water might be expected to have a boiling point close to 200 degrees lower than that observed, and life would not have evolved at all on Earth.

Multiple intermolecular hydrogen bonds are very important in other aspects of biology. To take a couple of examples,

● DNA is normally a double-stranded macromolecule with two polynucleotide chains, held together by hydrogen bonding between the bases, forming a DNA molecule.

● The shapes adopted by protein molecules are governed by hydrogen bonding interactions.

These cases will be examined in greater detail in Chapter 7, but Figure 2.29 shows a schematic representation of the key features associated with the linking of the two helical strands of DNA, in which a hydrogen bonded to (electronegative) nitrogen on a base on one helical strand, is associated through a hydrogen bond with the lone pair of either oxygen or nitrogen on a base of the other helical strand.

Strength of hydrogen bonds

Comparing the strengths of different interactions, it is clear that the term hydrogen bonding is somewhat misleading, as the interactions are significantly weaker than, for example, covalent bonds. Table 2.8 lists the strengths of these interactions.

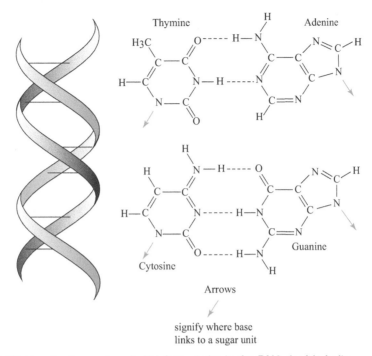

Figure 2.29 How hydrogen bonds link base pairs in the DNA double helix.

Table 2.8 **Comparison of the strengths of different interactions**

Type of interaction	Interaction energy (strength)/kJ mol^{-1}	Example of interaction
Ionic bond	400–4000	NaCl
Covalent bond	150–1100	Cl–Cl
Metallic bond	75–1000	Fe
Hydrogen bonding	10–40	$H_2O - H_2O$
Dipole–dipole	5–25	HCl – HCl
Dispersion	0.05–10	$F_2 - F_2$

Questions

2.1 Give the full electronic configuration and abbreviated electron configuration for each of the following:

(a) P
(b) K
(c) Se
(d) I

2.2 Give the charge for the most stable ion for each of the following elements:

(a) Mg
(b) O
(c) P
(d) Cs

2.3 Which of the following pairs of atoms and ions are isoelectronic?

(a) Cl^-, Ar
(b) Na^+, Ne
(c) Mg^{2+}, K^+
(d) Li^+, Ne
(e) O^{2-}, F^-

2.4 Explain which of the following electron configurations are not possible:

(a) $1s^2, 1p^1$
(b) $1s^2 2s^2 2p^2$
(c) $1s^2 2s^2 2p^6 2d^1$
(d) $1s^2 2s^3$

2.5 Explain which species in each of the following groups is found in nature:

(a) Na, Na^+, Na^-
(b) S^{2-}, S^-, S^+
(c) Cl, Cl^-, Cl^+

2.6 Using a periodic table rank the following sets of elements in order of decreasing first ionisation energy. Explain your answers.

(a) Kr, He, Ar
(b) Sb, Te, Sn
(c) K, Ca, Rb
(d) I, Xe, Cs

2.7 Use Lewis electron-dot diagrams and electron configurations to depict the formation of:

(a) K^+ and O^{2-} ions from the respective atoms and determine the formula of the compound
(b) Mg^{2+} and Cl^- ions from the respective atoms and determine the formula of the compound

2.8 Use the periodic table to rank the bonds in each set in order of decreasing bond length and bond strength:

(a) S—F, S—Br, S—Cl
(b) C=O, C—O, C≡O
(c) Si—F, Si—C, Si—O

2.9 For the following polar covalent bonds state which atom has a partial −ve charge and which a partial +ve charge:

(a) O—H
(b) F—N
(c) I—Cl

2.10 Rank the following bonds in order of increasing polarity:

(a) H—N, H—O, H—C
(b) Cl—F, Br—Cl, Cl—Cl

2.11 State which type of bonding (ionic, covalent, polar covalent, metallic) you would expect to find in samples of each of the following:

    (a) $KF(s)$
    (b) $N_2(g)$
    (c) $Na(s)$
    (d) $BrF_3(g)$
    (e) $BrO_2(g)$
    (f) $HCl(g)$
    (g) $Cr(s)$
    (h) $CaO(s)$
    (i) $S_8(s)$

2.12 Using VSEPR theory predict the shape of:

    (a) CCl_4
    (b) $BrCN$
    (c) PCl_3
    (d) OF_2
    (e) H_2S
    (f) BrF_3
    (g) I_3^-

2.13 Show how resonance may be used to rationalise the following bond lengths and predict the bond angles in each of the following:

    (a) CO_3^{2-} (carbonate): all C—O bonds are equal in length.
    (b) CH_3COO^- (acetate or ethanoate ion): both C—O bonds are equal in length.
    (c) Pyridine: C—N bonds are equal, C—C bonds are nearly equal in length.

2.14 For *each* of the following molecules draw a diagram showing the bond angles, the hybridisation of the constituent carbon, nitrogen or oxygen atoms, and the type (σ or π) of carbon–carbon, carbon–nitrogen or carbon–oxygen bonds present.

    (a) methylamine, CH_3NH_2
    (b) methanal (formaldehyde), $H_2C{=}O$
    (c) methanoic (formic) acid, HCO_2H

2.15 The structure of the thyroid hormone thyroxine is shown below. Redraw it, indicating the hybridisation state of all of the C, N and O atoms in the molecule.

71

2.16 What are the intermolecular forces in the following species:

(a) PCl_3
(b) $MgCl_2$
(c) H_2
(d) CH_3CH_2OH
(e) CH_3F
(f) CH_3NH_2

2.17 Oxygen and selenium are both in group 16. State why water forms H bonds but H_2Se does not.

2.18 State, with reasoning, which of the following pairs of substances has a higher boiling point:

(a) LiCl or HCl
(b) NH_3 or PH_3
(c) Xe or I_2
(d) NO or N_2

3 An introduction to the chemistry of carbon

Learning outcomes

By the end of this chapter students should be able to:

- identify functional groups within an organic molecule;
- provide a systematic name for simple organic molecules;
- give examples of, and distinguish between:

  (a) constitutional isomers, structural isomers and stereoisomers;
  (b) isomers and conformations;
  (c) geometrical isomers and enantiomers;

- assign the configuration prefix (E,Z) to a geometrical isomer;
- identify a plane of symmetry and chiral centres if present in a molecule;
- assign the absolute configuration prefix (R,S) to a molecular chiral centre.

3.1 Introduction

Carbon is the element that forms the basis of life on our planet, and this is the reason why the study of the chemistry of carbon compounds is known as organic chemistry. It is this fact that makes an appreciation of this area of chemistry fundamentally important to the study and practice of biology. This chapter will explore why carbon is such an important element by addressing a series of important questions.

- Why does it form so many compounds?
- How are the compounds named?
- What is the importance of molecular shape?
- How do we identify molecular shape?

On a wider scale, carbon chemistry has economic as well as biological importance. Some key aspects of carbon chemistry are identified below.

Carbon chemistry is a fundamental part of the following industries:

- pharmaceuticals
- agrochemicals

- food additives
- dyestuffs
- plastics.

It is a key component of the chemistry of biology.

Knowing the basics of the chemistry of carbon allows for an understanding of the structures of:

- peptides/proteins
- DNA/RNA
- carbohydrates
- fats.

Additionally it forms the basis for understanding the mechanisms of:

- catabolic (breakdown) pathways
- biosynthesis.

To fully appreciate the importance of organic chemistry to biology, it is necessary to appreciate the *structure and functions of organic compounds*. This phrase means:

- What shapes do organic compounds have?
- What are their component parts?
- How are they named?
- How do they behave? (That is, what reactions do they have and why/how do they behave that way?)

This chapter will concentrate on the first three of these topics.

There are, literally, millions of organic compounds. The reasons for this will be explored, and their biological importance will be examined.

Despite the complexity of many organic molecules, especially many of those involved in biological processes, there are underlying principles which allow for the prediction of both structure and behaviour at a molecular level. For example, using very simple principles, it is possible to explain:

- why DNA forms helices and not random spaghetti-like bundles;
- why proteins fold into unique compact 3D structures (see Figure 3.1 for a representation of the structure of myoglobin);
- how sugars and lipids are broken down (catabolised) and reassembled (synthesised) in living cells using defined pathways.

The structure shown in Figure 3.1 is represented in such a way as to emphasise the helical nature of the protein component of myoglobin. Proteins are polymers made up from chains of amino acids linked together, and by looking at the structural features of the individual amino acids it is possible to understand how these proteins adopt the helical structures.

Biological molecules are very large but, as referred to earlier, their behaviour can be explained by some general principles which are readily explained by much smaller molecules. Thus this chapter will explore the basic principles which determine organic molecule structure using (primarily) small molecules as examples, but larger biological molecules will be referred to as the chapter progresses.

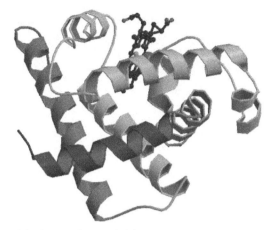

Figure 3.1 **Structural features of myoglobin.**

Properties of carbon

Carbon sits in the centre of the periodic table, and it has some important properties that reflect this pivotal position:

- Carbon forms strong covalent bonds with other elements (examples include H, N, O, S, P, halogens).
- Carbon forms strong covalent bonds with itself giving rise to chain, branched and ring structures.
- Carbon can form multiple bonds with itself and other elements, for example

$$-HC=CH-;\ -HC=O;\ -HC=N-;\ -HC=S;\ -C=N;\ -C\equiv C-,\ -C\equiv N$$

Box 3.1

The total number of bonds associated with a given carbon atom is always four

For example, compare the structures $-HC=CH-;\ -HC=O;\ -HC=N-;\ -HC=S;$ $-C=N;\ -C\equiv C-,\ -C\equiv N$ with a group of compounds known as alkenes. The former compounds have each got their C atoms surrounded by four others, a fact most easily seen within the ring systems (see the structure of cyclopropane, C_3H_6, below).

Cyclopropane, C_3H_6

Looking at the alkene C_2H_4, shown below:

$$H_2C=CH_2$$

(Continued)

Structurally this compound looks like this:

$$\begin{array}{c} H \\ \diagdown \\ H \diagup \end{array} C = C \begin{array}{c} H \\ \diagup \\ \diagdown H \end{array}$$

Note that in this case each C has 2 × (C—H) bonds, and then two bonds to the other C. *In both cases, each C still has four bonds, and this is found in all stable organic chemicals.*

3.2.1 Representations of organic structures

Because carbon has four bonds and, in alkanes, these are arranged tetrahedrally around even the simplest hydrocarbon molecules, this means they have a clear shape. Consider, for example, the compound pentane below.

It is important not to be deceived by different representations of organic molecules, and to remember the three-dimensional nature of carbon compounds. There are many different ways of presenting the same molecule, depending on what information is being delivered. For example, pentane is often seen represented in the following different ways:

$$CH_3 \cdot CH_2 \cdot CH_2 \cdot CH_2 \cdot CH_3 \quad or \quad CH_3 - CH_2 - CH_2 - CH_2 - CH_3$$

or

$$\begin{array}{ccc} & CH_2 & CH_2 \\ \diagup & \diagdown \diagup & \diagdown \\ CH_3 & CH_2 & CH_3 \end{array} \quad or$$

Similarly, for the molecule *trans*-but-2-ene, $CH_3 \cdot CH = CH \cdot CH_3$, there are many different ways that this is represented:

$$H_3C - \overset{\overset{\displaystyle H}{|}}{C} = \overset{\overset{\displaystyle }{C}}{\underset{\underset{\displaystyle H}{|}}{}} - CH_3 \quad or \quad \begin{array}{c} H \\ \diagdown \\ H_3C \diagup \end{array} C = C \begin{array}{c} CH_3 \\ \diagup \\ \diagdown H \end{array}$$

or

$$\begin{array}{c} H \\ \diagdown \\ Me \diagup \end{array} C = C \begin{array}{c} Me \\ \diagup \\ \diagdown H \end{array} \quad or$$

(*Note*: The term *trans* means that the two methyl groups are on opposite sides of the double bond. This is described more completely later, in Section 3.4.7 on geometric isomerism.)

In the cases above, both pentane and *trans*–but-2-ene are quickly, but reasonably accurately, represented by the final diagrams. These show the non-linear nature of the carbon skeletons, but are simplified because the hydrogens are implicit rather than explicitly shown.

In this text, given that different explanations may be enhanced by different representations, you will come across the different representational styles for drawing organic molecules.

3.3 Classification of organic molecules

One reason why there are so many organic molecules is that even relatively small numbers of atoms can join together in different ways. Take, for example, C_2H_6O – there are two compounds which share this formula, as shown below:

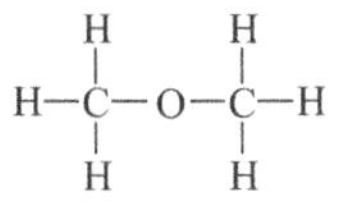

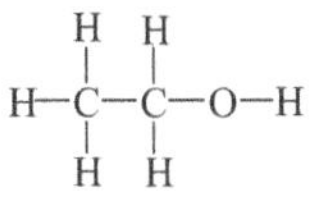

Compound **I** is an example of a category of compound called an ether, and compound **II** is ethanol, a member of the family of compounds called alcohols. These are different compounds from one another, differing in both their chemical and physical properties.

Before starting on the study of the structures and properties of organic compounds, it will be necessary to become familiar with the terminology (naming or nomenclature) used in discussing these compounds. Consider, for example, the amino acid, methionine. This is represented chemically by the structure shown below:

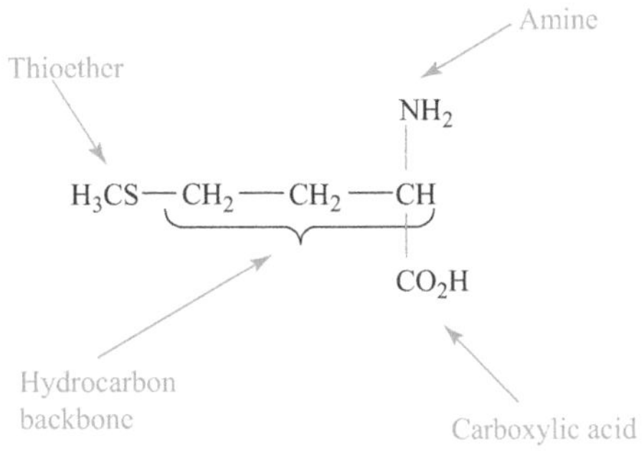

Which is more concisely represented as:

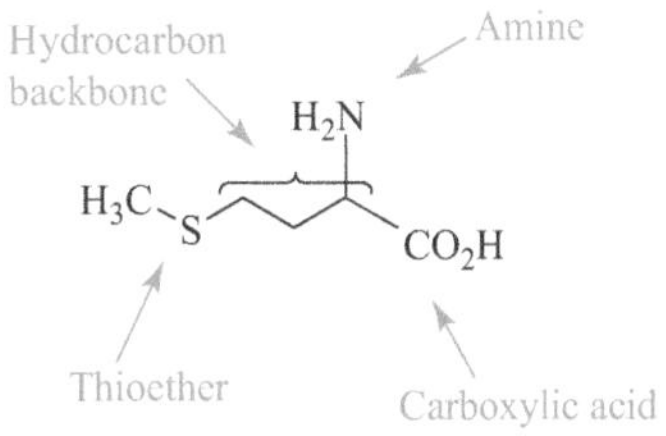

Looking at the representation of methionine shown above, there are a number of component parts to the molecule. The amine, NH_2, group, the carboxylic acid, CO_2H, group and the thioether between them control the principal chemical behaviour of this compound. Groups that control the properties of organic compounds in this way are called *functional groups*.

3.3.1 Nomenclature (naming) of organic compounds

Being able to name organic compounds according to a consistent set of rules is very important. Scientists need to be able to communicate with one another. There are systematic naming rules (referred to as IUPAC rules) used by practising chemists but, in the real world of chemical and biological sciences, trivial (non-systematic) names are in common usage, and it is equally important to be familiar with these.

Note: IUPAC is the abbreviation used for the International Union of Pure and Applied Chemistry, a body best known for standardising the rules of nomenclature in chemistry.

Trivial nomenclature bears no resemblance to structure. For example, consider the two compounds below.

Formaldehyde

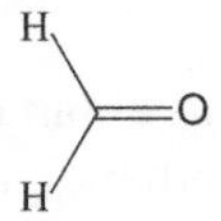

Acetaldehyde

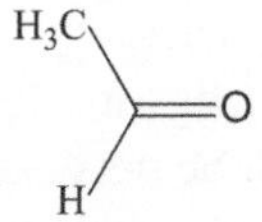

In these cases, the only way to deduce the structures from the names is to learn them. Table 3.1 lists some of the more common chemicals where trivial names are commonly used. Other common names will be introduced during the course of this text, particularly for biologically important molecules, for example amino acids or carbohydrates, where the systematic nomenclature becomes unwieldy.

3.3.2 Systematic nomenclature

This provides a more logically constructed system of naming, though occasionally parts are derived from a trivial base. We will start by considering the naming scheme for the simplest form of organic compounds, the saturated hydrocarbons; that is, hydrocarbons with no multiple bonds.

Straight chain alkanes

Most hydrocarbon names are based on the number of carbons forming the hydrocarbon chain (e.g. pentane signifies a five-carbon chain, hexane a six-carbon chain and so on). The four smallest hydrocarbon names (methane, ethane, propane and butane) have had their original trivial names carried through to the systematic usage.

Non-cyclic hydrocarbons, whether straight chain or branched chain, are described as acyclic.

Saturated acyclic hydrocarbons (known as paraffins), which have the general formula C_nH_{2n+2}, are called by the class name *alkanes*.

Table 3.1 **Some trivial names of everyday organic compounds**

Common (trivial) name	IUPAC name	Formula
Acetaldehyde	Ethanal	CH_3CHO
Acetic acid	Ethanoic acid	CH_3CO_2H
Acetone	Propanone	CH_3COCH_3
Acetyl chloride	Ethanoyl chloride	CH_3COCl
Acetylene	Ethyne	$CH{\equiv}CH$
Aniline	Aminobenzene	$C_6H_5NH_2$
Ethylene	Ethene	$CH_2{=}CH_2$
Formaldehyde	Methanal	$HCHO$
Formic acid	Methanoic acid	HCO_2H
Methyl acetate	Methyl ethanoate	$CH_3CO_2CH_3$
Methyl formate	Methyl methanoate	HCO_2CH_3
Phenol	Hydroxybenzene	C_6H_5OH
Toluene	Methylbenzene	$C_6H_5CH_3$

As mentioned above, the four lowest (smallest) members of this series have trivially based names, but the remainder have names based on a more systematic format, which is made up from two parts:

- the first part is variable and indicates the number of carbon atoms (using a Greek base);
- the second is the invariable suffix, '-ane'.

Table 3.2 gives some examples. These alkanes form part of an extensive series of compounds known as *a homologous series*, where a homologous series is defined as a series of compounds with the same general formula (in this case C_nH_{2n+2}). Thus methane is CH_4, ethane is C_2H_6 and so on.

Table 3.2 **Straight chain alkanes, general formula C_nH_{2n+2}**

Value of n	Name	Value of n	Name	Value of n	Name
1	Methane	8	Octane	15	Pentadecane
2	Ethane	9	Nonane	16	Hexadecane
3	Propane	10	Decane	17	Heptadecane
4	Butane	11	Undecane	18	Octadecane
5	Pentane	12	Dodecane	19	Nonadecane
6	Hexane	13	Tridecane	20	Eicosane
7	Heptane	14	Tetradecane	30	Triacontane

Table 3.3 **Alkanes and their related alkyl groups**

Parent alkane		Alkyl group	
CH_4	methane	CH_3-	methyl or Me
$CH_3 \cdot CH_3$	ethane	$CH_3 \cdot CH_2-$	ethyl or Et
$CH_3 \cdot CH_2 \cdot CH_3$	propane	$CH_3 \cdot CH_2 \cdot CH_2-$	propyl or Pr

Alkanes themselves can be useful materials. For example both propane and butane are known to many people as fuels, for small gas stoves and lighters respectively, and some larger chain alkanes act as sex pheromones in insects. Alkanes are, however, only the simplest forms of an enormous range and variety of compounds. For example, it is possible to react methane with the element chlorine to produce (amongst other things) the compound where one of the hydrogen atoms has been replaced by a chlorine, Cl, atom to form the compound CH_3Cl. The CH_3 part of this new compound is referred to as a methyl group, and other alkanes can have similarly related alkyl groups.

Alkyl groups are derived from the straight chain alkanes by the removal of a hydrogen atom from the terminal carbon atom. These groups are named by replacing the ending *-ane* of the name by *-yl* as shown thus for the first three members of the homologous series (Table 3.3).

Branched alkane structures

As described earlier, alkanes can have 'branched' structures, for example the structure below:

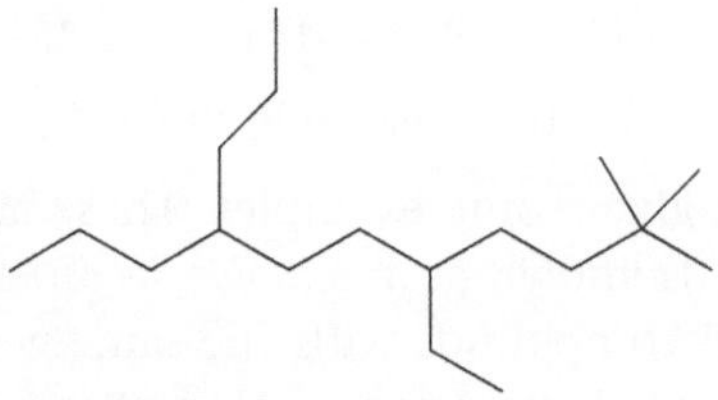

This is still an alkane (its formula is $C_{18}H_{38}$, which fulfils the criterion C_nH_{2n+2}), and even something relatively simple, with just four carbon atoms, has options.
Thus

(I)   (II)

Both I and II are C_4H_{10}, but they are not the same thing. They are chemically and physically distinct. So, if structure **I** is called butane, then what is structure **II** called?

The basic rules for naming alkanes that were described above are adapted and the rules outlined in Box 3.2 describe how they are applied to *give a name* to branched chain alkanes. Using this method, it is clear that structure **II** has a carbon chain of length 3, with a CH_3 group acting as a side chain at position 2. This would therefore be called 2-methyl propane, signifying that the methyl group is on the 2-position of a 3-carbon (propane) chain.

Box 3.2

> **Rules for naming branched chain alkanes**
>
> 1. Choose the longest chain of carbon atoms for the parent name.
> 2. Number the carbon atoms of this chain consecutively, starting from the end which gives the lower numbers to the side chains (which will, in the case of branched alkanes, be alkyl radicals).
> 3. If there are two or more different side chains, name these in alphabetical order (e.g. butyl, ethyl, methyl) but note that methyl, dimethyl and trimethyl are all considered to start with the letter m.

Example 3.1   Considering now the naming of the compound $C_{18}H_{38}$ shown earlier:

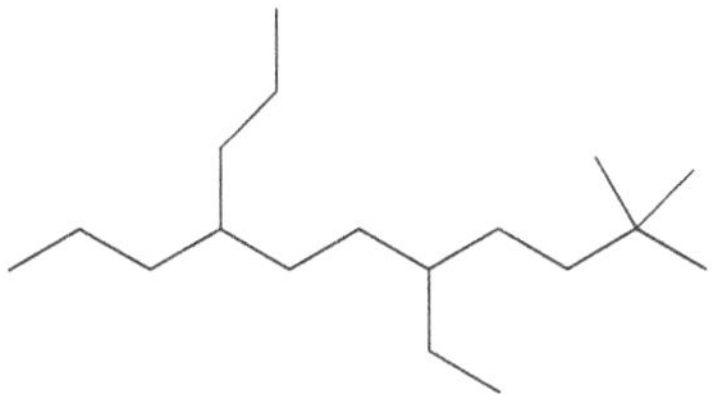

The longest chain is identified, and numbered, as shown below:

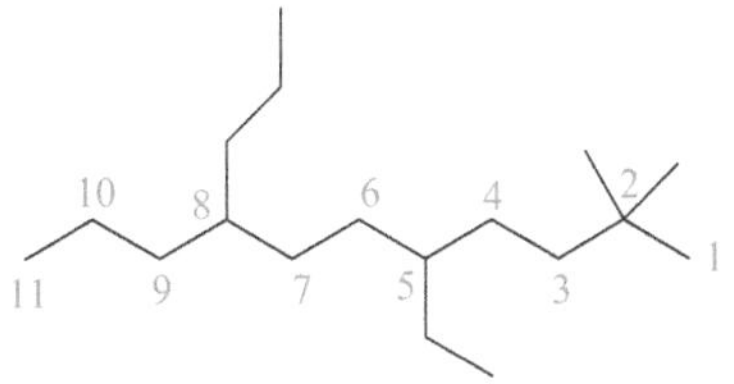

Note how the numbering starts at the right hand side, in order to ensure that the side groups start at the lowest number. The 2 position has 2 methyl groups on it, the 5 position has an ethyl group on it and the 8 position has a propyl group on it. Therefore, by noting that the 11–carbon chain is an undecane, then applying the rule stating that the side groups are listed alphabetically, this compound is named:

5-ethyl-2,2-dimethyl-8-propyl-undecane

Box 3.3

> **To convert a name to a structure**
>
> 1. Examine the name to decide how many carbon atoms are in the main chain, then write this down ignoring the hydrogen atoms or side chains.
> 2. Indicate the side chains with the appropriate groups of carbon atoms attached to the main chain at the right places.
> 3. Complete the structure by adding the necessary hydrogen atoms.

Example 3.2 Deriving a structure from a name (Box 3.3)

Starting with the name 5-ethyl-3-methylnonane, how is the structure constructed from this information?

1. It is a derivative of nonane, a 9-carbon chain:

   C—C—C—C—C—C—C—C—C

2. Indicate the locations and nature of the side groups. In this case there is a methyl at the 3 position and an ethyl at the 5 position:

   C—C—C—C—C—C—C—C—C
   $\qquad\;\;$ |$\qquad\quad$ |
   $\qquad\;\;$ C$\qquad\quad$ C—C

3. Add any H atoms to the C skeleton. Remember to see how many bonds each C atom has. If one C—C bond then add three H atoms; if two C—C bonds, then add two H atoms; and if three C—C bonds then add just one H atom. Thus:

   CH_3—CH_2—CH—CH_2—CH—CH_2—CH_2—CH_2—CH_3
   $\qquad\qquad\quad$ |$\qquad\qquad$ |
   $\qquad\qquad\;$ $CH_3$$\qquad\quad$ CH_2—CH_3

   This would normally be represented thus:

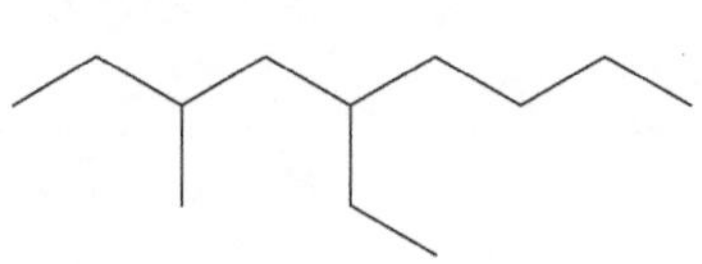

Unsaturated hydrocarbons

The alkanes, both straight chain and branched, are referred to as saturated hydrocarbons. In these, each carbon atom is joined to four other atoms, either carbon atoms or hydrogen atoms by single bonds only. There are also families of hydrocarbons where some of the carbon atoms are linked to other carbon atoms by either double bonds (known as alkenes) or triple bonds (known as alkynes), and these are referred to as unsaturated hydrocarbons.

Alkenes and alkynes

The naming of *unsaturated*, open chain (acyclic) hydrocarbons will now be considered. As mentioned above, hydrocarbons with one double bond are called *alkenes* and hydrocarbons with one triple bond are called *alkynes*. These points are summarised in Table 3.4.

Table 3.4   **Summary of acyclic hydrocarbon homologous series**

Family name	*General formula*	*Example*
Alkanes	C_nH_{2n+2}	H_3C—CH_3, ethane
Alkenes	C_nH_{2n}	H_2C=CH_2, ethene (trivial name ethylene)
Alkynes	C_nH_{2n-2}	HC≡CH, ethyne (trivial name acetylene)

Box 3.4

```
Naming of unsaturated compounds
```

Alkenes

The naming of alkenes is logically extended from the rules for alkanes, and the four rules below describe how alkenes, both simple and more complex, can be named.

1. Number the longest carbon chain (may have options if branched) that contains the double bond in the direction that gives the carbon atoms of the double bond the lowest possible numbers.
2. Indicate the location of the double bond by the location of its first carbon.
3. Name branched or substituted alkenes in a manner similar to *alkanes*.
4. Number the carbon atoms, locate and name substituent groups, locate the double bond, and name the main chain.

Alkynes

An alkyne is named in the same way but using *-yne* instead of -ene.

Multiple unsaturation

When there is more than one unsaturated centre in the molecule then the name must show their nature, number and positions. Alkene groups take precedence over alkyne groups. This means that:

- the main chain is that which contains the maximum number of alkenic centres;
- the main chain is numbered so as to give the lower numbers to the alkene group(s) *even if this results in the alkynic centre(s) having larger numbers.*

Example 3.3

Examples of naming of unsaturated compounds (Box 3.4)

Consider the compound shown below:

$$CH_3 \cdot CH_2 \cdot CH = CH \cdot CH_3$$

This is a 5–carbon chain, with the double bond starting at position 2 (not 3! – the compound is numbered to give the lowest number to the alkene group, so it is numbered from the right hand side as drawn), and as there are no branches it is given the name pent-2-ene.

Note, the left hand CH of the double bond is joined to $CH_3 \cdot CH_2-$, whilst the right hand CH is joined to $-CH_3$, so the structure could be rewritten using the abbreviations for methyl and ethyl, as shown:

$$Et \cdot CH = CH \cdot Me$$

Example 3.4

Moving on now to the branched chain alkene shown below.

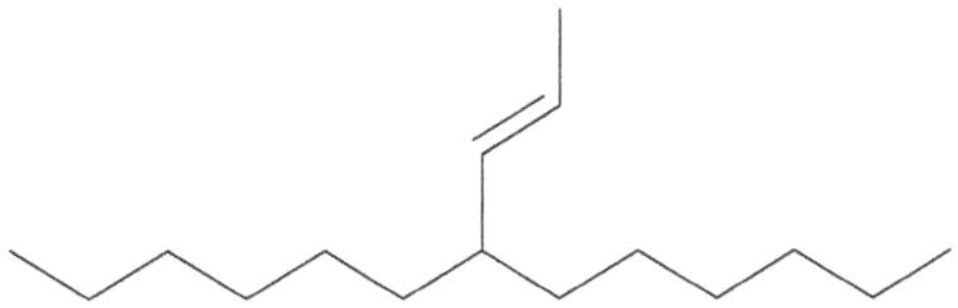

Here the longest chain is 13 carbon atoms long (the horizontal one, counting from left to right), but the longest chain *containing a double bond* is 10 carbon atoms long, as indicated below:

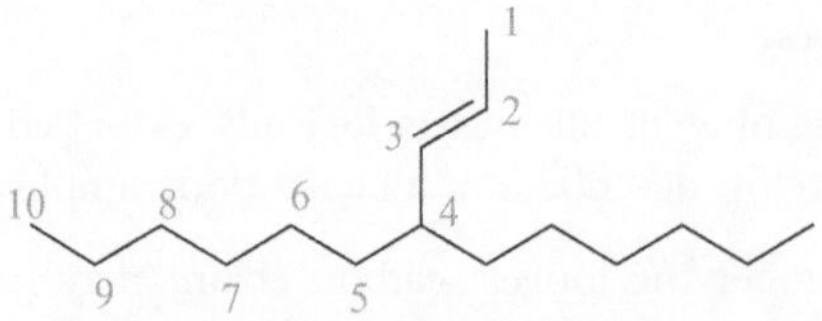

Note how the numbering gives the double bond the lowest number.The alkyl chain attached to the 4 position is a hexyl group. The compound is therefore named 4-hexyldec-2-ene.

Example 3.5   The compound $CH_3.C\equiv C.CH=CH.CH_2.CH_3$ is correctly named as hept-3-en-5-yne and *not* hept-4-en-2-yne, consistent with the numbering indicated below. This numbering ensures that the double bond is identified at the lowest numerical value, as required by the rules.

$$\overset{7}{H_3C}-\overset{6}{C}\equiv\overset{5}{C}-\overset{4}{CH}=\overset{3}{CH}-\overset{2}{CH_2}-\overset{1}{CH_3}$$

(Note that the above uses '–en-' rather than '–ene-' when there is still part of the name to be quoted.)

Note on abbreviating structures

Looking again at 4-hexyldec-2-ene, the structure can be shown as:

$$CH_3 \cdot CH_2 \cdot CH_2 \cdot CH_2 \cdot CH_2 \cdot CH_2 \cdot \underset{|}{CH} \cdot CH_2 \cdot CH_2 \cdot CH_2 \cdot CH_2 \cdot CH_2 \cdot CH_3$$
$$CH=CH \cdot CH_3$$

This is, however, already getting quite long to write out, so any way of abbreviating the drawing of the structures is welcome. In addition to the structural diagram:

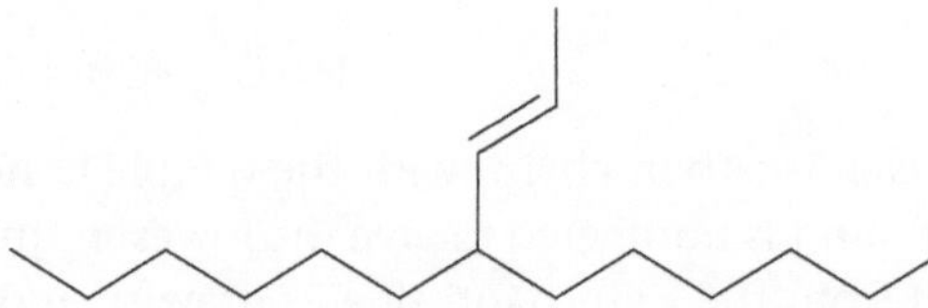

the standard way of doing so is to minimise the CH_2 chain by use of brackets (parentheses).

Thus the structure above could be rewritten as:

$$CH_3 \cdot (CH_2)_5 \cdot \underset{|}{CH} \cdot (CH_2)_5 \cdot CH_3$$
$$CH = CH \cdot CH_3$$

3.3.3   Introduction to the functional groups concept

Functional groups are so named because they are the parts of organic molecules where the main reactivity is found. It is not common to think of alkyl groups as functional groups, as they are quite unreactive and form the 'skeleton' of our compounds.

It is appropriate, however, to think of alkene and alkyne groups as functional groups, as they are relatively reactive.

We will now look at these, and at compounds of other types, and also consider some biologically important materials containing the functional groups in question. Compounds can contain just one functional group, but most that are important in a biological setting contain several functional groups.

Before continuing looking at how compounds are named, we will look at some of the biologically more important functional groups, with a view to being able to identify them in chemical structures.

Common functional groups in biological systems

Alkanes and alkyl groups

Because these contain only $C-C$ single bonds and $C-H$ bonds, they are not really regarded as functional groups, behaving more as the skeleton on which functional groups sit. Some examples are:

- petroleum components, for example methane, ethane;
- amino acid *alkyl* side chains;
- lipids, which contain long *alkyl* chains, e.g. palmitic acid (main fatty acid in butter)

$$CH_3 \cdot (CH_2)_{14} \cdot CO_2H$$

In many cases the properties of compounds are independent of the size/nature of alkyl groups; for example, the chemistry of alkyl halides is referred to in a generic sense. As such, it is useful to have a generic abbreviation for alkyl groups, and this is given by the letter R. Thus alkyl halides are often represented as RX, where R is the generic symbol for alkyl groups and X represents the halogens.

Alkenes and alkenyl groups

These contain $C=C$ bonds which (as will be seen later during the overview of reaction mechanisms in Chapter 8) are quite reactive, and so can be regarded as functional groups. Examples of compounds containing these, which are topical in terms of developing healthy eating regimes, are polyunsaturated fatty acids, which have multiple $C=C$ bonds, e.g. linolenic acid (a 45% component of corn oil):

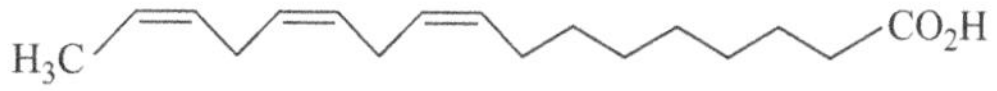

Alkynes and alkynyl groups

These groups contain carbon to carbon triple bonds, and although compounds containing this group are not widespread in nature, there are some, for example calicheamicin (below) and the related compound esperamicin, both of which are powerful antitumour agents (and both are derived from fungi) which cleave DNA.

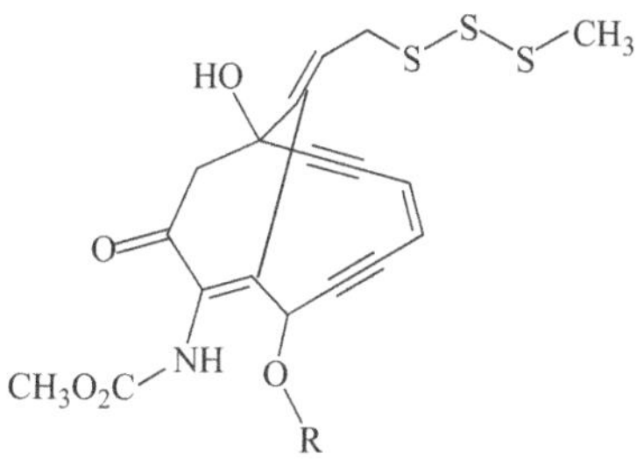

Arenes and aryl groups

These are aromatic groups, and in nature they usually are found as six-membered rings. The most famous example of an aromatic compound is benzene, the structure of which is

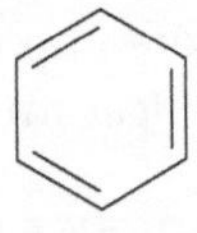

Aromatic rings are common, and the compound phenylalanine, which is a component part of the sweetener aspartame, is:

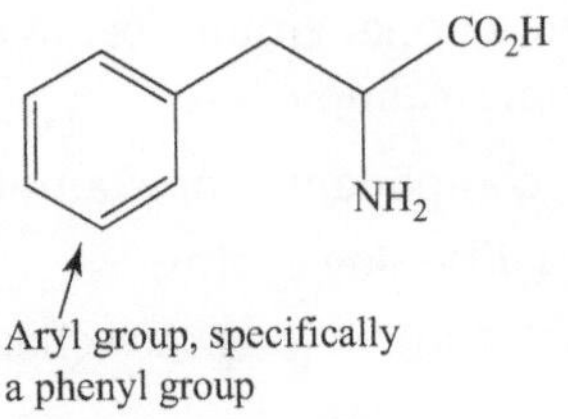

Aryl group, specifically a phenyl group

In addition to the aryl group, phenylalanine has two other functional groups, and these will be identified and described later.

Alcohols

The alcohols contain the group:

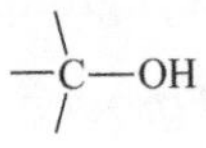

This is a very common functional group in nature, occurring, for example, in compounds such as carbohydrates (starches and sugars). Thus, for α-D-glucose:

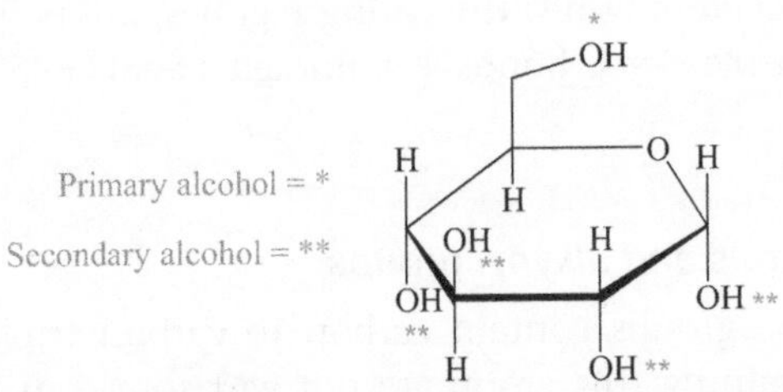

Primary alcohol = *

Secondary alcohol = **

Each carbon that is joined to an OH group is said to bear a hydroxyl group (or it may be said that they are hydroxylated). The glucose shown above can be said to be polyhydroxylated; it could be called a polyol.

The alcohols are given a further level of categorisation depending on how many H atoms are joined to the carbon of the C—OH bond. Thus:

- RCH_2OH is known as a primary alcohol (this includes CH_3OH, which is the situation where R = H);
- R_2CHOH is called a secondary alcohol; and
- R_3COH is called a tertiary alcohol.

The glucose structure shown above contains one primary alcohol group and four secondary alcohol groups.

Ethers

The ether group contains the unit:

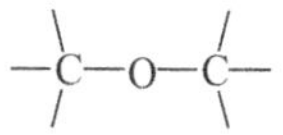

For example, diethylether would be $CH_3 \cdot CH_2 \cdot O \cdot CH_2 \cdot CH_3$, though the group is also found in the cyclic form of carbohydrates; for example, looking again at glucose:

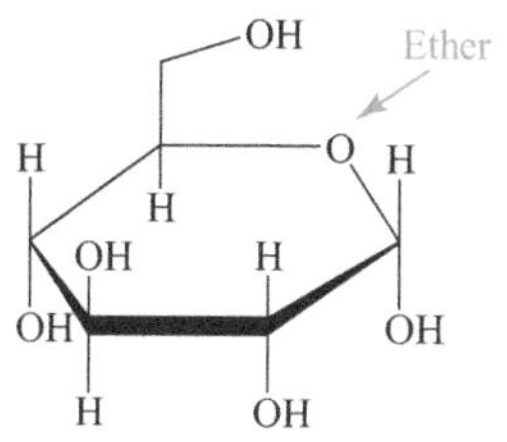

Halides

These contain the group:

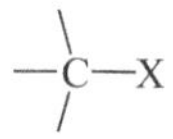

where X = F, Cl, Br or I.

Organic halides do occur in nature, but they are rare and are mainly found in metabolites produced by marine organisms. Exceptions include the thyroid hormone thyroxine.

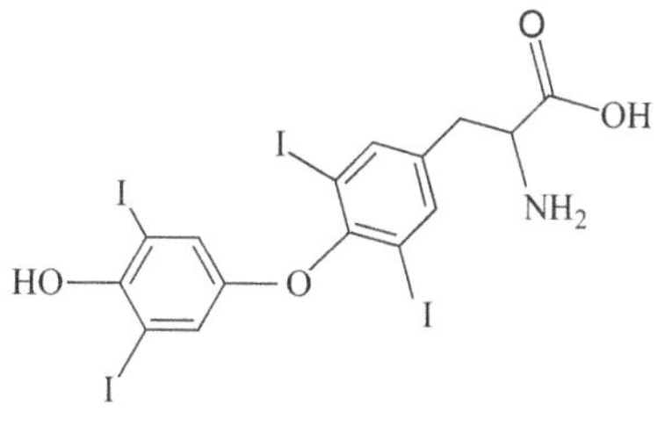

As with alcohols, there are primary, secondary and tertiary halides. Thus:

- RCH_2X is known as a primary alkyl halide (includes CH_3X);
- R_2CHX is called a secondary alkyl halide; and
- R_3CX is called a tertiary alkyl halide.

Amines

Amine compounds contain the grouping:

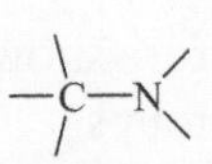

$$-\overset{\displaystyle |}{\underset{\displaystyle |}{C}}-N\overset{\diagup}{\diagdown}$$

Amines are subdivided into three types depending on the number of substituents on the nitrogen. Consider the methyl-substituted compounds as an illustration; thus:

CH_3NH_2 methylamine, a *primary* amine, two H atoms, one organic group

CH_3NHCH_3 dimethylamine, a *secondary* amine, two H atoms, two organic groups

$(CH_3)_3N$ trimethylamine, a *tertiary* amine, three organic groups.

Note how the use of the terminologies primary, secondary and tertiary differ with the amines when compared with the approach described above for alcohols and alkyl halides.

Amines are very common in nature. All of the amino acids are amines, as illustrated below, by returning to the previously quoted example of phenylalanine:

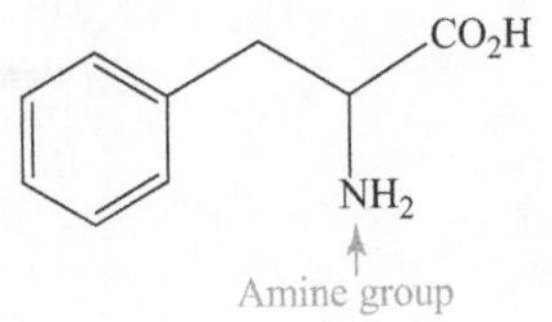

Amine group

Thiols

Thiols contain the grouping:

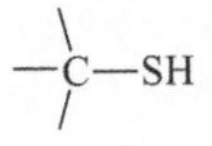

$$-\overset{\displaystyle |}{\underset{\displaystyle |}{C}}-SH$$

They are found in, for example, the amino acid cysteine:

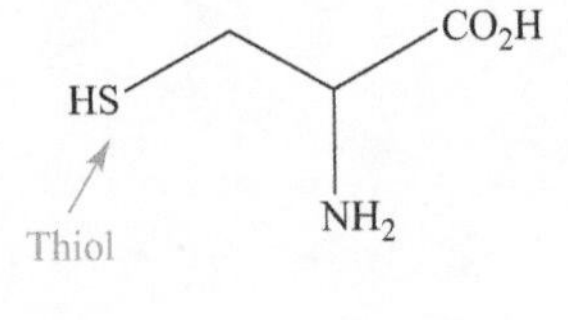

Thiol

Thioethers

These contain the grouping:

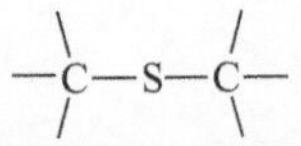

$$-\overset{\displaystyle |}{\underset{\displaystyle |}{C}}-S-\overset{\displaystyle |}{\underset{\displaystyle |}{C}}-$$

The amino acid methionine contains an example of a thioether:

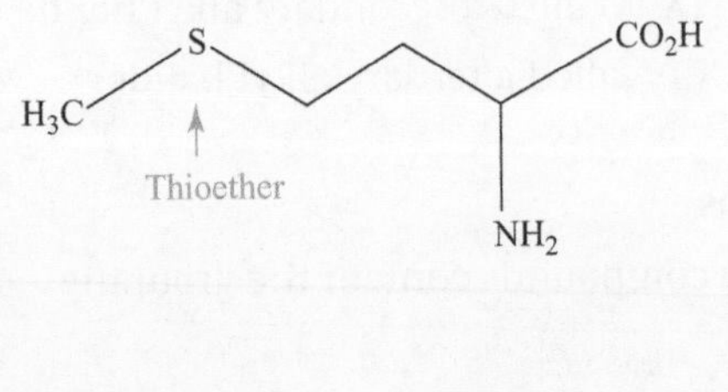

Thioether

Carbonyl functional groups

Carbonyl groups are those involving the entity:

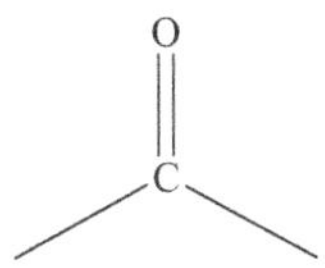

This encompasses a *very* large group of compounds and is subdivided and classified further according to the other atoms attached to the carbonyl carbon as summarised below:

- *Aldehydes and ketones:* These are very similar to each other, but there are enough differences in chemical behaviour (for example, as we will see later, ease of oxidation) for them to be considered as different:

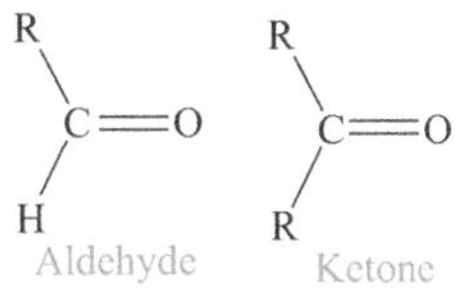

- *'Double' functional groups involving the carbonyl entity:* When two functional groups are bonded *directly* together each modifies the behaviour of the other, so that a new group with its own special properties is formed. The carbonyl group is usually one component and the range of compounds is known as 'carboxylic acid derivatives'.

- *Carboxylic acids:* $C{=}O$ *plus OH*. In these molecules there is very slight dissociation of the H linked to the O, and this is considerably more so than in just an alcohol. The presence of the carbonyl facilitates this through resonance dissociation:

Other biologically important compounds containing 'double functional groups' are:

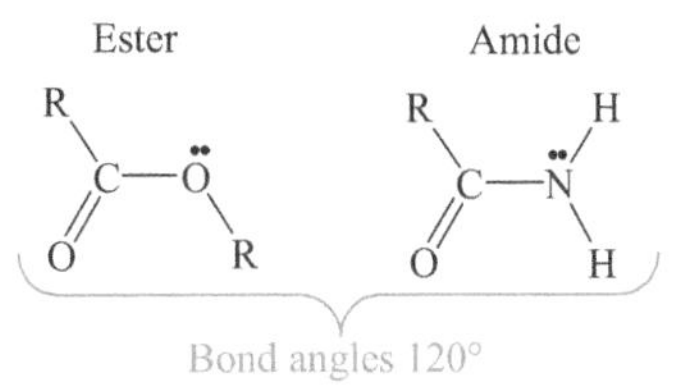

One point that is particularly noteworthy is that in an amide functional group, the carbon, the nitrogen and the atoms directly bonded to them are *coplanar* (all in the same plane) and that there is restricted rotation about the $C{-}N$ bond.

89

The different types of carbonyl groups occur extensively in nature. Referring again to the amino acids, and returning to the example already used, namely phenylalanine, the presence of a carboxylic acid group is indicated:

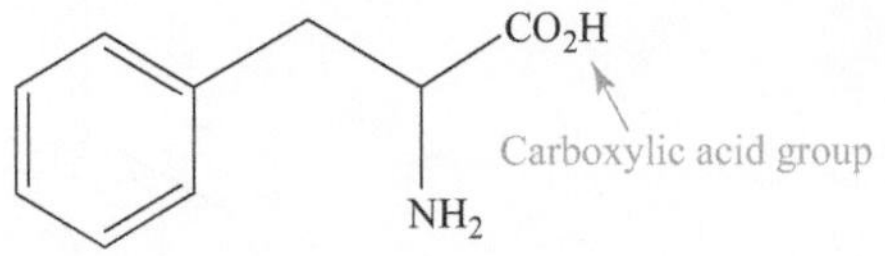

Furthermore, when two amino acids link together, the amine part of one amino acid joins with the carboxylic acid of the other, resulting in the formation of an amide group that links both amino acids together. The new compound is called a peptide (with two amino acids it is called a dipeptide), and the amino acid group is often referred to as a peptide link, or peptide bond. An example of a peptide is found in the well-known sweetener aspartame, which is a dipeptide based on two amino acids, namely phenylalanine and aspartic acid, as represented below:

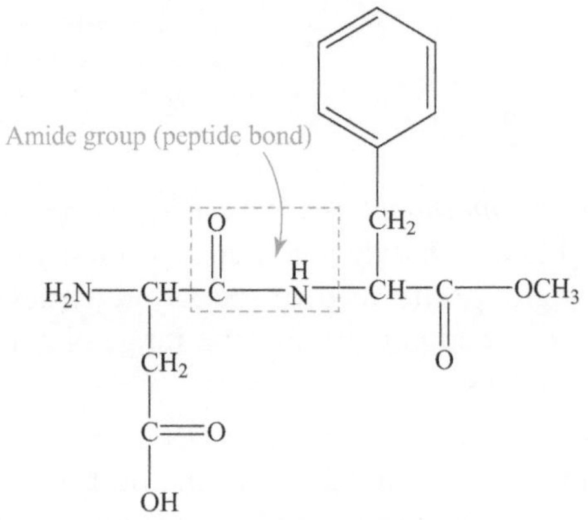

3.3.4 Naming of aliphatic compounds containing functional groups

Having learnt how to name both saturated and unsaturated hydrocarbons, and also having identified many of the key functional groups found in nature and in organic chemistry, the next logical step is to look at the formal naming of compounds containing many of these groups. Whilst most biological compounds are so big that trivial names are used for them, the interface between biology and chemistry is now so blurred that it is important for practising biologists to be familiar with the basics of language in chemistry.

Compounds containing *one type* of functional group will be discussed first, including cases in which this group is repeated several times in the same molecule. This functional group is indicated by a suitable prefix or suffix.

There are four important groups which, even when present as the only functional group, are designated by prefixes, as shown in the list below:

- **Halogen compounds** – chloro-, bromo-, iodo-, etc.
- **Ethers** – methoxy-, ethoxy-, etc.

- **Amines** – amino- (NH_2-), methylamino- (MeNH-), etc.
- **Nitro** – nitro- (NO_2)

Boxes 3.5 and 3.6 show how the rules described thus far are extended to incorporate these functional groups.

Box 3.5

```
Naming of molecules containing one type of functional group
```

In naming these compounds there are three important rules:

1. The main chain is now the one which contains the maximum number of *functional groups*.
2. The main chain is numbered so as to assign the lower numbers to the functional group.
3. The functional groups take precedence over unsaturated centres when selecting the main chain.

Box 3.6

```
Extending the naming rules
```

1. The main chain has the maximum number of functional groups. It is not now necessarily the longest, and a functional group takes precedence over unsaturated centres.
2. The main chain is numbered to give lower numbers to the functional groups.
3. If there seems to be a number missing in a name then the missing number is taken as one.
4. The C=O group is known as the *carbonyl group*.

Simple ethers are often named according to the two groups joined to the oxygen atom, e.g. CH_3—O—C_2H_5 is ethyl methyl ether. Simple halides are sometimes named thus, e.g. $CH_3CH_2CH_2$— Br is n-propyl bromide.

Other common functional groups are designated by a suffix which replaces the final 'e' of alkane. The full suffix should show the type of functional group, the number of that type, and its (their) position(s), as demonstrated in Table 3.5.

Table 3.5 **Naming of compounds with one functional group**

Functional group name	Functional group formula	Suffix	Systematic name
Alcohol	$(R \cdot OH)$	-ol	alkan**ol**
Aldehyde	$(R \cdot CHO)$	-al	alkan**al**
Ketone	$(R \cdot CO \cdot R')$	-one	alkan**one**
Acid	$(R \cdot CO_2H)$	-oic acid	alkan**oic acid**
Ester	$(R \cdot CO_2R')$	-oate	alkyl alkan**oate**
Acid chloride	$(R \cdot COCl)$	-oyl chloride	alkan**oyl chloride**
Amide	$(R \cdot CONH_2)$	-amide	alkan**amide**

In naming aldehydes and ketones it must be remembered that the carbon atom of the CHO or CO group(s) is included in the name and numbering of the main chain. Therefore, if we consider the ketone $CH_3 \cdot CH_2 \cdot CO \cdot CH_3$, the chain, including the CO carbon, is four carbon atoms long, and the CO carbon would be designated as being at position 2 in order to give it the lowest number in the chain. The name therefore would be *butan-2-one*. In the case of such simple ketones however, these are often named according to the groups bonded to the carbonyl group. For example:

$$CH_3 \cdot CH_2 \cdot CO \cdot CH_3$$

ethyl   methyl ketone

When naming carboxylic acids and their derivatives, it is important to remember that the carbon atom of the carboxyl group is included in the main chain also. The name of an ester is always made up of two words, one describing the acid component and the other describing the alcohol component of the ester.

This ester is methyl ethanoate:

Methyl (from methyl alcohol)

Ethanoate
(named after ethanoic acid)

The structure of organic molecules

An analysis of the structure of organic molecules can be divided into three main categories:

- composition, which considers the number/type of atoms;
- connectivity, which looks at how the atoms are linked; and
- shape, which involves considering the 3D spatial arrangement of the atoms (known as stereochemistry).

These aspects form the focus of this chapter, laying the foundations for later work.

3.4.1 Structural features of organic chemistry

Many features of organic and biological chemistry depend critically on the shapes of molecules. For example, as we will see later, how drugs interact with receptor sites is very dependent on the shape adopted by the drug.

Before going further it is wise to recall a few key facts and their consequences:

- Carbon forms four bonds.
- Carbon can form multiple bonds both with itself and with other atoms.
- Shapes of molecules are often controlled by numbers of electron groups (Lewis structures and VSEPR theory).

Therefore:

- If carbon has four single bonds (four 'electron groups'), then the four bonds will be tetrahedrally arranged about it.
- If carbon has one double and two single bonds, then the arrangement is based on three electron groups, and will be trigonal.
- If the carbon has one triple and one single bond, then the arrangement is based on two electron groups, and will be linear.

Developing these ideas, it follows that the three structure motifs shown below are those most commonly found in carbon chemistry:

As demonstrated already, organic compounds can adopt a great variety of shapes and structures, and there is a range of possible structures even for a given molecular formula.

This general area will now be explored further, with particular emphasis on *why this aspect of chemistry is so important in a biological context.*

3.4.2 Introduction to isomerism

Isomerism is defined as the existence of two or more compounds (isomers) having the same molecular formula but a different arrangement of atoms within the molecule.

There are a variety of different types of isomerism, and we will consider what these are and look at the criteria for distinguishing between the different types of isomerism. Consider Figure 3.2.

3.4.3 Structural/constitutional isomerism

Structural isomerism occurs where molecules have a given molecular formula, but the atoms are connected in different ways.

Some examples are:

- C_2H_6O – there are two compounds which share this formula. These are:

$$H_3C-O-CH_3 \quad \text{and} \quad CH_3-CH_2-OH$$

The left hand compound is an ether (dimethyl ether) and the right hand compound is an alcohol (ethanol).

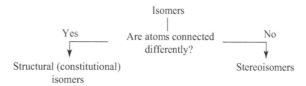

Figure 3.2 **Flow chart demonstrating different categories of isomerism.**

- C_4H_7N – has a number of different possible structural isomers, but to illustrate the point consider the two shown below:

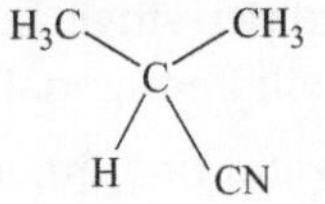

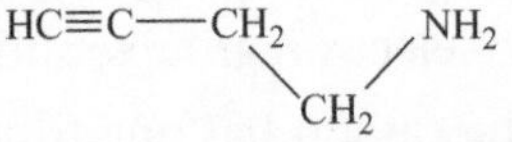

The left hand one of these is an alkyl cyanide, whereas the compound on the right contains two functional groups, an alkyne and a primary amine!

- C_4H_9NO – among the several options for this formula are:

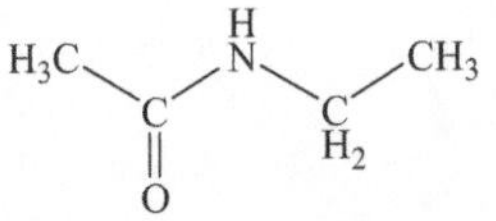

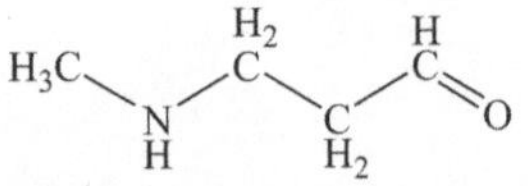

The left hand compound is an amide; the right hand one has two functional groups, a secondary amine and an aldehyde.

- C_6H_{14} – has five such isomers, and their structures are represented below:

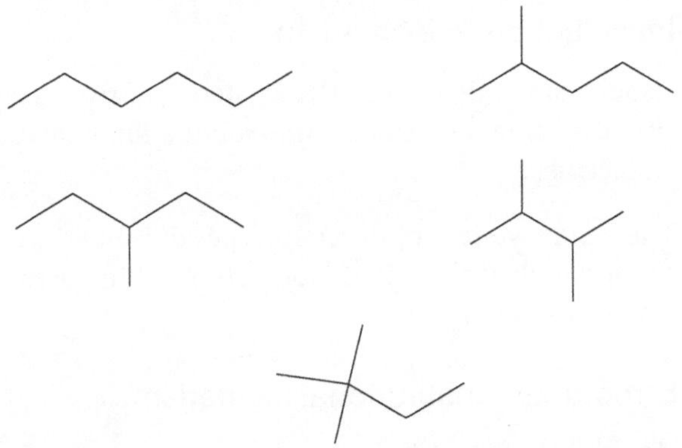

In all of the cases of structural isomerism, the different isomers will have different physical properties (for example, melting/boiling points) and will often be chemically different too.

Recalling Figure 3.2, we will now focus on the situations where the atom connectivities are the same, but differ in the relative spatial arrangements. This is *stereoisomerism*.

3.4.4 Introduction to stereoisomerism

This is often thought of as being separated into two categories, namely *conformational isomerism* and *configurational isomerism*, and the latter is often considered to be made up of a series of specific sub-areas, namely *geometric isomers, enantiomers and diastereomers*. Consider the flow diagram in Figure 3.3.

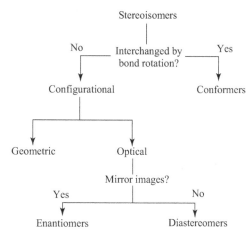

Figure 3.3 **Flow diagram for different categories of stereoisomerism.**

3.4.5 Conformation

Conformation is an exploration into how rotation around single bonds affects molecular flexibility and shape. An understanding of some basic principles will develop an understanding about why, as an example, steroids adopt the shapes they do. There are a few key facts:

- Most molecules have a large degree of flexibility associated with them due to bond rotation.
- Bond rotation arises often with $C-C$ single bonds (but is also the case with $C-N$ or $C-O$ single bonds).
- No such flexibility with multiply bonded systems (the rationale for this was described in Section 2.9).

Molecular shape depends:

- on the bonds between adjacent atoms;
- on the forces of attraction and repulsion between atoms of the molecule that are not joined directly by a covalent bond.

Consider ethane as an example. Figure 3.4 demonstrates a series of different ways of looking at ethane, the last three of which emphasise, in different ways, the shape of a molecule of ethane.

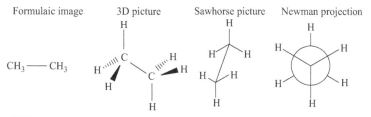

Figure 3.4 **Different representations of ethane.**

In Figure 3.4, the Newman projection may need some additional explanation. With a Newman projection we are, effectively, looking along the C—C bond of ethane head-on:

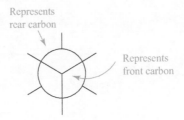

The positions of the two carbon atoms are indicated, and the groups attached to the carbons (Hs in this case) are shown coming out from them. Now consider the implications of rotation about the C—C bond, with particular reference to Figure 3.5:

● When the hydrogen atoms on adjacent carbon atoms are *eclipsed* the energy is at its highest – i.e. the situation is *less stable*.

● The energy of the *staggered* form is at a minimum, i.e. this is the preferred situation.

● The energy difference between the eclipsed and staggered forms of ethane is very small compared with the value of bond energy (ca.12 kJ mol^{-1}, compared with ca. 350–400 kJ mol^{-1}), and there is relatively free rotation about the C—C bond at room temperature.

In the case of butane, there is a slightly more complex situation, summarised by Figure 3.6 (where we are looking at the bond between C2 and C3). The least stable (highest energy) form is where the two methyl groups are eclipsed, but there are also energy maxima where methyl groups are eclipsed with H atoms (intuitively sensible, as methyl groups take up more space than H atoms).

This idea of bond rotation is also very important when considering cyclic alkanes.

Conformation in cyclic systems

There is a degree of freedom of rotation associated with single bonds (refer to the examples of ethane and butane discussed earlier), but the cyclic nature of the

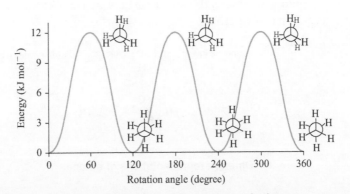

Figure 3.5 Relative energies of different conformations of ethane.

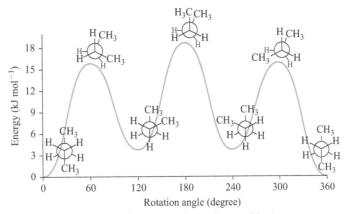

Figure 3.6 Relative energies of different conformations of butane.

systems inevitably results in restrictions. Many biologically important compounds, e.g. antibiotics, steroids and carbohydrates, have structures incorporating rings of three, four, five or more atoms, mostly carbon, joined by single bonds. Where an oxygen, sulfur or nitrogen atom is found in place of a carbon atom, as frequently occurs, it usually only results in minor variations in shape.

The conformations adopted by simple cyclic molecules are therefore particularly relevant when considering the shape of some of the more complex molecules.

As with the simple chain molecules discussed earlier, eclipsing strain and non-bonded interactions are important. However, ring information introduces extra factors affecting the stability of the molecule. Carbon atoms bonded to form a ring take up a regular form in which all the bond lengths and all the bond angles are equal.

For purely geometrical reasons the angles between the bonds forming the ring may differ markedly from the values normally found in chain structures. This introduces into the molecule an additional source of strain, known as *angle strain*. This becomes more pronounced as the bond angle deviates from the normal tetrahedral value of 109.5°.

Cyclopropane

The three carbon atoms forming the ring must be coplanar. This implies bond angles of 60° and a considerable degree of angle strain. To approach as nearly as possible a regular tetrahedral configuration around the carbon atoms, the hydrogen atoms lie above and below the plane of the ring.

Cyclobutane and cyclopentane

These have bond angles of 90° and 108° respectively (cyclopentane therefore has little angle strain).

Cyclopropane   Cyclobutane   Cyclopentane

97

In each of these flat molecules, however, there is a significant amount of eclipsing strain since the C—H bonds are lined up in the same way as in the eclipsed conformation of ethane. In ethane this eclipsing is reduced by rotation about a C—C bond, which moves the C—H bonds closer to a staggered arrangement:

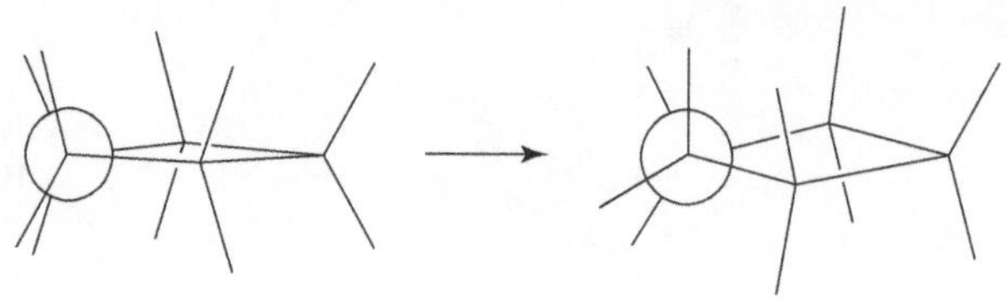

In these cyclic systems, however, any attempt to reduce eclipsing strain by bond rotations results in an increase in angle strain. In practice a balance is reached between the two opposing effects.

Cyclohexane, C_6H_{12}

Key features relating to the shape of cyclohexane are:

- A planar arrangement of carbon atoms would result in eclipsed bonds as before, *but also* in ring bond angles (120°) larger than in a normal saturated molecule.
- Attempts to relieve bond eclipsing by bond rotation therefore also results in relief of angle strain.

Two conformations exist for cyclohexane that are free from angle strain. One of these, known as the *chair conformation*, is also free of eclipsing strain. The ring is puckered and all C—H bonds are fully staggered:

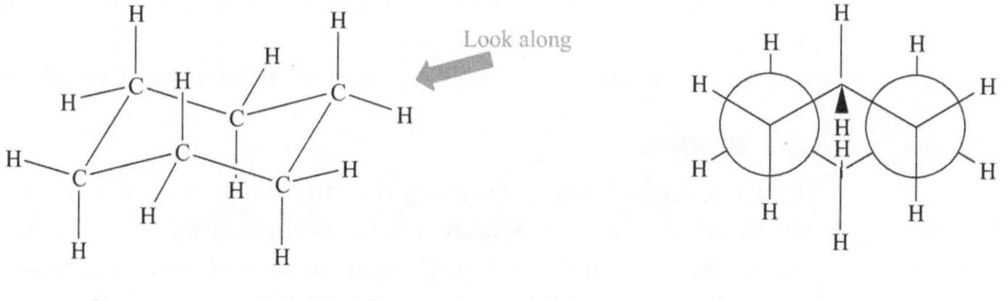

Chair conformation of cyclohexane

Unlike the conformation of cyclopentane this form is comparatively rigid since any rotation introduces strain in all angles and partial eclipsing in all bonds. This completely strain-free conformation of cyclohexane has several important features.

- Six of the hydrogen atoms lie alternately above and below the carbon framework so that their bonds are all parallel. The atoms are said to occupy *axial (a)* positions.
- The remaining six hydrogen atoms are distributed around the outside of the carbon framework so that the C—H bonds are parallel to the sides of the ring. These hydrogen atoms are said to occupy *equatorial (e)* positions.

- The internuclear distances between equatorial hydrogen atoms on adjacent carbon atoms is equal to the distance between axial and equatorial hydrogen atoms on adjacent carbon atoms, and also equal to the distance (250 pm) between axial hydrogen atoms on the same side of the ring.

Although rigid, this conformation can be converted into another chair conformation by bending the lower part of the chair up and the upper part down:

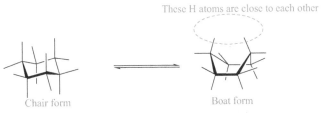

Interconversion of cyclohexane chair conformations

This involves passing through less stable arrangements in which some angles are distorted and some bonds eclipsed. *The result of this process is that those hydrogen atoms that originally occupied equatorial positions occupy axial positions and vice versa.*

If only the lower part is bent upwards the rigid chair conformation is converted into a flexible conformation which is free of angle strain, *although not free of strain resulting from bond eclipsing.* The flexibility arises because, unlike the chair form, bond rotations reduce the eclipsing of some bonds but increase the eclipsing of others and, in this instance, need not alter the bond angles. One conformation of the flexible form is known as the *boat conformation*:

These H atoms are close to each other

Chair form   Boat form

Interconversion of chair and boat conformations

Four of the C—C bonds are fully staggered but the other two are fully eclipsed. The two H atoms from the 'bow' and the 'stern' of the boat point in towards the centre of the molecule resulting in severe non-bonded interaction between these two atoms whose centres are only 184 pm apart. These additional strains mean the boat conformation is a lot less stable (about 30 kJ mol^{-1}) than the chair conformation and therefore in the majority of compounds containing six-membered rings nearly all the molecules are chair conformations.

Effects of substitution

Although for cyclohexane itself the two chair forms are identical, this is not true when one or more hydrogen atoms is replaced by a different atom or another group of atoms. The substitution of one hydrogen atom in the chair form of cyclohexane leads to two different chair conformations:

- In one of these the substituent occupies an axial position.
- In the other, an equatorial position.

99

As in cyclohexane these two conformations rapidly interconvert, and in a compound such as methylcyclohexane some molecules will have one conformation and some will have the other. The proportion of each will depend upon their relative stabilities.

The two conformations are not equally favourable since the presence of an atom or group larger than hydrogen may introduce additional strain into the molecule.

An equatorial methyl group introduces little, if any, strain. It lies at the end of two chains of four atoms which are arranged in exactly the same way as the four carbon atoms in the anti (strain–free) conformation of n-butane:

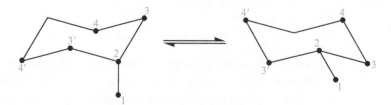

For molecules with, for example, an axial methyl group, each of four atoms corresponds to the skew conformation of n-butane. The resulting interactions, shown below, are often described as *1,3-diaxial interactions*.

Conformations are preferred in which the substituents occupy equatorial positions. If the substituent is very large then the axial conformation may be impossible and all molecules take the equatorial conformation. The importance of the stability associated with the equatorial position is reflected in the structures of the simple carbohydrates known as hexoses. One form is a six-membered ring in which one of the atoms is oxygen. The other five atoms in the ring are carbon atoms and each is also bonded to a hydrogen atom and other substituent, usually a hydroxyl group. Each substituent may occupy axial or equatorial positions and it is interesting that the most abundant, naturally occurring hexose, glucose, has all the groups positioned equatorially:

β-D-Glucose

The effect of double bonds in rings

The presence of a double bond in a six-membered ring, as in cyclohexene, results in a flattening of the chair conformation in the region of the double bond. This results in the molecule adopting the *half-chair* conformation. The sequence of carbon atoms C_6—C_1=C_2—C_3 (shown to the front in the diagram below) lie in a plane as in *cis*-but-2-ene, the two remaining carbon atoms C_4 and C_5 respectively lie above and below the plane of the other four carbon atoms. This molecule, unlike cyclohexane, is not strain–free:

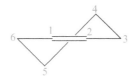

Half-chair conformation of cyclohexene

Conformation in fused ring systems

The conformation of molecules containing one or more fused ring systems, as is found in the class of organic compounds known as the steroids, follows directly from those of the disubstituted single ring compounds.

Decalins, $C_{10}H_{18}$, consist of two cyclohexane rings in which two adjacent carbon atoms are shared by both rings. They are disubstituted cyclohexanes in which the two substituents are joined together to form another ring. Two molecules, *cis*- and *trans*-isomers, are therefore possible and their conformations may be derived from those of the 1,2-dimethylcyclohexanes. In *trans*-decalin the second ring is formed from the two equatorial positions and also takes the chair form:

Trans-decalin

The two hydrogen atoms are on opposite sides at the ring junction which is said to be *trans*. The structure is very rigid, and is prevented from being transformed into a diaxial form by the second ring. Since all bonds are staggered and all angles are 109.5°, *trans*-decalin is strain–free. In *cis*-decalin the second ring also takes the chair form but it is formed from one axial and one equatorial position, and the ring junction is *cis*.

Cis-decalin

This molecule, unlike the *trans*-isomer, gives the impression of having the two rings almost at right angles and the structure is fairly flexible. It is, however, less stable than the *trans*-isomer since interactions arise between three pairs of hydrogen atoms which occupy axial positions on one or other of the rings.

Biological examples of ring systems

There are millions of biological examples. The hexoses referred to above are some. Below are a couple of other compound types known generally. *The steroids* are a family of biologically important compounds that all contain a basic skeleton of four fused rings. Three of these are six-membered and the remaining one is five-membered. In most, but not all, steroids the rings are *trans*-fused as in pregnanediol:

Pregnanediol

Owing to the inflexible nature of the ring system, the hydroxyl group attached to the first ring cannot move to the more stable equatorial position as this would require bond breakage.

Molecules in which the fused rings both contain less than six atoms always have *cis*-ring junctions. This means that the rings are not in the same plane and inaccurate representations may give an entirely misleading idea of the shape of the molecule. This is clear with *cis*-decalin above, and is well demonstrated by the skeleton of the penicillins:

i.e.

$$C_6H_5-CH_2-CONH$$

Penicillin G

3.4.6 Introduction to configurational isomerism

Having explored the concepts and relevance of conformation, the different forms of configurational isomerism will now be explored. The different types of configurational isomers are illustrated in Figure 3.7.

3.4.7 Geometrical isomerism

Geometrical isomerism arises where bond rotation is restricted or prevented. This situation is found in compounds containing double bonds (see Section 2.9.3) or

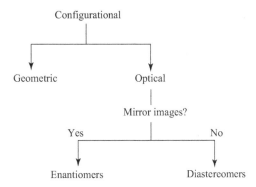

Figure 3.7 **Types of configurational isomers.**

in ring systems. In compounds with a double bond, if the two identical groups are joined to *one* of the doubly bonded carbon atoms then there are no geometrical isomers of the compound:

Changing the H and I on C1 around in the two structures (a) and (b) makes no difference – the two structures remain identical and are *not* geometric isomers.

If, however, *both* double bonded carbon atoms are joined to two non-identical groups then two different compounds are possible – these are known as *geometrical isomers:*

Changing the H and I around on C1 in these two does result in non-identical structures – these *are* geometric isomers.

The groups joined to one of the double bonded carbon atoms may be the same as those joined to the other, as in the example above (but they need not be). Where at least one of the groups at each doubly bonded carbon is the same (e.g. both carry an H atom) the nomenclature is straightforward:

● Where the identical groups lie on the *same* 'side' of the double bond the molecule is known as the *cis* isomer (in (x) above).

103

- Where the identical groups lie on the *opposite* 'side' of the double bond the molecule is known as the *trans* isomer (in (y) above).

A similar situation occurs with *ring* compounds. Consider as an example the two 3-membered rings below:

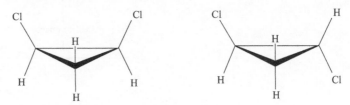

In the left hand ring, the chlorine atoms are on the same side of the ring as each other, whereas in the right hand example, the chlorine atoms are on opposite sides of the ring. The two forms are not interchangeable by bond rotation and so, as in the case of compounds x and y earlier, they are geometric isomers of one another.

- The left hand example could be referred to as *cis* (the two chlorine atoms are on the same side of the plane of the ring).
- The right hand example can be described as *trans* (the two chlorines are on opposite sides of the plane of the ring).

Geometrical isomers have *different physical properties* and frequently *different chemical reactivities*.

So far the usage of the terms *cis* and *trans* has been demonstrated, but in the situation where all four groups joined to the C=C double bond are different, a new method of naming compounds is needed.

E and *Z* nomenclature

A set of rules (Box 3.7) was devised in order to enable a higher 'priority' to be assigned to one of the two groups on the left hand carbon of the double bond, and similarly to one of the two groups on the right hand carbon of the double bond (assuming that the molecule is drawn with the C=C 'horizontal').

- If the two higher priority groups lie on the *same* side of the double bond, the molecule is designated the *Z*-isomer (from German *zusammen – together*)
- If the two higher priority groups lie on the *opposite* sides of the double bond, the molecule is designated the *E*-isomer (from German *entgegen(gesetzt) – opposite*).

Box 3.7

Rules for deciding priority

1. The group with the atom of highest atomic number directly attached to the double bond atom has highest priority.
2. Where groups are joined by the same kind of atom, e.g. both carbon, examine the second atoms in the groups. Here one higher atomic number atom takes priority over two or more lower atomic number groups. Thus O,H,H has higher priority than C,C,H.
3. A double bonded atom counts as two single bonded atoms; thus an aldehyde function —CH=O (i.e. O,O,H) takes priority over an alcohol —CH(Me)OH (i.e. O,C,H).

Example 3.6   For the isomers **(a)** and **(b)** shown below,

(a)   **(b)**

- on carbon 1, Br is of higher priority than H (from rule 1 above) and
- on carbon 2, CHO is of higher priority than CH_2OH (we cannot decide between them on the basis of rule 1 – the atom joined to the alkene double bond position is carbon in both cases, so we need to use both rules 2 and 3, where $C{=}O$ has a higher priority than $C{-}O$).

For structure **(a)**, the highest priority group on C_1 (the Br) is on the opposite side of the molecule from the highest priority group on C_2 (the CHO). Hence, compound **(a)** is an *E*-isomer. For compound **(b)**, the highest priority group on C_1 is on the same side of the molecule from the highest priority group on C_2. Hence, compound **(b)** is a *Z*-isomer.

3.4.8 Symmetry, chirality and optical isomerism

Most people understand what is meant when something is said to be 'symmetrical' without necessarily being able to define it precisely. There are actually many different ways in which things can be symmetrical:

- In terms of the *plane of symmetry* which slices an object into two halves that are mirror images of each other. A 'symmetrical' object may have one plane of symmetry (e.g. a mug) or more planes of symmetry (e.g. a cube).
- In terms of the *centre of symmetry*, which can exist as well as or instead of a plane (for example, a cube or a sphere).

All objects with these symmetry elements are identical with their mirror images and are said to be *achiral*. For example, in the cases below, each of which has planes of symmetry, all the molecules are identical with their mirror images (if you find it difficult to visualise, using a model kit often helps).

Chirality

Objects without planes or centres of symmetry (e.g. shoe, screw, golf club) are not identical with their mirror images. They exist in 'left-handed' and 'right-handed' forms and are said to be *chiral*.

Molecules also can be classified as achiral or chiral. Consider the molecule below:

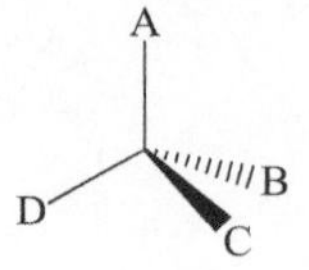

This represents a carbon atom with four different groups arranged tetrahedrally around it. This has no elements of symmetry about it whatsoever. Now compare it with its mirror image, as shown below:

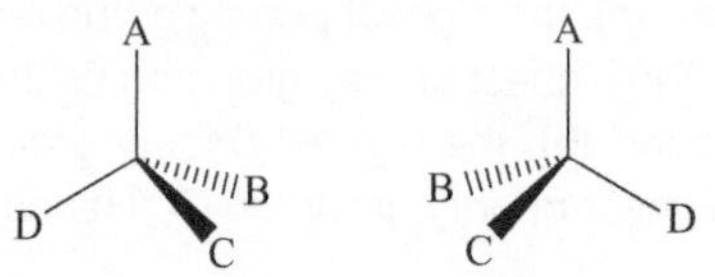

The two mirror image structures are *non-superimposable* (non-superimposable simply means it is impossible to overlay one directly over the other) and they are said to be chiral.

The two forms in this case are said to be *enantiomers. These 3D representations are sometimes difficult to visualise, and a set of molecular models can assist in this.*

A few real examples

Lactic acid, a product in the metabolism of glucose, is chiral and has the two forms shown below.

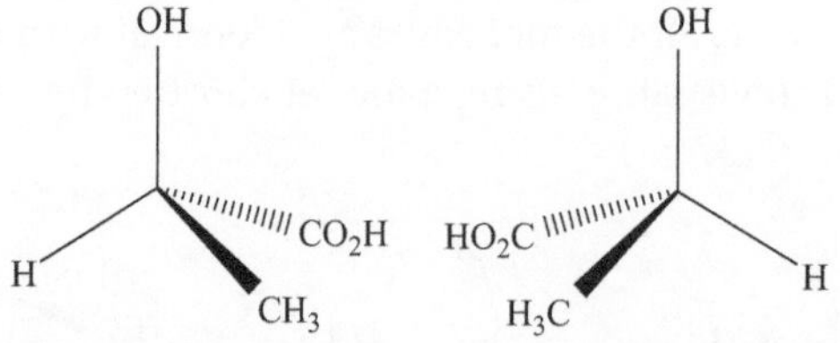

These two forms are mirror images which are non-superimposable, and so are chiral, and are enantiomers of one another.

The diagram below shows two further examples of enantiomers, one of them showing the two possible forms of alanine, an amino acid, and the other shows an enantiomeric pair containing two chiral centres, namely 1,2-dibromocyclohexane, which will be explored more fully later. The significance of the (+) and (–) terms will be described shortly.

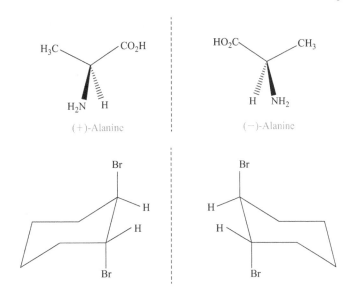

(+)-Alanine

(−)-Alanine

To summarise, if a molecule has one carbon atom with four different groups bonded to it, then that carbon atom becomes *chiral* and is known as a *chiral centre*. It is also (principally in the USA) known as a *stereogenic centre* or a stereo centre and is sometimes still referred to as an *asymmetric carbon atom*.

Properties of enantiomers

Enantiomers show *identical* chemical behaviour when reacting with achiral molecules (but usually react at very different rates with other chiral molecules). Enantiomers have *identical* physical properties *except* that in solution they rotate the plane of plane polarised light by the same amount in opposite directions. They are said to be optically active, and as such are referred to as optical isomers. This term does, however, include more than enantiomers, as we will see later.

- When the plane of an advancing beam of plane polarised light is rotated to the right (clockwise), the enantiomer is said to be dextrorotatory (+).

- When the plane of an advancing beam of plane polarised light is rotated to the left (anticlockwise), the enantiomer is said to be laevorotatory (–). Hence the labels in front of the alanine pictures above.

This is known as optical rotation, and the degree of optical rotation of an optically active compound is measured using a polarimeter.

- Achiral (i.e. non-chiral) compounds are not optically active.

- A mixture containing equal numbers of both (+) and (–) enantiomers has zero optical rotation and is known as a *racemic mixture*.

The *R/S* system

In the 1950s a rigorous system was devised by Cahn, Ingold and Prelog by which each chiral centre in a molecule could be defined unambiguously. In this way the *absolute configuration* of the molecule can be completely specified. Each chiral centre within a molecule, regardless of the number of centres, is defined as either *R* (from Latin *rectus*) or *S* (from Latin *sinister*) as follows:

1. Label each of the four groups bonded to the chiral centre according to their priority (1 = highest priority, 4 = lowest priority) using the same rules as for geometrical isomers, e.g.

2. View the molecule so that the group of lowest priority (4) points away from you – it is obscured by the chiral centre.

3. If the sequence 1...2...3 is clockwise (as above) the centre is labelled *R*.

4. If the sequence 1...2...3 is anticlockwise, the centre is labelled *S*.

In three dimensions:

(*R*)-isomer   (*S*)-isomer

Note: There is no automatic correlation between the absolute configuration (*R/S*) of an enantiomer and the direction in which it rotates the plane of plane polarised light (+/−).

Example 3.7

Prioritise groups in order 1–4

Group 4 put to the back.
1−2−3 goes anticlockwise.
It is an *S*-form.
Compound is *S*-alanine.

Example 3.8

Prioritise groups in order 1−4

Group 4 put to the back.
1−2−3 goes clockwise.
It is an *R*-form.
Compound is: *R*-3-bromo-2-hydroxymethyl-2-methylpropanal

Note: Many people struggle to visualise how to rearrange 3D images in the way necessary to achieve the determination of *R/S* configuration. The simplest way of

108

doing so is to make up a molecular model. This is not always possible, however, and there is a relatively simple 'by the book' way of redrawing structures to get the lowest priority group to the back. This is:

1. swap the position of the lowest priority group with the group currently sitting at the back and then
2. swap the positions of the other two groups.

Consider the case from above, namely *R*-3-bromo-2–hydroxymethyl-2-methylpropanal:

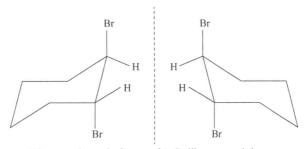

Box 3.8

Generalisations concerning molecules and chirality
1. If only one chiral centre is present, the molecule is chiral.
2. If more than one chiral centre is present, the molecule is not necessarily chiral. The molecule may have a plane of symmetry.
3. Some molecules are chiral even though they have no chiral centres – they do not have a plane of symmetry.

Molecules with more than one chiral centre

The example of 1, 2-dibromocyclohexane, shown earlier, identified that a compound can have more than one chiral centre and that enantiomeric forms can exist in such cases:

The enantiomeric forms of 1, 2-dibromocyclohexane

This idea will be explored more extensively using tartaric acid, which occurs naturally in many plants, as an example. The tartaric acids are examples of molecules that have two chiral centres. Since the groups bonded to both chiral centres are the same, there is the possibility that the chirality at one centre can exactly balance that at the other, resulting in a plane of symmetry in the molecule. Where this happens one centre is R, the other is S and the compound is known as a **meso** compound, e.g. **(x)** below. It is achiral.

If both centres in the molecule have the same absolute configuration then a pair of enantiomers results, as shown by (y) and (z) below.

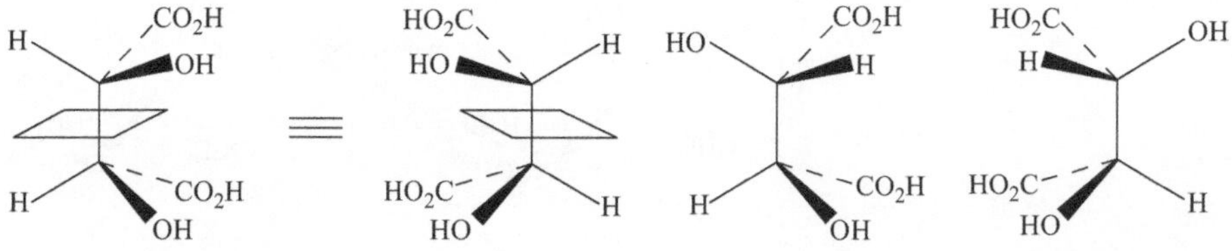

meso–tartaric acid
m.p. = 140 °C
(x)

S,S-(−)-tartaric acid
m.p. = 170 °C
(y)

R,R-(+)-tartaric acid
m.p. = 170 °C
(z)

Thus **(y)** and **(z)** are non-superimposable mirror images, hence they are enantiomers.

The D/L system

Often in biologically interesting systems, particularly sugars and amino acids, optically active compounds are referred to as either D- or L-isomers. This is largely a historical form of terminology, and the background will now be discussed. The system was introduced by Fischer over 100 years ago (1896), and was based on an assumption. He said: 'We will *assume* that (+)-glyceraldehyde has the actual 3-D structure shown below and we will call it the D configuration; its enantiomer will have the L configuration'.

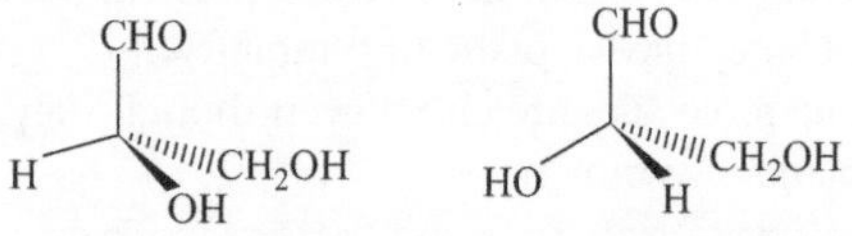

D-Glyceraldehyde

L-Glyceraldehyde

It took until 1951 before Fischer's assumption was shown to be correct. The configurations of other optically active compounds are referred to as these forms of glyceraldehydes. (This will be described in Chapter 7.) Nearly all naturally occurring monosaccharides have the D configuration and nearly all naturally occurring amino acids have the L configuration. Although not satisfactory when trying to deal with molecules with many chiral centres, this system is still used for these two classes of compounds, as emphasised below:

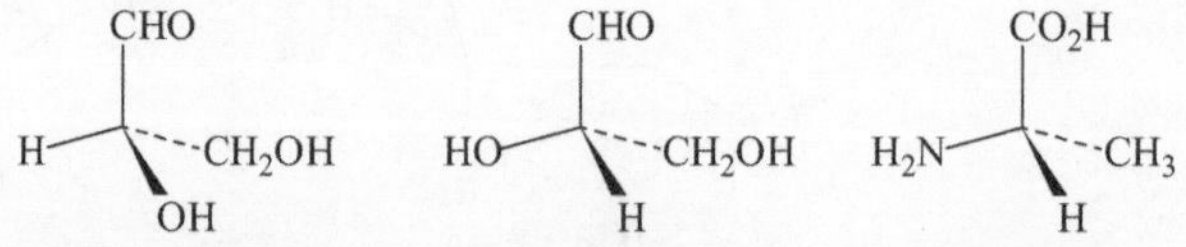

D-Glyceraldehyde

L-Glyceraldehyde

L-Alanine

Fischer formulae

Returning to D-glyceraldehyde, not only did Fischer propose the structure of this compound, but he also proposed a two-dimensional way of providing a three-dimensional representation of the structure. The approach is highlighted below and summarised in Box 3.9.

This is the generation of the Fischer projection for D-glyceraldehyde, and note the OH is on the right hand side. When discussing biomolecules in more detail, the method for how this approach is extended to assign D- and L-configurations to carbohydrate molecules and to amino acids will be shown.

Using this approach for L-glyceraldehyde:

In a Fischer projection:

- horizontal lines point towards you;
- vertical lines point away from you;
- hence, the stereocentre is the only atom in the plane of the paper.

Box 3.9

Rules of Fischer projections

1. They *can* be rotated about 180°, i.e.

2. They *cannot* be rotated through 90°

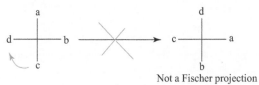

Not a Fischer projection

(in fact the two structures are enantiomers of each other).

3. They *cannot* be 'turned over', i.e.

Fischer projections are used extensively to represent 3D images, though particularly in biochemical books. As mentioned above, we will see them again (particularly in the discussion of biomolecules in Chapter 7), and these representations will be looked at more fully at that point.

3.4.9 Some examples of why shape is important

Example 3.9   Carvone

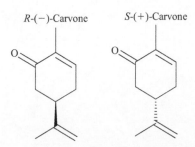

- *R*-(–)-Carvone is found in spearmint oil, and is responsible for the distinctive odour of spearmint.
- *S*-(+)-Carvone is found in the oil of caraway seeds, and is responsible for their odour.

How does this work?

Inside the nose are receptors, which are protein molecules. Receptors are chiral, and can therefore bind enantiomers specifically. The nose contains thousands of receptors, and each enantiomer of carvone interacts with different receptors, hence their different odours.

Example 3.10 L-Dopa – known in treatment of Parkinson's disease

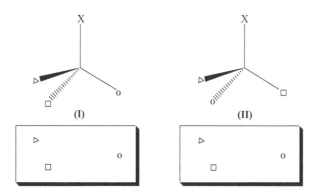

S-(−)-Dopamine
Used in treatment of Parkinson's

R-(+)-Dopamine
Contributes to side effects

Example 3.11 Amino acids

All naturally occurring amino acids are L-forms. It has been found that these all have S absolute configurations, *except* cysteine, which is R. (These will be described further in Chapter 7.)

Schematically the interaction between chiral molecules and receptors can be represented as shown:

X   X

(I)   (II)

The diagram illustrates clearly how one chiral form (I) can 'fit in' with the receptor sites whereas the other (II) is mismatched.

Example 3.12 Retinal – where *cis/trans* isomerism is important

The compound retinal is important in human vision. Retinal is derived from vitamin A (retinol), in the form shown below:

CH$=$O

This is called *trans*-retinal, and this is converted (enzymically) to the *cis*-form shown next:

113

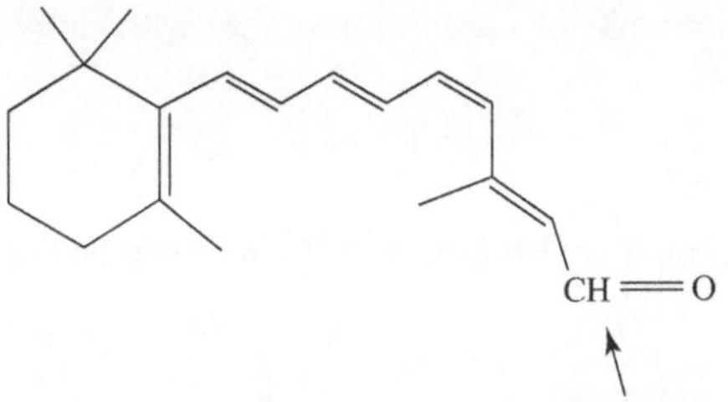

Aldehyde group
Reacts with protein (opsin)
to form rhodopsin

The aldehyde group of the *cis*-form reacts with a protein molecule called opsin to form rhodopsin. Light converts the *cis*-form to the *trans*-form, which is less securely bound to the opsin, thereby causing a response in the optic nerve, and the production of a visual image. Ultimately the uncomplexed *trans*-retinal is regenerated so that the whole cycle can begin again.

Hence the need for vitamin A, which is found in, for example, carrots.

Questions

Questions about nomenclature

3.1 Name the substances in the following structures:

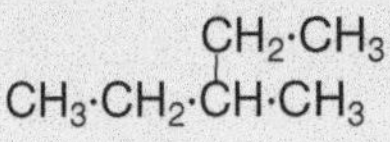

$$CH_3 \cdot CH_2 \cdot \overset{\displaystyle CH_2 \cdot CH_3}{\underset{}{CH}} \cdot CH_3$$

A

$$CH_3 \cdot CH_2 \cdot \overset{\displaystyle CH_3}{\underset{\displaystyle CH_3}{C}} \cdot CH_2 \cdot CH_3$$

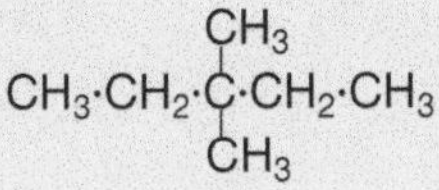

B

3.2 Write out the following structure in full and then name it:

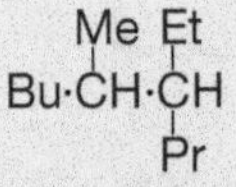

$$Bu \cdot \overset{\displaystyle Me}{\underset{\displaystyle Pr}{CH}} \cdot \overset{\displaystyle Et}{CH}$$

3.3 Write out the formulae for the following compounds:

hex-1-ene   4-methylhept-3-ene
A   **B**

3.4 Name the following:

$$CH_3CH_2C \equiv C \cdot CH(Et) \cdot CH = CH - C_2H_5 \qquad Et_2C = CH_2$$
A   **B**

3.5 For the following formulae, identify the functional groups:

$$CH_3CH_2CH_2 - O - CH = CH_2 \qquad CH_3 \cdot CH = CH \cdot CH(OH)CH_3$$

3.6 The following table gives the names and/or structural formulae of some organic compounds:

Name	Structure
Pentanal	$CH_3 \cdot CH_2CH_2CH_2 \cdot CHO$
4-Hydroxybutanal	
Pent-2-en-4-ynamide	$HC \equiv C \cdot CH = CHCONH_2$
	$CH_3 \cdot CH_2 \cdot CH_2 \cdot O \cdot CO \cdot CH_3$
Pent-3-en-2-ol	

Complete the table, then draw out the structures, and circle and name the functional groups.

Questions on conformation and geometric isomerism

3.7 Draw (a) the sawhorse and (b) the Newman perspective formulae for the preferred conformation of each of the following molecules, focusing your attention on the bonds indicated.

(i)
$$CH_3$$
$$CH_3\overset{|}{C}HCH_2CH_3$$
along the C2–C3 bond

(ii)
$$CH_3\overset{|}{C}HCH_2Br$$
along the C1–C2 bond

3.8 (a) Assign Z or E configurations to the following alkenes.

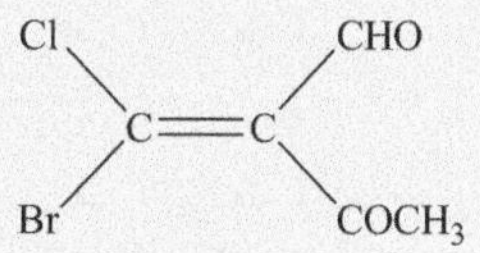

(I) 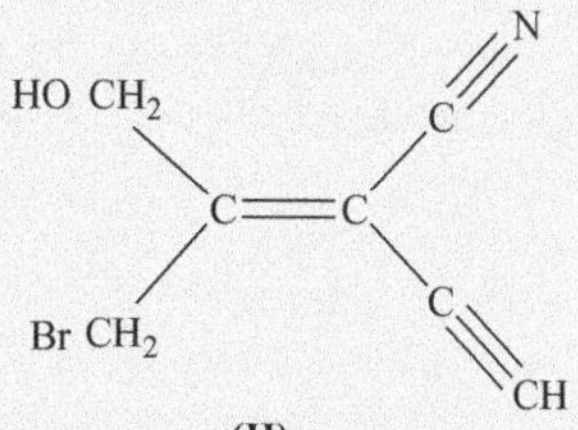 (II)

(b) State which of the following molecules can exist as geometric isomers, giving your reasons. Where geometric isomerism is possible draw both geometric isomers and assign their E or Z configuration.

(i) $ClCH = CHBr$

(ii) $Br_2C = CH_2$

(iii) $ICH = CH \cdot CH = CHEt$

(iv)

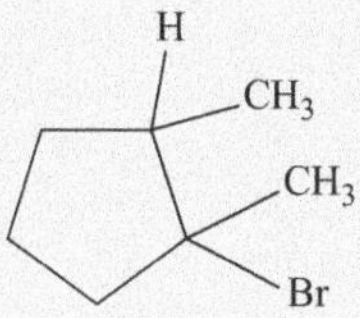

3.9 Answer the following concerning 1,3-dimethylcyclohexane. (Building a molecular model is helpful.)

(a) Draw clear diagrams showing the principal conformations of Z- and E-1, 3-dimethylcyclohexanes.

(b) For each geometrical isomer state which conformation has the lower energy, if applicable, giving your reasons.

(c) Does either isomer exhibit enantiomerism? If so, draw the possible pairs of enantiomers.

3.10 Give the absolute configuration (R or S) of the chiral centres in the following molecules.

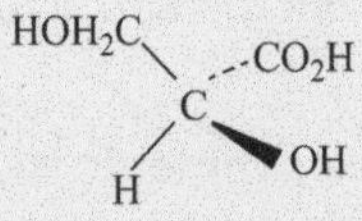

(a) 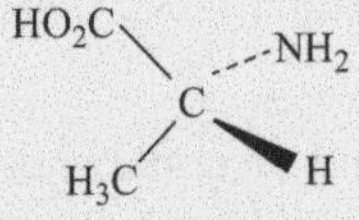   (b)

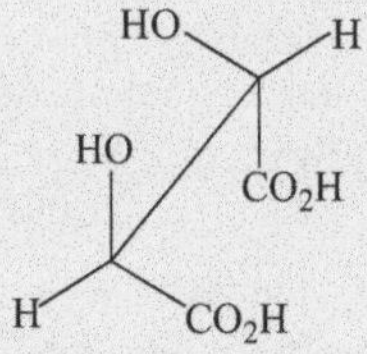

(c) 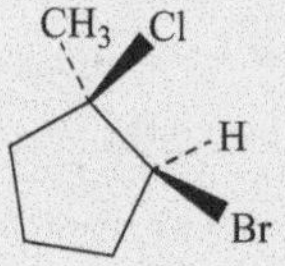   (d)

3.11 During glycolysis the metabolite glyceraldehyde-3-phosphate, shown below, is produced.

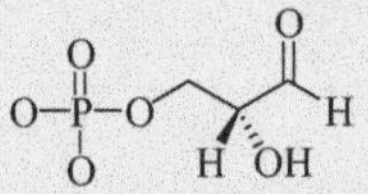

(a) Redraw this and circle and identify the aldehyde functional group.

(b) On the same drawing identify the hydroxyl group.

(c) Identify, by labelling it with an asterisk(*), the chiral centre in this molecule.

(d) Assign the chirality as R or S using the Cahn–Ingold–Prelog priority rules.

4 Energetics

Learning outcomes

By the end of this chapter students should be able to:

- recognise the different types of energy – potential energy, kinetic energy, heat, work etc.;
- appreciate that internal energy is a combination of heat and work, as summarised by the first law of thermodynamics: $\Delta U = q + w$;
- understand the nature of enthalpy change, ΔH, and be aware of its definition, and the relationship $\Delta H = \Delta U + P\Delta V$;
- know what is meant by the term *standard state*;
- be aware that enthalpy is a *state function,* with the consequences in terms of Hess's law and its applications;
- know the definition of entropy, know that it is a state function, and appreciate its role in determining spontaneity of a reaction, and to relate this to the second law of thermodynamics 'for any spontaneous process the entropy of the universe must increase';
- recognise that for a spontaneous reaction, the Gibbs free energy change $\Delta G < 0$;
- know that free energy is a state function;
- use the relationship $\Delta G = \Delta H - T\Delta S$;
- be aware of the difference between chemical and biological standard states.

4.1   Introduction

Why do biological processes work? What drives them forward? The area of science that provides the answers to these questions, and that this chapter covers, is traditionally known as *'thermodynamics'*, which derives from the ancient Greek for 'heat' and 'power'. This reflects the fact that for many years the problems that scientists focused on mostly were concerned with harnessing the heat from burning fuel to drive machinery.

Whilst this seems unlikely to be relevant to either chemistry or biology, an understanding of these principles is fundamental to the understanding of the chemistry associated with biology.

This chapter will explore:

- the nature of work and heat,
- the relationship between work and heat, and the energy of atoms and molecules,
- why chemical reactions 'go'.

Why does all this matter in biology? The fundamental processes in biology are also chemical reactions (for example the individual stages of glycolysis and of the Krebs cycle during the process of cellular respiration are examples of chemical reactions in an important biological context), and establishing the nature of the driving forces of these reactions permits a clearer insight into how and why they work. To illustrate the importance in the wider sense, the more that is known about these biological reactions, the better placed scientific researchers are for developing cures for, or at least developing methods to limit, medical conditions that afflict so many people.

4.1.1 The idea of energy

Everyone needs heat and energy, but what are they and how do we get them? *Food allows us to do 'energetic' things*, and it has a number of roles:

- It is a source of energy.
- It can be transformed into both work and heat.
- It drives the chemical processes on which life is based.

This chapter will explore the meaning of 'energy' and look at the relationship between the energy involved in chemical processes, as well as with heat and work, *all the time retaining an underlying focus on its role in biology*.

Biological processes involve quite complex molecules, but the principles that govern the processes involving these molecules are the same ones that are involved in much more simple chemical processes. Explaining the core principles can therefore be achieved by looking at some much simpler chemical examples, which will illustrate all of the key facts and ideas that are applicable throughout.

4.1.2 Energy: heat, work and the first law of thermodynamics

What is meant by the term 'energy'?

Energy can take many forms, broadly divided into *potential energy* and *kinetic energy*. The potential energy of an object is the energy it possesses as a consequence of its position. Examples include its height above the Earth, where it acquires energy as a consequence of the force of gravity. Electrons associated with an atom or with a bond also possess potential energy as a consequence of the interactions with nuclei or other electrons.

The kinetic energy of an object is the energy it possesses as a consequence of its motion – for instance the atoms or molecules in a gas, or an object dropped from a height, where the potential energy is exchanged for kinetic energy as the object falls. A ball rolling down a hill is a simple example (Figure 4.1).

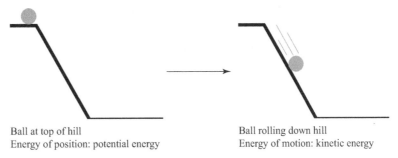

Ball at top of hill
Energy of position: potential energy

Ball rolling down hill
Energy of motion: kinetic energy

Figure 4.1 **Potential and kinetic energy.**

What is energy?

Energy is formally defined as 'the capacity to do work'. Work and heat provide mechanisms for the transfer of energy, so definitions of 'work' and heat are needed to gain insight into the concept of energy.

4.2 Temperature and heat

The temperature of a body is said to be a measure of its ability to transfer energy in the form of heat. If two bodies are brought into contact, energy in the form of heat is transferred from the higher temperature body to the lower temperature one, until both bodies are at the same temperature.

4.2.1 The nature of heat

For a very long time people did not understand the nature of 'heat'. Heat is now understood to be associated with random (thermal) motion of its constituent atoms or molecules (in a gas or liquid, the particles move with greater speed as heat is supplied, and in a solid they vibrate more vigorously).

Heat is sometimes said to be the energy associated with the disorderly movement of objects. This level of understanding is, however, a relatively recent situation, and a definition based on bulk properties was used in the past. Heat was defined in terms of the energy required to raise the temperature of an object by a particular amount. A little later it was shown that heat and work could be inter-converted.

Different substances have different responses when supplied with heat (for example, the filling of pies is much more likely to burn your mouth than a similar weight of pastry, and the filling is said to have a higher heat capacity). So, the definition was drawn up for a standard material, water, and this provided the basis for the former unit of energy called the calorie:

1 calorie (cal) is the energy required to raise the temperature of 1g of pure water by 1°C (to be precise, measured at the standard temperature of 25°C).

The calorie as a unit of energy has been officially discontinued for some time, though its use lives on in everyday life (*popular press, food industry, though the 'food calorie'* is, in fact, a *kilocalorie*). The official unit of energy now used is the joule (J). (See Box 4.3 for a more detailed look at units.) There is a simple conversion between the two units: $1\,cal \cong 4.18J$.

4.2.2 Heat capacity, C, and specific heat capacity, c

Heat capacity is defined as the amount of heat energy required to raise the temperature of a substance by 1 K. Heat capacity is, however, dependent on the amount of a material present, and a more useful way of thinking about heat capacity involves taking this into account. To summarise, therefore, if an object is heated or cooled, the amount of heat energy transferred, given the symbol q, depends on:

- the quantity of material, m;
- the magnitude of the temperature change, ΔT, where $\Delta T = T_{final} - T_{initial}$;
- the identity of the material involved (hence the example of the pie described above).

Specific heat capacity

The relationship between q and c is:

$$q = m \cdot c \cdot \Delta T \qquad [4.1]$$

From this equation, the definition of specific heat capacity can be derived, and it is 'the amount of energy, as heat, required to raise the temperature of a given amount of a substance (1 kg by convention, but 1 g is often used) by 1 kelvin'. Rearranging this equation gives:

$$c = \frac{q}{m \cdot \Delta T}$$

The specific heat capacity has units of $J\,K^{-1}\,kg^{-1}$. Specific heat capacity can also be quoted for molar quantities, and the units are $J\,K^{-1}\,mol^{-1}$.

4.2.3 Endothermic and exothermic processes

It is important to specify the direction of the flow of heat between a system and its surroundings:

- if a system gives out heat, its energy decreases and the heat flow, q, is negative; or
- if a system takes in heat during a reaction, the heat flow is positive.

These two cases are given special names:

- if $q < 0$ the process is exothermic;
- if $q > 0$ the process is endothermic.

Some key definitions, systems, surroundings, etc. are given in Boxes 4.1 and 4.2.

Box 4.1

Thermodynamic terms: system and surroundings

Before starting to discuss the subjects of energy, heat and work more extensively, it is important to ensure that certain terminologies are in place. When studying reactions in chemistry (or studying objects in physics) it is common practice to use the rather dry expression 'system' to represent the reaction being studied, with everything else called the 'surroundings'.

The system and the surroundings are separated by a boundary (in the case below, the beaker):

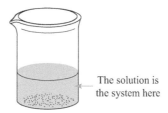

The solution is the system here

Everything that is not the system shown is the surroundings. Therefore:

System + surroundings = universe (i.e. absolutely everything)

Box 4.2

Types of thermodynamic system

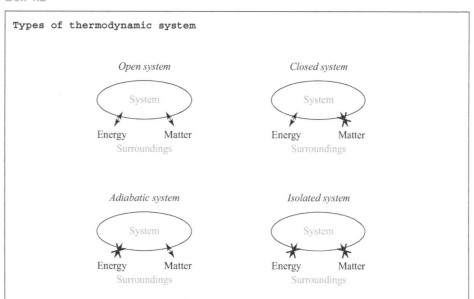

- In open systems energy and matter can be exchanged with the surroundings.
- In closed systems energy can be exchanged with the surroundings, but matter cannot.
- In adiabatic systems matter can be exchanged with the surroundings, but energy cannot.
- In isolated systems neither matter nor energy can be exchanged with the surroundings.

4.3 The first law of thermodynamics: introducing the concept of work

There are two general ways in which a system may change its energy: by the exchange of heat or work with its surroundings. This is summarised as:

$$\Delta U = q + w \qquad [4.2]$$

where:

- ΔU is the change in internal energy of the system;
- q is the heat change of the system;
- w is the work done by the system.

This relationship is called the *first law of thermodynamics*.

4.3.1 The nature of work

Sometimes work is said to be the energy associated with the orderly movement of objects (compare with heat). For example, a man pushing a car, or pushing a boulder (Figure 4.2), would be an example of work.

Work of expansion

The work of expansion against pressure is a common form of work in chemistry, and indeed in biology. If a chemical reaction produces a gas, it has to do work against atmospheric pressure, and mathematically this work is expressed by the following equation:

$$w = -P \cdot \Delta V \qquad [4.3]$$

Here

- 'P' is the pressure, commonly atmospheric pressure (about $100\,000\,\mathrm{N\,m^{-2}}$ or $100\,000\,\mathrm{Pa}$, or more exactly $101\,325\,\mathrm{Pa}$);
- ΔV is the change in volume.

Example 4.1 Suppose in the course of a chemical reaction carried out at atmospheric pressure 1 litre ($=1\,\mathrm{dm^3} = 10^{-3}\,\mathrm{m^3}$) of CO_2 is produced. The energy change of the system is given by:

$$\Delta U = w = -P\Delta V = -101\,325\,\mathrm{N\,m^{-2}} \times 10^{-3}\,\mathrm{m^3}$$

$$\approx -100\,000 \times 10^{-3}\,\mathrm{N\,m}$$

$$= -100\,\mathrm{J}$$

Figure 4.2 Representation of a man pushing a boulder up a slope.

There was no heat change involved, so $q = 0$, meaning '$\Delta U = q + w$' becomes '$\Delta U = w$'. (The symbol $\approx$ simply means 'approximately equal to'.)

The quantity of CO_2 produced has to be reported as cubic metres (m^3) in order to have consistency of units with the units of pressure. This is a relatively small amount of energy, but it is an energy penalty paid during respiration and production of CO_2.

4.3.2 Energy in the chemistry context

In chemistry, the principal way in which energy is exchanged in a process is through changes in bond energy. The stronger the bond between a pair of atoms, the more energy is required to break it (as discussed in Section 2.8 on bonding), and these bond energies are referred to as bond enthalpies.

Enthalpy

There are two general ways in which a system may change its energy: by the exchange of heat or work with its surroundings, and this is summarised as:

$$\Delta U = q + w$$

This relationship is called the first law of thermodynamics.

In many systems in chemistry and biology, the only work done is work of expansion which is given by $w = -P\Delta V$ (*the system is at constant pressure*). In such cases the first law relationship may be rewritten (by substituting 'w' with '$-P\Delta V$') as:

$$\Delta U = q - P\Delta V$$

4.3.3 The concept of enthalpy

ΔU means the change in the internal energy, so another way of saying the same thing is to consider the situation where a system has changed from state A to state B (represented schematically in Figure 4.3), and where we are considering the change in internal energy that results from this.

In Figure 4.3:

- The internal energies associated with the states A and B are, respectively, U_A and U_B.
- The heat change at constant pressure is referred to as q_p.
- The volumes of states A and B are V_A and V_B respectively.
- The system is at constant pressure, P.

State A   State B

Figure 4.3 **Change of state.**

Box 4.3

A note on the units of work and energy

The traditional way of thinking about work is expressed in the form of the equation:

$$w = F \cdot d \qquad \text{(work = force} \times \text{distance)}$$

where:

- w represents the work;
- F the force;
- d the distance moved in the direction of the force.

Note the use of symbols makes it quicker to perform calculations, otherwise it would be necessary to write out the quantities in long-hand each time.

In order for this work to be done, energy, given the symbol U (*note that in many other textbooks this energy, called internal energy, is given the symbol E*), is expended, and when the work is done, it is said that there has been an *energy change*, ΔU, where 'Δ' (the Greek letter delta) is shorthand for a change or difference in the value of the quantity.

The formal definition of internal energy is that it is the total energy of a system; it is a measure of the system's ability to do work.

$$\Delta U = w = F \cdot d \qquad \text{(assume no heat change, so } q = 0)$$

It is important to specify the direction of the distance moved in relation to the force. For example, consider a person lifting a weight:

- work is done *on* the weight, so energy is lost by the person doing the lifting;
- the weight gains energy.

From the perspective of the weightlifter, the force of gravity is downwards, and the distance the weight is moved is upwards, so:

- d is in a negative direction relative to F and
- w is negative from the viewpoint of the weightlifter.

Therefore, when the weight is raised,

- from the perspective of the system (the weightlifter), ΔU is negative ($\Delta U < 0$);
- from the perspective of the surroundings, ΔU is positive ($\Delta U > 0$).

Suppose it is necessary to calculate the amount of energy involved (rather than just its sign). The force due to gravity is given by:

$$F = m \cdot a$$

where:

- m is the mass of the object (in kg);
- a is the acceleration due to gravity (in m s^{-2}).

Combining the above, it is possible to work out the following units of work:

$$\text{Units (work)} = \text{units (F)} \times \text{units (d)} = (\text{kg m s}^{-2})(\text{m}) = \text{N m}$$

'N' (the newton) is a unit given to force which encompasses all the fundamental units, and:

$$1\,\text{N} = 1\,\text{kg m s}^{-2}$$

Work (and hence energy) is also given a unit that encompasses all the fundamental units. This is the joule (J), and:

$$1\,\text{J} = 1\,\text{N m} = 1\,\text{kg m}^2\text{s}^{-2}$$

On rewriting the first law of thermodynamics to reflect the idea that it is reflecting the difference between states A and B, then the result is:

$$\underbrace{U_B - U_A}_{\Delta U} = q_p - \underbrace{(PV_B - PV_A)}_{P\Delta V}$$

On rearranging, to put q_p on its own on one side of the equation:

$$\begin{aligned} q_p &= U_B - U_A + (PV_B - PV_A) \\ &= U_B - U_A + PV_B - PV_A \\ &= (U_B + PV_B) - (U_A + PV_A) \end{aligned}$$

$(U + PV)$ is now defined as a 'new' term called *enthalpy*, which is given the symbol H, so:

$$\begin{aligned} (U_B + PV_B) &= H_B \\ (U_A + PV_A) &= H_A \\ q_p = H_B - H_A &= \Delta H \end{aligned}$$

Substituting for q in $\Delta U = q - P\Delta V$ gives:

$$\Delta U = \Delta H - P\Delta V$$

Or, on rearranging:

$$\Delta H = \Delta U + P\Delta V$$

The formal definition of enthalpy is given in Box 4.4.

Box 4.4

General formal definition of enthalpy

The enthalpy change associated with a chemical reaction corresponds to the quantity of heat exchanged with the surroundings when the reactants are converted to products at constant pressure.

Enthalpy in terms of bond making and breaking

Consider the formation of water from its constituent elements, for which the stoichiometrically balanced equation is:

$$H_2\,(g) + \tfrac{1}{2}\,O_2\,(g) \rightarrow H_2O\,(g)$$

Both hydrogen and oxygen exist as *diatomic* molecules, so making water is going to involve:

- breaking $H-H$ and $O{=}O$ bonds;
- making $H-O$ bonds.

Typical bond energies (note they are per mole) are as follows:

H—H $432\,kJ\,mol^{-1}$

O=O $498\,kJ\,mol^{-1}$

H—O $464\,kJ\,mol^{-1}$

Remembering the balanced equation:

$$H_2\,(g) + \tfrac{1}{2}\,O_2\,(g) \rightarrow H_2O\,(g)$$

ΔH = difference between bond enthalpy of reactants and bond enthalpy of products. Energy is needed to break bonds, and conversely bond formation will release energy. Each of the processes associated with the reaction above will now be discussed.

Energy required to break bonds

There is 1 mole of H_2, so the system needs 432 kJ to break 1 mole of
H—H bonds $= +432\,kJ$

There is 0.5 mole of O_2 involved in the reaction, so 498/2 kJ is needed
to break O=O bonds $= +249\,kJ$

Therefore the total energy required to break bonds is (432 + 249) $= +681\,kJ$

Energy from making bonds

1 mole of water is made, but each molecule contains two O—H bonds,

Energy released $= 2 \times 464\,kJ$ $= -928\,kJ$

Therefore:

- overall, more energy is given out on bond formation than is needed to break bonds (the process is exothermic!);
- the difference is $247\,kJ\,mol^{-1}$;
- because the process is exothermic, the enthalpy change $\Delta H = -247\,kJ\,mol^{-1}$.

Key points

- The process of *forming* the bonds is the reverse of *breaking* the bonds.
- The bond enthalpy applies to the process of breaking bonds and requires an input of energy to the bond – which from the viewpoint of the system is a *positive* quantity; therefore the process of forming the bonds has a *negative* enthalpy.
- For the case of the reaction between oxygen and hydrogen, the total bond enthalpy of the products is greater than the bond enthalpies of the reactants, and this enthalpy is released by the system. In the case of the reaction between hydrogen and oxygen, it does this very rapidly indeed in the form of heat.

Because the enthalpy of a reaction is equal to the heat of the reaction carried out at some constant pressure then, as with the two cases of heat flow, a reaction is exothermic or endothermic if ΔH is respectively negative or positive.

(Experimentally, enthalpy changes are measured using the technique known as calorimetry. This technique involves the combustion of a measured quantity of material, with the heat given out measured through measurement of the rise in temperature of a thermally insulated water bath.)

Box 4.5

State functions

A state function depends only on the current state of the system, and is independent of the path by which that state was reached.

- Pressure, volume, temperature and *enthalpy* are examples in the context of this course (others, to be described later, are entropy and Gibbs free energy).

- A more everyday illustration would be to consider the journey between two cities (say Edinburgh and London). No matter which path is followed to go from one city to the other, whether it be a direct rail journey (route 1 on map), or whether it be via another location (for example, Dublin, route 2 + 3 on map), the distance between Edinburgh and London remains constant.

Box 4.6

Standard states

A standard state of a substance (element or compound) is defined as the most stable form of that substance that exists at a pressure of 1 bar (1 bar = $100\,000\,\mathrm{N\,m^{-2}}$, approximately atmospheric pressure) and at a specified temperature (often taken to be 298 K, but it need not be).

Box 4.7

Definitions

Standard enthalpy of formation

The enthalpy of formation of one mole of a particular material from its constituent elements in their standard states (pure and generally in the state of lowest enthalpy) at a particular temperature, commonly (*though not necessarily*) 298 K; these are given the symbol: $\Delta H^{\theta}_{f,298}$.

Note:

- $\Delta H^{\theta}_{f,298}$ for elements is 0 by definition.
- Most $\Delta H^{\theta}_{f,298}$ values are negative, signifying that formation of compounds from elements is usually exothermic.
- The superscript θ signifies standard conditions.

Standard enthalpy of combustion, $\Delta H^{\theta}_{c,298}$

$\Delta H^{\theta}_{c,298}$ is the enthalpy change when one mole of a substance is totally combusted in oxygen at 298 K and 1 atm.

[Consequences of definitions above:

$\Delta H^{\theta}_{f,298}$ values are based on the formation of 1 mole of compound. If n moles are formed, then the enthalpy change will be $(n \times \Delta H^{\theta}_{f,298})$.

Similarly, on considering heats of combustion, then burning more than one mole (e.g. n moles) will involve a greater energy change, i.e. $(n \times \Delta H^{\theta}_{c,298})$].

Standard enthalpy of reaction, ΔH^{θ}_r

The enthalpy associated with a reaction in which pure reactants in their standard states are converted into pure products in their standard states.

4.3.4 Examples of enthalpy changes in biological processes

Energy and enthalpy play a key role in biological processes, because organisms need energy to do work. In biosynthesis, that work is the production of molecules essential to life processes.

One form of energy is provided by light from the Sun in photosynthesis, which leads to the formation of relatively complex molecules from water and carbon dioxide:

$$6CO_2\,(g) + 6H_2O\,(l) \rightarrow C_6H_{12}O_6\,(s) + 6O_2\,(g)$$

This process requires an energy input of 2858 kJ mol^{-1} (i.e. $\Delta H^{\theta} = +2858$ kJ mol^{-1}). The reverse process is related to the combustion of the molecule $C_6H_{12}O_6$, known as glucose (and also as dextrose or blood sugar). This can be measured readily in a calorimeter, and the heat given up by the combustion process is -2858 kJ mol^{-1}.

It would not be much use to the function of organisms if the only way in which the energy could be released was by producing heat. It is possible, however, to couple the energy of processes with other chemical transformations, and there is a key intermediary system to transfer such energy, which is the pair of molecules ADP and ATP (see Figure 4.4). 'A' stands for adenosine (the combination of cyclic organic

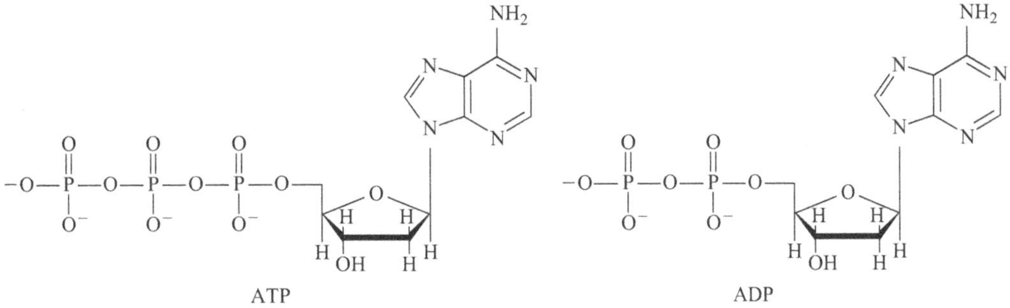

Figure 4.4 **Structures of ATP and ADP.**

fragments at the right of the figure), 'D' or 'T' stands for 'di' or 'tri', and 'P' is 'phosphate'.

The conversion of ATP to ADP in the presence of water involves breaking a phosphate bond:

$$ATP + H_2O \rightarrow ADP + phosphate$$

This has an enthalpy change of $-29\,kJ\,mol^{-1}$ and, if it is coupled with other molecules, the energy that is released can be used to form bonds that require energy. In other words it can lead to the formation of new molecules of higher energy, and perhaps a different biological role.

The regeneration of ATP from ADP, which is the reverse process and therefore needs an input of energy, is driven by reactions that involve the breakdown of glucose, and indeed the energy released in just one glucose molecule degrading drives the formation of many molecules of ATP.

Whilst it is possible to determine the magnitude of these enthalpy changes using a calorimeter, and measuring the heat given off (in effect, directly measuring the enthalpy), there are many examples of reactions that cannot easily be followed in a calorimeter, for instance the rearrangement of key biological molecules to produce forms with different structures and function.

4.3.5 The determination of enthalpies: Hess's law

Fortunately, because enthalpy is a state function (see Boxes 4.5–4.7), it is often possible to work out the enthalpy of a particular process in terms of enthalpies of processes that are much easier to measure. To illustrate this, consider the rearrangement of a simple hydrocarbon molecule: hexane from the straight-chain (n) form, to the branched isomer 2-methyl-pentane (C_6H_{14}):

In this example the objective is to determine the enthalpy of step A. This value depends only on the nature of the reactants and products, and is independent of the route taken to get from reactants to products.

This principle is called Hess's law. It is a consequence of the fact that the enthalpy of a species is a *state function*.

Hess's law calculation using heats of combustion

If the same amount of oxygen is introduced to both sides of the process, A, the enthalpy change is unaltered. It does, however, open up a new route that the reaction can take, which is through burning either molecule. In both cases the product is CO_2 and water, and though just enough O_2 is added in the diagram below to ensure complete combustion, if more had been introduced (if oxygen were in excess), some of it would have remained unused, and would not have altered the energetics of the process:

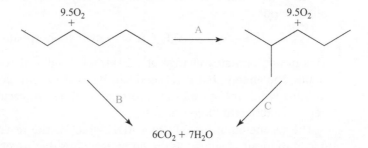

The enthalpy change associated with process A (ΔH_A) plus the enthalpy change associated with process C (ΔH_C) is then the enthalpy of process B (ΔH_B):

$$\Delta H_A + \Delta H_C = \Delta H_B$$

where:

● ΔH_B is the enthalpy of combustion of n-hexane; and

● ΔH_C is the enthalpy of combustion of 2-methyl-pentane.

In this case it is possible to say:

$$\Delta H_A = \Delta H_{combustion}(\text{n-hexane}) - \Delta H_{combustion}(\text{2-methyl-pentane})$$

Both of the quantities on the right hand side are readily measured by calorimetry, and have been recorded in the literature, and these are:

● $\Delta H_{combustion}(\text{n-hexane}) = -4163.2\,\text{kJ mol}^{-1}$

● $\Delta H_{combustion}(\text{2-methyl-pentane}) = -4157.3\,\text{kJ mol}^{-1}$

So, the enthalpy of conversion of hexane to 2-methyl-pentane, calculated using literature enthalpy of combustion values, is:

$$\Delta H_A = -4163.2 - (-4157.3) = -5.9\ \text{kJ mol}^{-1}$$

Hess's law calculation using heats of formation

An alternative form of energy cycle that is commonly used to calculate unknown enthalpy changes involves the production of the species involved from their constituent elements. So, for the isomerisation of n-hexane to 2-methyl-pentane (C_6H_{14}) again use Hess's law:

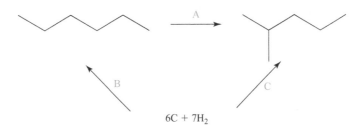

Here, the enthalpy change of process A (ΔH_A) plus the enthalpy change of process B (ΔH_B) equals the enthalpy change of process C (ΔH_C):

$$\Delta H_A + \Delta H_B = \Delta H_C$$

Now, recalling that ΔH_f is the enthalpy of formation of 1 mole of the molecule(s) in question from their constituent elements (see earlier for the more formal and complete definition), then,

- ΔH_B is the enthalpy of formation of hexane;
- ΔH_C is the enthalpy of formation of 2-methyl-pentane.

So,

$$\Delta H_A = \Delta H_C - \Delta H_B$$

which can be rewritten as

$$\Delta H_A = \Delta H_f \, (2\text{-methyl-pentane}) - \Delta H_f \, (\text{hexane})$$

These enthalpy of formation data are again readily obtainable from literature sources, and in these cases the figures are:

- $\Delta H_f \, (2\text{-methyl–pentane}) = -204.6 \, \text{kJ mol}^{-1}$
- $\Delta H_f \, (\text{hexane}) = -198.7 \, \text{kJ mol}^{-1}$

And so, the enthalpy of conversion of hexane to 2-methyl-propane, calculated using the literature enthalpy of formation data is:

$$\Delta H_A = -204.6 - (-198.7) = -5.9 \, \text{kJmol}^{-1}$$

Enthalpy values may depend on temperature, on the purity of a material, and on the phase in which it exists (for example, in the case of carbon, whether it is present as graphite, diamond or some form of fullerene).

Enthalpy of reaction: the general case

Let us consider a chemical reaction, for example the hypothetical reaction in which 'a' moles of a compound A react with 'b' moles of a compound B to yield 'c' moles of a compound C and 'd' moles of a compound D, which is shown by:

$$a \, A + b \, B \rightarrow c \, C + d \, D$$

The enthalpy change for this general reaction is given by:

$$\Delta H^\ominus = \left[c\Delta H_f^\ominus(C) + d\Delta H_f^\ominus(D)\right] - \left[a\Delta H_f^\ominus(A) + b\Delta H_f^\ominus(B)\right]$$

which generalises to

$$\Delta H^\ominus = \Sigma\Delta H_f^\ominus(\text{products}) - \Sigma\Delta H_f^\ominus(\text{reactants}) \qquad [4.5]$$

The symbol Σ means 'the sum of'.

Example 4.2 The application of Hess's law

Determine the enthalpy of combustion of methanol, given the enthalpies of formation of reactants and products.

The enthalpy of combustion of methanol is defined by the equation:

$$CH_3OH\,(g) + {}^3\!/_2\,O_2\,(g) \rightarrow CO_2\,(g) + 2\,H_2O\,(g)$$

Now consider a cycle, showing the combustion of methanol, along with the heat of formation equations for methanol, carbon dioxide and water:

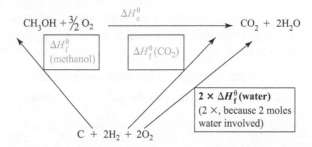

Using the approach outlined in the previous section, on considering the options going *from* the elements in their natural states *to* CO_2 and H_2O, then in order to balance the cycle, the sum of the enthalpy values going directly must equal the sum of the enthalpy values going indirectly. This is summarised as:

$$\Delta H_c^\ominus + \Delta H_f^\ominus(\text{methanol}) = \Delta H_f^\ominus(CO_2) + \left[2 \times \Delta H_f^\ominus(\text{water})\right]$$

(Note the multiplier applied to the production of water.)

Rearranging the equation leads to:

$$\Delta H_c^\ominus = \Delta H_f^\ominus(CO_2) + \left[2 \times \Delta H_f^\ominus(\text{water})\right] - \Delta H_f^\ominus(\text{methanol})$$

Values for standard molar enthalpies of formation are:

For CO_2: $-393.5\,\text{kJ}\,\text{mol}^{-1}$
For H_2O: $-241.8\,\text{kJ}\,\text{mol}^{-1}$
For CH_3OH: $-201.5\,\text{kJ}\,\text{mol}^{-1}$

[O_2, as an element, has an enthalpy of formation of 0 and so has not been shown.]

So, putting the figures into the calculation,

$$\Delta H_c^\ominus = \Delta H_f^\ominus(CO_2) + \left[2 \times \Delta H_f^\ominus(\text{water})\right] - \Delta H_f^\ominus(\text{methanol})$$
$$= (-393.5\,\text{kJ}) + \left[2\,(-241.8\,\text{kJ})\right] - (-201.5\,\text{kJ})$$
$$= -675.6\,\text{kJ per mol of methanol}$$

An alternative approach to Hess's law calculations

Rather than drawing out the reactions in the form of cycles, which can prove cumbersome, it is often more convenient to perform these calculations using an algebraic method. This method will now be demonstrated by using it to solve the preceding example.

Example 4.3   Step 1   Show the equation of interest

The enthalpy of combustion of methanol, which is the required value, is defined by equation A:

$$CH_3OH\ (g) + \tfrac{3}{2}\ O_2\ (g) \rightarrow CO_2\ (g) + 2H_2O\ (g) \tag{A}$$

Step 2   Express, in equation form, all of the information that has been provided

In this case the enthalpy of formation values of CO_2, H_2O and CH_3OH:

$C\ (s) + O_2\ (g) \rightarrow CO_2\ (g)$	$\Delta H_f^\theta = -393.5\ \text{kJ mol}^{-1}$	B
$H_2\ (g) + \tfrac{1}{2}O_2\ (g) \rightarrow H_2O\ (g)$	$\Delta H_f^\theta = -241.8\ \text{kJ mol}^{-1}$	C
$C\ (s) + 2H_2\ (g) + \tfrac{1}{2}O_2\ (g) \rightarrow CH_3OH\ (g)$	$\Delta H_f^\theta = -201.5\ \text{kJ mol}^{-1}$	D

Step 3   Combine equations B, C and D to produce A

There are a couple of points to note:

- As equation A has CH_3OH on the left hand side, then equation D should be written in reverse, as $-D$, thus:

$$CH_3OH\ (g) \rightarrow C\ (s) + 2H_2\ (g) + \tfrac{1}{2}O_2\ (g)$$

- Equation A has 2 moles of water on the right hand side, so this needs to be accounted for when combining B, C and D.

The three equations are combined thus:

$$B + 2C - D$$

This yields:

$$C\ (s) + O_2\ (g) + 2H_2\ (g) + O_2\ (g) + CH_3OH\ (g) \rightarrow C\ (s) + 2H_2\ (g) + \tfrac{1}{2}O_2\ (g) + CO_2\ (g) + 2H_2O\ (g)$$

Now cancel the items that occur on both the right and left hand sides:

$$\tfrac{1}{2}O_2\ (g)$$
$$CH_3OH\ (g) + \cancel{C\ (s)} + O_2\ (g) + \cancel{2H_2\ (g)} + \cancel{O_2\ (g)} \rightarrow$$
$$\cancel{C\ (s)} + \cancel{2H_2\ (g)} + \cancel{\tfrac{1}{2}O_2\ (g)} + CO_2\ (g)2H_2O\ (g)$$

Which gives:

$$CH_3OH(g) + \tfrac{3}{2}\ O_2(g) \rightarrow CO_2(g) + 2H_2O(g)$$

So, putting the figures into the calculation,

$$B \quad + \quad 2C \quad - \quad D$$

$$\Delta H_c^\theta = (-393.5\ \text{kJ}) + 2\,(-241.8\ \text{kJ}) - (-201.5\ \text{kJ})$$
$$= -675.6\ \text{kJ per mol of methanol}$$

133

4.4 Spontaneous processes, entropy and free energy

Thus far the discussions have looked at the concepts of energy and how these relate to chemical processes. These will now be extended to develop the ideas of thermo-dynamics to answer the question of 'why does a reaction go?'

A reaction that 'goes' is said to be a 'spontaneous' reaction, and although it is often the case that many spontaneous processes, including chemical reactions, are exothermic, this is not always true, so there must be a different criterion for a reaction to be spontaneous.

- Examples of spontaneous processes that involve no energy change:
  - The expansion of a gas from one container into an evacuated chamber (Figure 4.5). This costs no energy but will always happen.
  - The flow of heat from a hot object to a cold object within an insulated container. For example, a pack of cola will always get colder if it is placed in an insulated container of ice even though the total amount of heat inside the box remains effectively constant.
- Examples of spontaneous processes that are *endo*thermic:
  - Dissolution of certain salts (e.g. NH_4NO_3) in water chills it – heat is drawn from the surroundings to the system.
  - Evaporation of volatile liquids cools the surface on which they lie. For example, ethyl bromide, C_2H_5Br, used in local anaesthesia, evaporates rapidly when placed on skin, and cools it. For this process, $\Delta H_{vap} = +27.4$ kJ mol^{-1} (ΔH_{vap} is the enthalpy of vaporisation).

The common factor in each of these cases is a greater dispersal of either energy (as heat) or of matter.

Dispersal of matter also increases on going from solid to liquid to gas phases of matter. The flow of heat from one body to another cooler body always results in greater dispersal of heat. In both cases the degree of dispersal (or of disorder) is quantified by the *entropy*, S, which has units of energy/temperature or JK^{-1}.

Entropy, S, like enthalpy, is a *state function*, independent of history, so its value can be determined for a given substance (chemical composition), in a given state (gas, liquid, solid, solution…) and under external factors such as temperature or pressure. (*Note that values are given at some standard temperature (usually 298K) and pressure and purity.*)

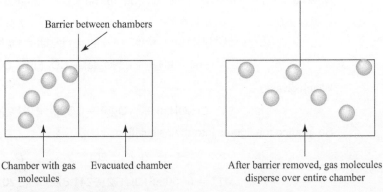

Figure 4.5 **Expansion of a gas.**

A few generalisations can be made:

- Entropy values increase as the volume of gas increases.
- Entropy values increase as the state of the material changes from solid to liquid to gas.
- Entropy values increase in the *surroundings* if heat, q, is passed *from* the *system*.

Box 4.8 contains information about entropy and defines the standard entropy change for a reaction.

Box 4.8

Some information and definitions

- At the absolute zero of temperature, the entropy of a perfectly crystalline solid is zero (this is referred to as the third law of thermodynamics).
- The standard molar entropy value for a substance is the entropy content of one mole of the substance at a pressure of 1 bar*, and is recorded as an absolute value, being shown as S^θ.
- ΔS_{surr} is the entropy change in the surroundings, ΔS_{sys} is the entropy change in the system.
- $\Delta S_{sys} + \Delta S_{surr}$ = entropy change in the universe.
- Also, the definition for ΔS_{sys} for a chemical reaction is:

$$\Delta S_{sys} = \text{total entropy of products} - \text{total entropy of reactants}$$
$$\Delta S^\theta = \Sigma \Delta S_f^\theta(\text{products}) - \Sigma \Delta S_f^\theta(\text{reactants})$$

*Remember 1 bar = $100\,000\,\mathrm{N\,m^{-2}}$.

It is possible to make a precise prediction that if the heat transfer from the system to the surroundings is done very (infinitely) slowly, under what are referred to as reversible conditions, then:

$$\Delta S_{surr} = -q_{sys}/T = -\Delta H_{sys}/T$$

(because, for a reaction at constant pressure, $q_{sys} = \Delta H_{sys}$).

In any real system it is impossible to achieve the idealised reversible conditions, and the result of this is some of the heat is effectively wasted, so the entropy change is less than the reversible case:

$$\Delta S_{surr} < -\Delta H_{sys}/T$$

We will now explore how this helps to predict whether a reaction is spontaneous.

4.4.1 The second law of thermodynamics

One way of stating this law is: 'the entropy of the universe increases in any spontaneous process'. This can be represented as:

$$\Delta S_{sys} + \Delta S_{surr} > 0$$

It is unrealistic to try to monitor 'the universe', or indeed ΔS_{surr}. It has been seen, however, that this term arises from the transfer of heat from the system, so on substituting for ΔS_{surr} with $-\Delta H_{sys}/T$ in the relationship:

$$S_{sys} + \Delta S_{surr} > 0$$

this then gives

$$\Delta S_{sys} - \Delta H_{sys}/T > 0$$

(From now on the focus will be on the properties of the system, so the subscript 'sys' will be dropped, and its usage will be taken to be implicit.)

Thus the following relationships, which are variants of one another, all hold:

$$\Delta S - \Delta H/T > 0 \text{ or } T\Delta S - \Delta H > 0 \text{ or } \Delta H - T\Delta S < 0$$

The person who developed this concept, J. Willard Gibbs (1839–1903), gave his name to a quantity that accounts for this balance and this is called the free energy or, more commonly, the *Gibbs free energy*:

$$\Delta G = \Delta H - T\Delta S \qquad\qquad [4.7]$$

In summary:

- If $\Delta G < 0$, the reaction goes. It is said to be spontaneous.

- If $\Delta G > 0$, the reaction does not go. It is said to be non-spontaneous.

Gibbs free energy is often defined as the maximum energy available to do useful work, and the difference between it and the enthalpy value is the energy 'discarded' as entropy in the system. This relationship forms the basis of predicting whether any chemical or fundamental biological process can go or not.

Note that it does not predict how fast a process will go. (This is the subject of Chapter 9, on reaction rates.)

It is now possible to produce Table 4.1, which summarises the impact of enthalpy and entropy changes on whether or not a reaction is spontaneous. So, for any reaction, if ΔH and ΔS are known (details can usually be derived from standard tables), then it is possible to calculate ΔG, and thereby determine whether the reaction 'goes' – whether it is spontaneous or not. Box 4.9 contains a definition and description of the standard free energy change for a reaction.

Table 4.1 **Thermodynamic properties and reaction spontaneity**

ΔH	ΔS	ΔG	*Spontaneous*
Negative	Positive	Negative	Yes
Positive	Negative	Positive	No
Negative	Negative	Depends on relative values of ΔH, ΔS	Favoured at lower T
Positive	Positive	Depends on relative values of ΔH, ΔS	Favoured at higher T

Example 4.4   The boiling point of water

It is possible to calculate the boiling point of water by using standard thermodynamic parameters (taken from the *CRC Handbook**) for the transition of water to steam, which is given by:

$$H_2O \, (l) \rightarrow H_2O \, (g)$$

Box 4.9

Standard free energy change, ΔG^{θ}

The value of ΔG^{θ} simply refers to the difference in free energy of the reactants and products in their standard states, in the proportions in which they appear in the equation for the reaction to form one mole of product. It can be calculated from:

$$\Delta G^{\theta} = \Delta H^{\theta} - T\Delta S^{\theta}$$

Just as there is a standard enthalpy for formation given by ΔH_f^{θ}, so there is a standard free energy of formation, the symbol for which is ΔG_f^{θ}. ΔG_f^{θ} is the standard free energy of formation of a compound from its elements in their reference states.

The standard reaction free energy may be calculated from ΔG_f^{θ} values of the reaction components. Thus, for a reaction:

$$a\,A + b\,B \rightarrow c\,C + d\,D$$

The standard free energy change for this reaction is given by:

$$\Delta G^{\theta} = c\Delta G_f^{\theta}(C) + d\Delta G_f^{\theta}(D) - a\Delta G_f^{\theta}(A) - b\Delta G_f^{\theta}(B)$$

which is summarised by:

$$\Delta G^{\theta} = \Sigma \Delta G_f^{\theta}(\text{products}) - \Sigma \Delta G_f^{\theta}(\text{reactants})$$

But ΔG^{θ} only applies to a reaction going from pure unmixed reactants to pure unmixed products. It is far more usual to wish to know about the thermodynamics of mixtures and reactions that do not form pure products, or which only reach an equilibrium ratio of reactants to products. (This will be explored more completely in the next chapter.)

The enthalpy change is $\Delta H = +44.0\,\text{kJ mol}^{-1}$ and the entropy change is $\Delta S = 117.9\,\text{J K}^{-1}\,\text{mol}^{-1}$. At the boiling point, T_b, the process of going from liquid to gas crosses over from $\Delta G > 0$ to $\Delta G < 0$; that is, from a non-spontaneous to a spontaneous process. In other words, at T_b, $\Delta G = 0$.

Remembering the relationship $\Delta G = \Delta H - T\Delta S$, then assuming at the point where $T = T_b$, the free energy change $\Delta G = 0$, this relationship can be rewritten as:

$$0 = \Delta H - T_b\Delta S$$

This rearranges to give:

$$\Delta H = T_b\Delta S$$

and hence can give T_b,

$$T_b = \frac{\Delta H}{\Delta S}$$

*The CRC Handbook of Chemistry and Physics, also known as the 'Rubber Book', is a repository of physical and chemical data.

Putting in the values for ΔH and ΔS, *noting that ΔH is quoted in kJ and ΔS in J, therefore it is necessary to convert both to the same units, usually to J. So,*

$$T_b = 44\,000/117.9 = 373.1K \, (= 100\,°C)$$

Example 4.5 The decomposition of the mineral calcium carbonate, $CaCO_3$

This decomposition is represented by the reaction:

$$CaCO_3 \, (s) \rightarrow CaO \, (s) + CO_2 \, (g)$$

This reaction can be studied at two different temperatures, namely 'room' temperature, 25°C/298 K, and a very high temperature, 2000°C/2273 K, using the standard thermodynamic values for enthalpy and entropy of $\Delta H = +173.0\,kJ\,mol^{-1}$ and $\Delta S = 160.5\,J\,K^{-1}\,mol^{-1}$ respectively. As indicated in the preceding example, it is important to be consistent with the units.

At 'low' temperature (e.g. 298 K):

$$\Delta G = \Delta H - T\Delta S = 173\,000 - (298)160.5 = +125.2\,kJ\,mol^{-1}$$

so it does not go (it is non-spontaneous).

If the system is warmed to a high temperature, e.g. 2273 K:

$$\Delta G = \Delta H - T\Delta S = 173\,000 - (2273)160.5 = -191.8\,kJ\,mol^{-1}$$

so it *does* go.

Example 4.6 Dissolution of NH_4NO_3 in water at 298 K

This process is endothermic, with an enthalpy of solution, $\Delta H = +26.4\,kJ\,mol^{-1}$ and an entropy change of $\Delta S = 108.2\,J\,K^{-1}\,mol^{-1}$. Therefore:

$$\Delta G = \Delta H - T\Delta S = 26\,400 - (298)108.2 = -5843\,J\,mol^{-1} = -5.8\,kJ\,mol^{-1}$$

In other words, this is compatible with dissolution at room temperature.

4.4.2 Free energy and ATP: coupling of reactions

During the discussion of enthalpy (Section 4.3.4), the hydrolysis of ATP to produce ADP was discussed, as summarised below:

$$ATP + H_2O \rightarrow ADP + phosphate \qquad \Delta H^\theta = -29\,kJ\,mol^{-1}.$$

The standard free energy change for this process is $\Delta G^\theta = -31\,kJ\,mol^{-1}$. In order that organisms do not exhaust their supply of ATP, this process needs to be reversed, that is:

$$ADP + phosphate \rightarrow ATP + H_2O$$

However, as the re-creation of ATP is the reverse of its hydrolysis, the standard free energy change, ΔG^θ, for the process is $+31\,kJ\,mol^{-1}$, which appears to go against the second law of thermodynamics, since this indicates the reaction is not spontaneous. *This is overcome by coupling this unfavourable reaction to a favourable one such that the overall free energy change is negative.*

One such favourable reaction is the conversion of glucose to lactic acid. Lactic acid is responsible for a burning sensation in the muscles involved with intense exercise. The conversion of glucose to lactic acid is summarised below:

$$C_6H_{12}O_6 \rightarrow 2CH_3CH(OH)CO_2H \qquad \Delta G = -197 \, kJ \, mol^{-1}$$

In practice, 1 mole of glucose degrades together with the regeneration of two moles of ADP, with the overall process being summarised by:

$$2ADP + 2 \, phosphate + C_6H_{12}O_6 \rightarrow 2ATP + 2H_2O + 2CH_3CH(OH)CO_2H$$

The free energy change for the whole process is calculated from the simple summation below, which is a Hess's law calculation, applied this time to Gibbs free energy (recall Gibbs free energy is a state function):

$$C_6H_{12}O_6 \longrightarrow CH_3CH(OH)CO_2H \qquad \Delta G_{(I)} = -197 \, kJ \, mol^{-1}$$

and

$$2ADP + 2 \, phosphate \longrightarrow 2ATP + 2H_2O \quad 2\Delta G_{(II)} = (2 \times 31) \, kJ \, mol^{-1}$$

gives

$$C_6H_{12}O_6 + 2ADP + 2 \, phosphate \longrightarrow 2ATP + 2H_2O + CH_3CH(OH)CO_2H \quad \Delta G_{(III)}$$

Therefore:

$$\Delta G_{(III)} = \Delta G_{(I)} + 2\Delta G_{(II)}$$

$$\Delta G_{(III)} = -197 \, kJ \, mol^{-1} + (2 \times 31) \, kJ \, mol^{-1} = -135 \, kJ \, mol^{-1}$$

(*A detailed analysis of how this happens will be given in Chapter 12 on metabolism.*)

4.4.3 Biological example: thermodynamic rationale of micelle behaviour

Micelles result from an aggregation of a particular type of a group of compounds called lipids, discussed briefly here. (A more complete look at lipids will follow in Chapter 7.)

The sodium salts of long-chain carboxylic acids (soaps) are almost completely miscible in water. Except in extremely dilute solutions, however, they dissolve as micelles, which are spherical clusters of carboxylate ions that are dispersed throughout the aqueous phase. The non-polar (and thus hydrophobic, which means water-hating) alkyl chains remain in the non-polar environment (in the interior of the micelles) while the polar (and therefore hydrophilic, or water-loving) carboxylate groups are exposed to the polar environment of the aqueous medium. A diagrammatic representation of both a soap and a cross-section of a spherical micelle is shown in Figure 4.6.

In dilute solutions the hydrophobic tails of the long chain acids are surrounded by 'cages' of water molecules, then as the concentration of the acid salt increases, the micelles are formed. In thermodynamic terms,

- the process is generally very slightly endothermic (ΔH is usually about 1–2 kJ mol^{-1});
- the ΔS for micelle formation is generally negative (micelles are more organised than a collection of individual molecules).

139

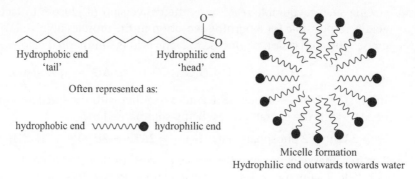

Figure 4.6 **Micelle formation.**

This would imply the process would not be spontaneous, *but*

● the ΔS for the solvent is large and positive, as a result of the breaking up of the cages of water that has taken place, thereby resulting in an overall significant entropy increase.

Box 4.10

```
More about standard states
```

Reminder: standard states

A *standard state* of a substance (pure element or compound) is defined as the most stable form of that substance that exists at a pressure of 1 bar (approximately atmospheric pressure) and at a specified temperature (often taken to be 298 K, but it need not be).

Extension 1: Compounds in solution

A standard state of a substance (pure element or compound) in solution is defined as the most stable form of that substance that exists at a pressure of 1 bar, at a specified temperature (usually 298 K) and at a concentration of 1 mol L^{-1}.

Extension 2: Biological standard state

If a solution contains H^+ ions (usually shown as H_3O^+), then the standard state would dictate that the concentration, $[H_3O^+]$, would be $1 \, mol \, L^{-1}$.

Biological processes usually occur at approximately neutral pH, which corresponds with a hydrogen ion concentration, $[H_3O^+]$, of $10^{-7} \, mol \, L^{-1}$, and so the *biological standard state* is quoted as being defined as 'the most stable form of that substance that exists at a pressure of 1 bar, at a specified temperature (usually 298 K) and at a concentration of 1 mol L^{-1}, except for the hydrogen ion concentration, $[H_3O^+]$, which has a value of $10^{-7} \, mol \, L^{-1}$.'

Standard state symbols

So far standard changes in parameters such as enthalpy, entropy and free energy have been represented by using the symbol θ as a superscript; for example, standard free energy changes have thus far been represented by $\Delta G^{\oplus}$.

For biological standard states, however, the symbol used is $\oplus$ (or sometimes θ'), again as a superscript. Therefore, for example, the standard free energy change under biological standard state conditions would be $\Delta G^{\oplus}$.

4.1 An athlete during training does 800 kJ of work during their warm-up on an exercise bike, whilst losing 150 kJ of heat to the surroundings at 37 °C (310 K).

(a) Calculate the change in the athlete's internal energy.

(b) Calculate the entropy change in the athlete's surroundings.

4.2 The standard enthalpy of combustion of maltose, represented below, is $-5640.1 \text{ kJ mol}^{-1}$. The equation below is an incomplete representation of this.

$$C_{12}H_{22}O_{11} \text{ (s)} + O_2 \text{ (g)} \rightarrow CO_2 \text{ (g)} + H_2O \text{ (l)}$$

(a) Complete this reaction equation.

(b) Given that CO_2 and H_2O are the only products, calculate the enthalpy of formation of maltose.

Values for standard molar enthalpies of formation are:

for CO_2, $-393.5 \text{ kJ mol}^{-1}$

for H_2O, $-285.8 \text{ kJ mol}^{-1}$.

4.3 Which of the following processes is likely to be accompanied by an increase or a decrease in the entropy value? In each case provide a rationale for your answer.

(a) The evaporation of propanone (acetone) from a bottle of nail polish remover.

(b) The formation of ice on a cold window.

(c) The photosynthesis of glucose: $6CO_2 \text{ (g)} + 6H_2O \text{ (l)} \rightarrow C_6H_{12}O_6 \text{ (s)} + 6O_2 \text{ (g)}$

4.4 Below is the equation showing the reaction between nitrogen pentoxide and water, to produce nitric acid at 310 K:

$$H_2O \text{ (l)} + N_2O_5 \text{ (g)} \rightleftharpoons 2HNO_3 \text{ (l)}$$

The following table shows selected thermodynamic data for this process.

	H_2O (l)	N_2O_5 (g)	HNO_3 (l)
ΔH_f^{θ} (kJ mol^{-1})	+285.8	+11.3	−174.1
S^{θ} (J K^{-1} mol^{-1})	+69.9	+355.7	+155.6

(a) Use the data from the table to calculate the value of ΔS^θ for the reaction.

(b) Use the data to calculate the standard enthalpy change, ΔH^θ, for this reaction.

(c) Hence calculate ΔG^θ. Is the reaction spontaneous?

4.5 When the mineral $CaCO_3$ is heated, it decomposes as follows:

$$CaCO_3 \text{ (s)} \rightarrow CaO \text{ (s)} + CO_2 \text{ (g)}$$

The enthalpy change, ΔH, for this process $= 179.1\,kJ\,mol^{-1}$, and the entropy change, ΔS, for the process $= 161\,J\,K^{-1}\,mol^{-1}$.

At low temperatures the free energy change, ΔG, for this reaction is positive and at high temperatures it becomes negative. There is therefore a temperature, T_d, at which decomposition becomes spontaneous.

(a) What is the free energy change for the reaction at the temperature at which the reaction changes from being non-spontaneous to spontaneous?

(b) Rearrange the equation $\Delta G = \Delta H - T\Delta S$ to express the value of T_d in terms of ΔH and ΔS.

(c) Calculate T_d (assume that the values of ΔH and ΔS do not change significantly with temperature).

5 Equilibria: how far does a reaction go?

Learning outcomes

By the end of this chapter students should be able to:

- understand what is meant by the term equilibrium;
- know how to determine the reaction quotient Q for a reaction in terms of the composition of the system, and know how it relates to the equilibrium constant K;
- appreciate the relationship between the equilibrium constant for a reaction and the free energy change associated with the reaction, $\Delta G = -RT \ln K$;
- appreciate the importance of non-ideal behaviour and the necessity to describe the true equilibrium in terms of 'activities';
- calculate the composition of a system at equilibrium from the starting concentrations and the equilibrium constant;
- state Le Chatelier's principle and understand its significance.

5.1 Introduction

The sign of the free energy change defines whether a reaction goes: it does not, on its own, say how far the reaction will go. Whilst some reactions go to completion (an *irreversible reaction*), there are many chemical and biological processes that do not do so, and reactions of this kind produce mixtures of reactants and products. These are *reversible reactions* and produce *equilibrium mixtures*. Consider these examples of each type of reaction:

- An irreversible reaction might be the reaction that occurs when magnesium is burned in oxygen to produce magnesium oxide (often demonstrated in schools, as the magnesium burns with a very bright glow). The magnesium is usually a metallic strip (known as magnesium ribbon), and the magnesium oxide is a white powder.

$$2Mg \text{ (s)} + O_2 \text{ (g)} \rightarrow 2MgO \text{ (s)}$$

Note here how the arrow $\rightarrow$ indicates the direction of the reaction.

- A reversible reaction occurs between an organic acid and an alcohol to form an ester. Taking the example of the reaction between ethanol (CH_3CH_2OH) and ethanoic acid (CH_3COOH), which produces the ester, ethyl ethanoate, and water, to illustrate this.

$$CH_3COOH\ (l) + CH_3CH_2OH\ (l) \rightleftharpoons CH_3COOCH_2CH_3\ (l) + H_2O\ (l)$$

In this case, starting with just the acid and the alcohol, at a given temperature, the reaction will proceed to a point at which there is a mixture of all four components, namely the acid, the alcohol, the ester and the water. Similarly, on mixing the ester and water, a reaction occurs producing, eventually, a four-component mixture of the acid, the alcohol, the ester and water. In each case the reaction appears to have stopped, but in fact the rate of the reaction going in one direction is being matched by the rate going in the opposite direction. When this happens the system is said to be at *equilibrium*. Note here how the symbol $\rightleftharpoons$ indicates the reversible nature of the reaction.

5.2   Developing the idea of equilibrium: the equilibrium constant

Every equilibrium reaction is described by an equilibrium constant, which is given the symbol K. This is usually measured in terms of the concentrations of the species involved when at equilibrium (though for gaseous reactions partial pressures are used – see Section 5.3.3). As mentioned above for the reaction between an organic acid and an alcohol, for a given temperature the same equilibrium position is reached whether starting with the reactants or the products.

If this is generalised by taking the hypothetical reaction where a moles of a compound A react with b moles of a compound B to yield c moles of a compound C and d moles of a compound D, as illustrated:

$$a\text{A} + b\text{B} \rightleftharpoons c\text{C} + d\text{D}$$

The equilibrium constant for this reaction is given by:

$$K = \frac{[C]^c[D]^d}{[A]^a[B]^b} \qquad [5.1]$$

The key points about this are:

- For a given temperature, the value of K describes the composition of an equilibrium mixture and is characteristic of a particular reaction.
- Even though K is a constant, the equilibrium composition for a reaction does not need to be the same.

Considering the second point for the general reaction above, where the values of a, b, c and d are all 1, that is:

$$\text{A} + \text{B} \rightleftharpoons \text{C} + \text{D}$$

If this reaction had an equilibrium constant of $K = 1$, then clearly if $[A] = [B] = [C] = [D] = 1$ mol L^{-1}, the system would be at equilibrium, because

$$K = \frac{[C][D]}{[A][B]} = \frac{1 \times 1}{1 \times 1} = 1$$

However, the value of K would also be 1 if the composition was $[A] = 10$ mol L^{-1}, $[B] = 0.1$ mol L^{-1}, $[C] = 50$ mol L^{-1} and $[D] = 0.02$ mol L^{-1}, thus:

$$K = \frac{[C][D]}{[A][B]} = \frac{50 \times 0.02}{10 \times 0.1} = 1$$

Both mixtures would be at equilibrium! The equilibrium constant depends on the initial amounts of material.

5.2.1 Calculation of equilibrium constants and concentrations

Returning to the general reaction:

$$aA + bB \rightleftharpoons cC + dD$$

where:

$$K = \frac{[C]^c[D]^d}{[A]^a[B]^b}$$

This means that if the equilibrium concentration of at least one of the species is known, along with starting concentrations, then it will be possible to use that information to determine the value of the equilibrium constant.

Similarly, where the value of the equilibrium constant is known, along with the starting concentrations, then it will be possible to calculate the equilibrium concentrations. The procedure for these calculations makes use of constructing a so-called ICE table (from Initial, Change, Equilibrium quantity values). This approach is outlined by the examples below.

Example 5.1 Calculation of equilibrium composition

0.5 moles of hydrogen (H_2) and 0.5 moles of iodine (I_2) were reacted in a 0.50 litre container at 425°C, in the presence of a catalyst, to produce hydrogen iodide (HI), until equilibrium was established. The equilibrium constant for the reaction is 54. Calculate the equilibrium concentrations of the species present.

Step 1   Show the equilibrium

$$H_2\ (g) + I_2\ (g) \rightleftharpoons 2HI\ (g)$$

Step 2   Itemise the known concentrations

As there are 0.5 moles of both H_2 and I_2, and the volume is 0.5 L, their concentrations are, therefore, $C = n/V = 0.5/0.5 = 1.0$ mol L^{-1}.

	H_2	I_2	HI
Initial	1.0	1.0	0
Change			
Equilibrium			

Step 3 Itemise the known changes in concentrations

There are $2x$ moles of HI produced for every x moles of H_2 and I_2 used.

	H_2	I_2	HI
Initial	1.0	1.0	0
Change	$-x$	$-x$	$+ 2x$
Equilibrium			

Step 4 Show the equilibrium concentrations in terms of the changes made

	H_2	I_2	HI
Initial	1.0	1.0	0
Change	$-x$	$-x$	$+ 2x$
Equilibrium	$(1.0-x)$	$(1.0-x)$	$2x$

Step 5 Calculate the value of x

The first step is to write the expression for the equilibrium constant using the concentrations above.

$$K = \frac{[HI]^2}{[H_2][I_2]} = 54$$

Then feed in the concentration values:

$$\frac{(2x)^2}{(1 - x)(1 - x)} = 54$$

which can be rewritten as:

$$\frac{(2x)^2}{(1 - x)^2} = 54$$

If the square root of both sides is taken:

$$\frac{(2x)}{(1 - x)} = \sqrt{54} = 7.35$$

and so following the calculations through:

$$2x = 7.35(1 - x)$$
$$2x = 7.35 - 7.35x$$
$$9.35x = 7.35$$
$$x = 0.79$$

Therefore the equilibrium concentrations are:

$[H_2] = [I_2] = (1.00 - 0.79) = 0.21 \text{ mol L}^{-1}$

and

$[HI] = 2 \times 0.79 = 1.58 \text{ mol L}^{-1}$

Example 5.2 60.0 g of ethanoic acid (CH_3CO_2H) and 46.0 g of ethanol (CH_3CH_2OH) were reacted at 50°C until equilibrium was established. By this time 12.0 g water and 58.7 g ethyl acetate ($CH_3CO_2C_2H_5$) were produced. What is the equilibrium constant for this reaction at this temperature?

Step 1 Show the equilibrium

$$CH_3CO_2H + CH_3CH_2OH \rightleftharpoons CH_3CO_2C_2H_5 + H_2O$$

Step 2 Set up the ICE table, starting with the initial values

	CH_3CO_2H	CH_3CH_2OH	$CH_3CO_2C_2H_5$	H_2O
Mol mass (g mol^{-1})	60.0	46.0	88.0	18.0
No. of moles: Initially	1	1	0	0

Step 3 Calculate the equilibrium quantities

Recalling that the stoichiometry of the reaction shows that 1 mole of both of the reactants produces 1 mole of the products.

At equilibrium, there are: 58.7 g ethyl acetate, which is (58.7/88.0) moles = 0.667 moles

Similarly, there are: 12 g water, which is 12.0/18.0   = 0.667 moles

Therefore, the number of moles of ethanoic acid   = 1 − 0.667   = 0.333

and the number of moles of ethanol   = 1 − 0.667   = 0.333

Step 4 Calculate K

Therefore as the equilibrium constant is given by:

$$K = \frac{[CH_3CO_2C_2H_5][H_2O]}{[CH_3CO_2H][C_2H_5OH]}$$

Then

$$K = \frac{0.667 \times 0.667}{0.333 \times 0.333} = 4$$

5.3   Equilibrium and energetics

5.3.1 Background

Having established that chemical reactions can exist at equilibrium, we need to explore why they do so. To understand how systems can achieve a state of equilibrium it is important to consider the way in which free energy changes as a chemical reaction progresses.

Returning to the hypothetical reaction:

$$aA + bB \rightarrow cC + dD$$

It will be assumed that the reaction as written is spontaneous (in other words, ΔG^{θ} is negative). The value of ΔG^{θ} is the *standard free energy change*, which simply refers

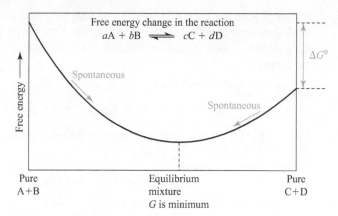

Figure 5.1 **Variation of free energy with composition.**

to the difference in free energy of the reactants and products in the proportions in which they appear in the equation for the reaction to form one mole of product:

$$\Delta G^\theta = G_{products} - G_{reactants} = (cG(C) + dG(D)) - (aG(A) + bG(B))$$
$$= cG(C) + dG(D) - aG(A) - bG(B)$$

However, the evolution of G with the extent of the reaction does not necessarily take the form of a straight line, and there is usually a minimum in G at some intermediate point, as illustrated in Figure 5.1.

Example 5.3   What happens when *a* moles of A are reacted with *b* moles of B?

The reaction proceeds until the free energy minimum is reached. As the system reaches this point, the free energy must rise for the reaction to continue, and the second Law of Thermodynamics states that no further progress is allowed.

At the minimum in the curve the reaction cannot move in either direction and the system is said to be *at equilibrium*, and contains a mixture of A, B, C and D.

Example 5.4   What happens if the reverse situation is applied, and *c* moles of C react with *d* moles of D?

C and D will react to some extent and the system will move to the left as indicated in Figure 5.1, until it too arrives at the equilibrium point comprising a mixture of A, B, C and D.

This process can therefore be regarded as being reversible and, as was mentioned earlier, this fact is indicated in the equation for the reaction by replacing the arrow ' → ' by a pair of arrows pointing in opposite directions: ⇌

Therefore the reaction is more correctly represented as:

$$aA + bB \rightleftharpoons cC + dD$$

What the graph shows is that there is a relationship between the free energy and the composition of the reaction mixture.

5.3.2 The reaction quotient

The reaction quotient Q provides the way of defining the reaction composition. Considering again the general reaction:

$$aA + bB \rightleftharpoons cC + dD$$

then Q is defined as

$$Q = \frac{[C]^c[D]^d}{[A]^a[B]^b} \tag{5.2}$$

Considering this with reference to a modified form of the diagram (Figure 5.2) relating free energy with reaction progress for the reaction between A and B, we can see that three situations can apply here:

1. If Q occurs to the left of the minimum G value, then the reaction will proceed in a forward direction until equilibrium is established.
2. If Q occurs when G is a minimum, the system is at equilibrium.
3. If Q occurs to the right of the minimum G value, the reaction will proceed in the reverse direction until equilibrium is established.

At the point where G is at a minimum, then the value of Q is given a special label, K, which is the equilibrium constant (as seen in Section 5.2).

For $Q < K$, the free energy change for the forward reaction is negative, and so the reaction between A and B is spontaneous until $Q = K$.

For $Q > K$, the free energy change for the reverse reaction is negative, and so the reaction between C and D is spontaneous until it reaches $Q = K$.

Therefore, at equilibrium, there is a relationship between the concentrations of the reactants and products, called the equilibrium constant, and this is given the symbol K. For the generalised reaction above this is:

$$K = \frac{[C]^c[D]^d}{[A]^a[B]^b}$$

The implication of this equation would be that equilibrium constants could have units (if the indices in the equilibrium expression did not cancel. In fact they do not

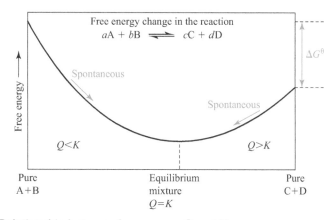

Figure 5.2 Relationship between free energy, Q and K.

have units, and the reasons for this are described in Box 5.1. (Equilibrium constants based on concentrations, as shown above, are often shown as K_c.)

Box 5.1

Activity, or effective concentration

When carrying out experiments in solutions in the laboratory, quantities of materials used are measured in terms of molar concentrations. In solutions where concentrations are greater than approximately 0.1 mol L^{-1}, there are significant degrees of interaction between solute species, where these exist as molecules or as ions. Because of these interactions, and particularly when dealing with solutions of concentrations greater than 0.1 mol L^{-1}, the effective concentration of the species that are available to react will be less than the actual concentration dissolved. This situation can be represented symbolically as shown:

$$a = \gamma \frac{c}{c^{\theta}}$$

where:

- a is the activity of the species in solution (in essence the effective concentration);
- γ is the activity coefficient, a 'fudge factor' by which the concentration is multiplied in order to yield the activity;
- c is the concentration of the species in solution, and c^{θ} is the standard state concentration, defined as 1 mol L^{-1}.

Because the definition of activity uses the ratio $\frac{c}{c^{\theta}}$, this has a consequence that the concentration units cancel out, and, as γ is just a number, then activity itself has no units.

Note that:

- at concentrations of 0.1 mol L^{-1} or lower, the activity coefficient, γ, is taken to be 1, and so the value of activity, a, is taken to be the numerical value of the concentration;
- for pure solids and liquids the activity, a, is taken to be 1.

Therefore, in the generalised example of the reaction:

$$aA + bB \rightleftharpoons cC + dD$$

then the value of Q is represented by:

$$Q = \frac{a_C^c \times a_D^d}{a_A^a \times a_B^b}$$

and therefore at the equilibrium point, it can be said that:

$$K = \frac{a_C^c \times a_D^d}{a_A^a \times a_B^b}$$

where a_C^c = activity of species C to the power of its stoichiometric value, c.
Similarly

a_D^d = activity of species D to the power of its stoichiometric value, d;

a_A^a = activity of species A to the power of its stoichiometric value, a;

a_B^b = activity of species B to the power of its stoichiometric value, b.

The equilibrium constant, K, is based on activities, and activity does not have units, so K does not have units.

5.3.3 Calculating equilibrium constants in the gas phase, using partial pressures; K_p

For reactions in the gas phase it is often more useful to measure quantities of gas in terms of pressure rather than concentration. Consider the general reaction:

$$aA\,(g) + bB\,(g) \rightleftharpoons cC\,(g) + dD\,(g)$$

If this reaction is carried out in a contained environment then for this the total pressure within the container, $P_{total} = p_A + p_B + p_C + p_D$, where p_A, p_B, p_C and p_D are the partial pressures of the components.

Partial pressure means the pressure that each component would exert if it were present in the container on its own.

The appearance of the equilibrium constant based on partial pressures is very similar to that of the equilibrium constant based on concentration, and is:

$$K_P = \frac{p_C{}^c \times p_D{}^d}{p_A{}^a \times p_B{}^b} \qquad [5.3]$$

Box 5.2

Remember: equilibrium is a dynamic process

It is important to recognise that when a system is at equilibrium, it is not in a static situation. What is happening is that the forward reaction and the reverse reaction are both occurring, but their rates are equal. In other words, at equilibrium:

Rate of forward reaction = rate of reverse reaction

5.4 The relationship between ΔG^{θ} and K

The more negative the value of ΔG^{θ}, the more the reaction goes towards the products and the higher the value of K. This is often referred to as a *product favoured* reaction.

Figure 5.3 shows how the position of the equilibrium varies with ΔG^{θ}. Note that with a negative value the equilibrium position favours an excess of products compared with reactants, whereas when the value of ΔG^{θ} is positive the equilibrium favours an excess of reactants compared with products.

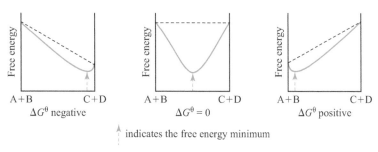

indicates the free energy minimum

Figure 5.3 The variation of equilibrium position with ΔG^{θ}.

As Figure 5.3 demonstrates, there is clearly a link between the value of the standard free energy change, ΔG^{θ}, for a reaction, and the position of the equilibrium constant for that reaction.

The mathematical expression for this relationship is:

$$K = e^{\frac{-\Delta G^{\theta}}{RT}} \qquad [5.4]$$

where:

- R is a fundamental constant, known as the gas constant, and equals 8.314 $J\,K^{-1}\,mol^{-1}$;
- T is the absolute temperature in kelvin.
- 'e' is just a number (= 2.718. . .) and it is the base used in natural logarithms.

Appendix 1, which gives some basic mathematical tools used in this book, provides an overview of the importance of e in science, and some examples of its usage, in particular its relationship with many of the laws of nature. These have logarithmic forms, but the base involved is e (again see Appendix 1), and the logarithms that use this base are called natural logarithms. Natural logarithms, $\log_e$, are usually represented by the symbol ln (this is how the natural logarithm button appears on most calculators).

As a brief reminder, if $y = e^x$ then $\ln y = x$.

Note too that $\ln (e^x) = x$, so, in the case of the relationship between K and ΔG^{θ}, namely

$$K = e^{\frac{-\Delta G^{\theta}}{RT}}$$

Taking natural logs of both sides gives:

$$\ln K = \ln e^{\frac{-\Delta G^{\theta}}{RT}}$$

which, using the log rule $\ln x^a = a \ln x$, rearranges to:

$$\ln K = -\frac{\Delta G^{\theta}}{RT} \times \ln e$$

Also from the rules of logarithms, $\ln e = 1$, so this means that:

$$\ln K = -\frac{\Delta G^{\theta}}{RT}$$

Multiplying both sides by $-RT$ gives the result:

$$\Delta G^{\theta} = -RT \ln K \qquad [5.5]$$

5.4.1 A more detailed look at reaction quotient Q and equilibrium constant, K

Focusing now on the response of the different species in a system at equilibrium to changes in concentration, the equation:

$$\Delta G^{\theta} = -RT \ln K$$

relates ΔG^{θ} to the equilibrium constant K.

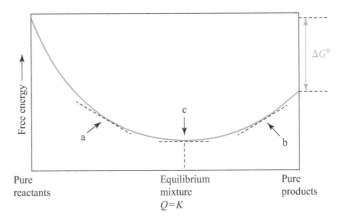

Figure 5.4 **Variation of free energy change with composition.**

But from Figure. 5.2, showing how free energy varies with reaction progress, it is clear that the free energy, G, of the system at any time depends on the concentrations of the mixture at that time.

The approach that has been used already can be extended to incorporate Q and a new term, ΔG_r, which represents the free energy change at any point during the reaction. This is illustrated in Figure 5.4.

ΔG_r is the free energy change associated with the reaction at any point of the process, and is given by the slope of the tangent to the curve at that point, as indicated at three different points labelled a, b and c. At point a, the slope of the tangent is negative, indicating the value of ΔG_r is negative at that point, and hence the forward reaction is spontaneous. At point b, the slope of the tangent is positive, indicating the value of ΔG_r is positive at that point, and hence the forward reaction is non-spontaneous (though the reverse reaction will be spontaneous). At point c the slope of the tangent is 0, indicating the value of ΔG_r is 0 at that point, and hence the reaction is at equilibrium.

There is a mathematical relationship that can be found between the values of ΔG_r for a particular composition, the value of ΔG^θ and this relationship involves the composition of the reaction mixture as represented by the reaction quotient, Q. The relationship is:

$$\Delta G_r = \Delta G^\theta + RT \ln Q \qquad [5.6]$$

Box 5.3 demonstrates the significance of this relationship. Because at the point of equilibrium, where $Q = K$, the free energy change $\Delta G_r = 0$, then the relationship can be rewritten as:

$$0 = \Delta G^\theta = +RT \ln K$$

This can be rearranged to:

$$\Delta G^\theta = -RT \ln K$$

153

Important points

- The relationship:

$$\Delta G_r = \Delta G^\theta + RT \ln Q$$

means that even if a reaction has a positive standard free energy change, ΔG^θ, it can still progress spontaneously until $Q = K$, because the reaction free energy change, ΔG_r, can be negative until the minimum free energy value is reached, even if the standard free energy change, ΔG^θ, is positive.

- Rearranging $\Delta G^\theta = -RT \ln K$

gives

$$\ln K = -\Delta G^\theta / RT$$

This means that the value of $\ln K$ (and hence of K itself) depends on the temperature.

Box 5.3

Significance of the relationship between reaction free energy and Q

For the hydrolysis of ATP in a cell:

$$ATP^{4-} + H_2O \rightleftharpoons ADP^{3-} + HPO_4^{2-} + H^+$$

Taking the equation relating free energy change and reaction composition, namely:

$$\Delta G_r = \Delta G^\theta + RT \ln Q$$

then putting in the relevant concentrations for Q, it can be expanded to:

$$\Delta G_r = \Delta G^\theta + RT \ln \frac{[ADP^{3-}][HPO_4^{2-}][H^+]}{[ATP^{4-}][H_2O]}$$

where

$$Q = \frac{[ADP^{3-}][HPO_4^{2-}][H^+]}{[ATP^{4-}][H_2O]}$$

This looks very messy, but a number of assumptions can be made that will simplify this:

- H_2O need not be included; it is the solvent and therefore has an effective concentration of 1 M.

- The pH $= 7.0$ and does not change during the reaction, so $[H^+] = 10^{-7}$ M.

- Under these conditions the $[H_2O]$ and $[H^+]$ terms can be 'incorporated' into ΔG^θ to give the biological standard free energy change, $\Delta G^\oplus$.

- There is a 'biological' equilibrium constant, related to $\Delta G^\oplus$, which is designated as $K^\oplus$.

The value of $\Delta G^\oplus$ that is based on these biological standard conditions is -30.5 kJ mol^{-1}. The equation relating free energy to composition, using biological conditions, can be rewritten as:

$$\Delta G_r = \Delta G^\oplus + RT \ln \frac{[ADP][P_i]}{[ATP]}$$

For simplicity, we have substituted ADP for ADP^{3-}; ATP for ATP^{4-} and P_i (inorganic phosphate) for HPO_4^{2-}. Under biological standard conditions, the concentrations of each of these species is 1 M, and any variation from these concentrations will result in ΔG_r having a value different from that of $\Delta G^{\oplus}$

Consider now the significance of operating concentrations found in a cell. If in skeletal muscle cells the [ATP] is measured to be about $10 \times$ [ADP] and $[P_i] = 0.11$ mM (mM = millimolar) (1 mM = 0.001 mol L^{-1}) then

$$\Delta G_r = \Delta G^{\oplus} + RT \ln \frac{[ADP][P_i]}{[ATP]}$$

gives, on putting in the numbers (remembering to convert mM to mol L^{-1}):

$$\Delta G_r = \Delta G^{\oplus} + RT \ln \frac{[1][0.00011]}{[10]}$$

Although the concentrations of ATP and ADP are not known, the ratio between the species is known, so it is fine to put the numbers in as shown.

$$\Delta G_r = \Delta G^{\oplus} + RT \ln (1.1 \times 10^{-5})$$
$$= -30\,500 + (8.314 \times 310 \times -11.42)$$
$$= -59\,933 \text{ J mol}^{-1}$$
$$= -59.9 \text{ kJ mol}^{-1}$$

Organisms can control how much energy they obtain from ATP by maintaining concentrations of the various species at certain levels.

5.5 Disturbing an equilibrium

An equilibrium reaction can be affected in a number of different ways, for example adding or removing reactants or products, changing the pressure (in a gas phase reaction) or changing the temperature at which the reaction is carried out. The effect of disturbances to chemical equilibria was considered by the noted French chemist Henry Louis Le Chatelier, whose approach was summarised in the renowned 'Le Chatelier's principle'.

5.5.1 Statement of Le Chatelier's principle

When a system in chemical equilibrium is disturbed by a change in temperature, pressure or the concentration of a reaction component, the system will shift the equilibrium position so as to counteract the effect of the disturbance. This is classified in Box 5.4.

Box 5.4

Comments on Le Chatelier's principle

If a reaction is at equilibrium, then *adding reactants, or removing products*, will result in the reaction quotient, Q, decreasing it, because

$$Q = \frac{[C]^c[D]^d}{[A]^a[B]^b}$$

and if [A] and/or [B] increases (or [C] and/or [D] reduces), then Q must decrease.

Because the relationship between reaction free energy and Q is given by:

$$\Delta G_r = \Delta G^\theta + RT \ln Q$$

then it follows that if Q decreases, so does $\ln Q$, and hence so does ΔG_r, and hence the *forward reaction* becomes favoured (spontaneous).

Removing reactants, or adding to the product side, will cause an increase in the value of Q, and hence will cause an increase in the value of ΔG_r, and hence the *reverse reaction* becomes favoured.

5.5.2 Le Chatelier's principle and the effect of temperature on equilibria

Reactions are usually accompanied by at least a small change in temperature. In other words they will either be exothermic (if heat is released) or endothermic (if heat is taken in). Consider the equilibrium:

$$aA + bB \rightleftharpoons cC + dD$$

- If the forward reaction is exothermic then it could be rewritten as:

$$aA + bB \rightleftharpoons cC + dD + \text{heat}$$

- If the forward reaction is endothermic then it could be rewritten as:

$$aA + bB + \text{heat} \rightleftharpoons cC + +dD$$

Taking the case where the *forward reaction is exothermic* and additional heat is supplied, then heat can be regarded as one of the products, so in order for the system to re-establish equilibrium then the relative proportions of reactants and products will need to change, resulting in relatively more reactant and relatively less product.

Taking the case where the *forward reaction is endothermic* and additional heat is supplied, then heat can be regarded as one of the reactants, so in order for the system to re-establish equilibrium then the relative proportions of reactants and products will need to change, resulting in relatively more product and relatively less reactant.

Key features of the relationship between K and temperature are summarised in Boxes 5.5 and 5.6.

Box 5.5

The value of K is temperature-dependent

- For an exothermic reaction, as temperature increases, K decreases.
- For an endothermic reaction, as temperature increases, K increases.

Box 5.6

Dependence of equilibrium constant, K, on temperature, T; a more mathematical look

Taking the relationship between standard free energy, enthalpy and entropy changes:

$$\Delta G^{\theta} = \Delta H^{\theta} - T\Delta S^{\theta}$$

and that between standard free energy change and equilibrium constant:

$$\Delta G^{\theta} = -RT \ln K$$

then it follows that:

$$-RT \ln K = \Delta H^{\theta} - T\Delta S^{\theta}$$

This can be rearranged very readily (divide both sides by RT) to give:

$$\ln K = -\frac{\Delta H^{\theta}}{RT} + \frac{\Delta S^{\theta}}{R}$$

(remember R is the gas constant).

This relationship is called the van't Hoff isochore, and it is in the form of a straight line graph, where a plot of $1/T$ as the x axis and $\ln K$ as the y axis will give a gradient of $-\frac{\Delta H^{\theta}}{R}$ and an intercept on the y axis of $\frac{\Delta S^{\theta}}{R}$.

It is found that both ΔH^{θ} and ΔS^{θ} do, in fact, change with T, but in general not quickly when compared with the factor $1/T$ in the equation, so to a first approximation it can be assumed that they are independent of T. This means that a graphical plot of $\ln K$ vs $1/T$ would produce a straight line such as:

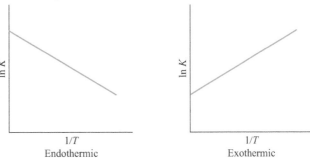

This is really just a manifestation of Le Chatelier's principle.

(i) If a reaction is exothemermic ($\Delta H^{\theta} < 0$), then $\ln K$ is smaller for a smaller value of $1/T$. As $1/T$ is the inverse of T then, as T rises, so $\ln K$ becomes smaller (meaning K becomes smaller).

(ii) If a reaction is endothermic the opposite situation holds, and K rises as T falls.

In words, the equilibrium point shifts to minimise the effect of changing the reaction conditions, which is consistent with Le Chatelier's principle.

5.5.3 Examples involving Le Chatelier's principle

Why does washing dry better on a windy day?

This is an everyday example of how disturbing an equilibrium can be useful. The equilibrium in question is:

$$\text{water} \underset{\text{cool}}{\overset{\text{heat}}{\rightleftharpoons}} \text{water vapour}$$

Even at everyday temperatures in mid-latitude countries (for example the UK) this equilibrium will be established, albeit with the equilibrium being substantially to the left (in favour of the reactant). When damp washing is hung outside on a breezy day the equilibrium is constantly being disturbed as the water vapour is blown away (the product is removed), thereby continually encouraging the forward reaction.

Effect of temperature on the dissociation of dinitrogen tetroxide, N_2O_4

The equilibrium in question here is:

$$N_2O_4 \rightleftharpoons 2NO_2$$

The reaction as shown is endothermic (ΔH is positive). The NO_2 in this equilibrium is coloured brown, whereas the N_2O_4 is colourless.

Consider now the following experiment (shown in Figure 5.5).

Figure 5.5 Samples of N_2O_4 at different temperatures.
(Charles D. Winters/Science Photo Library)

- A sample of N_2O_4 is introduced to three tubes, such that the amount of colour in each is the same at room temperature.
- One of the tubes is heated to about 80°C (on left), one is cooled to 0°C (on right) and one is kept at room temperature (middle), which is about 20°C.

The sample at 0°C is seen to have lost some of its colour when compared with the sample at room temperature, whereas that at 80°C has a much more intense colour.

5.6 Energetics and equilibria in the biological context

5.6.1 Calculating $\Delta G^\oplus$ from experimentally determined compositions (via K values)

Consider the reaction of phosphate with glucose to produce glucose-6-phosphate (abbreviated to p-glucose).

| Equilibrium Molar concentrations | [0.0080] | [0.0125] | [2.78 × 10⁻⁵] |

Using the equilibrium molar concentrations in the reaction as shown, the value of the equilibrium constant, K, can be calculated as:

$$K = \frac{[\text{p-glucose}]}{[\text{glucose}][\text{phosphate}]}$$

$$= \frac{2.78 \times 10^{-5}}{0.0080 \times 0.0125}$$

$$= 2.78 \times 10^{-1}$$

As the equilibrium constant is known, it is therefore possible to calculate the standard Gibbs free energy change associated with the process:

$$\Delta G^\oplus = -RT \ln K$$

$$= -8.314 \, \text{J K}^{-1}\text{mol}^{-1} \times 310 \, \text{K} \times \ln (2.78 \times 10^{-1})$$

$$= -8.314 \, \text{J K}^{-1}\text{mol}^{-1} \times 310 \, \text{K} \times (-1.28)$$

$$= 3299 \, \text{J mol}^{-1}$$

$$= +3.3 \, \text{KJ mol}^{-1}$$

This means that the reaction is non-spontaneous as written, and the equilibrium therefore lies to the left.

5.6.2 Calculating equilibrium compositions from $\Delta G^{\oplus}$

The standard free energy change for the equilibrium for citrate^{3-}/isocitrate^{3-} is +6.65 kJ mol^{-1}. This can be used to calculate the ratio of the species. Thus, for the process:

$$\text{Citrate}^{3-} \rightleftharpoons \text{Isocitrate}^{3-}$$

This reaction has a $\Delta G^{\oplus}$ value of +6.65 kJ mol^{-1} at 37°C (37°C = 310 K).

The first stage is to use the relationship between the standard free energy change and equilibrium constant, in order to calculate the value of K:

$$K = e^{-\Delta G^{\oplus}/RT}$$

(where $R = 8.314$ J K^{-1} mol^{-1}, $T = 310$ K and $\Delta G^{\oplus} = +6650$ J mol^{-1}; *note units of $\Delta G^{\oplus}$ in J for consistency with units of R*).

Putting in the values:

$$K = e^{-\{6650/(8.314 \times 310)\}}$$

$$= e^{-2.58} = 0.0758$$

From the stoichiometry of the reaction then it follows:

$$K = [\text{isocitrate}^{3-}] \, / \, [\text{citrate}^{3-}]$$

$$= 0.0758$$

Rearranging this shows:

$$[\text{isocitrate}^{3-}] = 0.0758[\text{citrate}^{3-}]$$

and so

$$[\text{citrate}^{3-}] = 13.2 \, [\text{isocitrate}^{3-}]$$

5.6.3 Macromolecule–ligand interactions

Biological macromolecules are usually taken to mean large carbohydrates (e.g. starches); proteins (e.g. enzymes); nucleic acids (e.g. DNA) or lipids. Molecules such as these function by interaction with other species, for example enzyme–inhibitor interactions, haemoglobin–oxygen interactions.

The concept of equilibria is applicable to many such intermolecular interactions, and in biology these are referred to as *macromolecule–ligand interactions*. A ligand can be a macromolecule or it can be a small molecule. When a macromolecule interacts with a ligand, a *complex* is said to be formed.

Formation of the complex (or binding/association of molecules) is characterised by an equilibrium constant (K_A), known as an association constant. The reverse reaction, the dissociation of such a complex, is characterised by a dissociation constant (K_D).

If the macromolecule is given the symbol M, and the ligand with which it interacts is called L, then the process of interaction can be summarised by the equilibrium:

$$M + L \underset{K_D}{\overset{K_A}{\rightleftharpoons}} ML$$

Where

$$K_A = \frac{[ML]}{[M][L]} \text{ and } K_D = \frac{[M][L]}{[ML]}$$

K_A is the association constant; the equilibrium constant for the forward reaction. K_d is the dissociation constant; the equilibrium constant for the reverse reaction. From the above it is clear that the relationship between the two constants is

$$K_D = K_A^{-1}$$

In biology it is the dissociation constant K_D that is often used, since, for 1:1 binding, its dimension is that of a concentration, that makes it something that can easily be related to the concentration of compounds involved.

Dissociation constants indicate how 'strong' the complex is, and it is usual to refer to nano, micro, millimolar affinities (where the values of K_D are 10^{-9}, 10^{-6}, 10^{-3} M, known as nano-, micro-, millimolar). As $K_D = K_A^{-1}$ this means that the corresponding K_A values are 10^9, 10^6 and 10^3. From these values it is clear that the formation of the complex is much more favoured for nanomolar rather than millimolar affinities. Calorimetric techniques exist that measure the heat (or enthalpy) associated with complex formation. For weakly bound complexes, the same experimental methods can provide equilibrium constants, K.

As $\Delta G^\theta = -RT \ln K$, then, using experimentally determined values of K, it is possible to determine the free energy change, ΔG^θ, for the complexation process. Given that enthalpy values can also be found experimentally, then it is possible to determine values of ΔS^θ.

Analysis of such data provides valuable insights into the nature of complexes and their formation. Studies have been carried out which compare the interaction between the tetrameric protein streptavidin (which is derived from the bacterium *Streptomyces avidinii*) and biotin (vitamin B7), with the interaction between streptavidin and the dye HABA (2-(4′-hydroxyphenylazo)benzoic acid). This is summarised thus:

$$\text{Streptavidin + biotin} \rightleftharpoons \text{[streptavidin–biotin]} \qquad \text{(I)}$$

$$\text{Streptavidin + HABA} \rightleftharpoons \text{[streptavidin–HABA]} \qquad \text{(II)}$$

The thermodynamic parameters that resulted from the experiments involving these two complexation reactions are shown in Table 5.1.

Table 5.1   **Thermodynamic parameters from streptavidin complexation reactions**

	ΔH^θ kJ mol^{-1}	ΔS^θ JK^{-1}mol^{-1}	ΔG^θ kJ mol^{-1}	K_D
Biotin	-133.9	-192	-76.6	3.7×10^{-14}
HABA	$+7.1$	$+98$	-22.1	1.3×10^{-4}

The free energy change associated with biotin binding is determined by enthalpy terms, while that associated with HABA binding is dominated by entropy. There are a number of conclusions that can be derived from analysis of the data:

- The large negative ΔH associated with the biotin complex formation is consistent with the formation of (relatively strong) hydrogen bonds.

- A positive ΔH (as seen in the HABA complexation) or even a small negative value would be consistent with the formation of van der Waals interactions or expulsion of structured waters from the binding site *(therefore removing H bonds between H_2O and a biomolecule and also those between H_2O molecules)*.

- A negative ΔS is suggestive of the creation of a more ordered system.

- A positive ΔS is suggestive of the expulsion of structured water molecules from the binding site.

If only a macromolecule and its ligand are being considered, then the entropy of the system should decrease (going from a disordered state with two individual molecules to one more ordered molecule of a complex). This does not, however, take account of any displacement of water molecules that are bound to the macromolecule surface and the consequence that in releasing them to the bulk water, complex formation can result in an increase of the overall entropy.

5.6.4 Haemoglobin–oxygen

One of the many examples of equilibria in biological systems (already alluded to at the start of Section 5.6.3), is the uptake and transport of oxygen in the blood, where it is bound to the protein haemoglobin, Hb:

$$Hb + O_2 \rightleftharpoons Hb \cdot O_2$$

the equilibrium constant for which is:

$$K_A = \frac{[Hb \cdot O_2]}{[Hb][O_2]}$$

where:

- $[Hb.O_2]$ = the concentration of the haemoglobin–oxygen complex;
- $[Hb]$ = the concentration of free haemoglobin;
- $[O_2]$ = concentration of oxygen.

At a given temperature, K is constant, so if the concentration of oxygen is reduced (as happens, for example, when a climber is on a high mountain), the concentration of $Hb \cdot O_2$ must be reduced in proportion, and there is less oxygen in the bloodstream. One way to compensate for this effect is to raise the concentration of Hb, which the body does naturally to a limited extent, and the concentration of $Hb \cdot O_2$ can then rise again by the same proportion.

This acclimatisation is not always sufficient for the body to function properly, and bottled oxygen may then be needed. The key issue here is that K remains constant, so any factor that alters one of the terms on the top of the expression, must also alter one of the terms on the bottom by the same proportion. *This behaviour is an example of Le Chatelier's principle.*

5.7   Revisiting coupled reactions

In the previous chapter on energetics (Section 4.4.2) the rationale behind how the regeneration of ATP from ADP was made feasible by coupling it with the process of breaking down glucose (the glycolysis pathway) was described. In Section 5.6.1, it was demonstrated that the reaction of phosphate with glucose to produce glucose-6-phosphate was non-spontaneous, with a standard free energy change of $+3.3$ kJ mol^{-1}. The equilibrium constant, K, for this reaction was 2.78×10^{-1}.

This addition of a phosphate group to the 6 position of glucose is, however, the first step in the glycolysis pathway, so it is known that it 'goes', but what is the driving force? The driving force comes from the coupling of this reaction with the process of hydrolysing ATP to ADP which (as shown in Section 4.4.2) is spontaneous, with a standard free energy change of -31 kJ mol^{-1}.

Putting the reactions together, therefore, gives:

$$\text{(I)}$$

$$\text{ATP} + \text{H}_2\text{O} \longrightarrow \text{ADP} + \text{HPO}_4{}^{2-}$$

$$\text{(II)}$$

$$\text{(III)}$$

For reaction I, $\Delta G^{\ominus}{}_{(I)} = +3.3$ kJ mol^{-1}, so $K_{(I)} = 0.278$.
For reaction II, $\Delta G^{\ominus}{}_{(II)} = -31$ kJ mol^{-1}, so $K_{(II)} = 1.67 \times 10^5$.
For reaction III, $\Delta G^{\ominus}{}_{(III)} = -27.7$ kJ mol^{-1}, so $K_{(III)} = 4.64 \times 10^4$.

Reaction (III) = reaction (I) + reaction (II)

As $\Delta G^{\ominus}$ is a state function, Hess's law applies.

$$\Delta G^{\ominus}{}_{(III)} = \Delta G^{\ominus}{}_{(II)} + \Delta G^{\ominus}{}_{(I)} = -31 + 3.3 = -27.7 \text{ kJ mol}^{-1}$$

Recalling the relationships $\Delta G^{\ominus}{}_{(II)} = -RT \ln K_{(II)}$ and $\Delta G^{\ominus}{}_{(I)} = -RT \ln K_{(I)}$. This leads to:

$$\Delta G^{\ominus}{}_{(III)} = -RT \ln K_{(II)} + (-RT \ln K_{(I)})$$

$$= -RT(\ln K_{(II)} + \ln K_{(I)})$$

163

$$= -RT \ln (K_{(II)} \times K_{(I)})$$

because: $$\Delta G^{\oplus}_{(III)} = -RT \ln K_{(III)}$$

this leads to: $$-RT \ln K_{(III)} = -RT \ln (K_{(II)} \times K_{(I)})$$

and so: $$K_{(III)} = K_{(II)} \times K_{(I)}$$

Putting the numbers in: $$K_{(III)} = 1.67 \times 10^5 \times 0.278$$

$$= 46\,426\ (= 4.64 \times 10^4)$$

Self test

Use value of $\Delta G^{\oplus}_{(III)}$ to calculate $K_{(III)}$. The answer will be very slightly different because of rounding approximations.

Consequences

When equilibrium reactions are combined, their equilibrium constants are multiplied together to calculate the equilibrium constant for the combined process. This means that it is important to recognise that the way a reaction is written will affect the value of K. For example, consider the different ways in which the production of ammonia can be represented:

$$N_2 + 3H_2 \rightleftharpoons 2NH_3$$

for which the equilibrium constant is called K_1, or

$${}^1/_2\,N_2 + {}^3/_2\,H_2 \rightleftharpoons NH_3$$

for which the equilibrium constant is called K_2.

For the first reaction $$\Delta G^{\theta}_1 = -RT \ln K_1$$

For the second reaction $$\Delta G^{\theta}_2 = -RT \ln K_2$$

But, because the amount of material (the number of moles) involved in reaction 1 is twice that involved in reaction 2, then the energy changes involved in reaction 1 will be twice those involved in reaction 2. That is:

$$\Delta G^{\theta}_1 = 2 \times \Delta G^{\theta}_2$$

Therefore: $$-RT \ln K_1 = 2 \times -RT \ln K_2$$

So, dividing both sides by $-RT$,

$$\ln K_1 = 2 \times \ln K_2$$

The rules of logarithms mean this can be rearranged to:

$$\ln K_1 = \ln (K_2)^2$$

meaning that: $$K_1 = K_2{}^2\,(=K_2 \times K_2)$$

The value of the equilibrium constant depends on how the reaction is written in terms of the numbers of moles involved!

Box 5.7

The relationship between ΔG^{θ} and K at 25°C		
Value of ΔG^{θ} *(kJ mol^{-1})*	K	*Physical significance*
500	3×10^{-88}	For all practical purposes reactions with these values do not proceed in the forward reaction. Only reactants present at 'equilibrium'
100	3×10^{-18}	
40	10^{-7}	Both reactants and products have measurable presence at equilibrium; both forward and reverse reactions occur to significant extent
0	1	
-40	10^{7}	
-100	3×10^{17}	For all practical purposes reactions with these values proceed to completion. Only products present at 'equilibrium'
-500	4×10^{87}	

Questions

5.1 For the equilibrium   A (g) $\rightleftharpoons$ B (g) + C (g), the value of K_c at 298 K is 2.0. For each case below, state what will happen to the initial concentration of A in a reaction vessel of volume 1 L if initially:

(a) 1 mole of each of A, B and C are added to the reaction vessel,
(b) 2 moles of each of A, B and C are added to the reaction vessel,
(c) 3 moles of each of A, B and C are added to the reaction vessel.

5.2 Consider the equilibrium

$$H_2(g) + I_2(g) \rightleftharpoons 2HI(g) \quad \Delta H = -12.6 \,\text{kJ mol}^{-1}.$$

What happens to the equilibrium concentration of H_2 if:

(a) more I_2 is added to the container?
(b) more H_2 is added to the container?
(c) more HI is added to the container?
(d) the temperature is raised?
(e) a catalyst is added?

5.3 Write out the expression for the equilibrium constant of the reaction:

$$\text{L-glutamate} + \text{pyruvate} \rightleftharpoons \alpha\text{-ketoglutarate} + \text{L-alanine}$$

The reaction is catalysed by the enzyme L-glutamate-pyruvate transferase with $K_c = 1.11$.

The reaction components are placed in a vessel with this enzyme in the following concentrations: L-glutamate] $= 3.2 \times 10^{-4}$ M, [pyruvate] $= 1.5 \times 10^{-4}$ M, [α-ketoglutarate $= 9 \times 10^{-3}$ M, [L-alanine] $= 8 \times 10^{-3}$ M. Will the reaction proceed towards products or reactants?

5.4 The enzyme phosphoglycerate mutase catalyses the following reaction:

$$\text{2-phosphoglycerate} \rightleftharpoons \text{3-phosphoglycerate}$$

The equilibrium constant for this reaction is 5.8 at 298 K and 5.45 at 310 K.

(a) Is the process, as written, product or reactant favoured?
(b) From the values of the equilibrium constants at the two temperatures, is the process likely to be exothermic or endothermic? State your reasoning.
(c) Calculate the values of ΔG^θ for each temperature.
(d) Calculate ΔH^θ and ΔS^θ for this reaction. (You may assume that ΔH^θ and ΔS^θ are independent of temperature over the range 298 to 310 K.)
(e) What effect does the presence of the enzyme have on the equilibrium?

5.5 The hydrolysis of the disaccharide, sucrose, into its constituent monosaccharides, glucose and fructose, is shown below:

$$\text{Sucrose} + \text{H}_2\text{O} \rightleftharpoons \text{glucose} + \text{fructose}$$

At 25 °C, the standard free energy change $\Delta G^\theta = -29.3$ kJ mol^{-1}. Use this information and, as appropriate, the standard enthalpies of formation listed below, to work out the following:

(a) The enthalpy change for the reaction.
(b) The entropy change for this reaction.
(c) The equilibrium constant for the reaction.

Is this reaction driven by the change of enthalpy or entropy? Explain your reasoning.

Use the following standard enthalpies of formation (values in kJ mol^{-1}).

Sucrose: −2222

Glucose: −1268

Fructose: −1266

H$_2$O: −285.8

5.6 A protein exists in one of either two conformations, folded (F) or unfolded (U), with the transitions between the two conformations taking place reversibly according to the equilibrium,

$$F \rightleftharpoons U$$

Spectroscopic measurements show that the fraction of protein in the *unfolded* conformation, x_U, and the equilibrium constant, K, varied with temperature as shown in the table:

$T\,(^oC)$	x_U	K
63	0.21	0.27
74	0.78	3.55

Using these data,

(a) Write down an equation for the equilibrium constant, K, in terms of x_U, and show how the values quoted in the table are consistent with this equation.
(b) For both of the temperatures, calculate the standard Gibbs free energy change, ΔG^θ, accompanying the transition from folded to unfolded conformations.
(c) Calculate the standard entropy change, ΔS^θ, and the standard enthalpy change, ΔH^θ, accompanying the transition between folded and unfolded conformations. Assume that ΔS^θ and ΔH^θ do not change in this temperature range.
(d) Comment on these results.

6 Aqueous equilibria

Learning outcomes

By the end of this chapter students should be able to:

- define the terms acid and base;
- understand how the principles of chemical equilibrium explain the properties of acids and bases;
- know the meaning of the terms K_W, K_a and K_b and be able to relate them to one another, and hence to define the terms pK_W, pK_a and pK_b;
- define pH, understand the pH scale, and calculate pH values for both strong and weak acid and base solutions;
- understand the meaning of the term *buffer solution*;
- state the Henderson–Hasselbalch relationship, and demonstrate how it is used in preparing buffer solutions;
- describe solubility as an equilibrium system.

6.1 Introduction

6.1.1 Why is this important in biology?

Water is essential to life, forming about 80% of the mass of cells. In living organisms many substances exist in solution, with water being the solvent in which they are dissolved. Additionally, many biological processes are very dependent on pH for their functioning, and in order to understand fully what the term pH means, it is necessary to understand something of the behaviour of water itself.

6.1.2 The importance of pH and pH control

The pH of solutions affects our everyday lives in many ways. Below are a few:

- Physiological
  - Drug delivery – many medications, prescribed or not, function badly (or not at all!) in acid solutions. The digestive fluids in the stomach are very acidic and so will preclude some medications from being delivered orally. Alternative delivery mechanisms are often necessary (pharmacology).

- Blood is almost neutral, at pH 7.4. A deviation of 1 pH unit would result in death!
- Enzyme function is very pH dependent.
- Ecological
  - Acid rain – arises from assimilation of fossil fuel emissions in rainwater, it enters lakes and streams, with detrimental effects on native fish stocks.
  - Agriculture/horticulture – plants grow best in soil of appropriate pH, so there is often a need to maintain suitable pH for specific crops via fertilisation.
  - Water purification in sewage works must be carried out at optimum pH.

The principal aims of this chapter will then be:

- to introduce the idea of self-ionisation of water;
- to define the terms acid and base, emphasising the biological context;
- to show how the principles of chemical equilibrium are used to understand acids and bases;
- to define the terms pH, pK_a and pK_b;
- to illustrate the importance of pH control in a biological context;
- to define the term buffer solutions;
- to illustrate how buffer solutions can be prepared.

6.2   Self-ionisation of water

Water is generally regarded as molecular H_2O, and to a large extent that is an accurate picture. Water does, nevertheless, ionise to a *very* small extent and this can be represented by the equilibrium:

$$H_2O + H_2O \quad \rightleftharpoons \quad H_3O^+(aq) + OH^-(aq)$$

The equilibrium constant for this reaction is represented by:

$$K = \frac{[H_3O^+][OH^-]}{[H_2O]^2}$$

The concentration of water, $[H_2O]$, is regarded as a constant, and so the equilibrium is usually shown as:

$$K_w = [H_3O^+][OH^-] = 1.0 \times 10^{-14} \tag{6.1}$$

where K_w is referred to as the self-ionisation constant for water, and in the above case is quoted for 298 K. At 1.0×10^{-14} the value of K_w is very small, meaning that the equilibrium is *very* much to the left.

Given that the stoichiometry of the ionisation produces one mole of H_3O^+ for every one mole of OH^- produced, then in this equilibrium mixture:

$$[H_3O^+] = [OH^-]$$

Therefore, in water,

$$[H_3O^+] = \sqrt{10^{-14}} = 10^{-7}\,mol\,L^{-1} \qquad [6.2a]$$

and hence

$$[OH^-] = \sqrt{10^{-14}} = 10^{-7}\,mol\,L^{-1} \qquad [6.2b]$$

These are important and will be referred to later.

6.3　Acids and bases

6.3.1　What do the terms acid and base mean?

Historically one of the fundamentally most important definitions of the terms acid and base was proposed by the Swedish scientist Svante Arrhenius. The Arrhenius definitions are:

- An acid is a material that dissolves in water to release a proton or hydrogen ion (H^+).
- A base is a material that dissolves in water to release a hydroxide ion (OH^-).

For example, hydrogen chloride is an acid, and in solution in water it ionises to become hydrogen ions (really H_3O^+ – in aqueous solutions H^+ does not exist as such, but is 'hydrated' to form H_3O^+) and chloride ions. This is summarised in the equation:

$$HCl + H_2O \quad \rightarrow \quad H_3O^+\,(aq) + Cl^-\,(aq)$$

Sodium hydroxide is a base, which dissociates when dissolved in water to produce hydroxide ions, as shown:

$$NaOH\,(aq) \quad \rightarrow \quad Na^+\,(aq) + OH^-\,(aq)$$

There are, however, problems associated with the Arrhenius approach outlined above, namely that it is restricted to just aqueous solutions. Hydrogen chloride, however, was shown to display acid behaviour in ammonia:

$$HCl + NH_3 \quad \rightarrow \quad NH_4^+\,(aq) + Cl^-\,(aq)$$

This illustrated there was a need for a more broadly based definition, one that incorporates Arrhenius's and expands on it. In 1923 the scientists Johannes Nicolaus Brønsted and Martin Lowry simultaneously produced proposals which developed the ideas of acid–base behaviour to account for acid–base behaviour in non-aqueous situations, and their proposals are now usually referred to as the Brønsted–Lowry theory.

The Brønsted–Lowry definitions are summarised very briefly as follows:

- An acid is a proton donor.
- A base is a proton acceptor. (In water a base will release OH^-.)

These definitions can be represented in chemical equilibrium format as shown.

$$\text{For an acid:} \quad HA + H_2O \quad \rightleftharpoons \quad H_3O^+ + A^-$$
$$\text{For a base:} \quad B + H_2O \quad \rightleftharpoons \quad BH^+ + OH^-$$

The new definition for bases means that compounds such as ammonia, which in itself does not contain any OH^- groups, can act as a base, thus:

$$NH_3 + H_2O \quad \rightleftharpoons \quad NH_4^+ + OH^-$$

For biological systems this is the definition that will be used. (A general representation of an acid is HA; a general representation of a base is B.)

6.3.2 Properties of acids

- Acids release a hydrogen ion into water (aqueous) solution.
- An acid and a base combine to make a salt and water (neutralisation). A salt is any ionic compound comprising the anion of an acid and the cation of a base.
- Acids corrode active metals. Usually when an acid reacts with a metal, it produces a compound with the cation of the metal and the anion of the acid and hydrogen gas.
- Acids turn blue litmus to red. Litmus is the oldest known pH indicator (several others exist). It is red in acid and blue in base.
- Acids taste sour. Stomach acid is hydrochloric acid. Acetic acid is the acid ingredient in vinegar. Citrus fruits such as lemons, grapefruit, oranges and limes have citric acid in the juice. Sour milk, sour cream, yogurt and cottage cheese have lactic acid from the fermentation of the sugar lactose.

6.3.3 Properties of bases

- Bases release a hydroxide ion into water solution. (Or, in the Brønsted–Lowry model, cause a hydroxide ion to be released into water solution by accepting a hydrogen ion in water – see the example of ammonia above.)
- Bases neutralise acids in a neutralisation reaction.
- Bases denature proteins. This accounts for the 'slippery' feeling on hands when exposed to a base. Strong bases that dissolve in water well, such as sodium or potassium lye, are very dangerous because a great amount of the structural material of human beings is made of protein. Serious damage to flesh can be avoided by careful use of strong bases.
- Bases turn red litmus to blue.

6.3.4 Strong acids and strong bases

Acids

The common acids that are almost 100% ionised, called strong acids, are:

HNO_3 – nitric acid

HCl – hydrochloric acid

H_2SO_4 – sulfuric acid

$HClO_4$ – perchloric acid

HBr – hydrobromic acid

HI – hydroiodic acid

What does this mean?

It means that the general equation

$$HA + H_2O \quad \rightleftharpoons \quad H_3O^+ + A^-$$

goes completely to the right, suggesting essentially full ionisation, and can realistically be represented thus:

$$HA + H_2O \quad \rightarrow \quad H_3O^+ (aq) + A^- (aq)$$

Other acids are incompletely ionised, existing mostly as the un-ionised form. Incompletely ionised acids are called weak acids. Most organic acids (the carboxylic acids) are weak acids. *Weak acids will be described later.*

Sulfuric acid is the only one in the list of strong acids above that is diprotic. This term means that it has two ionisable hydrogens per formula (or two moles of ionisable hydrogen per mole of acid). Sulfuric acid ionises in two steps.

The *first* time a hydrogen ion splits off of the sulfuric acid, it acts like a *strong* acid. The *second* time a hydrogen splits away from the sulfate ion, it acts like a *weak* acid. Thus:

Ionisation step 1:

$$H_2SO_4 + H_2O \quad \rightarrow \quad H_3O^+ + HSO_4^- \qquad \text{loss of one } H^+$$

Ionisation step 2:

$$HSO_4^- + H_2O \quad \rightleftharpoons \quad H_3O^+ + SO_4^{2-} \qquad \text{loss of second } H^+$$

The other acids in the list are called monoprotic, having only one ionisable proton per formula. Thus, in the general case:

$$HA + H_2O \quad \rightarrow \quad H_3O^+ + A^-$$

Phosphoric acid, H_3PO_4, is a weak acid. Phosphoric acid has three hydrogen ions available to ionise and lose as a proton, and so phosphoric acid is triprotic.

Any acid with two or more ionisable hydrogens is described as *polyprotic*.

Bases

Below is a short list of strong bases, ones that completely ionise into hydroxide ions and a conjugate acid. All of the bases of group I and group II metals except for beryllium are strong bases. The bases of group II metals, magnesium, calcium, barium and strontium, are strong, but all of these bases have somewhat limited solubility. Magnesium hydroxide has a particularly low solubility. Potassium and sodium hydroxides are both commonly available in laboratories.

LiOH – lithium hydroxide

NaOH – sodium hydroxide

KOH – potassium hydroxide

RbOH – rubidium hydroxide

CsOH – caesium hydroxide

$Mg(OH)_2$ – magnesium hydroxide

$Ca(OH)_2$ – calcium hydroxide

$Sr(OH)_2$ – strontium hydroxide

$Ba(OH)_2$ – barium hydroxide

The bases of group I metals are called monobasic, whereas those of group II metals are called dibasic, referring to the number of moles of OH^- groups released per mole of base dissolved.

$$\text{Monobasic:} \quad MOH\ (aq) \quad \rightarrow \quad M^+\ (aq) + OH^-\ (aq)$$

$$\text{Dibasic:} \quad M(OH)_2\ (aq) \quad \rightarrow \quad M^{2+}\ (aq) + 2OH^-\ (aq)$$

Many of the basic organic compounds (and some inorganic materials) have an amino group ($-NH_2$) rather than an ionisable hydroxyl group. The amino group attracts a proton (hydrogen ion) to become ($-NH_3^+$). By the Brønsted–Lowry definition, an amino group definitely acts as a base, and the effect of removing hydrogen ions from water molecules is the same as adding hydroxide ions to the solution. Therefore:

$$RNH_2 + H_2O \quad \rightleftharpoons \quad RNH_3^+ + OH^-$$

6.4 Acid–base equilibria

As intimated earlier, the behaviour of weak acids and bases is described in terms of equilibria, and it is worth looking at that in a bit more detail.

6.4.1 Behaviour of weak acids

Weak acids are not completely dissociated in aqueous solution, and in a general sense the process being described is summarised thus:

$$HA + H_2O \quad \rightleftharpoons \quad H_3O^+ + A^-$$

HA is referred to as a weak Brønsted acid, with the equilibrium lying to the left hand side, meaning that the acid remains largely undissociated.

(Remember, an equilibrium is a dynamic process, and in the case shown above, the rate of dissociation of the acid HA, is matched by the rate of re-protonation of the A^- ion.)

Recalling that for any equilibrium it is possible to define an equilibrium constant, then for this equilibrium the equilibrium constant K is given by:

$$K = \frac{[H_3O^+][A^-]}{[HA][H_2O]}$$

In this case, because the reaction is in aqueous solution, the concentration of water is regarded as a constant (technically the activity is 1), and so the equilibrium is usually shown as:

$$K_a = \frac{[H_3O^+][A^-]}{[HA]} \tag{6.3}$$

in which K_a is referred to as the *acid dissociation constant*.

Consider now the specific case of a carboxylic acid (carboxylic acids are weak acids), where R = CH$_3$; that is, ethanoic (acetic) acid:

$$CH_3CO_2H + H_2O \;\rightleftharpoons\; H_3O^+ + CH_3CO_2^-$$

Ethanoic acid is a weak Brønsted acid, and as such the equilibrium lies to the left hand side.

The acid dissociation constant for ethanoic acid would be given by:

$$K_a = \frac{[H_3O^+][CH_3CO_2^-]}{[CH_3CO_2H]}$$

where $K_a = 1.8 \times 10^{-5}$.

Values of K_a, particularly for the weaker acids, are usually very small numbers and it is more common to use a more simplified number to report the degree of dissociation of weak acids. A very useful way of reporting this constant is as the pK_a value.

The pK_a value and its relationship with K_a

The pK_a is defined as the negative logarithm (to the base 10) of the equilibrium constant K_a (a detailed look at the mathematics behind logarithms is given in Appendix 1):

$$pK_a = -\log K_a \qquad [6.4]$$

If the value of pK_a is known, then this equation can be rearranged to find K_a, thus:

$$K_a = 10^{-pK_a}$$

(These conversions can be easily done with calculators.)

What does all of this mean?

Remembering that the dissociation of a weak acid is represented by the equation:

$$HA + H_2O \;\rightleftharpoons\; H_3O^+ + A^-$$

From this the K_a value is given by:

$$K_a = \frac{[H_3O^+][A^-]}{[HA]}$$

Then combining what has been seen about the relationship between K_a and pK_a, it can be concluded that:

- Smaller K_a values mean weaker acids (they do not dissociate to the same extent as stronger acids).
- Smaller K_a values give larger pK_a values, so larger pK_a values mean weaker acids.

Example 6.1 The K_a value of ethanoic acid is 1.8×10^{-5}. Calculate the pK_a.

Use p$K_a = -\log K_a$, so:

$$pK_a = -\log K_a$$
$$= -\log (1.8 \times 10^{-5})$$

The value for this can be found by using a calculator, and it is found that

$$pK_a = 4.745 \,(\text{often reported as } 4.75)$$

Example 6.2 The pK_a value for benzoic acid (also called formic acid, found in ant bites) is 4.20. Calculate the value of K_a.

$$pK_a = -\log K_a$$

so

$$K_a = 10^{-pK_a}$$

Putting in the numbers,

$$K_a = 10^{-4.20}$$
$$= 6.31 \times 10^{-5}$$

6.4.2 Behaviour of weak bases

In an aqueous solution of a weak base, B, the proton transfer from water to base is not complete, and there an equilibrium is set up:

$$B\ (aq) + H_2O \rightleftharpoons BH^+\ (aq) + OH^-\ (aq)$$

Again, as this is an example of a chemical equilibrium, then there will be an equilibrium constant, as indicated:

$$K = \frac{[BH^+][OH^-]}{[H_2O][B]}$$

Because the reaction is in aqueous solution, the concentration of water is regarded as a constant, and so the equilibrium is usually shown as:

$$K_b = \frac{[BH^+][OH^-]}{[B]}$$

in which K_b is sometimes referred to as the *base dissociation constant*.

As with the acid dissociation constant K_a, there is a common way of reporting K_b, which is as the pK_b value, and again this is a logarithmic relationship, where the pK_b is defined as the negative logarithm (to the base 10) of the equilibrium constant K_b.

$$pK_b = -\log K_b \qquad\qquad [6.5]$$

If the value of pK_b is known, then this equation can be rearranged to find K_b, thus:

$$K_b = 10^{-pK_b}$$

What does all of this mean?

- Smaller K_b values mean weaker bases (not as effective at de-protonating the water).
- Smaller K_b values give larger pK_b values, so larger pK_b values mean weaker bases.

6.5 Dissociation of acids and bases: conjugate acids and bases

Recalling that the equilibrium for an acid dissociation in water is represented thus:

$$HA + H_2O \rightleftharpoons H_3O^+ + A^-$$

175

then HA is acting as an acid (a proton donor) and water is acting as a base (a proton acceptor).

Looking now at the *reverse reaction*, that is:

$$H_3O^+ + A^- \ \rightleftharpoons \ HA + H_2O$$

then H_3O^+ is acting as an acid (a proton donor) and A^- is acting as a base (proton acceptor). In an equilibrium like this:

- HA and A^- are referred to as a *conjugate acid/base pair;*
- H_2O and H_3O^+ are a *base/acid conjugate pair.*

The equilibrium constant for this reaction can be stated:

$$K_a = \frac{[H_3O^+][A^-]}{[HA]}$$

Looking now at the behaviour of the conjugate base of HA, that is A^-, dissolved in water and acting as a base:

$$H_2O + A^- (aq) \ \rightleftharpoons \ HA(aq) + OH^- (aq)$$

The equilibrium constant K_b for this is given by:

$$K_b = \frac{[HA][OH^-]}{[A^-]}$$

Multiplying these values of K_a and K_b together (in other words, multiplying the equilibrium constant for the dissociation of an acid, HA, in water by the equilibrium constant for the protonation of the conjugate base of that acid by water) gives:

$$K_b \times K_a = \frac{[H_3O^+][A^-]}{[HA]} \times \frac{[HA][OH^-]}{[A^-]}$$

$$= [H_3O^+][OH^-]$$

since the [HA] and [A^-] values cancel. (*The principles behind this approach were covered in Section 5.7.*)

Recall equation [6.1]:

$$K_w = [H_3O^+][OH^-]$$

Therefore, for a conjugate acid/base pair

$$K_w = K_a \times K_b$$

Taking logarithms of both sides of this equation gives:

$$\log K_w = \log (K_a \times K_b)$$

and (using the rules of logs given in Appendix 1), this becomes:

$$\log K_w = \log K_a + \log K_b$$

Multiplying both sides of this equation by -1 gives:

$$-\log K_w = -(\log K_a + \log K_b)$$

$$= -\log K_a + (-\log K_b)$$

Now recall the definitions,

$$pK_a = -\log K_a$$

$$pK_b = -\log K_b$$

and introduce another definition along the same lines, such that

$$pK_w = -\log K_w$$

It is known (Section 6.2) that:

$$K_w = 1.0 \times 10^{-14}$$

so it logically follows that

$$pK_w = 14$$

and then

$$pK_w = pK_a + pK_b = 14 \qquad\qquad [6.6]$$

The *significance* of this is that if the pK_a for an acid in water is known, then it is possible to work out the pK_b for its conjugate base pair, as they always sum to 14.

Example 6.3 The dissociation of ethanoic acid in water is given by the equilibrium:

$$CH_3CO_2H \quad + \quad H_2O \quad \rightleftharpoons \quad H_3O^+ \quad + \quad CH_3CO_2^-$$

Ethanoic acid	Ethanoate
Acid	Conjugate base

For this equilibrium the pK_a value is 4.75. Therefore, for the ethanoate anion, which is the conjugate base for this acid,

$$pK_b = 14.00 - 4.75 = 9.25$$

Example 6.4 Ammonia behaves as a base, as represented in the equilibrium:

$$NH_3 \quad + \quad H_2O \quad \rightleftharpoons \quad NH_4^+ \quad + \quad OH^-$$

Ammonia	Ammonium
Base	Conjugate acid

For this equilibrium the $pK_b = 4.76$. Therefore, for the ammonium ion, NH_4^+, which is the conjugate acid of this base,

$$pK_a = 14.00 - 4.76 = 9.24$$

6.6 Acids and bases in aqueous solution: the concept of pH

6.6.1 Definition

Many advertisements for skin care products refer to them as being 'pH balanced'. This infers that the product is specially designed to minimise skin damage. But what do they mean by pH?

Recall the definition of an acid, which states that an acid is a proton donor. It is important, therefore, to have a way of referring to how much H_3O^+ a given solution of an acid contains. To this end the strength of a solution of an acid is measured by the concentration of hydrogen ions it releases into the solution, represented by $[H_3O^+]$.

Concentration values are not always easy numbers to handle involving, as they frequently do, powers of ten, so a scale of measurement based on logarithms is used, and this is *the pH scale*.

The definition of pH is:

$$pH = -\log_{10}\left[H_3O^+\right] \qquad\qquad [6.7]$$

Key features about the pH scale

- The pH scale is in easy numbers – a lot of lab work is done in the range from 0 to 14 (*but the scale can have pH > 14 or pH < 0*).

- pH does not have units.

- Because pH is the negative logarithm of the concentration of H_3O^+, then the lower the pH value, the more acidic the system (i.e. the greater the concentration of H_3O^+).

- If the pH is reduced by 1 unit, the concentration of H_3O^+ is increased by 10 times.

Using the pH scale

Given that there is a mathematical relationship between $[H_3O^+]$ and pH, then it is important to be able to follow the principles of how to interconvert them. This is illustrated by the following examples.

Example 6.5 An acid solution has a pH of 2. Calculate the hydrogen ion concentration, $[H_3O^+]$. Remembering the general rule of logarithms, that if

$$y = a^x \;(y \text{ equals } a \text{ to the power of } x)$$

then

$$x = \log_a y \;(x \text{ equals the log to the base } a \text{ of } y)$$

In the case of this question, the key factor is to remember that $pH = -\log_{10}[H_3O^+]$.

$$pH \quad = \quad -\log_{10}\left[H_3O^+\right]$$

$$x \qquad\qquad \log_a \quad y$$

Therefore rewriting the equation $x = \log_a y$, using the values from the question thus:

- $x = 2$ (the pH)
- $a = 10$ (pH values are derived from logarithms to the base 10)
- $y = [H_3O^+]$

This leads to:

$$2 = -\log_{10}[H_3O^+]$$

which is the same as:

$$-2 = \log_{10}[H_3O^+]$$

Therefore,

$$[H_3O^+] = 10^{-2}\,\text{mol}\,L^{-1}.$$

Example 6.6 An acid solution has a pH of 8.50. Calculate the hydrogen ion concentration $[H_3O^+]$.

Solving this question requires the same approach as that of Example 6.5. Therefore, as the definition of pH is:

$$pH = -\log_{10}[H_3O^+]$$

and this can be rearranged into the form

$$\left[H_3O^+\right] = 10^{-pH}$$

Then as the pH $= 8.50$, then

$$\left[H_3O^+\right] = 10^{-8.50}\,\text{mol}\,L^{-1}$$

This is fine, but numbers raised to fractional indices are not easy to visualise. It is easy, using a calculator, to convert $10^{-8.50}$ to a decimal number, and this gives:

$$10^{-8.50} = 3.16 \times 10^{-9}$$

Therefore,

$$\left[H_3O^+\right] = 3.16 \times 10^{-9}\,\text{mol}\,L^{-1}.$$

Example 6.7 A solution has a hydrogen ion concentration of $2.3 \times 10^{-7}\,\text{mol}\,L^{-1}$. Calculate its pH.

$$pH = -\log[H_3O^+] = -\log(2.3 \times 10^{-7})$$

so

$$pH = 6.625$$

6.6.2 What happens when acids are dissolved in water?

Because, in biological systems, any acids are found in water, then it is important to consider what happens when an acid is dissolved in water. Firstly it is important to remember the key features of pure water itself.

Pure water undergoes self-ionisation:

$$H_2O + H_2O \rightleftharpoons H_3O^+\,(aq) + OH^-\,(aq)$$

For this process at 25 °C (298 K) the equilibrium constant could be represented thus:

$$K_w = [H_3O^+][OH^-] = 1.0 \times 10^{-14}$$

179

K_w is referred to as the self-ionisation constant for water.

In this equilibrium mixture:

$$[H_3O^+] = [OH^-]$$

Therefore, in pure water,

$$[H_3O^+] = \sqrt{10^{-14}} = 10^{-7}\,\text{mol L}^{-1}$$

and similarly

$$[OH^-] = 10^{-7}\,\text{mol L}^{-1}$$

This means that in water, which is referred to as neutral because $[H_3O^+] = [OH^-]$,

$$pH = -\log[H_3O^+] = -\log 10^{-7} = 7$$

So, *neutral* is defined as *pH 7*.

6.6.3 What happens when the water equilibrium is disturbed?

This involves applying Le Chatelier's principle. In the context of the water equilibrium, remember that $K_w = [H_3O^+][OH^-]$. This means that:

- In all cases $[H_3O^+][OH^-] = 1.0 \times 10^{-14}$.
- Addition of an acid to pure water would increase the amount of H_3O^+. In order to concur with Le Chatelier's principle there would need to be a corresponding decrease in the amount of OH^-. *This leads to pH usually in the range 0–7.*
- Adding a base to pure water would decrease the amount of H_3O^+, but would increase the amount of OH^-. *This leads to pH usually in the range 7–14.*
- If equal amounts of acid and base are present, $[H_3O^+] = [OH^-]$. Therefore pH = 7.

Because $-\log_{10}[H_3O^+] = pH$, then similarly

$$-\log_{10}[OH^-] = pOH \qquad [6.8]$$

As

$$[H_3O^+][OH^-] = 1.0 \times 10^{-14}$$

so

$$\log_{10}([H_3O^+][OH^-]) = -14$$

Following the rules of logarithms, which state that $\log_{10}(A \times B) = \log_{10} A + \log_{10} B$, then it is clear that:

$$\log_{10}[H_3O^+] + \log_{10}[OH^-] = -14$$

Recalling the definitions of pH and pOH, then if both sides of the equation are multiplied by -1:

$$-\log_{10}[H_3O^+] + (-\log_{10}[OH^-]) = 14$$

In other words:

$$pH + pOH = 14 \qquad [6.9]$$

Example 6.8 An aqueous solution has a pH of 5. Calculate the pOH and then the $[OH^-]$.

$$pH + pOH = 14$$

and hence

$$pOH = 14 - 5 = 9$$

Therefore

$$-\log_{10}[OH^-] = 9$$

which is the same as

$$\log_{10}[OH^-] = -9$$

so

$$[OH^-] = 10^{-9}\,mol\,L^{-1}$$

Box 6.1

What about different temperatures?

The formation of H_3O^+ and OH^- ions from water is an endothermic process. In other words the forward reaction absorbs heat. Recalling Le Chatelier's principle, if a change is made to the conditions of an equilibrium process, the position of equilibrium moves to counter the change made. For an endothermic reaction this means that adding heat favours the forward reaction (as shown in Section 5.5.2). Therefore, at higher temperatures K_W will have higher values.

Take, for example, a temperature of $310\,K$

At $310\,K$, $K_w = [H_3O^+][OH^-] = 2.5 \times 10^{-14}$. As $[H_3O^+] = [OH^-]$, so

$$[H_3O^+] = \sqrt{(2.5 \times 10^{-14})}$$
$$= 1.58 \times 10^{-7}$$

This corresponds to pH = 6.8. This is still neutral, however, because $[H_3O^+] = [OH^-]$.

This means pH neutrality is temperature-dependent.

6.6.4 Calculating pH values for acids

The pH is a measure of the amount of H_3O^+ in a solution, from the definition:

$$pH = -\log_{10}[H_3O^+]$$

For a strong acid: if, for example, 0.30 moles of a strong acid (such as HCl) is dissolved in water, and the solution made up to a volume of 1 L, then full dissociation is expected:

$$HA + H_2O \rightarrow H_3O^+ + A^-$$

Thus the pH of the resulting solution would be:

$$pH = -\log_{10} 0.30 = 0.523$$

But, what if the acid dissolved is a weak acid? To illustrate this, consider the pH of a 0.30-mol L^{-1} solution of ethanoic acid, CH_3CO_2H, at 298 K.

$$CH_3CO_2H \ (l) + H_2O \ (l) \quad \rightleftharpoons \quad H_3O^+ \ (aq) + CH_3CO_2^- \ (aq)$$

The K_a for ethanoic acid at 298 K is 1.8×10^{-5}. For this, the equilibrium (acid dissociation) constant is:

$$K_a = \frac{[H_3O^+][CH_3CO_2^-]}{[CH_3CO_2H]}$$

The dissociation of ethanoic acid produces 1 mole of ethanoate, $CH_3CO_2^-$, for every one mole of H_3O^+. Therefore the concentrations $[H_3O^+]$ and $[CH_3CO_2^-]$ at equilibrium are equal, and will be given the value c.

The value of K_a is a very small number, so the values of $[H_3O^+]$ and $[CH_3CO_2^-]$, namely c, are very small in relation to the starting concentration of acid, $[CH_3CO_2H]$, which was 0.30 mol L^{-1}. This means that the equilibrium concentration of acid, which will be $(0.30 - c)$ mol L^{-1} can be approximated to 0.30 mol L^{-1} (see Box 6.2 to see the result if no approximation is made and Box 6.3 for a generalised view of the weak acid case).

Summarising these data in an ICE table (for approach, see Section 5.2.1):

	[CH$_3$CO$_2$H]	[H$_3$O$^+$]	[CH$_3$CO$_2$$^-$]
Initial state	0.30	0	0
Change	$-c$	$+c$	$+c$
At equilibrium	$(0.30 - c) \approx 0.30$	c	c

In order to calculate the value of pH, it will be necessary initially to calculate the value of the hydrogen ion concentration, $[H_3O^+]$. In other words, calculate the value of c. This involves using the expression for K_a,

$$K_a = \frac{[H_3O^+][CH_3CO_2^-]}{[CH_3CO_2H]}$$

The equilibrium values for the concentrations are then introduced:

$$K_a = \frac{c \times c}{0.30} = \frac{c^2}{0.30}$$

and on rearranging this gives:

$$c^2 = 0.30 \times K_a$$

So,

$$c^2 = 0.30 \times (1.8 \times 10^{-5}) = 5.4 \times 10^{-6}$$

and hence

$$c = 2.3 \times 10^{-3}$$

In other words, on dissolving 0.30 moles of acetic acid in 1 L of water, then 2.3×10^{-3} moles of H_3O^+ will be produced.

Therefore,

$$pH = -\log_{10}[H_3O^+] = -\log_{10}(2.3 \times 10^{-3}) = 2.64$$

Box 6.2

What if the approximation $(0.30 - c) \approx 0.30$ had not been made?

Starting with

$$K_a = \frac{[H_3O^+][CH_3CO_2^-]}{[CH_3CO_2H]}$$

Introducing the equilibrium concentration values

$$K_a = \frac{c \times c}{(0.30 - c)} = \frac{c^2}{(0.30 - c)}$$

On rearranging this gives:

$$(0.30 - c) \cdot K_a = c^2$$

Then

$$0.30K_a - cK_a = c^2$$

So

$$c^2 + cK_a - 0.30K_a = 0 \quad \text{(This is a quadratic equation!)}$$

Therefore

$$c = \frac{-K_a \pm \sqrt{K_a^2 - (4 \times 0.30 \times K_a)}}{2}$$

$$= \frac{-(1.8 \times 10^{-5}) \pm \sqrt{(1.8 \times 10^{-5})^2 - (-4 \times 0.30 \times \{1.8 \times 10^{-5}\})}}{2}$$

$$= \frac{-(1.8 \times 10^{-5}) \pm \sqrt{(3.24 \times 10^{-10}) - (-2.16 \times 10^{-5})}}{2}$$

$$= \frac{-(1.8 \times 10^{-5}) \pm 4.65 \times 10^{-3}}{2}$$

$$= \frac{4.63 \times 10^{-3}}{2}$$

$$= 2.31 \times 10^{-3}$$

(Note that there was also a second possible answer to the calculation above, but this would have led to a negative concentration, which is clearly impossible.)

The value of c calculated above is almost the same as that calculated when the approximation was used, and therefore gives rise to the same pH value. This means that the approximation used was valid.

Box 6.3

Generalised approximation for the pH of a weak acid

For the generalised weak acid equilibrium:

$$HA + H_2O \rightleftharpoons H_3O^+ + A^-$$

The acid dissociation constant is given by:

$$K_a = \frac{[H_3O^+][A^-]}{[HA]}$$

For this, $[H_3O^+] = [A^-]$, so it changes to:

$$K_a = \frac{[H_3O^+]^2}{[HA]}$$

which rearranges to:

$$K_a \times [HA] = [H_3O^+]^2$$

This then gives:

$$[H_3O^+] = \sqrt{K_a \times [HA]}$$

which leads to

$$pH = -\log_{10}(\sqrt{K_a \times [HA]})$$

(often shown as $pH = -\log_{10}(K_a^{1/2} \times [HA]^{1/2})$).

6.7 The control of pH: buffer solutions

6.7.1 Background

In nature, and in the laboratory, there is a need to be able to limit the effect of acids and/or bases when they are added to systems of interest. *Buffer solutions* achieve this goal.

A buffer causes solutions to be resistant to changes in pH on addition of small amounts of either a strong acid or a strong base. For a buffer to work it must have an acid component to react with any OH^- added, and it must have a base component to react with any H_3O^+ added. To this end, buffers are usually prepared from conjugate acid–base pairs. That is:

- a weak acid and its conjugate base pair (e.g. ethanoic acid/ethanoate);
- a weak base and its conjugate acid pair (e.g. ammonia/ammonium).

Addition of base to a buffer solution

Consider what would happen in the case where there is a buffer comprising a weak acid and its conjugate base pair, for example ethanoic acid/ethanoate. If base, OH^-, were added to the ethanoic acid/ethanoate buffer, the ethanoic acid would react with the base as shown:

$$CH_3CO_2H\ (aq) + OH^-\ (aq) \quad \rightleftharpoons \quad H_2O + CH_3CO_2^-\ (aq)$$

For this reaction, the equilibrium constant, K, is 1.8×10^9, because OH^- is a much stronger base than the $CH_3CO_2^-$ ion.

Addition of an acid to a buffer

If acid, H_3O^+, were added to the buffer, then the ethanoate would react with the acid as shown:

$$H_3O^+ \ (aq) + CH_3CO_2^- \ (aq) \quad \rightleftharpoons \quad H_2O + CH_3CO_2H \ (aq)$$

In this case, the equilibrium constant, K, is 5.6×10^4, again quite large, because H_3O^+ is a much stronger acid than is CH_3CO_2H.

The above explains, qualitatively, how buffers work. In real systems it is found that buffers tend to have pH ranges over which they are effective, and the reasons for this will now be looked at in more detail, with a view to working out how to make up buffer solutions.

6.7.2. Theoretical aspects of buffers

Recall the general representation of the dissociation of an acid:

$$HA + H_2O \quad \rightleftharpoons \quad H_3O^+ + A^-$$

The acid dissociation constant for this is:

$$K_a = \frac{[H_3O^+][A^-]}{[HA]}$$

Now, this can be easily rearranged, firstly by multiplying both sides of the equation by [HA], to give:

$$K_a[HA] = [H_3O^+][A^-]$$

Then on dividing both sides of the equation by $[A^-]$:

$$\frac{K_a[HA]}{[A^-]} = [H_3O^+]$$

This equation has K_a on one side of it and $[H_3O^+]$ on the other. As described earlier, for numerical simplicity both of these are often represented by their logarithmic equivalents, shown as pK_a and pH respectively. This equation can be made to contain pK_a and pH by taking negative logarithms (to the base 10) of both sides, as shown:

$$-\log_{10}\frac{K_a[HA]}{[A^-]} = -\log_{10}[H_3O^+]$$

This can be rewritten as:

$$-\log_{10}K_a + \left(-\log_{10}\frac{[HA]}{[A^-]}\right) = -\log_{10}[H_3O^+]$$

Putting in the symbols for pK_a and pH this becomes:

$$pK_a - \log_{10}\frac{[HA]}{[A^-]} = pH$$

This is usually shown in books the other way round, as:

$$pH = pK_a - \log_{10}\frac{[acid]}{[base]}$$

Note: A$^-$, which is the conjugate base of HA, usually exists as a salt, and hence the [A$^-$] is often referred to in texts as [salt]. This is not applicable, however, in cases where the acid component is a salt, such as the ammonium/ammonia buffer. It is, therefore, more helpful to use the term [base] for the concentration of the base component of an acid/base conjugate pair and [acid] for the concentration of the acid component of an acid/base conjugate pair.

From the rules of logarithms (see Appendix 1) the following relationships can be defined:

$$-\log_{10}\frac{[acid]}{[base]} = +\log_{10}\frac{[base]}{[acid]}$$

Incorporating this into the equation gives us:

$$pH = pK_a + \log_{10}\frac{[base]}{[acid]} \qquad [6.10]$$

This is known as the *Henderson–Hasselbalch equation.* It is important to learn how to use this!

Note: If [base] = [acid], then

$$\frac{[base]}{[acid]} = 1$$

In this case, log 1 = 0, and so pH = pK$_a$.

6.7.3 General strategy for making buffer solutions

In many experiments it is necessary to work over particular pH ranges, and in preparing buffer solutions then it is helpful to take into account the pK$_a$ values of the weak acids that could potentially be used (ideally the acid pK$_a$ should be within 1 of the target pH). This is because of the relationship, in the Henderson–Hasselbalch equation, between pH and pK$_a$. Tables 6.1 and 6.2 show pK$_a$ values for weak acids and weak bases respectively. Except for the value for TRIS in Table 6.1, all of the values refer to a temperature of 298 K. For TRIS it refers to 293 K.

Table 6.1 **Some pK$_a$ values for weak acids**

Acids	pK$_a$
Benzoic, $C_6H_5CO_2H$	4.20
Carbonic (1), H_2CO_3	6.37
Carbonic (2), HCO_3^-	10.298
Citric, $C_6H_8O_7 + H_2O \rightleftharpoons C_6H_7O_6^- + H_3O^+$	3.13
Citric (2), $C_6H_7O_7^- + H_2O \rightleftharpoons C_6H_6O_6^{2-} + H_3O^+$	4.76
Citric (3), $C_6H_6O_7^{2-} + H_2O \rightleftharpoons C_6H_5O_6^{3-} + H_3O^+$	6.40
Ethanoic (acetic), CH_3CO_2H	4.75
Methanoic (formic), HCO_2H	3.75
Phenol, C_6H_5OH	9.99
Phosphoric (1), $H_3PO_4 + H_2O \rightleftharpoons H_2PO_4^- + H_3O^+$	2.12

(Continued)

Table 6.1 (Continued)

Acids	pK_a
Phosphoric (2), $H_2PO_4^- + H_2O \rightleftharpoons HPO_4^{2-} + H_3O^+$	7.21
Phosphoric (3), $HPO_4^{2-} + H_2O \rightleftharpoons PO_4^{3-} + H_3O^+$	12.67
TRIS	8.30

TRIS = tris (hydroxylmethyl) aminomethane.

Table 6.2 Some pK_a values for weak bases

Bases	pK_b	pK_a
Ammonia	4.75	9.298
Aniline	9.37	4.63
Dimethylamine	3.13	10.87
Ethylamine	3.19	10.81
Ethylenediamine (1)	4.07	9.93
Ethylenediamine (2)	7.57	6.43
Pyridine	8.82	5.18
Trimethylamine	4.13	9.87

Thus, if an experiment needed to be performed under conditions where the pH was regulated at, say, pH = 4.5, then the most likely buffer would be made of ethanoic acid/ethanoate (benzoic acid/benzoate would be an alternative). Alternatively, experiments carried out at values close to pH = 9 are often performed in NH_3/NH_4^+ buffers.

To understand how effectively buffers work, consider the following examples.

Example 6.9 Addition of acid to water

What happens to the pH when 10^{-3} moles of hydrochloric acid are added to 1 litre of pure water?

The new 1-litre solution will contain 10^{-3} moles of HCl; in other words it will have a concentration [HCl] = 10^{-3} mol L^{-1}. Now, knowing that HCl is a strong acid and fully dissociates in water:

$$[H_3O^+] = 10^{-3} \, mol \, L^{-1}$$

Therefore, as

$$pH = -\log\left[H_3O^+\right]$$

then:

$$pH = -\log 10^{-3} = 3$$

Now remember, at 298 K the pH of pure water is 7, so adding 1 mL of 1.00 mol L^{-1} hydrochloric acid to almost 1 litre of pure water would change the pH from 7 to 3.

187

Example 6.10 Addition of acid to a buffer solution

What happens to the pH of a 1-litre ethanoic acid/ethanoate buffer made up with $0.700 \, mol \, L^{-1}$ ethanoic acid and $0.600 \, mol \, L^{-1}$ sodium ethanoate (the K_a value for ethanoic acid $= 1.8 \times 10^{-5}$), when 1 mL of $1.00 \, mol \, L^{-1}$ hydrochloric acid is added? (*This is a direct comparison with the effect of HCl on water.*)

Step 1 In order to be able to study the effect of adding acid to the buffer solution, the first step is to calculate the pH of the buffer solution before the acid is added.

As the K_a for ethanoic acid is known, then the value of pK_a is given by:

$$pK_a = -\log \, (1.8 \times 10^{-5})$$
$$= 4.745$$

Using the Henderson–Hasselbalch equation it is possible to relate pK_a to pH by:

$$pH = pK_a + \log_{10} \frac{[base]}{[acid]}$$

and from above it is known

$$pK_a = 4.745$$
$$[base] = 0.600 \, mol \, L^{-1}$$
$$[acid] = 0.700 \, mol \, L^{-1}$$

Therefore, putting the numbers into the equation gives:

$$pH = 4.745 + \log_{10} \frac{0.600}{0.700}$$
$$= 4.745 + \log_{10} 0.857$$
$$pH = 4.678$$

Step 2 The second stage is to consider the effect of adding 10^{-3} moles of H_3O^+ to the buffer. (This is because the HCl is fully ionised in solution.) The consequence of adding the HCl is that the base, namely ethanoate, will react with the H_3O^+ to give ethanoic acid:

$$H_3O^+ \, (aq) + CH_3CO_2^- \, (aq) \quad \rightleftharpoons \quad H_2O + CH_3CO_2H \, (aq)$$

($K = 5.6 \times 10^4$ – equilibrium very much to right.)

Adding 10^{-3} moles of H_3O^+ to 1 litre effectively means that the concentration of acid added is $0.001 \, mol \, L^{-1}$.

The consequence of this is that all of this acid will react with $CH_3CO_2^-$ resulting in the following changes to concentrations of the buffer species:

$$[CH_3CO_2^-] = 0.600 - 0.001 = 0.599 \, mol \, L^{-1}$$

and

$$[CH_3CO_2H] = 0.700 + 0.001 = 0.701 \, mol \, L^{-1}$$

Taking account of the effect of these revised concentrations on the pH, by feeding them in to the Henderson–Hasselbalch equation [6.10], using the new figures for this, which are:

$$pK_a = 4.745, [base] = 0.599 \, mol \, L^{-1} \, \text{and} \, [acid] = 0.701 \, mol \, L^{-1}$$

Then,

$$pH = 4.745 + \log_{10} \frac{0.599}{0.701}$$

$$= 4.745 + \log_{10} 0.857$$

$$= 4.678 \text{ (effectively unchanged compared with step 1)}$$

Conclusion from examples 6.9 and 6.10

The addition of a small amount of a strong acid to pure water caused a significant change to the pH of the water.

The addition of a small amount of a strong acid to a buffer solution caused an insignificant change to the pH of the buffer.

Example 6.11 What ratio of acid:base is required to make an ethanoic acid/ethanoate buffer of pH 4.30? (pK_a ethanoic acid $= 4.75$ to 2 decimal places.)

From the Henderson–Hasselbalch equation:

$$pH = pK_a + \log_{10} \frac{[\text{base}]}{[\text{acid}]}$$

Feed in the numbers known:

$$4.30 = 4.75 + \log_{10} \frac{[\text{base}]}{[\text{acid}]}$$

Therefore

$$\log_{10} \frac{[\text{base}]}{[\text{acid}]} = -0.45$$

$$\frac{[\text{base}]}{[\text{acid}]} = 10^{-0.45} = 0.355$$

Therefore ratio of acid:base is the inverse of this, and is

$$= 1/0.355$$

$$= 2.82{:}1$$

Example 6.12 How many pH units away from its pK_a is a buffer whose base:acid ratio is 4.43?

From the Henderson–Hasselbalch equation:

$$pH = pK_a + \log_{10} \frac{[\text{base}]}{[\text{acid}]}$$

This can be rearranged by subtracting pK_a from both sides of the equation,

$$pH - pK_a = \log_{10} \frac{[\text{base}]}{[\text{acid}]}$$

Therefore, as

$$\frac{[\text{base}]}{[\text{acid}]} = 4.43$$

189

then

$$pH - pK_a = \log_{10} 4.43 = 0.646$$

Example 6.13   Calculate the ratio of ammonium chloride to ammonia that is required to make a buffer solution with a pH of 9.00. The K_a for ammonium ion, NH_4^+, is 5.6×10^{-10}.

$$NH_3(aq) + H_2O(l) \rightleftharpoons OH^-(aq) + NH_4^+(aq)$$

From the Henderson–Hasselbalch equation:

$$pH = pK_a + \log_{10} \frac{[base]}{[acid]}$$

where $[NH_4^+] = [acid]$ and $[NH_3] = [base]$. Now, K_a for the ammonium ion is 5.6×10^{-10}, so the pK_a for ammonium ion is:

$$pK_a = -\log 5.6 \times 10^{-10} = 9.25$$

The target pH is 9.00, so

$$9.00 = 9.25 + \log_{10} \frac{[NH_3]}{[NH_4^+]}$$

thus,

$$\log_{10} \frac{[NH_3]}{[NH_4^+]} = -0.25$$

and hence

$$\frac{[NH_3]}{[NH_4^+]} = 10^{-0.25} = 0.56$$

So, the ratio of $NH_4^+ : NH_3 = 1/0.56 = 1.78:1$

Summary of buffer solution capability

A solution is buffered against the addition of a:

- strong acid so long as a significant amount of the base component of the conjugate pair remains;
- strong base so long as a significant amount of the acid component of the conjugate pair remains.

6.8   Polyprotic acids

In contrast to a simple monoprotic acid like ethanoic acid, with only one equilibrium between the acid and base, a polyprotic acid contains more than one acidic hydrogen. For a polyprotic acid, n acidic hydrogens will exist in solution in equilibrium with n base forms (for a total of $n + 1$ species).

For example, when phosphoric acid ($n = 3$, a triprotic acid) is dissolved in solution, the following equilibria are established among the four species H_3PO_4 (phosphoric acid itself), $H_2PO_4^-$ (dihydrogen phosphate anion), HPO_4^{2-} (hydrogen phosphate anion), and PO_4^{3-} (phosphate anion):

$$H_3PO_4 + H_2O \rightleftharpoons H_2PO_4^- + H_3O^+ \qquad K_a = 7.58 \times 10^{-3}$$
$$H_2PO_4^- + H_2O \rightleftharpoons HPO_4^{2-} + H_3O^+ \qquad K_a = 6.17 \times 10^{-8}$$
$$HPO_4^{2-} + H_2O \rightleftharpoons PO_4^{3-} + H_3O^+ \qquad K_a = 2.14 \times 10^{-13}$$

Note that each successive pair of species is linked in an independent equilibrium with H_3O^+ and that each equilibrium therefore has a unique equilibrium constant. In order to understand how polyprotic acid solutions respond to chemical changes (such as addition of OH^- or H_3O^+), it would be necessary to predict the concentrations of the four phosphate species, as well as H_3O^+ and OH^-, under a variety of conditions. This is beyond the scope of this text, but we will consider one or two points arising from them, and summarise, descriptively, the potential use of this.

Looking at the three equilibria it is clear that

- H_3PO_4 acts as a weak acid in equilibrium 1, and $H_2PO_4^-$ is its conjugate base.
- $H_2PO_4^-$ acts as a weak acid in equilibrium 2, and HPO_4^{2-} is its conjugate base.
- HPO_4^{2-} acts as a weak acid in equilibrium 3, and PO_4^{3-} is its conjugate base.

This means that, of the four species, two of them, $H_2PO_4^-$ and HPO_4^{2-} can act both as acids and as bases – they are *amphoteric*. H_3PO_4 acts only as a weak acid. PO_4^{3-} is only a base.

Table 6.3 shows the dominant acid–base pair at various pH ranges. Thus, by combining appropriate quantities of the acid–base pairs from these equilibria, it is possible to define the pH of a solution.

Table 6.3   **Dominant phosphate species as a function of pH**

pH range	Dominant species
0–4.7	H_3PO_4, $H_2PO_4^-$
4.7–9.7	$H_2PO_4^-$, HPO_4^{2-}
9.7–14	HPO_4^{2-}, PO_4^{3-}

Other examples:

Sulfuric acid

$$H_2SO_4 + H_2O \rightarrow HSO_4^- + H_3O^+ \quad K_a \text{ very large (essentially 100\% ionised)}$$
$$HSO_4^- + H_2O \rightleftharpoons SO_4^{2-} + H_3O^+ \quad K_a = 1.2 \times 10^{-2}$$

(Note that the first step acts as a strong acid, and the second step acts as a weak acid.)

Carbonic acid

$$H_2CO_3 + H_2O \rightleftharpoons HCO_3^- + H_3O^+ \qquad K_a = 4.3 \times 10^{-7}$$
$$HCO_3^- + H_2O \rightleftharpoons CO_3^{2-} + H_3O^+ \qquad K_a = 5.6 \times 10^{-11}$$

Box 6.4

Blood as a buffer

Much of body chemistry relies on buffer systems to maintain blood pH in the range 7.35–7.45. The main one in human blood is the carbonic acid/hydrogen carbonate system (referred to in Section 6.8). The acid component of the buffer is H_2CO_3, and its conjugate base is HCO_3^-. In the equilibrium:

$$H_2CO_3 + H_2O \rightleftharpoons HCO_3^- + H_3O^+$$

● The quantity of H_2CO_3 is controlled by respiration.

● The amount of HCO_3^- is controlled by excretion through urination.

CO_2 produced by the tricarboxylic acid (TCA) cycle diffuses across capillary walls into the blood and is responsible for the build-up of carbonic acid:

$$CO_2 + H_2O \;\rightleftharpoons\; H_2CO_3$$

The K_a value of 4.30×10^{-7} for carbonic acid results in a pK_a value of 6.37.

In order to maintain a pH range of 7.35–7.45, a normal ratio of base:acid of about 20:1 is needed. Deviation from this causes difficulty. For example, asthma sufferers (or people with other respiratory conditions such as emphysema) get a build-up of acid, resulting in the blood pH falling, called *respiratory acidosis*, which, in extreme cases, results in shock and ultimately death.

Correspondingly, if the bicarbonate:carbonic acid ratio is in excess of 20:1 then the result is alkalosis.

6.9 Salts

6.9.1 Introduction

A salt is formed when an acid and a base are mixed. Table 6.4 summarises the characteristics of different acid–base reactions.

Table 6.4 **Salt formation from acid–base reactions**

Acid	Base	Salt pH
Strong	Strong	pH = 7
Weak	Strong	pH > 7
Strong	Weak	pH < 7
Weak	Weak	Depends on which is stronger

The pH of salts

By comparing the outcome from dissolving salts made from a strong acid/strong base with those made from a weak acid/strong base, it is possible to see why the salts have different pH values when dissolved in water.

Salts of a strong acid/strong base

Using the example of HCl and NaOH, the products of the reaction between these are NaCl and H_2O. If NaCl is dissolved in water, then full dissociation occurs according to the following:

$$2H_2O(l) + Na^+(aq) + Cl^-(aq) \rightleftharpoons H_3O^+(aq) + Cl^-(aq) + Na^+(aq) + OH^-(aq)$$

Now, $Na^+(aq) + Cl^-(aq)$ are present on both sides of the equation, so the net change is represented by:

$$2H_2O(l) \rightleftharpoons H_3O^+(aq) + OH^-(aq)$$

This is the equation representing the auto-dissociation of water. The auto-dissociation of water has an equilibrium constant, represented by K_w of 1.00×10^{-14}.

In other words, a reaction between a strong acid and a strong base produces a neutral solution with pH = 7.

Salts of a weak acid/strong base

Considering the salt made from the weak acid, CH_3CO_2H, and the strong base, NaOH, which is sodium ethanoate, the equilibrium resulting from its dissolution in water is given by:

$$2H_2O(l) + CH_3CO_2Na(aq) \rightleftharpoons CH_3CO_2^-(aq) + H_3O^+(aq) + Na^+(aq) + OH^-(aq)$$

$$\Updownarrow$$

$$H_2O(l) + CH_3CO_2H(aq)$$

The salt fully dissociates in water, with the result that the four ions $CH_3CO_2^-$, H_3O^+, Na^+ and OH^- are released into solution. Of these, the $CH_3CO_2^-$ and H_3O^+ exist in equilibrium with CH_3CO_2H and H_2O, with the equilibrium very much towards CH_3CO_2H, which is a weak acid, so remains largely undissociated in solution. This results in a relatively basic solution as the amount of OH^- will be greater than the amount of H_3O^+.

Conclusion: the salt made from the reaction of a weak acid with a strong base gives a basic solution (pH > 7) whose precise pH depends on pK_b of the salt.

6.9.2 Titrations

The reactions between acids and bases can be monitored by performing experiments called titrations.

Strong acid/strong base

Consider what would happen on adding a strong base to a known volume of a strong acid of known concentration; for example, the addition of $10^{-1} \, mol \, L^{-1}$ NaOH to 25 mL of $10^{-1} \, mol \, L^{-1}$ HCl. The equipment used to perform such an experiment is illustrated in Figure 6.1.

On plotting the results of the titration, showing the volume of base added along the x axis and the pH along the y axis, the picture would look something like

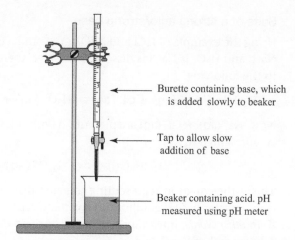

Figure 6.1 Equipment used to perform a titration experiment.

Figure 6.2. What the graph demonstrates is that the pH changes only very slowly until the amount of base is very nearly that of the acid.

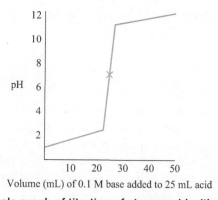

Volume (mL) of 0.1 M base added to 25 mL acid

Figure 6.2 Example graph of titration of strong acid with strong base.

The pH starts at 1.00 (corresponding to $[H_3O^+] = 0.100 \, mol \, L^{-1}$). When 24.50 mL of NaOH have been added the pH has increased to 3.00. When 24.75 mL has been added the pH increases to 3.3, and by the time the amount of NaOH added rises to 24.95 mL, the pH has risen to 4.00. The pH reaches 7.00 when the amount of NaOH added equals 25.00 mL. The mixture is neutral. The pH rises rapidly until the base has very little acid left to react with, and thereafter the increase tails off dramatically.

Strong acid/weak base

This type of reaction produces a titration plot like that shown in Figure 6.3 and, as referred to in the discussion about salts, the resultant is an acid salt, with a pH < 7, evident from the end point indicated by the × on the plot.

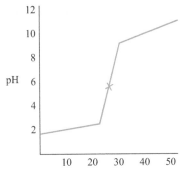

Figure 6.3 **Example graph of titration of strong acid with weak base.**

Strong base/weak acid

This type of reaction produces a titration plot like that shown in Figure 6.4, and, in this case, the resultant is a basic salt, with a pH > 7, evident from the end point indicated by the × on the plot.

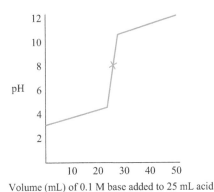

Figure 6.4 **Example graph of titration of strong base with weak acid.**

6.10 Introducing solubility

When a solid is added to a liquid it is often the case that the solid appears to vanish. It is, of course, dissolving. We see examples of this every day, for example when adding sugar to our hot drinks, or if we add salt to boiling water when cooking vegetables. What is happening in these situations?

6.10.1 Solubility of sugar and salt

Dissolving sugar

Consider firstly the situation of sugar (more correctly called sucrose, formula $C_{12}H_{22}O_{11}$). When this is added to water the intermolecular interactions between different sucrose molecules are broken, and this means that the individual molecules are released into solution (dissolved). This process can be regarded as an equilibrium:

$$C_{12}H_{22}O_{11}(s) \rightleftharpoons C_{12}H_{22}O_{11}(aq)$$

Eventually, if enough sugar is added to water, then a saturated solution is obtained, in which the rate at which the sugar dissolves is matched by the rate at which the molecules reform as solid. The solubility of sugar increases with increasing temperature, from about 2000 g per litre at 298 K to about 4200 g per litre by 363 K. Sugar is, therefore, regarded as being extremely soluble.

Solubility of salt (sodium chloride)

Sodium chloride is an ionic compound and, as such, when it dissolves in water it does so by splitting into solvated ions, as indicated in the equation:

$$NaCl(s) \quad \rightleftharpoons \quad Na^+(aq) + Cl^-(aq)$$

The driving force for this is that the energy released by solvating the ions is greater than the energy needed to break the ionic interactions holding the positive and negative ions together in the ionic lattice. The solubility of sodium chloride is approximately 36 g per litre at 298 K, rising to approximately 38 g per litre by 363 K. Salt is regarded as being soluble, giving a solution with solubility, if expressed in moles per litre, of approximately $0.62\,mol\,L^{-1}$ at approximately room temperature.

6.10.2 Insoluble ionic compounds: the concept of solubility product

Although most ionic compounds are relatively soluble in water, a few have only very limited solubility. One relatively insoluble ionic compound is silver chloride, AgCl, which has a solubility of 0.0019 g per litre at 298 K. Writing out the equilibrium resulting from dissolution of AgCl in water gives:

$$AgCl(s) \rightleftharpoons Ag^+(aq) + Cl^-(aq)$$

The equilibrium constant for this dissolution could be written out as:

$$K = \frac{[Ag^+][Cl^-]}{[AgCl]}$$

In this expression, because the solubility is so low, the denominator represents a solid material. As referred to earlier, this means that its effective concentration (activity) is 1, and hence the equilibrium can be rewritten as shown:

$$K_{SP} = [Ag^+][Cl^-]$$

In this the equilibrium constant has been represented with a subscript 'SP', which represents the term *solubility product*.

It is possible to calculate the value of the solubility product at 298 K from the information provided above. The solubility of AgCl is 0.0019 g per litre. The relative molecular mass of AgCl is 143.32, so the number of moles of AgCl that can dissolve in 1 litre is $(0.0019/143.32) = 1.326 \times 10^{-5}$ moles.

Therefore, because AgCl in solution completely dissociates into Ag^+ and Cl^- ions, then:

$$[Ag^+] = [Cl^-] = 1.326 \times 10^{-5}\,mol\,L^{-1}$$

Therefore, the solubility product is:

$$K_{SP} = [Ag^+][Cl^-] = 1.326 \times 10^{-5} \times 1.326 \times 10^{-5} = 1.757 \times 10^{-10}$$

This is usually rounded to 1.8×10^{-10}.

What if the ionic compound is not a 1:1 salt?

This can be demonstrated by considering the dissolution of the salt lead chloride at a temperature of 293 K, where lead has a valency of 2. The dissolution of lead chloride in water can be represented by:

$$PbCl_2(s) \rightleftharpoons Pb^{2+}(aq) + 2Cl^-(aq)$$

for which $K_{SP} = [Pb^{2+}][Cl^-]^2$

If the molar solubility of the lead chloride is given by S, then it follows that, since the stoichiometry of the dissolution shows that one molecule of $PbCl_2$ gives one ion of Pb^{2+} and two ions of Cl^-, then $[Pb^{2+}] = S$ and $[Cl^-] = 2S$. This leads to:

$$K_{SP} = [Pb^{2+}][Cl^-]^2 = S \times (2S)^2 = 4S^3$$

Lead chloride has a solubility of 9.9 g per litre at 293 K, so this must first be converted to $mol\,L^{-1}$. The relative molecular mass of $PbCl_2$ is 278, so the number of moles of $PbCl_2$ that will dissolve in a litre of water is $(9.9/278) = 0.0356$, or, in other words, $S = 3.56 \times 10^{-2}\,mol\,L^{-1}$. Therefore,

$$K_{SP} = 4S^3 = 4 \times (3.56 \times 10^{-2})^3 = 1.8 \times 10^{-4}\,(\text{at } 293\text{ K})$$

6.10.3 The common ion effect

Sometimes it is necessary to precipitate a sparingly soluble salt. Because the ions are in dynamic equilibrium with the solid salt, then it is possible to use Le Chatelier's principle: if a second salt that contains one of the same ions is added to a saturated solution, then the solubility of the other ion is decreased and the salt precipitates. The decrease in solubility is called the *common ion effect*.

Let us return to the earlier example of a saturated solution of silver chloride in water, for which the equilibrium is:

$$AgCl(s) \rightleftharpoons Ag^+(aq) + Cl^-(aq)$$

where

$$K_{SP} = [Ag^+][Cl^-] = 1.8 \times 10^{-10}$$

and the molar solubility of AgCl in water is $1.3 \times 10^{-5}\,mol\,L^{-1}$.

On adding sodium chloride to the solution, the concentration of Cl^- ions increases. Le Chatelier's principle states that the system responds by tending to minimise the effect of adding more Cl^- ions, so some of the Cl^- ions will form solid AgCl by taking some Ag^+ ions out of solution. Because there are now fewer Ag^+ ions in solution, the solubility of AgCl is lower in a solution of NaCl than it is in pure water. A similar effect occurs whenever two salts have an ion in common.

Box 6.5

Acidity and basicity: extending the concept to non-aqueous systems

As you may recall, the Brønsted–Lowry definitions are summarised very briefly as:

- an acid is a proton donor;
- a base is a proton acceptor. (In water a base will release OH^- anyway.)

A broader definition, also developed in the 1920s, extended the acid–base concept to non-protic systems (systems that did not necessarily involve H). This approach was named after the scientist who proposed it, Gilbert N. Lewis. The proposal was:

- A Lewis acid is a substance that accepts electron pairs.
- A Lewis base is a substance that donates electron pairs.

A Lewis base (nucleophile) must have a lone pair of electrons, therefore all Lewis bases are also Brønsted bases.

Lewis acids (electrophiles) include Brønsted acids, the proton itself, all positive ions, and any molecule that can accommodate an additional electron pair. Therefore in the reaction between ammonia and boron trifluoride,

$$NH_3 + BF_3 \rightarrow F_3B \cdot NH_3$$

NH_3 – Lewis base, has an electron pair to donate.

BF_3 – Lewis acid (electron-deficient, the B has only six valence electrons, accepts two to make up its octet).

Questions

6.1 Define the terms:

    (a) Brønsted–Lowry acid
    (b) Brønsted–Lowry base
    (c) Strong acid/strong base
    (d) Weak acid/weak base

6.2 State the definitions of pH and pOH. Using these definitions, and any other relevant information, calculate the pH of the following aqueous solutions:

    (a) $0.05\,mol\,L^{-1}$ HNO_3
    (b) $8 \times 10^{-4}\,mol\,L^{-1}$ HCl
    (c) $0.0075\,mol\,L^{-1}$ HCl
    (d) $0.05\,mol\,L^{-1}$ NaOH

6.3 A weak acid can be defined by the equilibrium

$$HA + H_2O \rightleftharpoons H_3O^+ + A^-$$

    (a) State what K_a would be for this reaction in terms of the concentrations of the species present.
    (b) Using the definition of K_a, derive the Henderson–Hasselbalch expression.
    (c) What is the pH of a 1.00-litre solution made up from 0.010 moles of propanoic acid ($K_a = 1.3 \times 10^{-5}$) and 0.020 moles of sodium propanoate?

6.4 A solution of 0.04 moles of butanoic acid, $CH_3(CH_2)_2COOH$, in 0.80 L of water dissociates as shown by the following equilibrium:

$$CH_3(CH_2)_2COOH + H_2O \rightleftharpoons CH_3(CH_2)_2COO^- + H_3O^+$$

The acid dissociation constant for this is $K_a = 1.48 \times 10^{-5}$

(a) Write down an expression for the acidity constant, K_a, for this process in terms of the concentrations of the reagents.

(b) Calculate the pH of the solution.

(c) Calculate the pH of the solution when 0.02 moles of sodium butanoate are added to the butanoic acid solution.

(d) Describe qualitatively what happens to the pH of the solution in (iii) when 10^{-5} moles of a strong acid such as HCl is added to it.

6.5 (a) What would be the pH of 1 litre of water to which 0.01 moles of HCl had been added?

(b) Using the value $K_b(NH_3) = 1.78 \times 10^{-5}$, calculate the pH of a solution containing 0.15 mol L^{-1} NH_3 and 0.25 mol L^{-1} NH_4Cl.

(c) What would be the effect of adding 0.01 moles of HCl to 1 litre of the solution in part (b)?

(d) Comment on the result from (c), and compare with the data from (a).

6.6 H_3PO_4 is a triprotic acid.

(a) Draw out the three acid–base equilibria that illustrate this.

(b) For each equilibrium identify the phosphorus-containing species that act as an acid, and which acts as its corresponding conjugate base.

7 Biomolecules and biopolymers

Learning outcomes

By the end of this chapter students should be able to:

- discuss the classification of lipids, and discuss their importance (micelles, bilayers);
- know about the nomenclature, structure and properties of important monosaccharides, disaccharides and polysaccharides;
- describe the key features of the structure of the 20 essential amino acids, the formation and properties of the peptide bond, protein folding, secondary, tertiary and quaternary protein structure, and protein denaturing;
- describe the nomenclature, composition and structure of the nucleic acids, the differences between RNA and DNA, and the nature of the DNA double helix.

7.1 Introduction

This chapter will take a more detailed look at the common types of molecules found in biological systems, with the main emphasis on those of major importance in studies of cell biology, namely lipids, carbohydrates, proteins and nucleic acids. Many of the molecules crucial to life are composed of much smaller building blocks. In other words they are *polymers*. The term polymer derives from the Greek *poly* (many) *meros* (parts). Polymers are generally a very simple molecule (the *mer* part) which is chemically combined (*poly*merised) many times until it attains the final desired properties as a polymer. Many modern materials are synthetic polymers; for example, nylon, polythene and plastics in general. It is also appropriate, however, to think of many naturally occurring compounds as polymers, often referred to as biopolymers. To give a couple of examples:

- starches are polymers of simple carbohydrate units;
- proteins are polymers of simple amino acid units.

The aim of this chapter is, therefore, to ensure familiarisation with important biological molecules including:

- lipids;
- the monomer units of proteins, polysaccharides and nucleic acids;
- structural features in biopolymers.

It is not the object of this chapter to explore the biology associated with these biomolecules in any detail, except as a means of illustrating the key structural features that emerge from this overview.

7.2 Lipids

Lipids is a general name given to a group of biomolecules which are insoluble in water. The name lipid comes from the Greek word *lipos*, for fat. The group of compounds called lipids all have large hydrocarbon components, with very limited amounts of polar functional groups. Some of them contain ester or amide functional groups (such as waxes, triacylglycerols, glycophospholipids, sphingomyelins or glycophospholipids), whereas others do not (such as steroids or eicosanoids). As a consequence of this, they exhibit a variety of structural types (see Figure 7.1 for examples of lipid structures). In this book the main focus will be on looking at fats and at steroids.

7.2.1 Fats, oils and fatty acids

Natural fats/oils, for example vegetable fats and oils, are not chemically homogeneous, containing instead a mixture of materials such as fatty acids and mono-, di- and triglycerides (esters of glycerol with one, two or three fatty acids respectively).

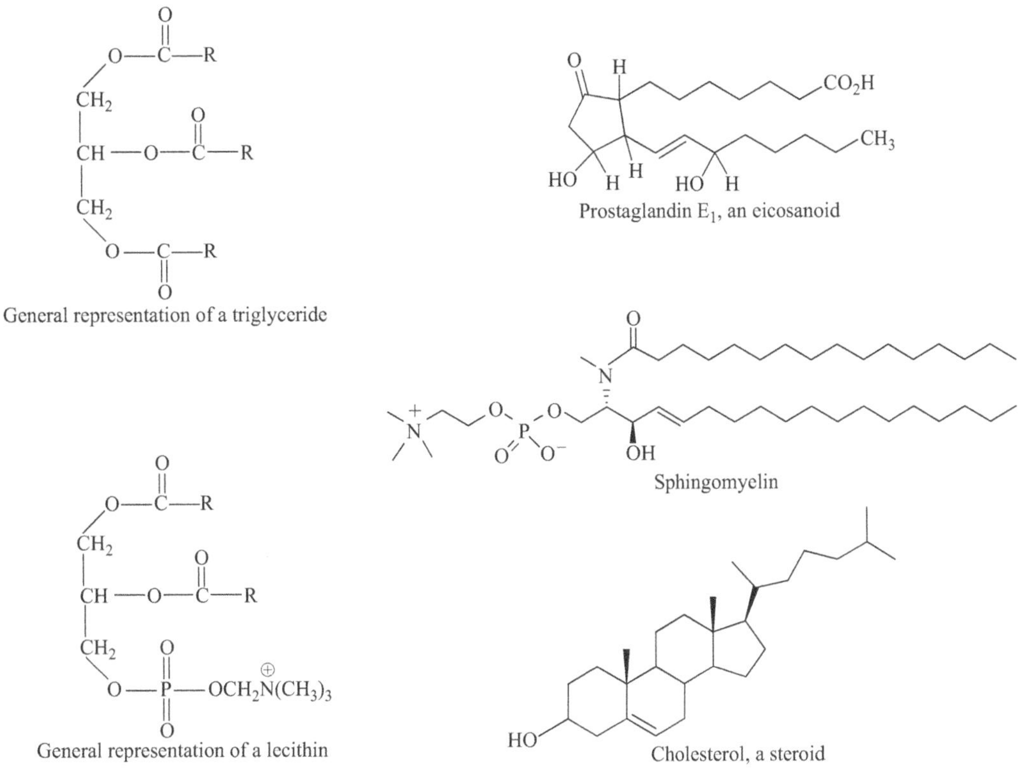

General representation of a triglyceride

Prostaglandin E_1, an eicosanoid

Sphingomyelin

General representation of a lecithin

Cholesterol, a steroid

Figure 7.1 **The range of different lipid structures.**

201

In chemical terms, *fats* are mostly made up of long chain esters of the triol glycerol:

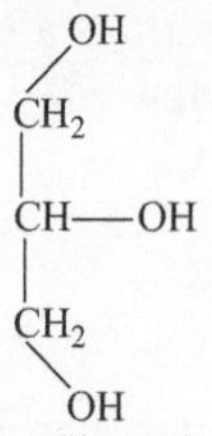

Glycerol

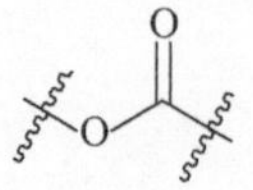

Ester group

Glycerol (often called glycerine) is a water-soluble clear oil. The most common fats are glycerol esters of long chain *fatty acids*. Fatty acids are long-chain carboxylic acids with a long unbranched aliphatic tail (chain), which is either *saturated* or *unsaturated*.

Saturated fatty acids

These have the general formula $CH_3(CH_2)_nCO_2H$, and some examples are:

- $n = 10$ ($C_{12}H_{24}O_2$), lauric acid (systematic name dodecanoic acid) has a melting point of 44 °C, and is:

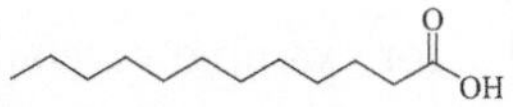

- $n = 14$ ($C_{16}H_{32}O_2$), palmitic acid (systematic name hexadecanoic acid) has a melting point of 63 °C, with the structure:

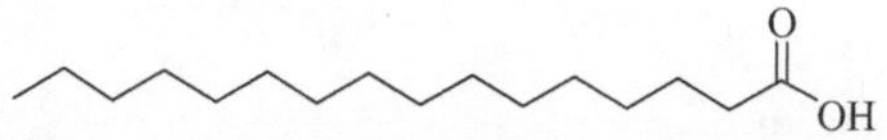

- $n = 16$, stearic acid, which has the molecular formula $C_{18}H_{36}O_2$ (systematic name octadecanoic acid), has a melting point of 70 °C and is:

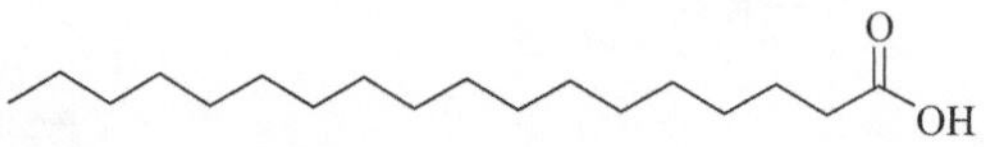

Unsaturated fatty acids

These are compounds where the chain contains one or more double bonds (the unsaturation). Examples include the following.

Oleic acid

This has the formula $C_{18}H_{34}O_2$, contains one $C=C$ double bond and, as such, exists in either *cis* or *trans* forms.

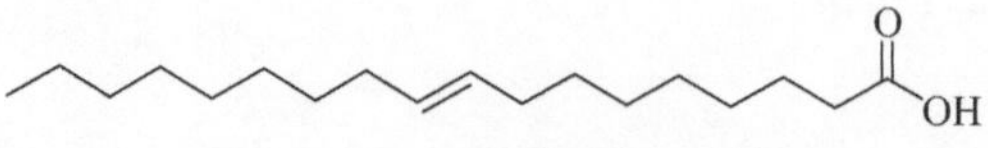

Trans-oleic acid

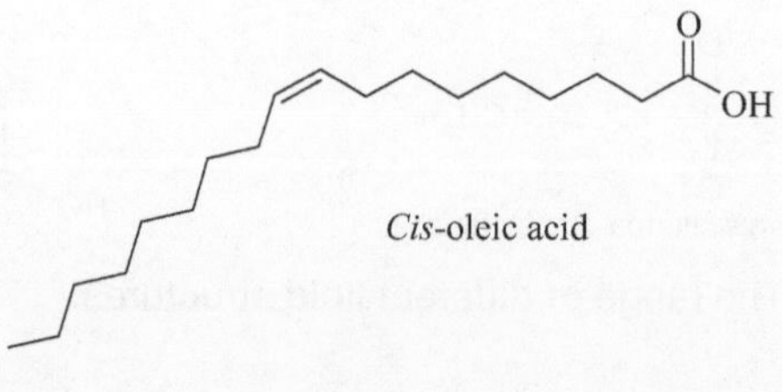

Cis-oleic acid

The physical properties of these are significantly different from one another. The *trans* isomer has a melting point of 45 °C whereas the *cis* isomer has a melting point of just 13 °C. The reason for this is that the bent nature of the *cis* isomer makes it more difficult for the molecules to pack together compared with the *trans* isomer.

Unsaturated fatty acids in general have one or more double bonds in the carbon chain, and like any alkene each double bond may be *cis* or *trans*.

Linoleic acid

9,12-Octadecadienoic acid is commonly called linoleic acid, and has a melting point of −9 °C. It is U-shaped because of the presence of two *cis* double bonds:

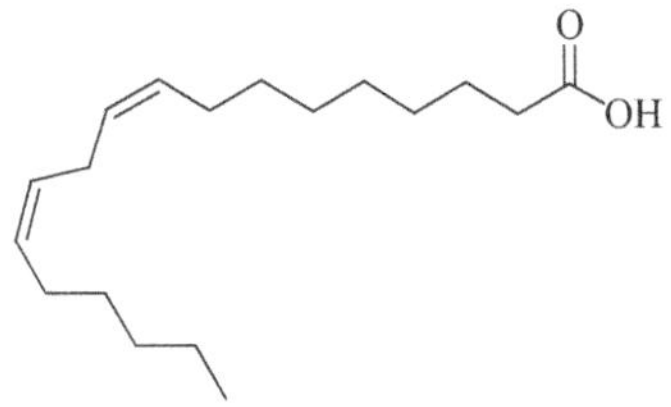

Recalling the nature of alkane structures (as discussed in Chapter 3), it follows that the carbon chains of saturated fatty acids can adopt many conformations. They are usually extended (straight) as this reduces steric repulsions between neighbouring groups. This means that they can pack efficiently and, because the intermolecular (van der Waals) attractions are comparatively strong, they have relatively high melting points.

In unsaturated fatty acids the *cis* configuration about the double bond puts a rigid bend in the chain. This interferes with packing, inhibiting the intermolecular interactions and thereby decreasing the melting points of unsaturated fatty acids.

Naturally occurring fatty acids almost always have *cis* double bonds, and *trans* fats may occur, but these are the products of hydrogenation where a food manufacturer adds hydrogen across one or more of the double bonds in a vegetable oil to raise the melting point. Solid commercial cooking fats are manufactured by *partial* hydrogenation of vegetable oils (achieved using a Ni catalyst). The resulting solid (for example, margarine) has a longer shelf-life as it is less prone to oxidation. There are problems, however, in that the process results in an unnatural *trans* (rather than *cis*) arrangement about remaining double bonds, and there is increasing evidence that *trans* fats are associated with heart disease.

7.2.2 Triglyceride fats

Because esters are the result of a reaction between an alcohol and a carboxylic acid, if glycerol is reacted with three carboxylic acid molecules, a triester results. For example, glycerol tripalmitate is formed from glycerol and palmitic acid. It is common in mammalian fat, and is quite unusual in that it is one of the few triesters that have three identical saturated fatty acid esters:

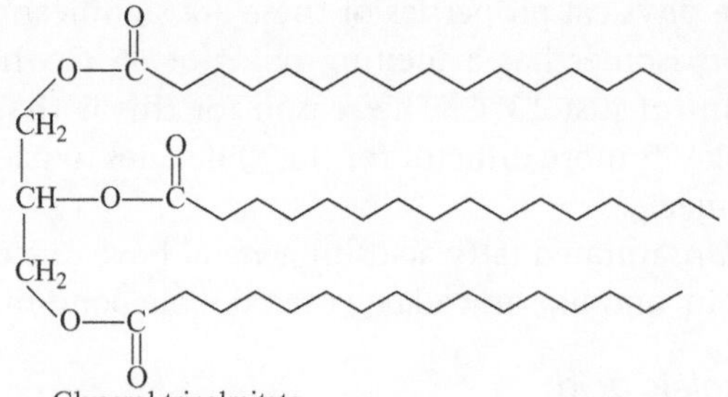

Glycerol tripalmitate

Fats are usually formed, however, from glycerol and a mixture of saturated and unsaturated long-chain alkanoic acids. The presence of the unsaturated fatty acids means the presence of one or more double bonds in the chain, which produces a 'kink' in the chain which means that the molecules cannot pack as tightly. This results in lower melting points. Compare, for example, the structure of the glycerol tripalmitate (above) with that of the dipalmitate oleate (below):

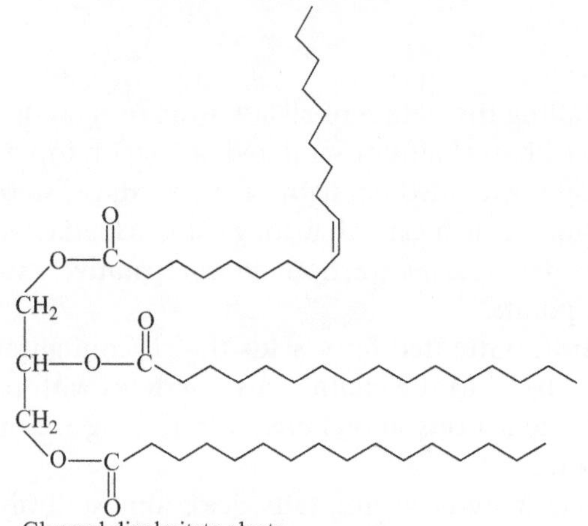

Glycerol dipalmitate oleate

7.2.3 Uses of fats: micelles

Fats can be hydrolysed to the component salts of the fatty acids and to glycerol by lipases. This can also be carried out in a laboratory environment by heating with alkali. This process is called *saponification* and it is the way soap is manufactured.

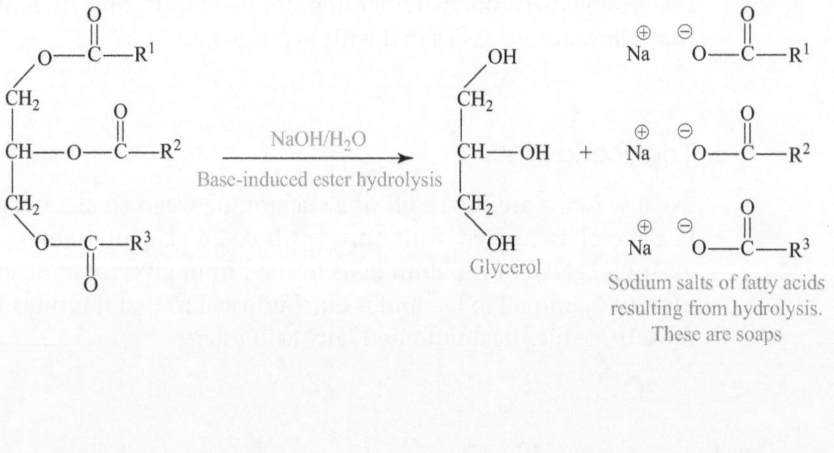

Sodium salts of fatty acids resulting from hydrolysis. These are soaps

The glycerol produced has various applications, in anti-freeze, paints, moisturising cosmetics and tobacco. As its tri-nitro derivative, it is important in the manufacture of dynamite. The structure of the soaps (which can be called sodium carboxylates as they are sodium salts of carboxylic acids) is crucial to their behaviour.

Consider, for example, sodium stearate:

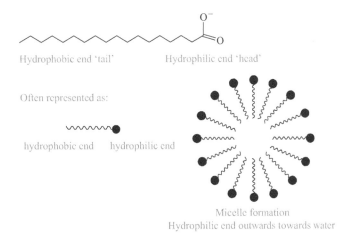

The sodium salts of long-chain carboxylic acids (soaps) are almost completely miscible in water. However, except in extremely dilute solutions, they dissolve as micelles, which are (almost) spherical clusters of carboxylate ions that are dispersed throughout the aqueous phase. The non-polar, and thus hydrophobic (which means water-hating), alkyl chains remain in the non-polar environment (i.e. in the interior of the micelles) while the polar, and therefore hydrophilic (water-loving), carboxylate groups are exposed to the polar environment of the aqueous medium. Pictorially this is shown in Figure 7.2 (as seen earlier in Chapter 4).

During the washing of clothes, for example, agitation breaks down grease into these micelles which have negatively charged surfaces. The grease droplets consequently repel each other and they remain suspended in the water as they can no longer coalesce and settle back on the clothes. Subsequently, the suspended droplets will be flushed down the drain along with the water. During this process, the sodium ions are free to move independently in the water.

Of more direct and obvious relevance to biology, micelles are now being researched for use as drug delivery vehicles as they exhibit excellent properties to enable controlled release of a substance. Oral administration of medicine is by far the easiest and most convenient route of delivery, especially in the case of chronic (long-term) therapies. Oral drug formulation presents significant difficulties, however, because the medication both has to be delivered intact and has to act at the optimum rate. The gastrointestinal tract presents many problems for a drug, from

Figure 7.2 **Micelles.**

structural barriers (mucus layer, microvilli, etc.) to physiological factors (the wide range of pH, the activity of enzymes, etc.). In the case of drugs which have poor water solubility, the dissolution time in the gastrointestine may be too long for it to be effective. This means that the desire for oral delivery poses a major challenge for effective delivery of poorly water-soluble medication.

7.2.4 Phospholipids

Phospholipids are triglycerides where one of the acids is phosphoric rather than carboxylic, and are found frequently as lecithins and cephalins. They generally form sheets rather than spheres (like the micelles do) and are important structural components forming the lipid bilayers of cell membranes. In the membrane the charged groups (the polar heads) are directed to the outside of the cell and inwards towards the cytoplasm (see Figure 7.3). They also have metabolic roles in nerve tissue.

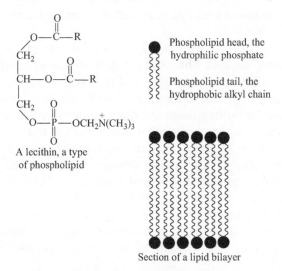

Figure 7.3 **Phospholipids and phospholipid bilayers.**

7.2.5 Waxes

These are esters of long-chain fatty acids, not with glycerol but with long-chain alcohols (which are themselves derived from fatty acids by reduction). Beeswax contains several of these, amongst which is the straight chain palmitate ester $C_{15}H_{31}COOC_{30}H_{61}$.

7.2.6 Steroids

These are a separate class of lipids, including cholesterol, the bile acids and the steroid hormones. Cholesterol (and its fatty acid esters) plays a structural role in (stiffening) membranes. Cholesterol itself, as shown, has eight chiral centres, though other steroids can have many more.

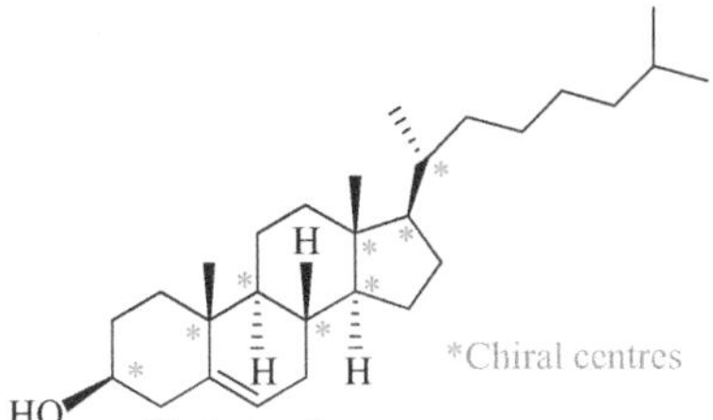

Cholesterol
*Chiral centres

The basic steroid structure is:

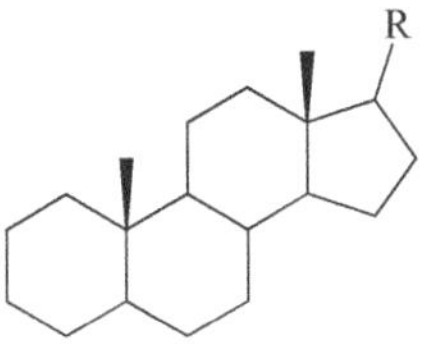

The structures as drawn above are, however, slightly misleading. The shapes of aliphatic ring systems are usually non-planar (as mentioned in Chapter 3), which has great significance in the structure of steroids. Considering cholesterol, for example, the molecule can be represented thus:

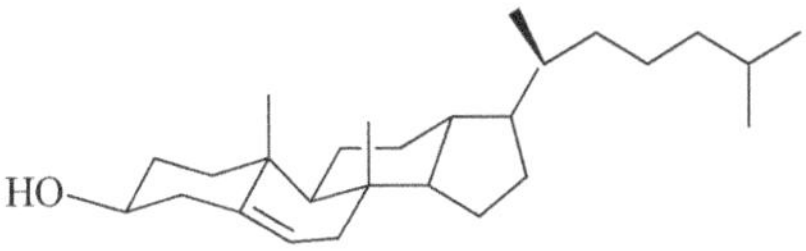

Cholesterol is the raw material for producing the sex hormones, which include the androgens (which are the male hormones and include testosterone) and the oestrogens (female) and also adrenocortical hormones such as aldosterone (which regulates sodium balance). It may also be dangerous, as too much cholesterol in the blood can clog the arteries, particularly those that supply the heart.

7.3 Carbohydrates

The group of compounds known as carbohydrates received their generic name as early observations found that they had the formula $(C \cdot H_2O)_n$. This inferred that they might be 'hydrates of carbon', though the arrangements of the atoms show that it has little to do with water molecules.

The carbohydrates are, in fact, a class of molecules ranging from simple sugars (*monosaccharides*) to complex *polysaccharides* which contain several million monosaccharide monomers. Generally, they are classed as *simple* or *complex*. The simple sugars (monosaccharides) are carbohydrates like glucose or fructose which cannot be converted into smaller sugars by hydrolysis. Complex carbohydrates are composed of two or more simple sugars linked together. Normal household sugar is sucrose which is made up of a glucose and a fructose linked together and is classified as a disaccharide.

Starch and cellulose are two common carbohydrates. Both are macromolecules with molecular weights in the hundreds of thousands. Both are described as polymers (hence '*polysaccharides*'); that is, each is built from repeating units, monomers, much as a chain is built from its links. Polysaccharides are important cell constituents; for example, cellulose is used in the cell walls of plants.

The principal energetic pathway of the cell is glycolysis, which is the breakdown (catabolism) of glucose. (The mechanism for this will be looked at more extensively in Chapter 12.) Glucose is stored in mammalian cells as the polysaccharide glycogen and in plant cells as starch. Other sugars play structural roles in cell matrices and membranes and addition of sugar units to proteins (glycoproteins) modulates the properties of the proteins.

7.3.1 Monosaccharides

These are the simplest carbohydrate molecules. They are linear polyalcohols with an aldehyde or ketone group, and they range in size from three carbon (trioses) to seven carbon (heptoses) in chain length. Thus the trioses, glyceraldehyde and dihydroxyacetone (both of which can be designated as $\{C_3(H_2O)_3\}$), are the simplest of the monosaccharides.

D-Glyceraldehyde   Dihydroxyacetone

It was by a lucky guess that the D-molecular geometry was assigned to (+)-glyceraldehyde in the late nineteenth century, an assumption which was confirmed by X-ray crystallography in 1951.

The simplest carbohydrates are also known as sugars or saccharides (from the Greek, *sakcharon* = sugar) and the name ending of most sugars is -*ose*. For example, glucose is the principal sugar in blood and fructose is the sugar found in fruits and honey.

Monosaccharides are classified according to:

- the number of carbon atoms present in the molecule;
- whether they contain an aldehyde or keto group.

Thus, a monosaccharide containing three carbon atoms is called a triose; one containing four carbon atoms is a tetrose; for five carbon atoms we have pentose; and a hexose contains six carbon atoms.

In addition, monosaccharides may be classified as either *aldoses* or *ketoses*, where:

- the -*ose* part indicates it is a carbohydrate;
- the *ald-* prefix indicates that the carbonyl is present as an aldehyde;
- the *ket-* prefix indicates that the carbonyl is present as a ketone.

Ribose
an *aldo*pentose

Glucose
an *aldo*hexose

Fructose
a *keto*hexose

7.3.2 Carbohydrate stereochemistry

For largely historical reasons, the D-form and the L-form of glyceraldehyde serve as configurational standards for all monosaccharides. A monosaccharide whose highest number stereocentre (the penultimate carbon as drawn in the Fischer representation) has the same configuration as D-glyceraldehyde is designated as a D-sugar (conversely for an L-sugar). By the Fischer convention, acyclic forms of monosaccharides are drawn vertically with the aldehyde or keto group at or nearest the top (the idea of D- and L-configurations, and of Fischer views were introduced in Chapter 3).

Nature, fortunately, uses only sugars which have the *R*-configuration at the penultimate carbon in the chain, these being with the OH group on the right. In the Fischer convention this centre is designated as having the D-configuration, and monosaccharides with this configuration are therefore said to belong to the D-series. Thus, for example, D-erythrose and D-glucose are naturally occurring, but L-erythrose and L-glucose are not.

However, nature does not restrict the stereochemistry at any of the other centres, meaning they can be variable, so even though one stereocentre is limited, there exists a series of diastereometric monosaccharides. Thus there are two natural D-tetroses, four natural D-pentoses and eight D-hexoses, as shown in Figure 7.4.

Ketoses, as the name implies, have a *ketone* instead of an aldehyde in their structure. Although there are many possible ketose structures, there are only a few which are of biological significance. The most important is D-fructose which is metabolically closely related to D-glucose.

7.3.3 Cyclisation in sugars

Pentoses and hexoses are not stable in the open chain form. They rapidly cyclise to form *cyclic hemiacetals* (the formation of these is examined in detail in Section 8.10.3), and the reaction occurs when an aldehyde (or ketone) reacts with an alcohol. For glucose this means one of the alcohol units at one end of the chain interacting with the aldehyde unit, as shown in Figure 7.5.

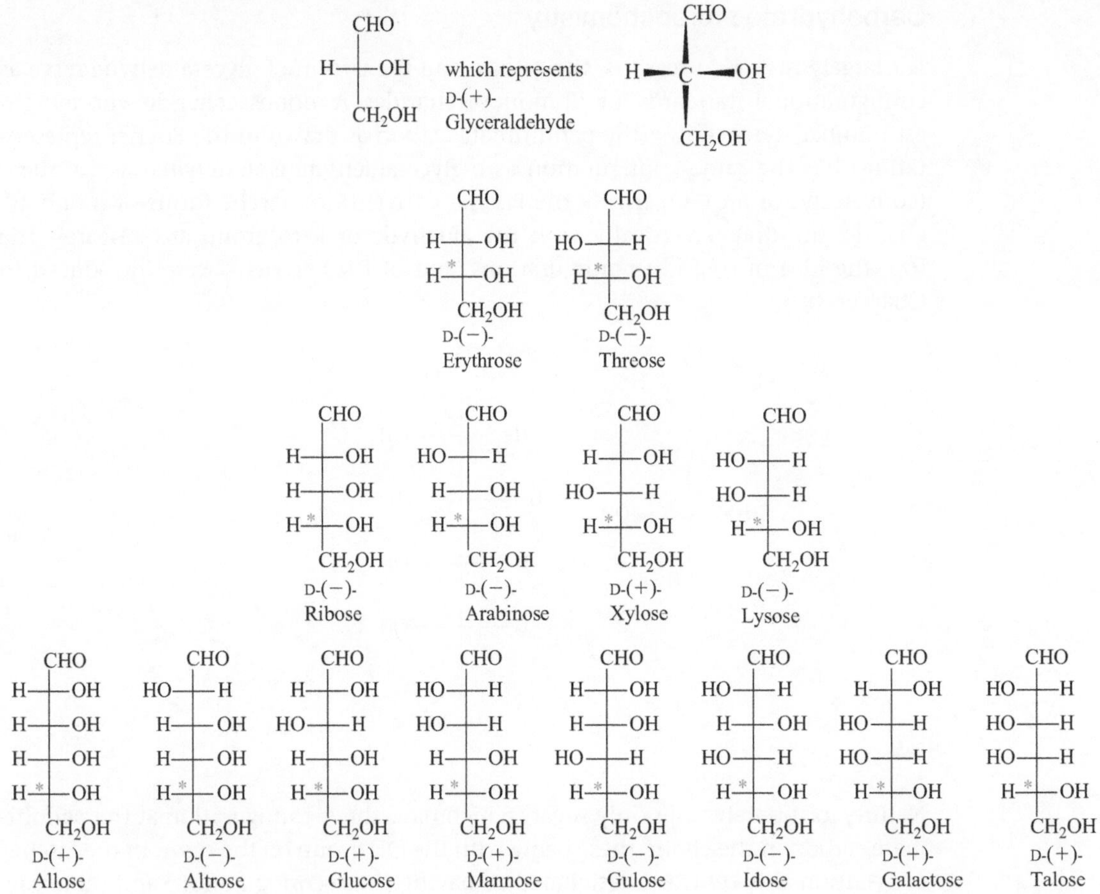

CHO
H——OH   which represents   H——C——OH
CH₂OH   D-(+)-   CH₂OH
              Glyceraldehyde

CHO   CHO
H——OH   HO——H
H——*OH   H——*OH
CH₂OH   CH₂OH
D-(−)-   D-(−)-
Erythrose   Threose

CHO   CHO   CHO   CHO
H——OH   HO——H   H——OH   HO——H
H——OH   H——OH   HO——H   HO——H
H——*OH   H——*OH   H——*OH   H——*OH
CH₂OH   CH₂OH   CH₂OH   CH₂OH
D-(−)-   D-(−)-   D-(+)-   D-(−)-
Ribose   Arabinose   Xylose   Lysose

CHO   CHO   CHO   CHO   CHO   CHO   CHO   CHO
H——OH HO——H H——OH HO——H H——OH HO——H H——OH HO——H
H——OH H——OH HO——H HO——H H——OH H——OH HO——H HO——H
H——OH H——OH H——OH H——OH HO——H HO——H HO——H HO——H
H——*OH H——*OH H——*OH H——*OH H——*OH H——*OH H——*OH H——*OH
CH₂OH CH₂OH CH₂OH CH₂OH CH₂OH CH₂OH CH₂OH CH₂OH
D-(+)-   D-(−)-   D-(+)-   D-(+)-   D-(−)-   D-(−)-   D-(+)-   D-(+)-
Allose   Altrose Glucose Mannose Gulose   Idose   Galactose Talose

Figure 7.4 The family of D-aldoses. The structures are all drawn as Fischer projections. The bonds to left and right come out of the paper, whereas bonds up and down are pointing away (into the paper). The most oxidised carbon is always at or near the top. The D-configuration corresponds to the sugars where the penultimate carbon (marked*) has the OH on the right, which corresponds to the *R*-configuration

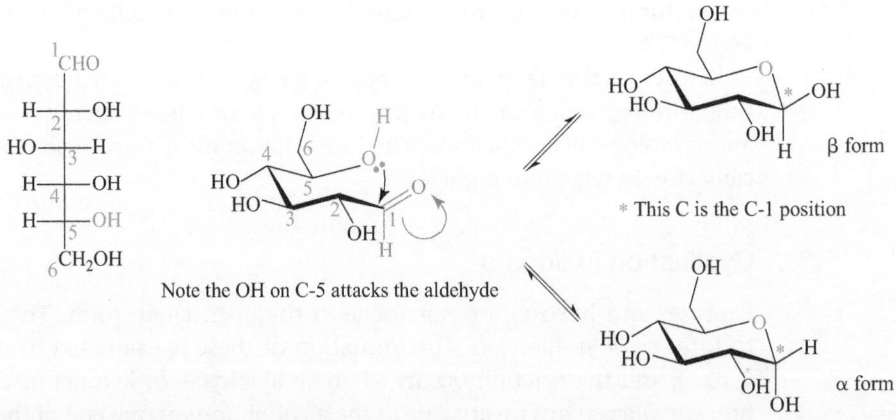

¹CHO
H——OH   2
HO——H   3
H——OH   4
H——OH   5
⁶CH₂OH

Note the OH on C-5 attacks the aldehyde

β form

* This C is the C-1 position

α form

Figure 7.5 Cyclisation of glucose.

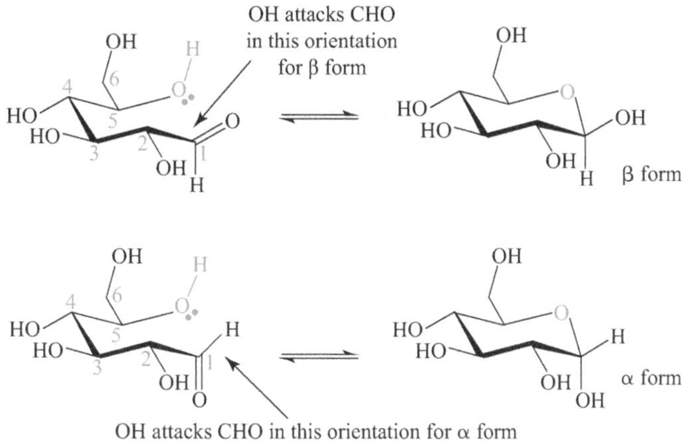

Figure 7.6 **Formation of α and β forms of glucose.**

The reason why there are both α and β forms in the cyclic structure arises from the free rotation about the bond linking C-1 and C-2, as demonstrated in Figure 7.6. The different ways of representing carbohydrate structures are summarised in Box 7.1.

The cyclisation of glucose, as shown in Figure 7.6, is an example of a cyclisation of an aldohexose to form a six-membered ring. This type of cyclic structure is called a *pyranose*, named because of the relationship of the ring structure to the molecule *pyran* (though in fact the monosaccharides are structurally related to the fully saturated tetrahydropyran).

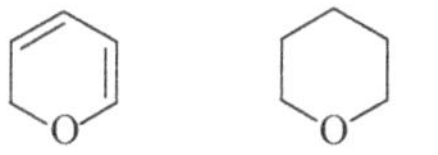

Pyran   Tetrahydropyran

The cyclisation of glucose (and of the other aldohexoses) tends to produce pyranoses as shown above. They can also, however, cyclise by attack of the OH associated with the C-4 on the aldehyde, to produce a five-membered ring called a *furanose* (Figure 7.7).

Dissolution of glucose results in an equilibrium being set up, in which the major components of the equilibrium mixture are the pyranose forms of glucose (though the furanose forms and the open chain form are present at any time). Furanose forms are, however, the main structural type resulting from the cyclisation of the aldopentoses, such as ribose, and also of the ketohexoses, such as fructose (Figure 7.8).

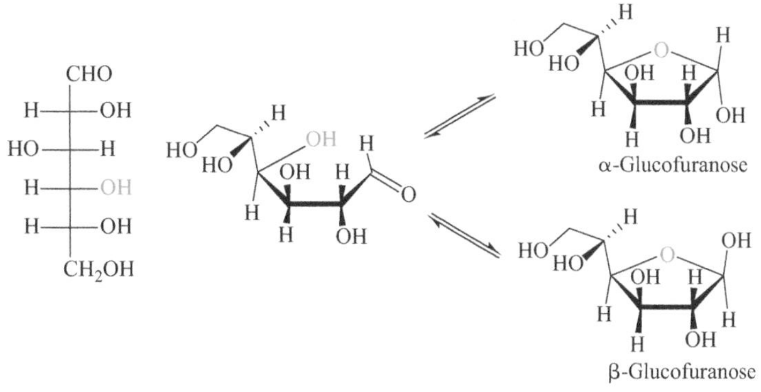

Figure 7.7 **Formation of a glucofuranose.**

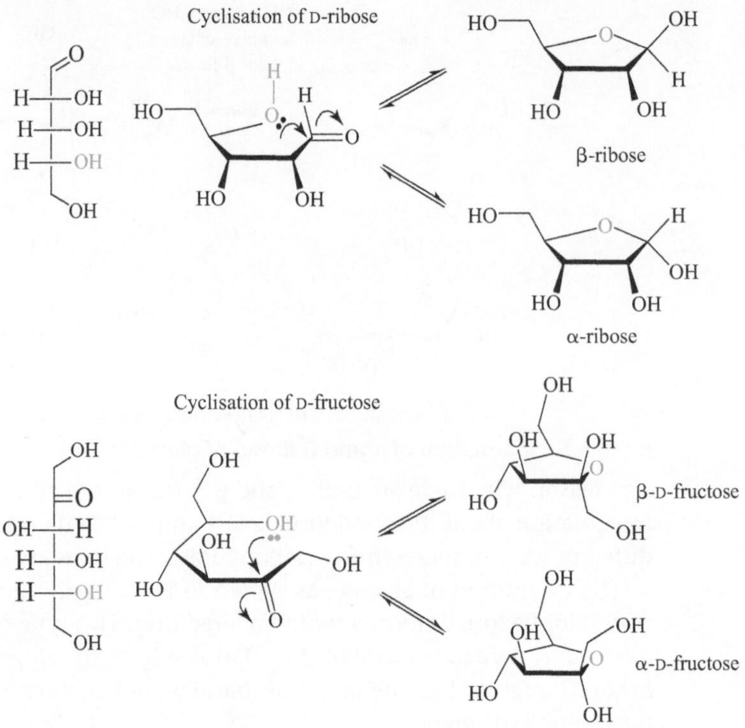

Figure 7.8 Cyclisation to furanose forms from D-ribose (an aldopentose) and D-fructose (a ketohexose).

Box 7.1

Drawing sugar structures

Fischer formulae tend to be a bit awkward to use when describing the cyclic forms. They also require drawing a 'bent bond'. There are two options:

● Haworth formulae (used a lot in traditional biochemical textbooks and particularly useful for extended polysaccharide structures);

● perspective formulae.

These two appear quite similar when used to illustrate five-membered rings, but are quite different for six-membered rings. Below are these three representations (Fischer, Haworth and perspective) as applied to the α-cyclic form of glucose:

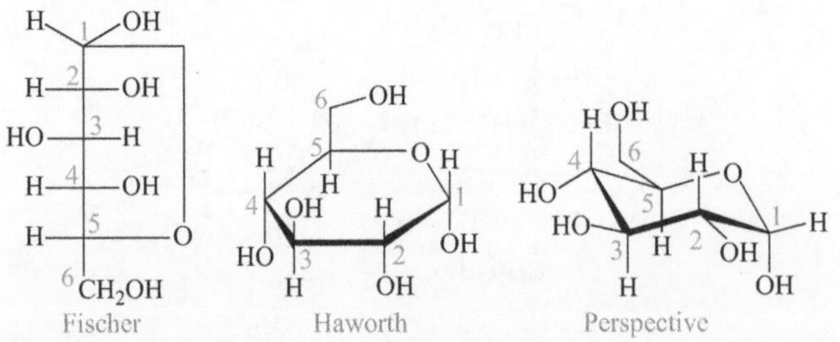

The numbering scheme is shown in each case to allow the different representations to be related to each other.

7.3.4 Disaccharides and polysaccharides

The joining of monosaccharides involves a corresponding loss of water; in other words it is a condensation reaction. Below, for example, is shown the formation of the disaccharides cellobiose and maltose, which arise as a result of, respectively, β-1,4- and α-1,4-linkages (sometimes called glycosidic bonds) between glucose units.

More extensive chains may be made up in a similar way. For example, cellulose is made from a series of glucose molecules linked together with β-1,4-linkages, and cellulose can be made up from hundreds, or even thousands, of glucose units.

By contrast the polysaccharide amylose is made up from a series of glucose molecules linked together with α-1,4-linkages:

213

Cellulose is the structural material that forms the primary cell wall in green plants. The hydroxyl groups that point out from the chain structure of cellulose can interact, through hydrogen bonds, with other cellulose chains. This contributes to the strength of cellulose. Despite this strength, it is permeable to water and to solutes, ensuring that these materials can transport into, and out of, cells.

Amylose is found as a component of starch (somewhere between 10% and 20% by mass), along with *amylopectin*, which is itself another glucose-derived polysaccharide, and which makes up the remaining 80–90% of the starch. The ratio of the two polysaccharides depends on the source; for example, long grain rice has a higher amylose content than does medium or short grain varieties. In the case of amylopectin the glucose units are linked by 1,4-α-linkages as in amylose, but the chain also has branches off, arising from α-1,6-linkages:

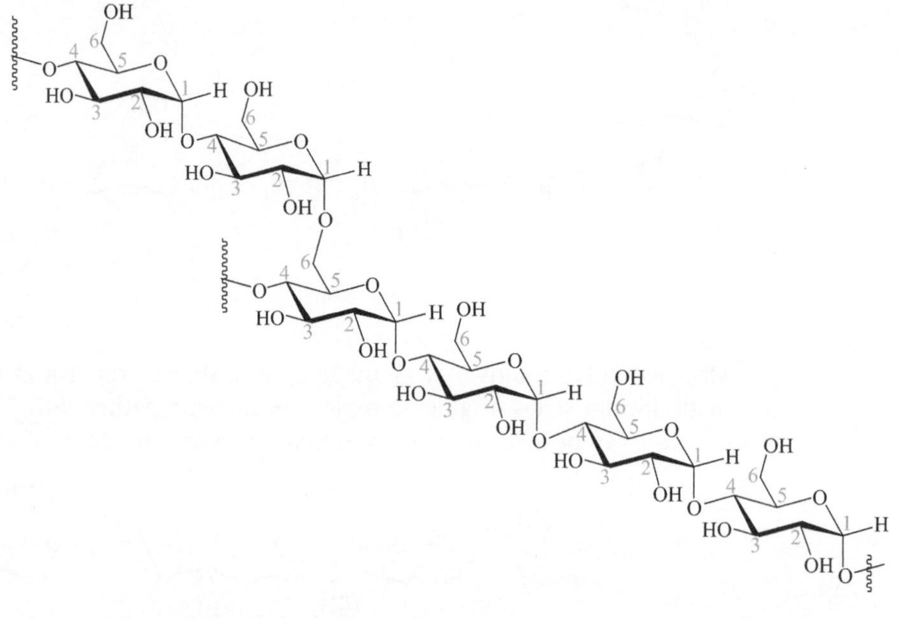

In amylopectin the branches come off the main 'amylose' chain, and they are typically separated by about 10–20 glucose units.

The examples shown so far have involved joining together of the same type of sugar unit, illustrated by glucose dimers and polymers. It is, however, possible for different sugars to combine together to form disaccharides and polysaccharides.

Lactose is a disaccharide found in milk, and it is a dimer composed from D-galactose and D-glucose:

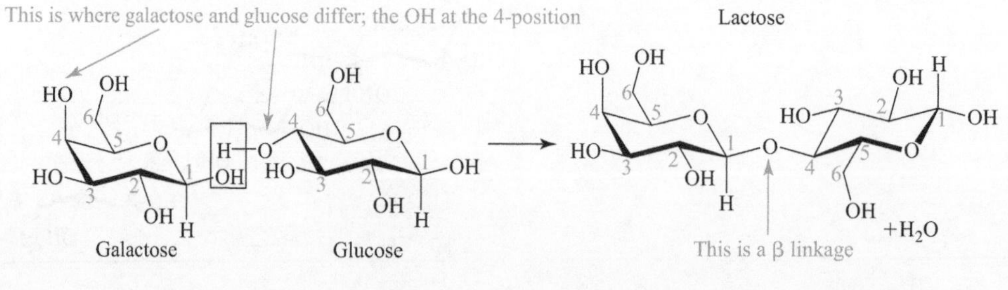

This is where galactose and glucose differ; the OH at the 4-position

Galactose   Glucose

Lactose

This is a β linkage

$+H_2O$

Lactose intolerance

Lactose digestion in mammals is facilitated by the enzyme lactase, and the production of this enzyme tends to fall off as mammals age beyond infancy. Humans differ from other mammals in this, however, and the production of lactase is often maintained in humans. There are significant regional variations, however, with a high percentage of adult northern European populations retaining lactase production, whereas many African and east Asian populations have lost this. For this reason dairy products form a significant part of northern European diets, but play little part in east Asian diets.

7.4 Amino acids, peptides and proteins

7.4.1 Introduction

Any molecule that contains both an amine and a carboxylic acid could be called an amino acid (aa). Those that can form proteins all have the amino group attached to the α-carbon, the one next to the carbonyl carbon. The generalised form of these amino acids is shown in three different ways: linear, Fischer and stereo forms.

Therefore this category of compounds could more correctly be referred to as either α-amino acids or as 2-aminoalkanoic acids.

These differ only in the R group, and the range of these in nature is limited. There are 22 amino acids that are found on analysis of the hydrolytic breakdown of protein structures, and although this hydrolysis of naturally occurring proteins can yield up to 22 α-amino acids, only 20 are actually used by cells to synthesise proteins, as both hydroxyproline and cystine are synthesised only after the polyamide chain is intact.

Table 7.1 lists the 20 amino acids used to synthesise proteins. This set of 20 amino acids can be subdivided into the following groups:

- **Aliphatic/non-polar:** glycine, alanine, valine, leucine, isoleucine, proline
- **Aliphatic/neutral polar:** serine, threonine
- **Aromatic:** phenylalanine, tyrosine, tryptophan
- **S-containing:** cysteine, methionine
- **Acidic:** aspartic acid, glutamic acid
- **Amides:** asparagine, glutamine
- **Basic:** lysine, histidine, arginine

Table 7.1 gives the side chains (R groups) for the 20 most common amino acids, with their three- and one-letter abbreviations. An L–amino acid has the L–configuration at the α-carbon atom:

Table 7.1 **Common amino acids**

Name	Three letter code	One letter code	R group
Glycine	Gly	G	$H-$
Alanine	Ala	A	CH_3-
Valine	Val	V	$(CH_3)_2CH-$
Leucine	Leu	L	$(CH_3)_2CH-CH_2-$
Isoleucine	Ile	I	$CH_3-CH_2-C(CH_3)H-$
Serine	Ser	S	$HO-CH_2-$
Threonine	Thr	T	$HO-CH(CH_3)-$
Proline	Pro	P	 (whole molecule shown)
Phenylalanine	Phe	F	$C_6H_5-CH_2-$
Tyrosine	Tyr	Y	$p-HO-C_6H_4-CH_2-$
Tryptophan	Trp	W	
Methionine	Met	M	$CH_3-S-CH_2-CH_2-$
Cysteine	Cys	C	$HS-CH_2-$
Aspartic acid	Asp	D	HO_2C-CH_2-
Glutamic acid	Glu	E	$HO_2C-CH_2-CH_2-$
Asparagine	Asn	N	$H_2N-CO-CH_2-$
Glutamine	Gln	Q	$H_2N-CO-CH_2-CH_2-$
Histidine	His	H	
Lysine	Lys	K	$NH_2-CH_2-CH_2-CH_2-CH_2-$
Arginine	Arg	R	$HN=C(NH_2)-NH-CH_2-CH_2-CH_2-$

7.4.2 Acid–base behaviour of amino acids: zwitterions

The chemistry of the amino acids is complicated by the fact that the amine, $-NH_2$, group is a base and the $-COOH$ group is an acid. Species exist in their acidic forms at pH values lower (more acidic) than their pK_a value, so as the acid component ($-COOH$) of an amino acid has a pK_a value of about 2, whereas the protonated amino group ($-NH_3^+$) has a pK_a value of about 9, then at approximately neutral pH the acid will be deprotonated and the amine group will be protonated, thus:

$$H_2N \overset{\displaystyle CO_2H}{\underset{\displaystyle R}{\rule{0pt}{1em}}}\!\!\!-H \rightleftharpoons H_3\overset{+}{N} \overset{\displaystyle COO^-}{\underset{\displaystyle R}{\rule{0pt}{1em}}}\!\!\!-H$$

Zwitterion

Despite the fact that the zwitterion contains positive and negative charges, the net charge on the molecule remains as zero. Under acid or base conditions, however, this is not necessarily the case:

$$H_3\overset{+}{N}\overset{\displaystyle COOH}{\underset{\displaystyle R}{\rule{0pt}{1em}}}\!\!\!-H \rightleftharpoons H_3\overset{+}{N}\overset{\displaystyle COO^-}{\underset{\displaystyle R}{\rule{0pt}{1em}}}\!\!\!-H \rightleftharpoons H_2N\overset{\displaystyle COO^-}{\underset{\displaystyle R}{\rule{0pt}{1em}}}\!\!\!-H$$

Acid conditions, e.g. pH = 0   Zwitterion   Base conditions, e.g. pH ≥ 11

Given the information above, it is clear that an amino acid cannot exist as an uncharged compound, because the carboxylic acid group cannot exist in protonated form at pH values where the amino group is not protonated also.

7.4.3 The isoelectric point

The titration curve of a typical amino acid (alanine) is shown in Figure 7.9. This reflects the point made earlier that there are two pK_a values, namely that of the carboxylic acid group with a pK_a value of approximately 2.5, and that of the protonated amine group, with a pK_a value of approximately 9.5.

The mid-point, where the alanine is in the zwitterionic form, is called the *isoelectric point*, referred to as the p*I*. The p*I* corresponds to the pH where the overall

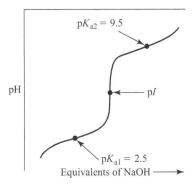

Figure 7.9 **Titration curve for alanine.**

structure is neutral, and it is the pH where the amino acid is least soluble in water. When there are extra ionisable groups in the side chains (for example, where the R group contains acids or amines) the situation is more complex. The amino acid lysine, for example, has two amino groups, each with a different pK_a. This increases the number of possible forms, but the form depends on pH.

Calculation of p*I* values

The actual value of the p*I* depends, therefore, on the detailed structure of an amino acid, including the side chain labelled as R in Table 7.1. To illustrate the point, consider the three amino acids glutamic acid, alanine and lysine, drawn below to represent the dominant ionic forms present at physiological pH (approximately pH = 7):

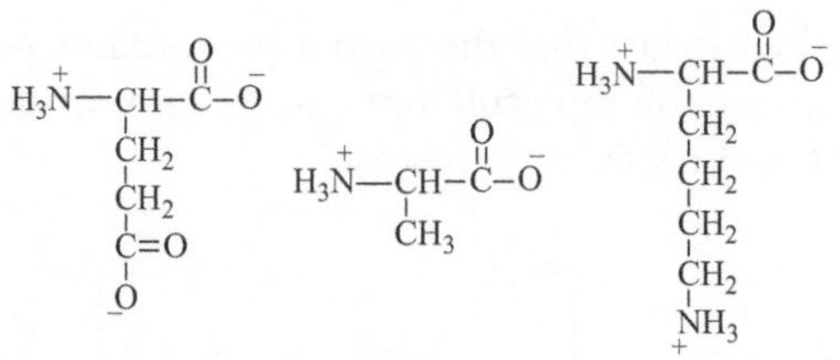

Glutamic acid   Alanine   Lysine

These show that, whilst the alanine has its charges balanced at pH values near neutral, glutamic acid has a net negative charge and lysine has a net positive charge at these pH values. The p*I* value is the average of the pK_a values of the protonated amine and the carboxylic acid group, and thus for alanine, the simplest, of those shown, it is $\dfrac{(9.5 + 2.5)}{2} = 6.0$.

Considering now the example of lysine, looking initially at the situation at physiological pH, and noting the pK_a values of the ionisable groups:

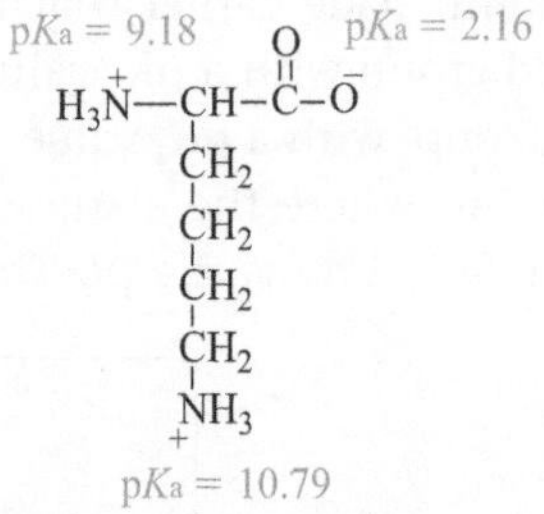

In order for this to become neutral, the amount of positive charge on the molecule must balance the negative. Starting at physiological pH, this can be achieved by increasing the pH (by adding base) until the point at which the molecule achieves overall neutrality. This happens at the average of the pK_a values of the two protonated amine groups; that is, $pI = \dfrac{(10.79 + 9.18)}{2} = 9.89$ (approximately 10.0). At this pH, all of the carboxylic acid groups are present as $-CO_2^-$ ions and the averaged population of the $-NH_3^+$ groups is equal to one.

Considering, finally, the example of glutamic acid, again taking an initial look at the situation at physiological pH, and noting the pK_a values of the ionisable groups:

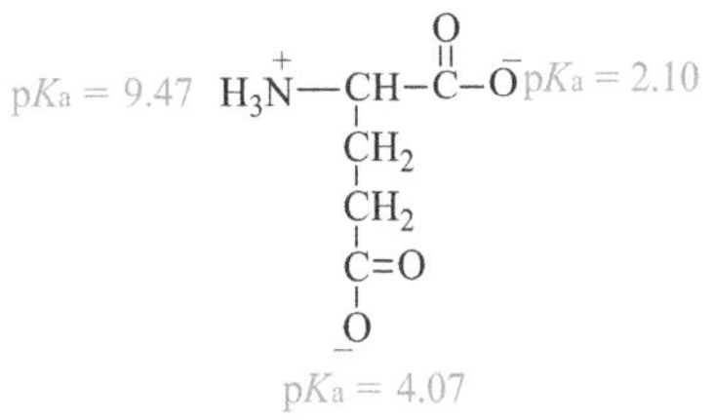

$$pK_a = 9.47 \quad \overset{+}{H_3N}-CH-\overset{\overset{O}{\parallel}}{C}-\bar{O} \quad pK_a = 2.10$$
$$\begin{array}{c} | \\ CH_2 \\ | \\ CH_2 \\ | \\ C=O \\ | \\ \bar{O} \end{array}$$
$$pK_a = 4.07$$

In order for this to become neutral, the amount of positive charge on the molecule must balance the negative. Starting at physiological pH, this can be achieved by decreasing the pH (by adding acid), thereby protonating at the carboxyl groups until the point at which the molecule achieves overall neutrality. This happens at the average of the pK_a values of the two carboxylic acid groups; that is, $pI = \dfrac{2.10 + 4.07}{2} \approx 3.1$ (approximately 3.1).

7.4.4 The stereochemistry of amino acids

The 'core' structure of the amino acids is shown below in zwitterion form:

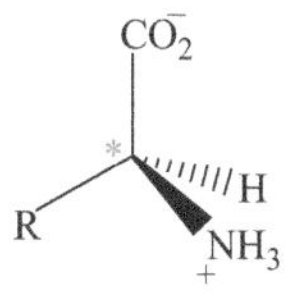

It is clear that the central carbon (marked with an orange asterisk) has four different groups on it, and hence it is chiral. As with carbohydrates, it is possible to represent the amino acids by a Fischer representation, and the Fischer representation of this structure is:

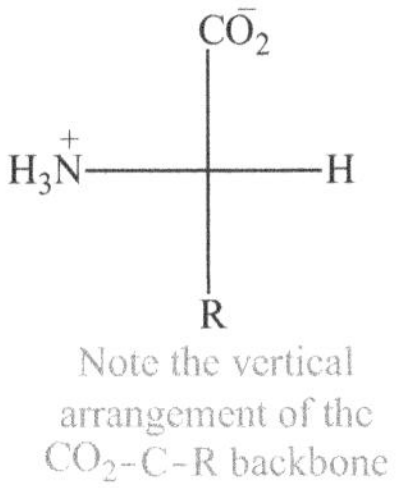

Note the vertical
arrangement of the
CO_2-C-R backbone

Comparison of this with the D-glyceraldehyde referred to earlier (Section 7.3.2) shows that the amino group is on the left hand side of the structure, which is the opposite side to that used for D-glyceraldehyde. This means the amino acid has an L-configuration.

219

The naturally occurring amino acids that make up protein structures almost always have the L-configuration. The one exception is glycine, which has two hydrogen atoms attached to the central carbon atom, and is not chiral as a result. The use of the D- and L-terminologies still exists, though largely for historical reasons. A more rigorous method (as intimated in Chapter 3) for determining absolute configuration in chiral systems is to use the Cahn–Ingold–Prelog system, whereby the chiral centre is identified as being either R- or S-.

Taking the example of alanine, the D- and L-forms of which are shown:

In alanine, the group priorities are:

$$-NH_3^+ > -CO_2^- > -CH_3 > H$$

Using this approach,

- For L-alanine, the 1–2–3 priority order goes in an anticlockwise direction when looking along the C^*—H bond.

- For D-alanine, the 1–2–3 priority order goes in a clockwise direction when looking along the C^*—H bond.

This means that:

- L-alanine corresponds with the S-configuration;
- D-alanine corresponds with the R-configuration.

It has already been mentioned that L-forms are almost universal and D-forms are rare, not participating in protein structures. It is found that, with only one exception, L configurations correspond with S and D configurations correspond with R. The exception referred to is the amino acid cysteine:

In cysteine, the group priorities are:

$$-NH_3^+ > -CH_2SH > -CO_2^- > H$$

Using this approach,

- For L-cysteine, the 1–2–3 priority order goes in a clockwise direction when looking along the C^*—H bond.

- For D-cysteine, the 1–2–3 priority order goes in an anticlockwise direction when looking along the C^*—H bond.

This means that:

- L-cysteine corresponds with the *R*-configuration;
- D-cysteine corresponds with the *S*-configuration.

7.4.5 Peptides and proteins

The amide group

Enzymes can cause α-amino acids to polymerise through the elimination of water, as shown:

$$H_3\overset{+}{N}-CH-\overset{O}{\overset{\|}{C}}-\overset{-}{O} \quad H_3\overset{+}{N}-CH-\overset{O}{\overset{\|}{C}}-\overset{-}{O}$$
$$\underset{CH_3}{\qquad} \qquad \underset{CH_3}{\qquad}$$

$$\downarrow -H_2O$$

$$H_3\overset{+}{N}-CH-\overset{O}{\overset{\|}{C}}-NH-CH-\overset{O}{\overset{\|}{C}}-\overset{-}{O}$$
$$\underset{CH_3}{\qquad} \qquad \underset{CH_3}{\qquad}$$

Amide or peptide link

The [—CO—NH—] (amide) linkage that forms between the amino acids is known as a peptide bond or peptide linkage. Amino acids when joined in this way (as opposed to being free) are called amino acid residues.

The polymers that contain 2, 3, a few (3–10) or many residues are called dipeptides, tripeptides, oligopeptides and polypeptides respectively. The term oligopeptide is frequently used to underline the fact that a peptide contains more than one different type of amino acid residue. The distinction between a peptide and a protein is largely semantic. Large natural oligopeptides (usually with >80 residues) with defined 3D structures are normally referred to as proteins, and proteins are molecules that contain one or more such polypeptide chains.

Two properties of the amide functional group are important:

- The amide is uncharged.
- The amide is essentially rigid and planar.

The planar rigid structure is a consequence of resonance with two resonance forms which can be drawn for the amide:

$$\underset{H}{\overset{\diagdown}{N}}-\overset{O}{\overset{\diagup}{C}} \quad \longleftrightarrow \quad \underset{H}{\overset{\diagdown}{\overset{+}{N}}}=\overset{\overline{O}}{C}$$

The consequence is that peptide structures are not as conformationally flexible as might have been imagined, and they can be considered as 'plates', formed by the amide functions, joined by bridges consisting of the α-carbon atoms of the amino acid residues. In addition, because most of them are chiral and have relatively large groups on them, so it substantially reduces the number of energetically favourable shapes that a large chain can adopt. So these factors contribute to the ultimate 3D structure of a protein.

Polypeptides are linear polymers, with one end of the polypeptide chain terminating in an amino acid residue that has a free protonated amine group (known as the *N-terminal residue*) while the other end terminates with a free carboxylate ($-CO_2^-$) group (called the *C-terminal residue*).

$$H_3\overset{+}{N}-CH-\overset{\overset{O}{\|}}{C}+HN-CH-\overset{\overset{O}{\|}}{C}+HN-CH-\overset{\overset{O}{\|}}{C}-\overset{-}{O}$$

R	R′	R″
Residue 1	Residue 2	Residue 3
N-terminal end		C-terminal end

7.4.6 Primary, secondary, tertiary and quaternary structures

Protein primary structure

The *primary structure* of a protein or peptide is merely the amino acid sequence. This is normally written and named from the N-terminus. The sequences are often shown with the amino acids represented by their one letter codes, which are shown in Table 7.1. This enables the sequences to be identified in as concise a way as possible.

One such example is the small (76 amino acid) protein ubiquitin, which is found in all eukaryotic cells (hence its name, from the word ubiquitous, which means found everywhere). The primary structure (or amino acid sequence) found in human cells is:

> MQIFVKTLTGKTITLEVEPSDTIENVKAKIQDKEGIPPDQQRLIFAGKQLEDGRTLSD
> YNIQKESTLHLVLRLRGG

In the arrangement above, each letter is the single letter code for an amino acid. The primary structure gives the ordering of the amino acid sequence, but the way in which peptide chains are arranged in three dimensions is also vitally important to the way in which a protein functions.

Protein secondary structure

The *secondary structure* of a protein is defined by the local conformation of its polypeptide backbone. Whilst the amide group itself is planar and has a fixed configuration, rotations of the groups attached to the amide nitrogen and the carbonyl carbon are relatively free and these rotations allow peptide chains to form different conformations. Structural forms known as β-sheets and α-helices result from the conformational changes within the polypeptide chain that allow the formation of regular folding patterns, usually in the form of pleated sheets and helices.

β-*Sheets*

There are two types of β-sheets, shown in Figure 7.10, namely:

- the parallel β-sheet.
- the antiparallel β-sheet.

Antiparallel arrangement

Parallel arrangement

Figure 7.10 β-sheet formation in proteins.

The *parallel β-sheet* is characterised by two peptide strands running in the same direction held together by hydrogen bonding between the strands, as indicated in Figure 7.10. The grey dotted lines represent hydrogen bonds between the strands (the hydrogen bonds form between the H on the amide nitrogen atoms and a lone pair on amide oxygen atoms). Without seeing the ends of the peptide it can be difficult to tell in which direction a peptide strand is running. One way to distinguish between whether two adjacent strands are running in the same direction (parallel) or opposite directions (antiparallel) is to count the number of atoms in the hydrogen bonded rings. In a parallel β-sheet each hydrogen bonded ring has 12 atoms in it. Also, the hydrogen bonds holding together the two peptide chains are not 180 degrees.

The *antiparallel β-sheet* has two peptide strands running in opposite directions held together by hydrogen bonding between the strands, as indicated in Figure 7.10. The orange dotted lines show hydrogen bonds between the strands. A feature characterising an antiparallel β-sheet is the number of atoms in the hydrogen bonded rings, with the number of atoms in hydrogen bonded rings in an antiparallel β-sheet alternating between 14 and 10. Another feature is that the hydrogen bonds in an antiparallel β-sheet are linear, and this probably accounts for antiparallel β-sheets being more stable than parallel β-sheets. In protein structural diagrams the β-sheets are usually depicted by arrows, as shown:

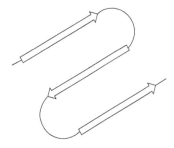

This is so that even fairly complex structural features can be summarised in an easy to recognise shorthand format.

The α-helix

Another very important secondary structural type that defines the folding and function of a protein is the α-helix. This structure is a right handed helix with 3.6 amino acid residues per turn. Each carbonyl group (C=O) in the nth residue

223

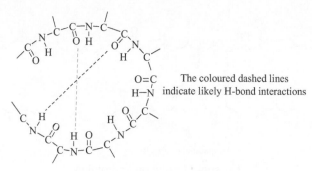

The coloured dashed lines indicate likely H-bond interactions

Figure 7.11 An α-helical arrangement of amino acids in a protein.

of the chain has a hydrogen bond to the NH hydrogen of an amide group in the $(n+4)^{th}$ amino acid residue in either direction, and the R-groups extend outwards from the axis of the helix. The N—H $\cdots$ O distance is approximately 280 pm. A schematic representation, viewed from the top, is shown in Figure 7.11. Note how this is an example of *intra*molecular hydrogen bonding, whereas it is more common to think of hydrogen bonds as an *inter*molecular phenomenon.

When α-helices are shown in protein structure diagrams they are usually shown as coils. Figure 7.12 shows a representation of a form of ubiquitin, and the presence of β-sheets and an α-helix are indicated. Not all of the protein is ordered in the form of β-sheets or α-helices, and those parts of the protein structure exist as *random coils*.

Protein tertiary structure

The *tertiary structure* of a protein is its 3D shape that arises from further foldings of its polypeptide chains and the additional folding superimposed on the coils of the α-helices. In other words there are interactions between the different β-sheets and/ or α-helices. Under normal environmental conditions they occur in one particular way, which is both characteristic of a particular protein and is also important for its function.

Various forces, such as ionic attraction (formation of 'salt bridges'), hydrogen bonding and hydrophobic interactions, acting between side chains of individual amino acids may be responsible for the establishment of the tertiary structure, and disulfide bridges sometimes provide covalent stabilisation.

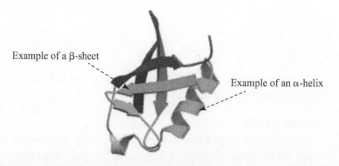

Example of a β-sheet

Example of an α-helix

Figure 7.12 Protein structure diagram of ubiquitin.

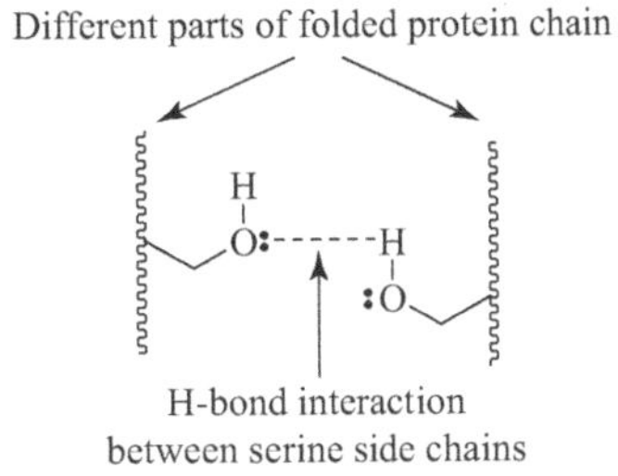

Different parts of folded protein chain

H-bond interaction
between serine side chains

Figure 7.13 **The contribution of hydrogen bonds to protein tertiary structure.**

Disulfide bridges

Tertiary structure is largely maintained by disulfide bonds. Disulfide bonds are formed between the side chains of cysteine by oxidation of two thiol groups (SH) to form a disulfide bond (S—S). This is also sometimes called a disulfide bridge.

Hydrogen bonds

Hydrogen bonding can exist between side chains, such as serine, because the side chain of the amino acid serine contains an —OH group and this provides the possibility of producing a hydrogen bond between two serine residues in different parts of a folded chain (Figure 7.13).

Similar hydrogen bonds involving —OH groups, or —COOH groups, or —$CONH_2$ groups, or —NH_2 groups in various combinations can be formed. Recall, however, that a COOH group and an —NH_2 group would form a zwitterion and produce a stronger ionic bond rather than a hydrogen bond.

Salt bridges

Salt bridges result from the neutralisation of an acid and amine on side chains, and this results in an ionic bond between the positive ammonium group and the negative acid group. This can happen for different combinations of acidic or amine amino acid side chains. For example, glutamic acid has a carboxylic acid group on its side chain that forms a salt with the amine on the lysine side chain, and this type of interaction is illustrated in Figure 7.14 (compare with the formation of zwitterions in Section 7.4.2).

Hydrophobic interactions

The non-polar groups on amino acid side chains mutually repel water and other polar groups with the result that there seems to be a net attraction of the non-polar groups for each other. Hydrocarbon alkyl groups on Ala, Val, Leu and Ile behave in this way. For most proteins the effect is to support folding that exposes the maximum number of polar (hydrophilic) groups to the aqueous environment and encloses a maximum number of non-polar (hydrophobic) groups within its interior.

225

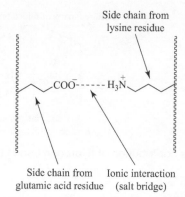

Figure 7.14 **The contribution of 'salt bridges' to protein tertiary structure.**

Quaternary structure

The *quaternary structure* refers to interactions between different protein monomers. The monomers may be the same or they may be different. For example, haemoglobin is a protein tetramer, made up of two different types of monomer (referred to as subunits), and each subunit occurs twice (see Figure 7.15).

7.4.7 Denaturing of proteins

The denaturing of proteins involves the disruption and/or destruction of its secondary and tertiary structures. It does not affect the primary structure, as the processes involved in denaturing are not sufficiently vigorous to break the amide bonds, but they will disrupt the α-helices and β-sheets in a protein, resulting in it uncoiling into a random shape. Usually denaturing a protein results in the precipitation or coagulation of the protein.

Typically ways in which denaturing can be effected are:

- heating a protein, which can disrupt hydrogen bonds and non-polar hydrophobic interactions (think about what happens to the white of an egg on cooking);
- alcohol will disrupt hydrogen bonds, hence its use as a hand wash;
- acids and bases disrupt salt bridges.

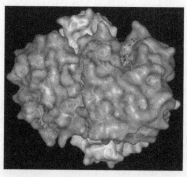

Figure 7.15 **Representation of haemoglobin.**

7.5 Nucleic acids

7.5.1 Introduction

The nucleic acids deoxyribonucleic acid (DNA) and ribonucleic acid (RNA) are the carriers and processors of the cell's genetic information. DNA molecules preserve hereditary information while RNA transcribes and translates this information, which allows the synthesis of all the proteins required in a cell. Recent developments in the understanding of genetic information, the manner in which it is copied and how it is decoded into the functioning proteins of the cell has resulted from the study of the nucleic acids, meaning that it is important to possess some appreciation of their components, namely the nucleotides and nucleosides.

The building blocks (monomers) of nucleic acid structures are the *nucleotides* (see Figure 7.16). The complete hydrolysis of a nucleotide produces three component parts:

- a five-carbon monosaccharide (sugar), either: ribose (RNA) or deoxyribose (DNA);
- a phosphate ion;
- a heterocyclic base: either a purine or a pyrimidine.

The monosaccharide

The central portion of the nucleotide is the monosaccharide, which is always present as a five-membered ring; that is, as a furanoside. The carbon atoms of the monosaccharide portion are numbered as 1', 2', 3' etc. As indicated in Figure 7.16, for nucleotides obtained from RNA, the sugar component is D-ribose, while that for DNA is 2'-deoxy-D-ribose, in which the —OH position at 2' is replaced by an H.

D-Ribose   2'-Deoxy-D-ribose

The heterocyclic base is attached through a linkage (known as an N-glycosidic linkage) to C1' and this is always a β anomer. The phosphate group is present as a phosphate ester attached to C5' (as shown in Figure 7.16) or C3'.

When there is no phosphate the compound is referred to as a *nucleoside*. For example, adenosine and deoxyadenosine are nucleosides.

Ribose   OH OH
Adenosine

Deoxyribose
(because OH at 2'
replaced by H)

OH
Deoxyadenosine

227

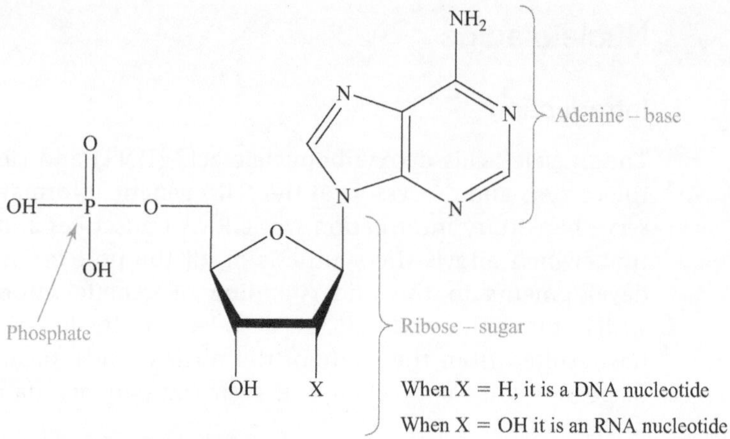

Figure 7.16 **A nucleotide monomer.**

Purines and pyrimidines

The nucleosides that can be obtained from DNA all contain 2-deoxy-D-ribose as their sugar component and one of the four heterocyclic bases: adenine, guanine, cytosine or thymine. For RNA there are also four bases, but these are either: adenine, guanine, cytosine or uracil. Two of these are *purines* and the others are *pyrimidines*.

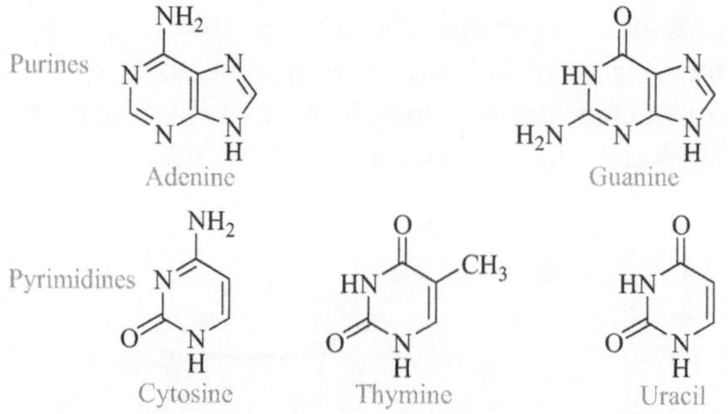

Overall, the purine or pyrimidine base links to the sugar via a dehydration reaction as shown:

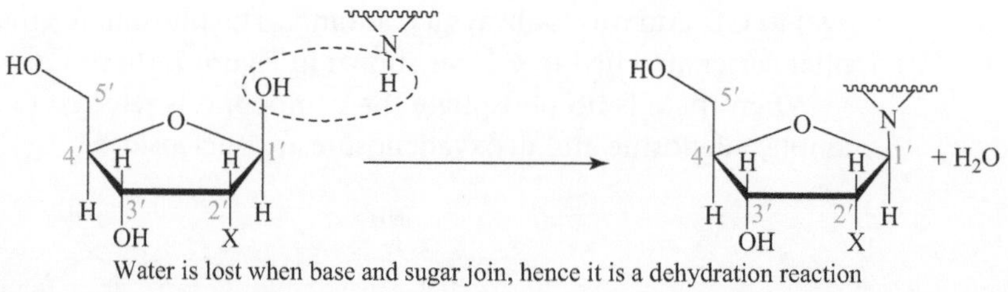

Water is lost when base and sugar join, hence it is a dehydration reaction

7.5.2 Primary structure of nucleic acids

Nucleotides bear the same relation to nucleic acids that amino acids do to proteins: in other words they are monomeric units. Whereas the connecting links in proteins

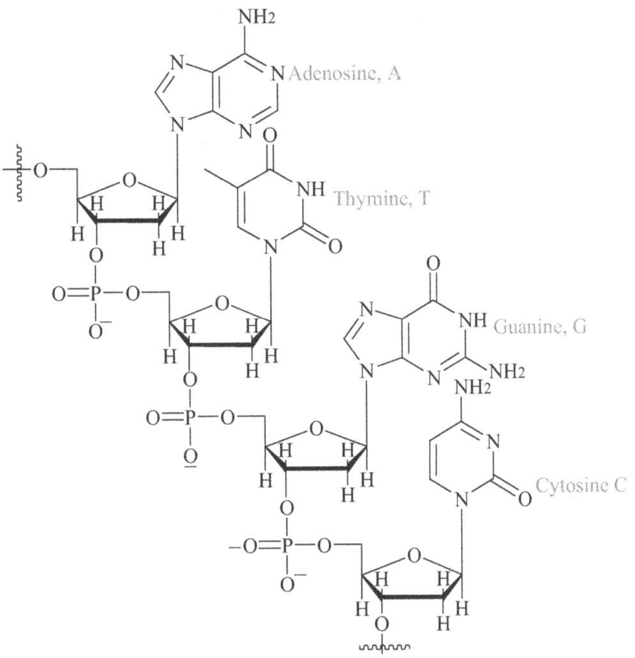

Figure 7.17 **Representation of a short sequence from a strand of DNA.**

are amide groups, in nucleic acids they are phosphate ester linkages. Phosphate esters link the 3'-OH of one ribose (or deoxyribose) with the 5'-OH of another ribose (or deoxyribose). The nucleic acid is, therefore, a long unbranched chain with a 'spine' of sugar and phosphate units with heterocyclic bases extending from the chain at regular intervals, as indicated for a short sequence illustrated in Figure 7.17.

The direction, and order, of the bases shown in Figure 7.17 are listed in the following abbreviated way:

$$5'A-T-G-C\ 3'$$

The numbers at the start and end indicate the position of the phosphate on the ribose at each end of the chain (compare this with the N- and C-terminal ends of peptides). The order shows the base sequence along the chain of DNA which encodes the genetic information. This is the primary structure.

7.5.3 Secondary structure in nucleic acids

The background to DNA

In 1953, James Watson and Francis Crick proposed a model for the secondary structure of DNA (which was verified by Rosalind Franklin and Maurice Wilkins using X-ray diffraction). Watson and Crick developed the work of Erwin Chargaff, who had noted the presence of regular features in the percentages of heterocyclic bases obtained from DNA.

- The total mole percentage of purines is approximately equal to that of the pyrimidines, which can be summarised as:

$$\frac{\%G + \%A}{\%C + \%T} \approx 1$$

- The mole percentages of adenine and of thymine are approximately equal (%A/%T ≈ 1), as are those of guanine and cytosine (%G/%C ≈ 1).

In fact, A will only hydrogen bond with T, and C will only hydrogen bond with G.

The X-ray diffraction data demonstrated that the secondary structure of DNA is a double helix, with the two nucleic acid chains being bridged by hydrogen bonds between base pairs on opposite strands. The double chain is wound into a helix with both chains sharing the same axis, as shown in Figure 7.18, which also shows the pairing of the bases, which takes place via hydrogen bonds, and occurs between adenine (A) and thymine (T) and also between guanine (G) and cytosine (C).

The purine and pyrimidine bases are on the inside of the helix (which positions the bases for pairing), and the phosphate and deoxyribose components are on the outside of the helix. The planes of the bases are almost perpendicular to the axis. There are, in fact, different double helix forms of DNA, known as A DNA, B DNA

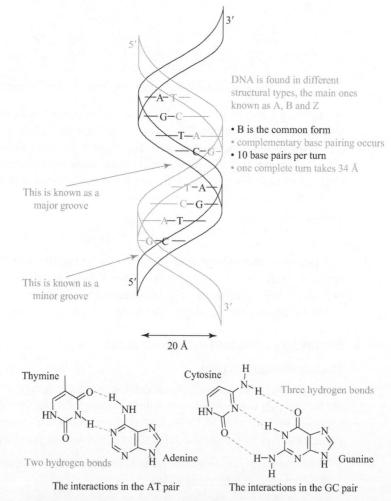

DNA is found in different structural types, the main ones known as A, B and Z

- B is the common form
- complementary base pairing occurs
- 10 base pairs per turn
- one complete turn takes 34 Å

This is known as a major groove

This is known as a minor groove

20 Å

Thymine

Two hydrogen bonds

Adenine

The interactions in the AT pair

Cytosine

Three hydrogen bonds

Guanine

The interactions in the GC pair

Figure 7.18 Representation of a DNA double helix, and the base pairing that holds it together.

and Z DNA. The most stable of these, and the form usually referred to, is the B form. There are a number of key points:

- It is the form favoured for random-sequence DNA under physiological conditions.
- The helices are right handed (this is the same as screw threads).
- The strands are *complementary in nucleotide sequence*, which means that a T on one strand is base-paired with an A on the other strand; similarly if one strand has a G, it pairs with a C on the other strand.
- The strands are *antiparallel*: a strand aligned $5' \rightarrow 3'$ will partner one aligned $3' \rightarrow 5'$.
- The major groove is wide and deep; the minor groove is narrow and deep.
- The DNA secondary structure is not rigid, but it is flexible.

DNA from a single human cell contains millions of base pairs, and extends in a single thread for almost 2 m long.

7.5.4 Structural features of RNA

There are three different types of RNA, known as messenger RNA (mRNA), ribosomal RNA (rRNA) and transfer RNA (tRNA), which each serve different biological functions.

Whilst DNA strands come together to form a double helix, this is not the case for RNA, which remains as single strands. Secondary structure does occur as a consequence of folding; for example, consider tRNA (which is the smallest of the three types, usually comprising 70 to 90 nucleotides). As can be seen in Figure 7.19, it folds itself in such a way as to resemble a cloverleaf.

Relative stability of DNA and RNA

The presence of the OH at the 2' position of the ribose part in the nucleotides that make up RNA has a profound impact on the stability of it relative to DNA, because the vicinal 2'-OH group makes the 3-phosphodiester bond susceptible to nucleophilic cleavage. This accounts for the need for DNA to be the deoxy form, because if it were

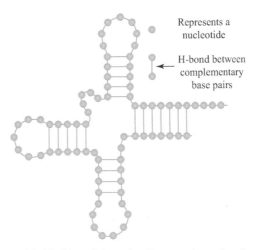

Represents a
nucleotide

H-bond between
complementary
base pairs

Figure 7.19 tRNA strand folded into 'cloverleaf' secondary structure.

231

broken down too readily then it could not carry hereditary information. RNA, on the other hand, needs to be broken down when its task is completed.

DNA replication

DNA replication is a process that occurs in living organisms, and it is the basis of biological inheritance. It creates two complete strands of DNA (one for each daughter cell) where only one existed before (from the parent cell). Shortly before the cell begins to divide, the double strand of DNA within the nucleus of the cell is 'unzipped' by an enzyme, which breaks the hydrogen bonds between the bases. The result is that two halves of the DNA now exist independently. There are free-floating nucleotides within the cell nucleus which begin to interact (through hydrogen bonding) with the bases of both strands. Because A only hydrogen bonds with T and C only with G, both strands form exact matches, meaning that complementary strands are formed along each chain and, in effect, each chain acts as a template for the formation of a complementary strand.

Once unwinding has completed there are two identical DNA molecules, and these molecules are passed on, one to each of the daughter cells. This is represented pictorially in Figure 7.20.

RNA transcription

This is the process by which a DNA sequence is copied, and a complementary RNA produced. It occurs in several steps, and these can be summarised thus:

1. Enzymes called RNA polymerases bind to specific regions of DNA called promoters.

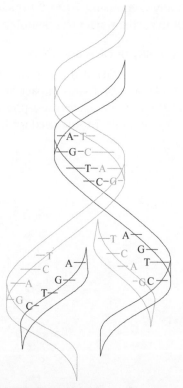

Figure 7.20 Representation of DNA replication.

2. When the polymerase binds to the promoter, the DNA chain separates, and one side of the DNA chain is then used for transcription. This chain is called the template.

3. The polymerase 'brings in' the RNA nucleotides that are complementary to those on the DNA chain. This is the transcription process. Note that in RNA uracil is used instead of thymine as the complementary base for adenine.

4. Transcription continues until termination, where the polymerase recognises the termination point of the DNA sequence.

The single strand of RNA (messenger RNA) thus produced is then used in the building up of proteins.

Questions

7.1 Look at the three structures below and answer the following questions.

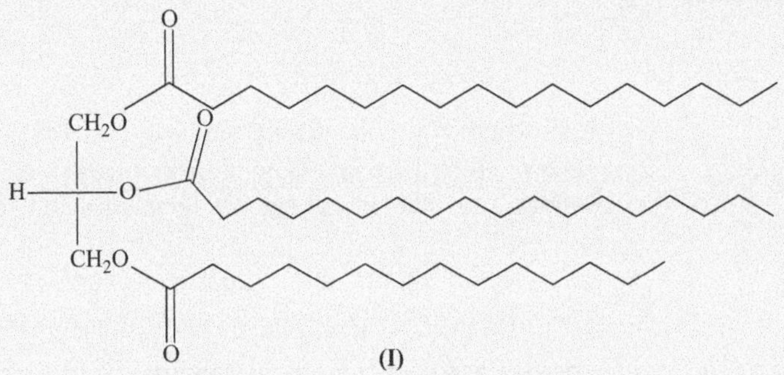

(I)

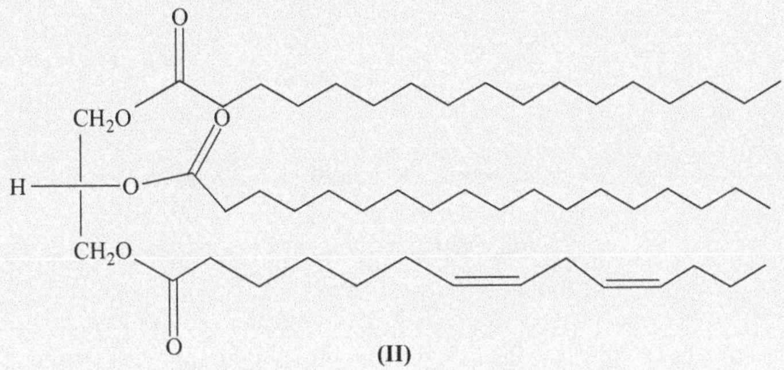

(II)

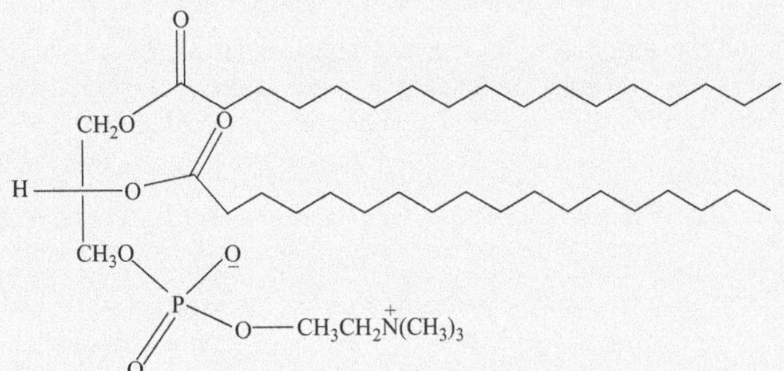

(III)

(a) Which of these are triglycerides? Identify the glycerol and fatty acid components.

(b) Compound **I** is a hydrogenated form of compound **II**. Which has the higher melting point?

(c) Is compound **III**, the phospholipid, chiral? If so identify any chiral centre(s).

(d) What behaviour would you expect this compound to exhibit when dispersed in water?

7.2 Below is a perspective representation of a monosaccharide. Draw the Fischer and Haworth representations of this structure and answer the following questions.

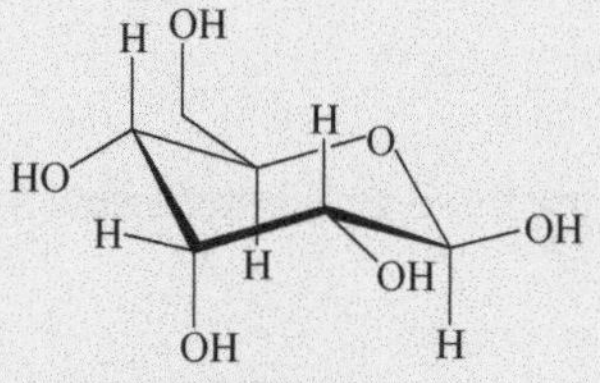

β-D-Allose

(a) State whether the ring is a pyranose or a furanose.

(b) State whether the sugar is an aldose or a ketose.

(c) Redraw the diagram above and indicate the locations of its five chiral centres.

(d) Comment on whether the sugar is a D- or an L-isomer.

(e) State whether the ring form is an α- or a β-anomer.

7.3 The molecule shown is a monosaccharide.

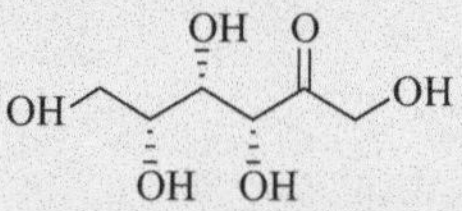

(a) State whether it is a pentose or a hexose.

(b) State whether it is an aldose or a ketose.

(c) Draw the Fischer formula for this molecule, and comment whether it belongs to the D- or L-chiral series.

(d) This compound does not exist to any significant extent as the linear form drawn here, but as a mixture of two cyclic furanose forms. Draw the structures for these explaining how they differ.

7.4 Amylose, which is a natural polysaccharide (soluble starch) derived from glucose, is a useful foodstuff for humans whereas another glucose-based polysaccharide, cellulose, is not. Comment on the structural differences between amylose and cellulose, and on why humans can digest one and not the other.

7.5 Peptides and proteins are polymers of amino acids. What is the name of the functional group that results when two amino acids are joined together?

7.6 Draw the two structures of the tetrapeptide Ala-Met-Ser-Gly, one with the Ala being at the C terminal end, the other where it is at the N terminal end.

7.7 Oxytocin is a nonapeptide hormone with the amino acid sequence cysteine-tyrosine-isoleucine-glutamine-asparagine-cysteine-proline-leucine-glycine. Categorise each of these amino acids as either non-polar, amide containing, sulfur containing or aromatic.

7.8 Describe the two main elements of secondary structure found in proteins, and the forces that are responsible for the stabilisation of these secondary structures.

7.9 Describe, and illustrate with diagrams as necessary, the tertiary structure of proteins.

7.10 RNA and DNA are polymers made of nucleotide building blocks.

   (a) State the three main components of nucleotides.
   (b) The nucleobases in DNA are guanine, adenine, cytosine and thymine. State which bases are complementary with each other.
   (c) Draw the hydrogen bond interactions between thymine and its complementary base in a DNA duplex.
   (d) State how the chemical composition of RNA differs from that of DNA.
   (e) Write down the names of the four bases found in RNA.

7.11 Describe the differences between nucleotides and nucleosides.

7.12 Draw a diagram to illustrate how nucleotides link together in DNA and RNA.

7.13 Write down the base sequence of DNA that is a complement to the DNA sequence: $3' - GCGTCA - 5'$.

8 Reaction mechanisms

Learning outcomes

By the end of this chapter students should be able to:

- recognise the most important functional groups in a molecule;
- predict the type(s) of reaction a molecule might undergo;
- understand the terms S_N1, S_N2, E1 and E2 reactions;
- identify whether a reagent is able to act as a nucleophile or as an electrophile;
- understand the importance of resonance effects;
- recognise the criteria for positive and negative inductive effects;
- draw out a reaction mechanism using curly arrows formalism.

8.1 Introduction

Chapter 3 introduced organic chemistry, focusing mainly on naming schemes of organic molecules, and in looking at the role that the shape of molecules plays in a biological setting. In this chapter the main objective will be concerned with examining what reactions occur frequently in biological settings, and, in particular, looking at how they happen, and why they go one way rather than another. In many cases the principles of stereochemistry covered earlier are important. A *chemical equation* is a form of accountancy and says nothing about what happens at a molecular level. For example, during a reaction some bonds are broken and others formed, which happens as a result of electrons being moved around, but the equation says nothing about the detailed ordering of these processes.

8.2 Organic reaction types

There are millions of different reactions of organic compounds and yet they *all* fall into one of four different types: addition, elimination, substitution and isomerisation.

8.2.1 Addition reactions

These are reactions in which two molecules combine, and they almost invariably require one of the two to have a multiple bond. For example, consider the reaction between ethene and hydrogen bromide:

$$H—Br \quad + \quad H_2C{=}CH_2 \quad \longrightarrow \quad H_3C—CH_2Br$$

The HBr adds 'across' the double bond, the H adding to one of the carbons associated with the double bond, and the Br to the other.

Sometimes a molecule can add to itself: under the right conditions acetaldehyde (ethanal) will do this:

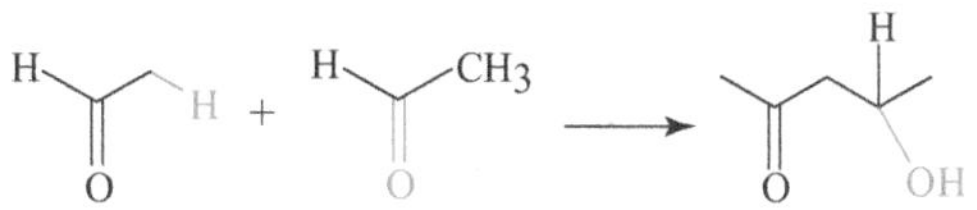

Here one C—H bond from one aldehyde molecule adds to the C=O in another (bonds involved indicated in orange). The reaction is known as the **aldol reaction**. This reaction and its reverse are found in many metabolic pathways, just involving different (more complex) aldehyde molecules.

8.2.2 Elimination reactions

These are, in essence, the opposite of addition reactions. In elimination reactions, a molecule loses some atoms to form a compound with a double bond or a ring. Consider, for example, the following, where two Br atoms are removed from adjacent carbon atoms, and a C=C bond results:

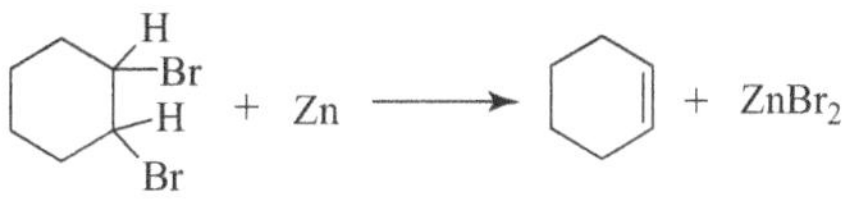

Enzyme-catalysed elimination of water (relevant atoms in orange) from citrate ions is one of the many processes involved in the citric acid (Krebs) cycle involved in the degradation of food and the production of energy:

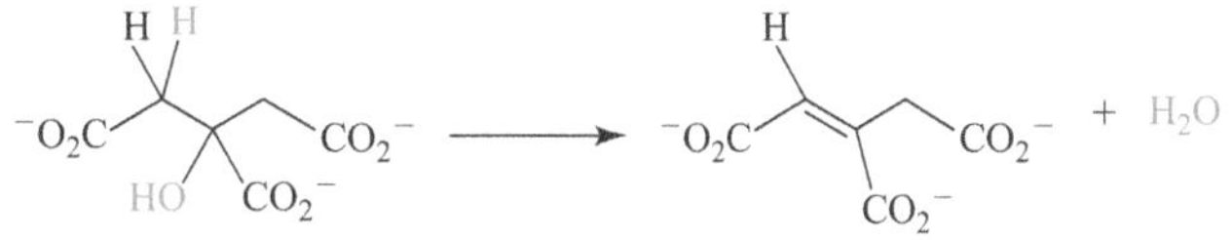

8.2.3 Substitution reactions

These involve replacement of one atom, or group of atoms, by another. Thus, in the reaction of bromoethane with the hydroxide ion:

$$HO:^- + H_3C—CH_2Br \rightarrow H_3C—CH_2OH + Br:^-$$

In this example the Br on the bromoethane has been substituted by the OH, resulting in the formation of an alcohol and the release of a bromide ion. The colon represents a lone pair of electrons.

237

8.2.4 Isomerisation reactions

Isomerisation reactions involve a single molecule undergoing electron reorganisation. These are usually initiated by heat or light.

This type of process occurs in the chemistry of vision (the conversion of *trans-* to *cis*-retinal, as described in Chapter 3).

8.2.5 Oxidation and reduction

In some of the four different reaction types described above, *oxidation* or *reduction* occurs as well.

- Oxidation is the incorporation of oxygen, or the loss of hydrogen or the loss of an electron.
- Reduction is the incorporation of hydrogen, or the loss of oxygen or the gain of an electron.

Example involving oxidations and reductions

Here is an example of transformations involving reversible oxidation and reduction processes.

8.3 Reaction mechanisms

To understand how and why these and, indeed, all reactions occur, it is necessary to look in detail at what the electrons are doing. The *reaction mechanism* is a description of how the electrons are reorganised in going from the starting materials to the products. This usually involves a series of steps, each involving no more than two molecules or molecular fragments.

The concepts of energetics and equilibria (introduced in Chapters 4 and 5) described how it was possible to establish whether a reaction was feasible, and also how much product was likely to form. This said nothing, however, about how fast a reaction might progress, or how rapidly equilibrium was achieved.

For example, reactions involving the combustion of alkanes, resulting in the production of CO_2 and H_2O, are spontaneous, so why don't alkanes just decompose automatically? It is because there is an energy barrier that needs to be overcome, called the *activation energy*.

Consider the situation in Figure 8.1, which shows how energy varies against the progress of the reaction for a one step reaction. Whilst the products are more stable than the reactants (they have a lower energy associated with them), and the reaction is therefore spontaneous according to the thermodynamic rules, we find that energy still needs to be put into the system initially in order to get 'over the hurdle' of the activation energy.

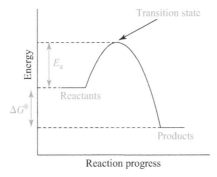

Figure 8.1 A reaction profile (E_a = activation energy; ΔG^θ = free energy difference between reactants and products).

Figure 8.2 demonstrates the situation that arises if the reaction proceeds via an intermediate material, rather than going directly from reactants to products. In this case the presence of a second transition state is noteworthy, as well as the presence of two activation energies:

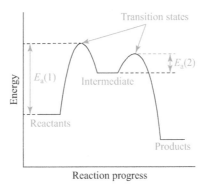

Figure 8.2 A reaction profile for a reaction involving an intermediate stage.

- $E_a(1)$ = activation energy of step 1;
- $E_a(2)$ = activation energy of step 2.

8.3.1 Catalysts

Very often, if the reactants and products are thermally stable, the energy provided can be in the form of heat. This then gives the reactant molecules enough kinetic energy to overcome the activation energy barrier. In some instances, however, either the material is insufficiently stable for this to happen or the activation energy barrier may be too high. In such situations a catalyst may sometimes help. Catalysts are substances that reduce the activation energy barrier, as indicated in Figure 8.3.

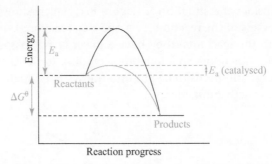

Figure 8.3 Comparison of the reaction profiles of a catalysed and an uncatalysed reaction.

Key definitions

A *reactive intermediate* is a molecular fragment that is formed during a chemical reaction but is too short-lived to be isolated and it reacts further. It is represented by an energy minimum on the reaction profile and all its bonds are fully formed.

A *transition state* is represented by an energy maximum on the reaction profile where some bonds are not completely broken and/or formed (sometimes called an *activated complex*).

There are two ways in which bonds can be broken: *homolysis* and *heterolysis*.

8.3.2 Homolysis

In homolysis there is a separation of the two electrons that make up the bond that is being broken. This involves bond breaking where the electrons in the bond separate so that one electron goes to each of the atoms originally bonded by them. This gives two neutral fragments which are atoms and/or free radicals.

$$Cl\text{—}Cl \longrightarrow Cl^{\bullet} + Cl^{\bullet}$$

In the above:

- A half-headed curly arrow of the type shown represents the movement of a single electron.
- Breaking any bond this way in a neutral molecule always gives two *neutral* fragments.

8.3.3 Heterolysis

In heterolysis the two electrons remain as a pair, and therefore during the process of bonds being broken, the electrons in the bond stay as a pair and *both* go with one of the atoms originally bonded by them. This gives charged fragments; in other words, ions.

$$H\text{—}Br \longrightarrow H^+ + Br^-$$

In the above:

- A *full-headed curly arrow* of the type shown represents the movement of a pair of electrons.
- The tail of the arrow is where the electrons are before the bond is broken.
- The head of the arrow is where the electrons finish up after the bond is broken.

Curly arrows are also used to show bond formation (see Box 8.1). Considering, for example, the ionisation shown above, it only occurs if water is present. In practice a lone pair of electrons on the oxygen atom of a water molecule forms a bond with the hydrogen of the HBr, at the same time as the H—Br bond breaks, thereby forming the hydronium ion, H_3O^+, as indicated below:

Most reactions of organic compounds involve heterolysis and the production of reactive intermediates that are either positively or negatively charged ions.

In *heterolytic reactions* there are two types of reagents and intermediates: *electrophiles* and *nucleophiles*.

An electrophile is a molecule or molecular fragment that seeks electron-rich areas of molecules (or alternatively could be regarded as being susceptible to attack by electron-rich groups). It has an empty orbital that can accept a pair of electrons and is usually positively charged. Examples of electrophiles include:

- H^+ (the proton)
- Br^+ (the bromonium ion)
- carbocations such as CH_3^+
- Lewis acids such as $AlCl_3$, which has a vacant orbital on the aluminium.

A nucleophile is a molecule or molecular fragment that seeks positive centres or electron-deficient regions of molecules. A nucleophile must have a *lone pair of electrons* and, whilst they are often negatively charged, there are many examples of neutral nucleophiles. Examples include:

- hydroxide, $HO:^-$
- bromide, $Br:^-$
- ammonia, $H_3N:$
- carbanions such as $^-:CH_3$.

Box 8.1

Curly arrows

Curly arrows are a shorthand method, used by organic chemists, to indicate the movement of electrons during the course of a reaction. To summarise some key features:

- Most reactions involve a 'pair of electrons' and their movement is depicted as a so-called 'curly arrow'.
- The tail of the arrow is where the electrons are before the bond is broken (from a lone pair or a bonding pair from a σ- or π-bond).
- The head of the arrow is where the electrons finish up after the bond is broken (new lone pair or a new bond).
- The number of atoms and charges at the beginning and end of a step must always be balanced!

(Continued)

Question

Why, in the dissociation of HBr in water, if two electrons are moving to the Br, does the H end up with just one + charge and the Br with just one − charge?

Summarised as:

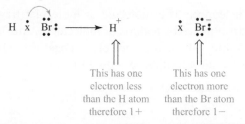

often shown as

Answer

The H—Br bond is covalent and, as such, one of the electrons in the bond can be associated with the contributing H atom and the other from the Br atom. Using a Lewis electron-dot approach the HBr can be represented thus:

$$H \times \quad \bullet \ddot{\underset{\bullet\bullet}{Br}} \colon \quad H \overset{\bullet}{\times} \ddot{\underset{\bullet\bullet}{Br}} \colon$$

H atom   Br atom   HBr molecule

In this representation the electron nominally from the H is represented by x, whereas the electron from the bromine is shown by the round dot. Showing now the ionisation process, it is clear that the result of the movement of the two electrons from the HBr bond to the Br is that the H now has one less electron compared with the H atom, and the Br now has one more electron than does the parent atom.

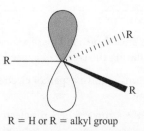

This has one electron less than the H atom therefore 1 +   This has one electron more than the Br atom therefore 1 −

8.3.4 Carbocations and carbanions; types and key points

Carbocation

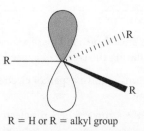

R = H or R = alkyl group

- It has six electrons around the central carbon, with two in each C—R bond.
- It is sp^2 hybridised, with a vacant (empty) p orbital.
- It is positively charged.
- Because it is sp^2 hybridised, it has a planar geometry.

Carbanion

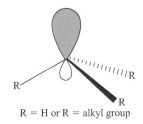

R = H or R = alkyl group

- It has eight electrons around the central carbon.
- It is negatively charged.
- There are two electrons in each C—R bond.
- The sp$_3$ orbital holds a lone pair of electrons.
- It is pyramidal due to the sp^3 character of the carbon.

<table>
<tr><td>8.4</td><td></td></tr>
</table>

8.4 Electronegativity and bond polarity

The electronegativity of the atoms involved in a bond determines the distribution of electrons in that bond. This means that the electrons are not shared equally, and they lie towards the more electronegative atom. Table 8.1 lists some key electronegativity values for elements that are frequently found in biomolecules. Covalent bonds in which the electrons are shared unequally are referred to as *polar covalent bonds*.

Table 8.1 **Electronegativity values for some important elements**

Element	Electronegativity
F	4.0
O	3.5
N	3.0
Cl	3.0
Br	2.8
S	2.5
C	2.5
I	2.5
H	2.1

When molecules are drawn, polar covalent bonds can be indicated by identifying the electronegativity of the atoms with the symbols δ+ and δ−. The symbol δ+ is located at the atom with the lowest electronegativity value, and the δ− is located at the more electronegative atom, to indicate where the greatest electron density is found. Thus the molecules ethanol and hydrogen chloride are often represented as shown:

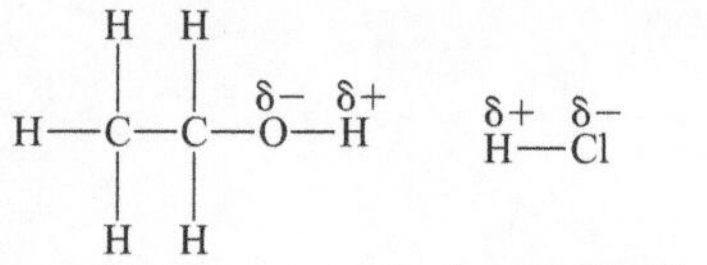

Ethanol   Hydrogen chloride

This approach is equally applicable where multiple bonds exist. For example, considering compounds containing the carbonyl functional group, then the oxygen is $\delta-$ and the carbon is $\delta+$, as shown:

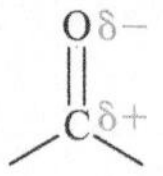

Indeed the electrons are less strongly held in a double bond, which means that there is a slightly greater degree of bond polarisation.

As we will see throughout the coming chapter, the existence of such partial charges is often used to rationalise why a mechanism will proceed one way rather than another. We will now examine some of the key mechanisms that are found in biological pathways, using simple chemical examples in the first instance, to illustrate how the ideas described so far in this chapter are used, and to develop our understanding of mechanisms.

8.5   Addition reactions

These reactions, as the name suggests, are reactions where one material is added to another to produce a product. Simple examples, involving alkenes, are demonstrated below:

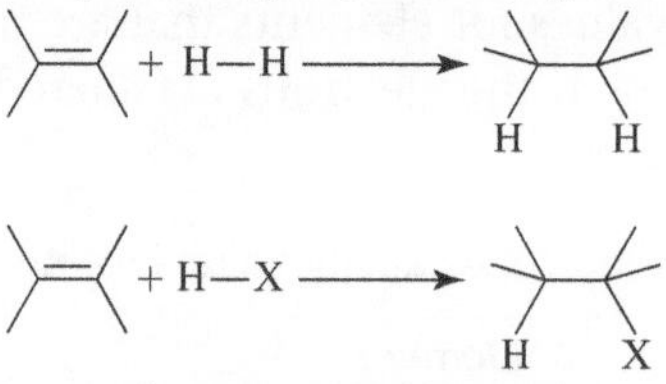

In each of the above a diatomic molecule has added to a double bond, one atom from the diatomic adding to one end of the double bond, and the other part of the diatomic adding to the other end of the double bond

More complex examples exist where two molecules of aldehydes (alkanals) add together (the aldol addition reaction). The general representation of this is:

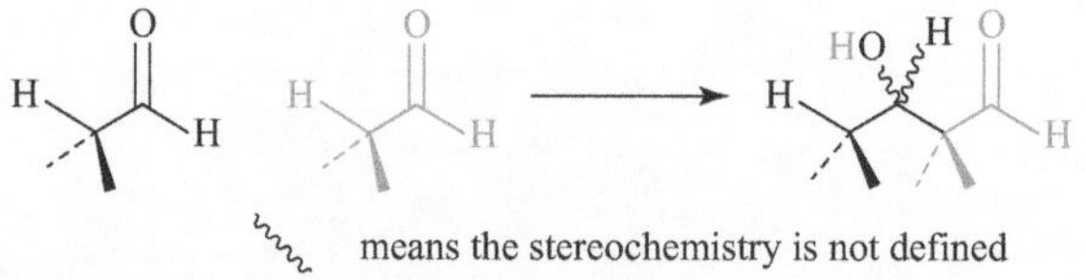

$\sim\sim\sim$ means the stereochemistry is not defined

8.5.1   Electrophilic additions to alkenes and alkynes

The equation for the reaction between hydrogen bromide and ethene (ethylene) is summarised by the equation:

$$H_2C={=}CH_2 + HBr \rightarrow C_2H_5Br$$

The resultant product is bromoethane (ethyl bromide), and the equation shown is merely an overall summary, and says nothing about the way the reaction actually proceeds. In trying to predict how this might take place, an important consideration is the role of the activation energy. The process will want to use as little energy as possible to get it started; in other words it will have as small an activation energy as possible.

Let us consider the nature of the two reagents involved in the reaction, namely the alkene and the hydrogen bromide, from the point of view of the electrons associated with each molecule. The alkene has electron-rich regions (the π-bond) above and below the plane of the atoms, whereas the HBr will have most of the electron density around the relatively electronegative bromine atom, as indicated:

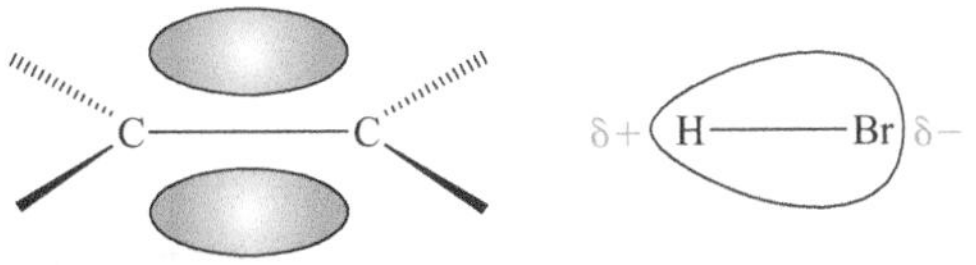

The easiest route will be for the H end of the HBr molecule to approach the π-electron region of the alkene: the π-electrons move to form a bond to the hydrogen, releasing both electrons of the HBr bond to the bromine. This becomes a bromide ion and there is positive charge on the alkene carbon which has not received the proton.

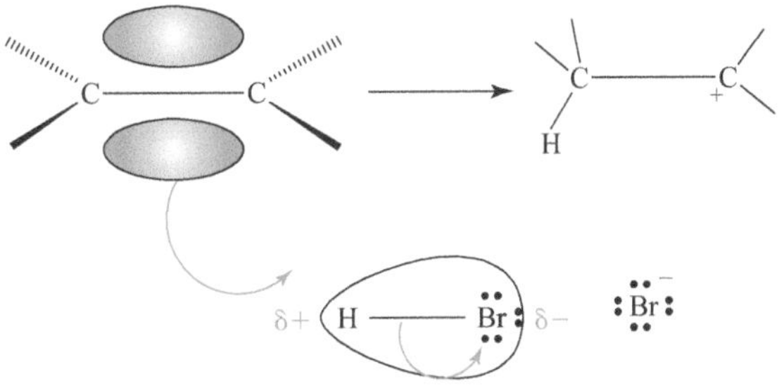

Normally the orbital shapes are not shown when drawing mechanisms, but it is more common to represent the above as shown in step 1 below. This is followed (step 2) by the reaction of the bromide ion (Br⁻) with the positively charged carbocation:

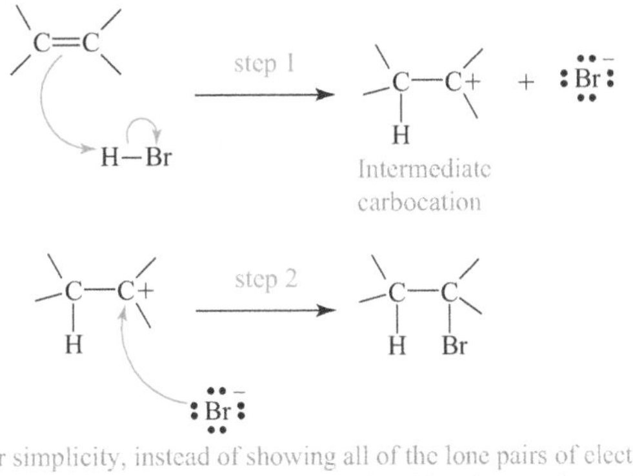

Intermediate
carbocation

Note that, for simplicity, instead of showing all of the lone pairs of electrons on species, it is often only those involved in bond making/breaking that are shown. Therefore the bromide ion could be represented by $\overset{..}{Br}$ rather than $:\overset{..}{\underset{..}{Br}}:$

Earlier it was shown that a 'reaction profile' gives a qualitative idea of the energy changes taking place as the reaction progresses. The reaction of HBr with an alkene involves the generation of an intermediate carbocation, and the profile for this reaction is described in Figure 8.4. This reaction is known as an *electrophilic addition*

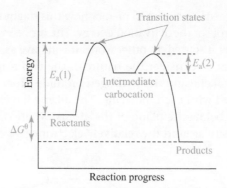

Figure 8.4 Reaction profile for alkene + HBr. (E_a(1) and E_a(2) = activation energies for steps 1 and 2 respectively; ΔG^θ = standard free energy change for the reaction.)

reaction since it starts by the addition of an electrophile H to the alkene. The mechanism as shown above illustrates how the reaction mechanism proceeds in a general sense, but if the alkene is unsymmetrical then the question arises about which carbon of the alkene is protonated. (In other words, to which carbon is the hydrogen added?)

8.5.2 Addition of HBr to asymmetrical alkenes

With an asymmetrical alkene, for example propene CH_3—CH=CH_2, there are two possible products as indicated below:

The mechanisms for the formation of these products are demonstrated below.

Possible route 1 involves H⁺ attack on C1

Secondary carbocation

Possible route 2 involves H⁺ attack on C2

Primary carbocation

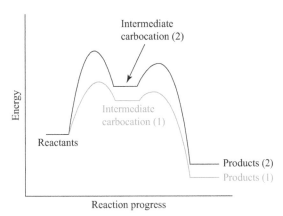

Figure 8.5 **Comparing the reaction profiles of different mechanisms.**

In practice, if the reagents are pure, then only route 1 is followed. This is because this route takes a lower energy path than route 2. The intermediate carbocation identified in route 1 has a lower energy than that for route 2 and so requires a lower activation energy for its formation (Figure 8.5). The activation energy for H^+ attack at C1 is lower because the cation to be formed has a lower energy. Carbocations tend to have a lower energy (be more stable) if the charge is spread (known as a delocalised charge) over a larger area rather than concentrated on one atom.

This is possible in two ways:

1. The charge can be spread where the carbon atom formally bearing the charge has alkyl groups attached to it. Alkyl groups have what is called a *positive inductive effect (called a +I effect)*, which means that they tend to push their electrons away from them if there is a positive centre nearby. This will partly reduce a positive charge on a neighbouring atom and the alkyl group becomes slightly positive, as illustrated:

$$H_3C\overset{+}{-}CH_2 \underset{\text{+ charge}}{\overset{\substack{\text{electrons pushed} \\ \text{from } CH_3 \text{ towards}}}{\Longrightarrow}} \overset{\delta+}{H_3C}\overset{(1-\delta)}{-}\overset{+}{CH_2}$$

So the more alkyl groups that are attached to the carbon bearing the charge, the lower will be the energy of the carbocation. (By contrast, electronegative atoms pull electrons towards themselves and are therefore said to have a *negative inductive effect; a −I effect.*)

2. Where there is a double bond joined to the carbon bearing the charge it is possible to draw another *canonical (resonance) structure* (see Box 8.2), as in the allyl ion shown next, so that the charge is located on a different atom. Neither of these is the true structure – the charge resides half on one carbon and half on the other and the pair of electrons, making the double bond shared by the three adjacent carbon atoms.

247

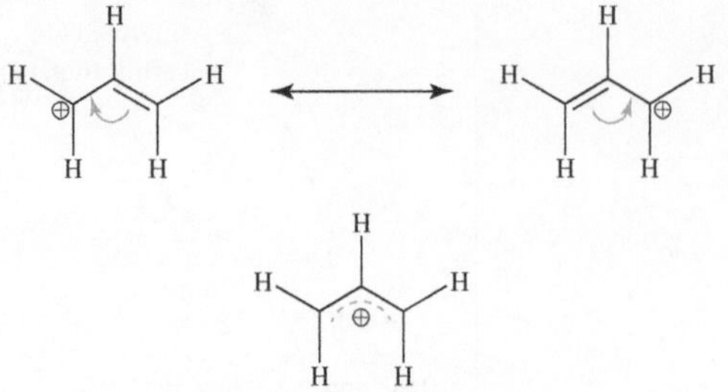

Delocalised: positive charge spread

Therefore, some carbocations will be more stable than others as a result of the spreading out of charge about the ion. The diagram below provides a rough idea of the relationship between carbocation type and relative stability:

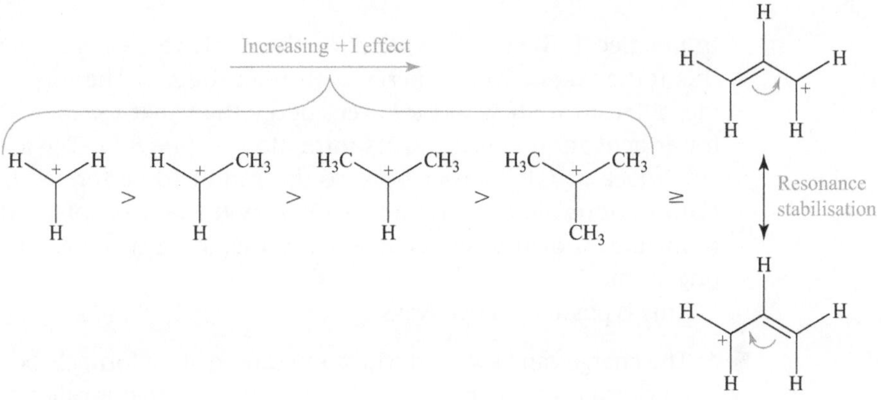

In the reaction of HBr with propene, the product from route 1 is formed rather than the product from route 2 and the reaction proceeds via the more stable carbocation because the activation energy for its formation is lower. This is not unreasonable since if two alternative routes are available, the lower energy one will be adopted (it has a lower hurdle).

This mechanism holds for all electrophilic addition reactions of alkenes and also to alkynes.

Box 8.2

Resonance effects

In Chapter 2 the concept of resonance was introduced, using both the carbonate (2−) ion and benzene to illustrate the idea. This concept will now be looked at again, using the electron-dot representation of *sulfur* dioxide to illustrate it.

$$\bar{O}\!-\!\overset{+}{S}\!=\!O \quad \longleftrightarrow \quad \bar{O}\!=\!\overset{+}{S}\!-\!O$$

(I)

(II)

- • represents the electrons nominally arising from the S
- x represents the electrons nominally arising from the O
- x represents the electron resulting in the negative charge

In sulfur dioxide, the two SO bonds are the same length and strength. This can be rationalised by using the representation given in **(I)** above. If this is represented in the Lewis electron-dot formulation, then the situation shown in **(II)** above results. In this it can be seen that:

- the S and the O atoms in the structures all have complete octets;
- the S contributes only a nominal five electrons to the structure, hence the positive charge (S is from group 16, and hence the atom has six valence electrons);
- the O that is singly bonded to the S has one additional electron and hence carries a negative charge (O is in group 16, and hence has six valence electrons).

We can now look at benzene again, showing how the delocalisation of electrons can be illustrated by use of curly arrows:

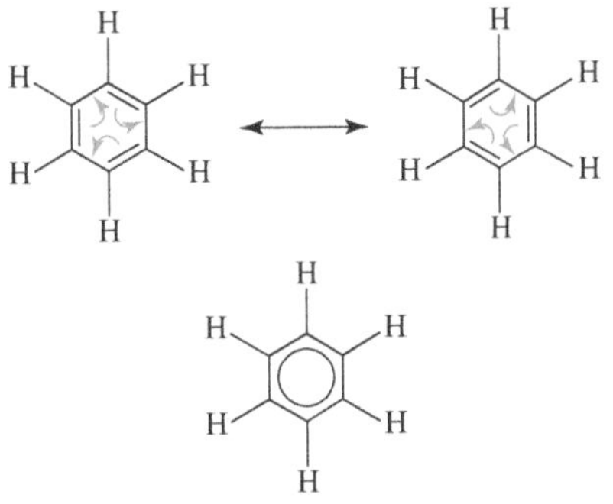

8.5.3 Addition of other electrophiles to alkenes

In the above reactions, the H—Br molecule can be regarded as having two components, namely an electrophile H^+ joined to a nucleophile $Br:^-$. This would be just one of a number of different molecules that could be represented as ENu. It is therefore possible to describe a whole series of reactions by a general reaction as shown:

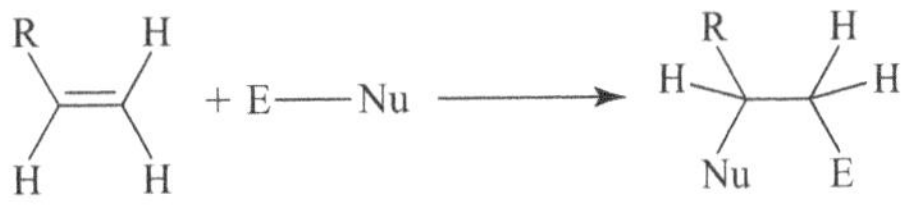

249

Examples

HCl and HI – as HBr

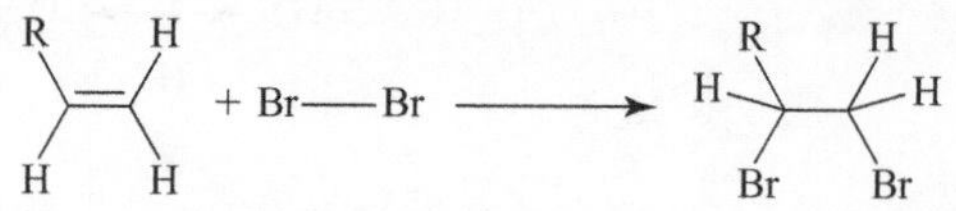

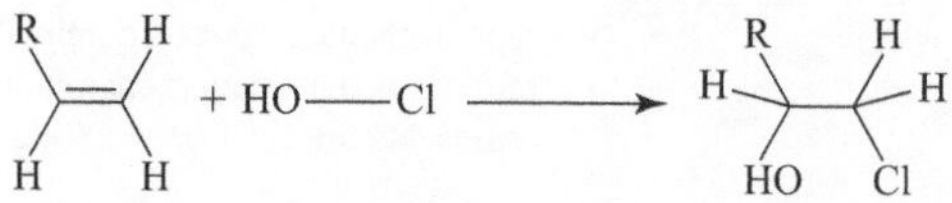

(Note here that the Cl atom acts as the electrophile)

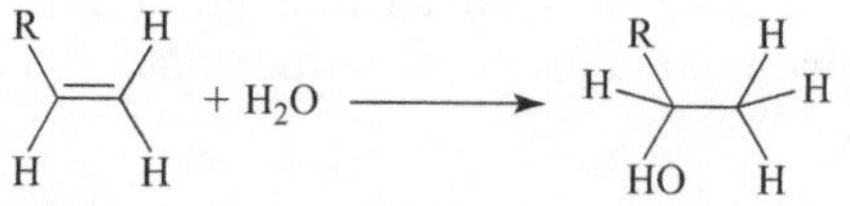

The last of the examples listed above, namely the addition of water to an alkene, only proceeds with the assistance of acid catalysis. In fact it takes place in dilute sulfuric acid solution, following the mechanism thus:

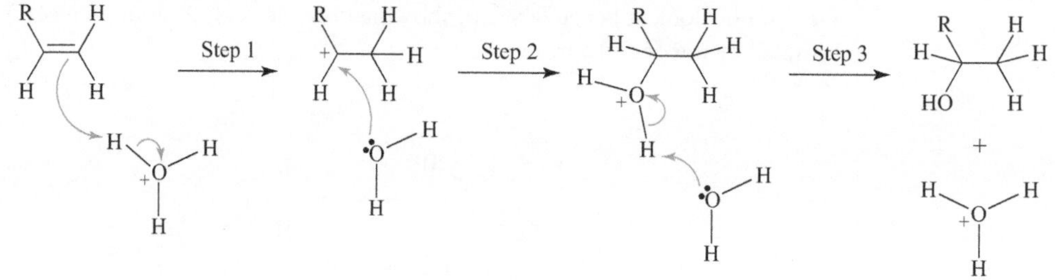

For brevity the H_3O^+ is often shown as H^+, so the above would therefore be redrafted as:

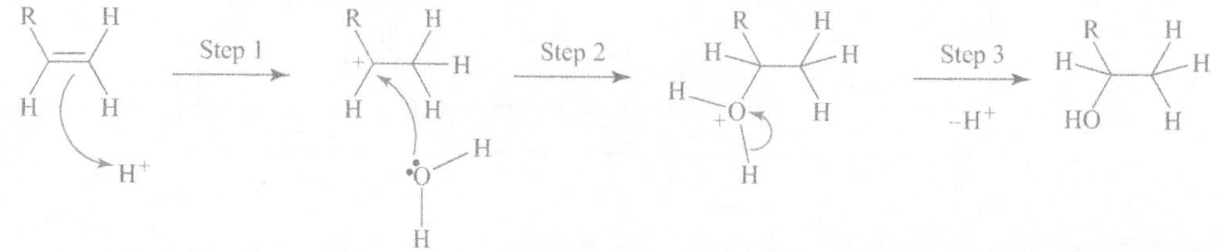

Step 1 involves H^+ addition to the alkene to form the most stable carbocation (in the example shown it is a secondary carbocation). Step 2 is the nucleophile (water in this case; it has a lone pair of electrons) attacking the positive carbon of the carbocation. Step 3 is the regeneration of the H_3O^+ catalyst, and the formation of the final product.

8.5.4 Electrophilic addition in biology

Part of the Krebs cycle involves the production of malate from fumarate. This, in essence, is the addition of water to the fumarate double bond, and the addition is facilitated by the use of the enzyme fumarate hydratase. The reaction is summarised below:

Fumarate + H_2O ⇌ (fumarate hydratase) Malate

8.5.5 Electrophilic addition without subsequent nucleophilic addition; loss of H⁺

In some cases, after the electrophile adds to the alkene the intermediate ion loses a proton instead of adding the nucleophile particularly if the electrophile is itself a carbocation. (An example of this occurs in one step in the biosynthesis of steroids, and this is described in Section 8.14.)

Box 8.3

```
Markownikoff's rule
```

This summarises the outcomes of the description of electrophilic addition of compounds of the type HX to asymmetrical alkenes. It states:

- When a hydrogen halide adds to an asymmetric alkene the addition occurs such that the halogen attaches itself to the carbon atom of the alkene bearing the least number of hydrogen atoms.
- When a hydrogen halide adds to an asymmetric alkene the addition occurs such that the halogen attaches itself to the carbon atom of the alkene bearing the greater number of carbon atoms.

8.5.6 Addition of HBr to conjugated dienes

Again we are using HBr as a representative molecule to illustrate the general principle of electrophilic addition. Conjugation results from a molecule having alternate double and single bonds (described in Chapter 2). Buta-1,3-diene is an example of a molecule exhibiting conjugation. The reaction of buta-1,3-diene with HBr results in the formation of two products, as summarised:

1, 4 addition   1, 2 addition

There are two possible sites at which the H⁺ could attack, resulting in two possible carbocation intermediates, but only one of them is protonated. This is because one of the carbocation intermediates exhibits resonance stabilisation.

251

Resonance stabilisation. Two canonical forms

[I]

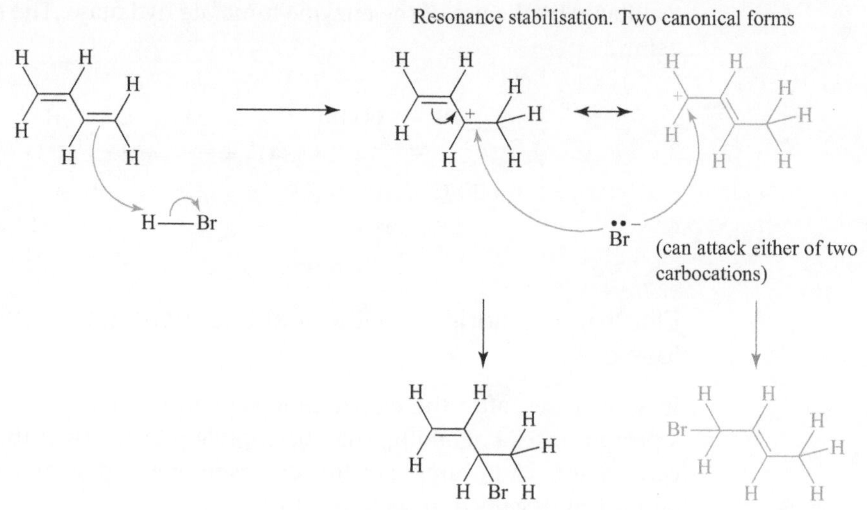

(can attack either of two carbocations)

[II]

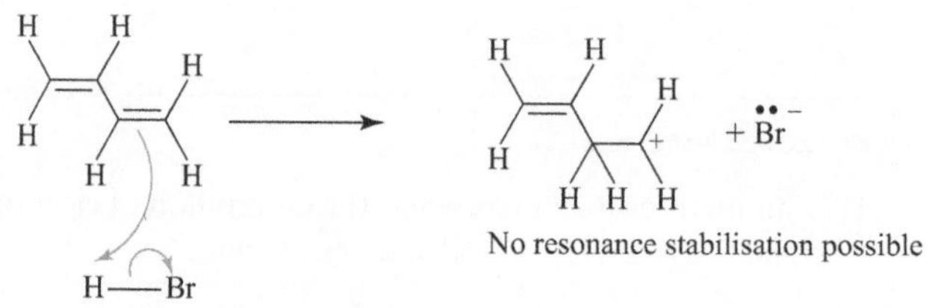

No resonance stabilisation possible

Because of resonance stabilisation [I] happens but [II] does not

- In [I] above, the resonance-stabilised intermediate carbocation leads to two products, as indicated by the attack of Br⁻ along either the 'black' route or the 'orange' route.

- In [II] above, the lack of resonance stabilisation in the intermediate means that it is higher energy than [I] and therefore the reaction will proceed via route [I].

8.5.7 Additions to alkynes

For all electrophilic additions except that of acidified water (H_3O^+) the addition mechanism is as for alkenes and happens twice. Consider, for example, the reaction of HBr with ethyne:

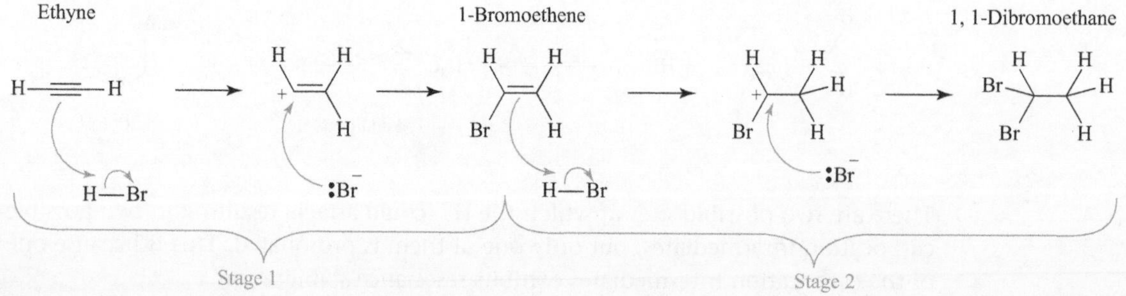

Ethyne   1-Bromoethene   1, 1-Dibromoethane

Stage 1   Stage 2

- Stage 1 involves an electrophilic attack of the HBr on the ethyne, to produce the brominated ethene.
- In stage 2 this alkene then undergoes an electrophilic attack by HBr, resulting in the formation of 1, 1-dibromoethane. The most stable carbocation in stage 2 is the one where the positive charge is on the most highly substituted carbon.

8.5.8 The addition of water (acidified) to an alkyne

This only happens once. The first product, the *enol* (called this because it has an alk*ene* group joined to an alco*hol* group) rearranges to a stable ketone before it has a chance to react again, as shown:

Product, an enol – one component of equilibrium

H transfers from O to=CH₂
Ketone is formed – very stable.
Equilibrium is established rapidly

8.6 Substitution reactions

So far the focus has been on addition reactions, and now we will look at a couple of other very important mechanistic pathways, namely substitution and elimination, starting with nucleophilic substitution.

8.6.1 Nucleophilic substitution at a saturated carbon atom

In a very generalised sense, reactions of this type can be summarised by the overall reaction:

$$Nu: + RX \rightarrow NuR + X:$$

One 'nucleophilic' group, X – known as the *leaving group*, is replaced by another, Nu, the nucleophile. Remember that nucleophiles are species that possess a lone pair of electrons that 'seek' a positive centre, and although they often carry a negative charge, it is not necessary that they do so.

Reactions of alkyl halides with aqueous sodium hydroxide are good examples of such reactions. In these the halide is the leaving group and the hydroxide, $:OH^-$, is the nucleophile. There are two possible mechanisms for nucleophilic substitution reactions: bimolecular and unimolecular.

8.6.2 Bimolecular nucleophilic substitution S$_N$2

The reaction of methyl iodide with aqueous sodium hydroxide provides an excellent example of this mechanism, as illustrated in Figure 8.6.

253

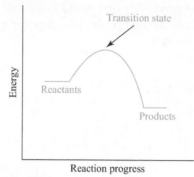

Reactants   A transition state   Products

The reaction profile for this is:

Figure 8.6 An S$_N$2 mechanism.

A number of key points arise from this mechanism:

- The attack of the nucleophile and loss of X are *concerted*, meaning they occur simultaneously.
- The nucleophile enters from the side opposite to the leaving group.
- There is no reactive intermediate.
- The reaction rate is proportional to both [HO$^-$] and to [CH$_3$I]. Extending this to a generalised case, the reaction rate is proportional to [Nu] and to [RX], usually represented as rate $\propto$ [Nu][RX], or rate = k[Nu][RX], where k is a constant called the rate constant (see Chapter 9). This dependency of the rate on the concentrations of the two reagents is the reason it is called bimolecular.
- The attacked carbon atom exhibits *inversion of configuration*, which is important if the starting material is chiral.

To illustrate the last point, we will consider a substitution of a chiral alkyl halide, namely *R*-2-bromobutane, which gives *S*-butan-2-ol on reaction with aqueous sodium hydroxide.

R-2-Bromobutane   Transition state   *S*-Butan-2-ol

8.6.3 Unimolecular nucleophilic substitution S$_N$1

The second mechanism that might be found in nucleophilic substitution reactions is referred to as an S$_N$1 mechanism, and differs from the S$_N$2 mechanism in a number of key ways:

- Whereas the S_N2 mechanism is a one-step process involving a single transition state, and hence no intermediate, the S_N1 mechanism is a two-step process involving an intermediate carbocation.
- Reaction rate depends only on the concentration of one species [RX], hence the name unimolecular.
- Starting with a single enantiomer gives both R and S products.

This is readily illustrated by using the example of the reaction of 2-bromo-2-methyl-propane with aqueous sodium hydroxide.

Step 1 involves a slow reversible ionisation, known as the rate-determining step (abbreviated as RDS).

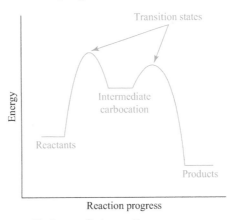

Step 2 involves a fast reaction of the intermediate carbocation with nucleophile. Attack on the carbocation intermediate is equally likely from 'above' (black) or 'below' (orange) (giving both R and S products in cases where three different groups are joined to the carbon bearing the leaving group):

Because the reaction is a two-step process involving a carbocation intermediate, the reaction profile is represented by Figure 8.7.

Figure 8.7 **The reaction profile for an S_N1 reaction.**

8.6.4 Determining which mechanism is followed

There is a relationship between the nature of the alkyl halide involved in a nucleophilic substitution and the mechanism that takes place, though it is possible for a reaction to follow both mechanisms. As an initial rule of thumb, however, Table 8.2 lists the relationship between likelihood of the mechanisms.

Table 8.2 **Overview of substitution mechanism vs structure type**

	CH_3X	RCH_2X	R_2CHX	R_3CX
S_N1 likelihood	No	No	OK	Good
S_N2 likelihood	Very good	Good	OK	No

There are exceptions to these simple guidelines however:

- Plenty of room around the carbon bearing the leaving group favours S_N2.
- A more stable carbocation favours S_N1.

Both S_N1 and S_N2 mechanisms require good nucleophiles. The strengths of nucleophiles are listed below in order of decreasing nucleophilic strength:

$$RS:^- > I:^- > RO:^- \sim HO:^- > Br:^- > Cl:^- \sim H_2O: > :CN^-$$

The mechanisms also need good leaving groups, which are listed below in order of decreasing leaving group ability:

$$I:^- > Br:^- \sim R_2S: > Cl:^- \sim H_2O:$$

One key point is that, in nucleophilic substitution reactions taking place at saturated carbons neither $RO:^-$ or $HO:^-$ act as leaving groups.

To remove HO it must first be protonated so that water becomes the leaving group. Heating an alcohol with aqueous HBr protonates the alcohol and also provides bromide ion as the nucleophile. It proceeds via an S_N1 mechanism. To illustrate this, consider the nucleophilic substitution of 2-methylpropan-2-ol.

8.7 Elimination reactions

In some cases an elimination reaction competes with nucleophilic substitution. The two most common elimination mechanisms are *bimolecular elimination*, labelled E2, and *unimolecular elimination*, labelled E1. Elimination reactions happen when the nucleophile is also a strong base, and acts as such. Overall the idea of elimination is summarised thus:

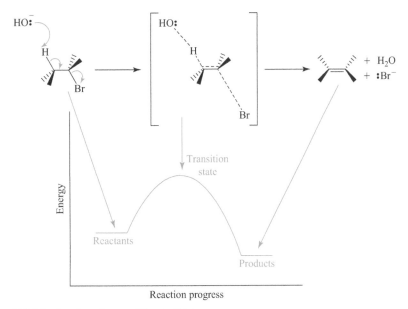

8.7.1 Bimolecular elimination, E2

There are certain requirements and features regarding E2 reactions, namely:

- There must be a H bonded to the C next to the C bearing the leaving group.
- The H—C—C—X group of atoms must lie in a plane with the H and X fully staggered.
- The reaction is concerted, meaning it goes through a single transition state, and no reactive intermediate is involved.

This is illustrated in Figure 8.8 for an alkyl bromide reacting with hydroxide. In the E2 elimination reaction the rate of reaction depends on the concentrations of both the alkyl halide and the base. This is represented by the relationship:

$$\text{Rate} = k[R_3CX][\text{base}]$$

8.7.2 Unimolecular elimination, E1

This mechanism has similarities with the unimolecular substitution, S_N1, mechanism in that the first step involves the formation of an intermediate carbocation. Thereafter, if the nucleophile is also a strong base, then rather than the reaction

Figure 8.8 **Mechanism for an E2 reaction.**

257

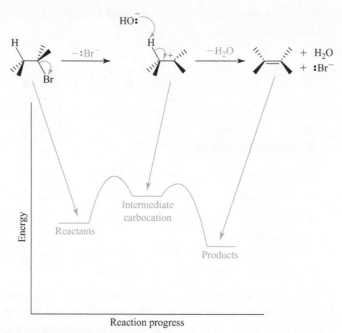

Figure 8.9 **Mechanism for an E1 reaction.**

proceeding solely as a substitution, there is a chance of an elimination. A general-ised representation of an E1 mechanism is illustrated in Figure 8.9.

As with the unimolecular substitution reaction, the reaction rate depends only on the concentration of one species, [RX], hence the name unimolecular elimination. Where a few products are possible the major product has the maximum number of alkyl groups joined to the double bond. For example, in the reaction between 2-bromo-2,3-dimethylbutane and sodium hydroxide, in addition to the possibility of substitution, where Br is replaced by OH, there are also two possible elimination products, as illustrated:

In this situation, where more than one elimination possibility exists, then the alkene formed will be the one with the highest number of substituents (known as Saytsev's rule).

Biological example of an S$_N$2 reaction

In many cases, *in vivo* methylation occurs via an S$_N$2 process, in which the nucleophile is often an amine and the leaving group is a sulfide RSR'. An important source of methyl groups in biological systems is *S*-adenosyl methionine (SAM). This is itself formed by an S$_N$2 process in which the sulfur of methionine acts as a nucleophile to displace the triphosphate group from ATP.

The formation of SAM is:

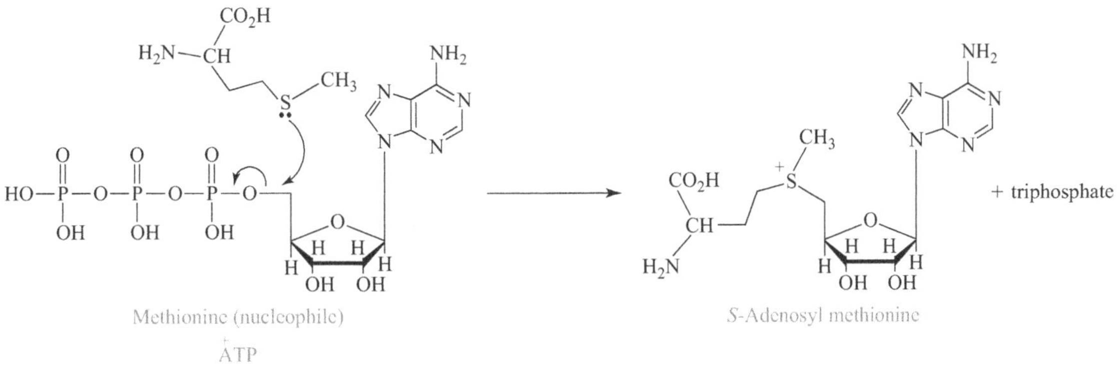

Methionine (nucleophile)

ATP

S-Adenosyl methionine

+ triphosphate

S-Adenosyl methionine can be thought of as the biological equivalent of iodomethane. The NH$_2$ group of norepinephrine acts as a nucleophile, to attack the methyl of SAM, and the dialkyl ammonium ion formed then loses a proton to give a new amine, adrenaline (epinephrine).

Norepinephrine is converted to adrenaline as shown:

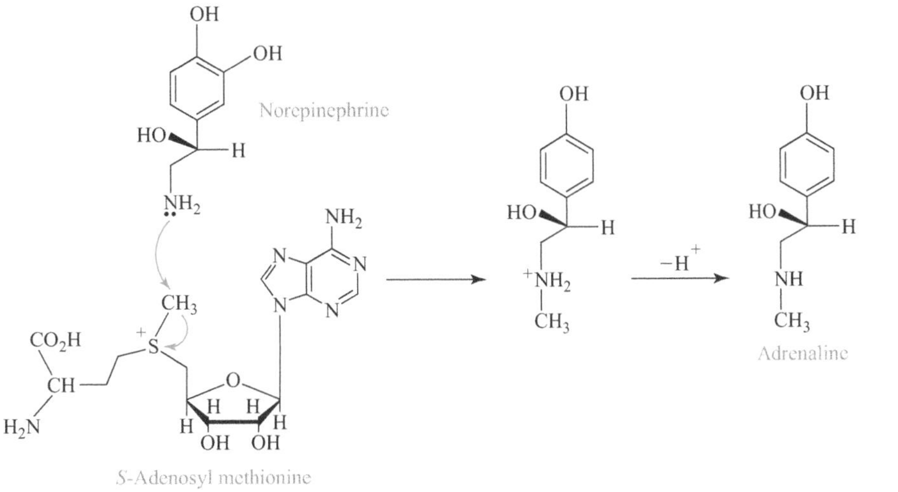

Norepinephrine

S-Adenosyl methionine

Adrenaline

8.9   Reaction mechanisms of carbonyl compounds

8.9.1   Introduction

The focus will now be switched to look at mechanisms involving carbonyl-containing functional groups. These functional groups are found in a wide variety of naturally occurring compounds. A few examples to illustrate this point include proteins, many fats, prostaglandins and the open chain forms of carbohydrates. The chemistry of carbonyl groups forms an integral part of the metabolic pathways, and so an understanding of the behaviour of these species is fundamental to biological chemistry.

8.9.2   Structure of the carbonyl group, $C{=}O$

The $C{=}O$ grouping has a similar structure to the $C{=}C$ grouping:

Alkene, $C{=}C$ Carbonyl, $C{=}O$

π (π^* not shown) π (π^* not shown)

$C{-}H$ and $C{-}C$ σ-bonds represented by——

$C{-}H$ and $C{-}O$ σ-bonds represented by——

sp^2 orbital holding lone pair of electrons

The carbon has, in each case, three σ-bonds, with the carbon being sp_2 hybridised. In the carbonyl grouping the π-bond arises from the overlap of p orbitals on C and O, with two lone pairs being found on the oxygen atom. This results in a planar $120°$ angle around the carbon atom. The double bond is therefore both shorter and stronger than a single $C{-}O$ bond, as illustrated in Table 8.3.

Table 8.3   **Comparison of $C{=}O$ and $C{-}O$**

	Double bond, $C{=}O$	**Single bond, $C{-}O$**
Length / pm	122	143
Strength / kJ mol^{-1}	732	385

Carbonyl groups encompass a very large group of compounds (as described in Chapter 3) and the compounds can be subdivided and classified further according to the other atoms attached to the carbonyl carbon. The RCO component of a carbonyl complex is referred to as the acyl group. *Aldehydes and ketones* are very similar to each other, but there are enough differences in chemical behaviour (for example ease of oxidation) for them to be considered as different:

Aldehyde   Ketone

There are a variety of 'double' functional groups involving the carbonyl entity; where two functional groups are bonded *directly* together each modifies the behaviour of the other, so that a new group with its own special properties is formed. The carbonyl group is usually one component and the range of compounds is known as 'carboxylic acid derivatives'.

In *carboxylic acids*, where there is C=O plus OH molecules there is very slight dissociation of the H linked to the O (as noted in Chapter 3), and this is considerably more so than in just an alcohol. The presence of the carbonyl facilitates this through resonance dissociation:

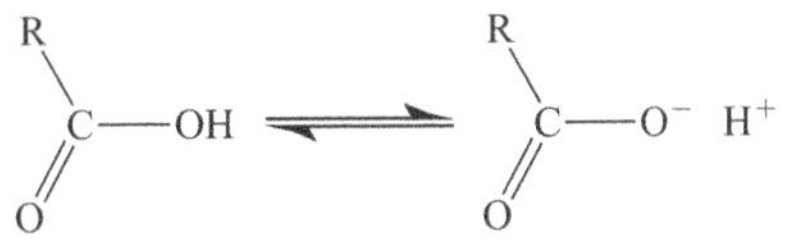

The other biologically important compounds containing 'double functional groups' that will be discussed in this chapter are:

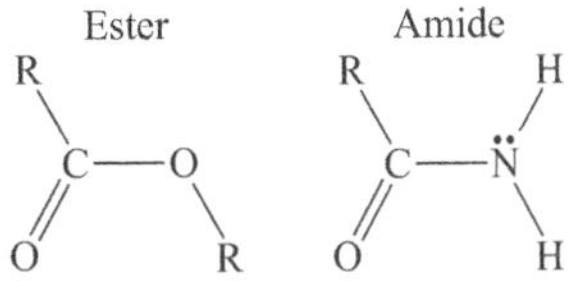

Oxygen is significantly more electronegative than carbon. Therefore the C=O bond is polarised, with the electrons in the C=O spending most of their time near the oxygen.

The carbonyl double bond is polarised with a partial positive charge on the C ($\delta+$) and a partial negative charge on the O ($\delta-$). This compares with the isolated C=C where the electrons are equally associated with each C.

8.9.3 Mesomeric effects

The effect of the polarisation of the C=O double bond resulting from this electronegativity difference between the two elements is often represented for a typical carbonyl thus:

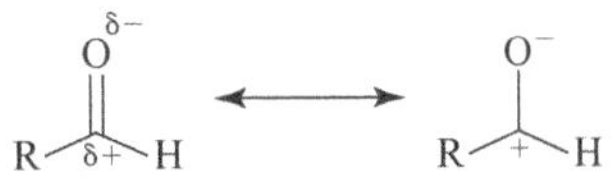

The movement of an electron pair gives rise to the alternative structures. These are so-called mesomeric forms. The 'true' structure lies between these two extremes.

There are consequences for the chemistry of carbonyl-containing compounds, and a couple of key points are:

● almost all the reactions involve a nucleophile (Nu:) attacking the carbon ($\delta+$) and the electrons in the π-bond moving to the oxygen;

● many of the reactions are reversible.

Initially, we will look at the aldehyde and ketone functional groups.

8.10 Reactions of aldehydes and ketones

8.10.1 Reaction of aldehydes and ketones with 'hydride'

To illustrate the idea of a nucleophilic attack on a carbonyl group, consider the reduction of a carbonyl compound using sodium borohydride:

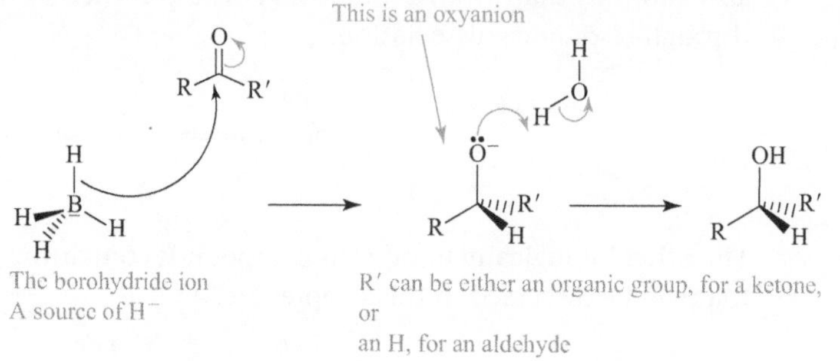

This is an oxyanion

The borohydride ion
A source of H⁻

R' can be either an organic group, for a ketone,
or
an H, for an aldehyde

This reaction is, effectively, the transfer of a *hydride ion* [H:⁻] from the borohydride ion to the C of the carbonyl group. The negatively charged *oxyanion* formed (in this example it is an alkoxide ion) reacts immediately with the solvent to form an alcohol. Water or methanol may be used as solvent.

The same reaction occurs whether the starting carbonyl compound is a ketone or an aldehyde, but *esters, acids and amides (and also some ketones) will not react with sodium borohydride.*

In vivo, nature has its own reducing agents which work the same way. For example NADH is a coenzyme which, among other things, converts pyruvic acid to lactic acid (this is a process that occurs in the muscle during vigorous exercise):

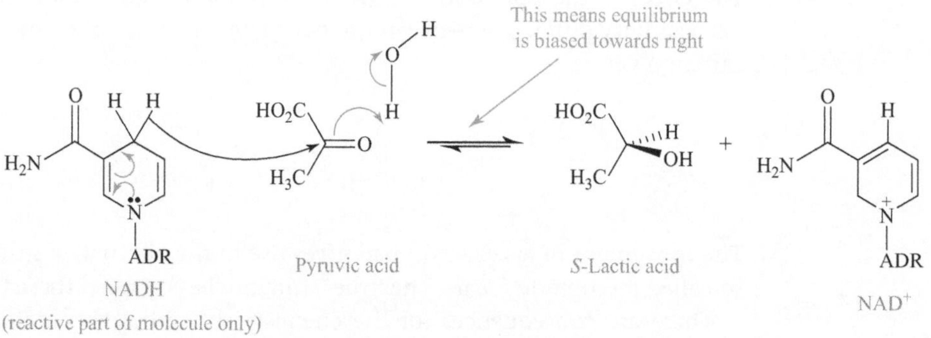

This means equilibrium is biased towards right

NADH
(reactive part of molecule only)

Pyruvic acid

S-Lactic acid

NAD⁺

8.10.2 Hydration of aldehydes and ketones

This is the reaction with water. For water-soluble aldehydes and ketones water will act as a nucleophile:

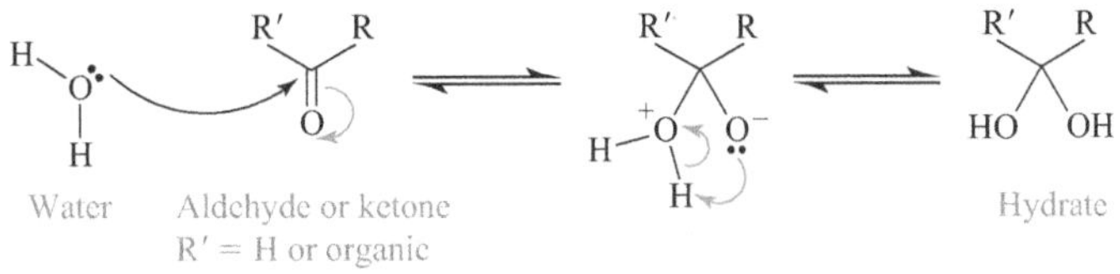

Water   Aldehyde or ketone   Hydrate
          R' = H or organic

In these reactions an equilibrium is set up, and which side the equilibrium lies on depends to some extent on the size of the groups, as illustrated in Table 8.4.

Table 8.4 **Dependence of equilibrium constant in hydration of aldehydes and ketones with nature of alkyl groups**

R	R'	$K = \dfrac{[hydrate]}{[carbonyl]}$
CH_3	CH_3	0.001
CH_3	H	1
H	H	2288
CCl_3	H	2000

With esters, acids and amides, the reaction with water alone has an equilibrium constant that is essentially zero; in other words there is almost no hydration, leaving entirely carbonyl compound. (However, a reaction can be forced if acid or base is present, as will be described later.)

8.10.3 Hemiacetal formation

Reaction of aldehydes with alcohols results in an equilibrium process with the product known as a *hemiacetal*. If a ketone is used as the starting material, then the mechanism is the same, but the product is called a *hemiketal*. The mechanism of hemiacetal/hemiketal formation is comparable with that of hydration, as shown:

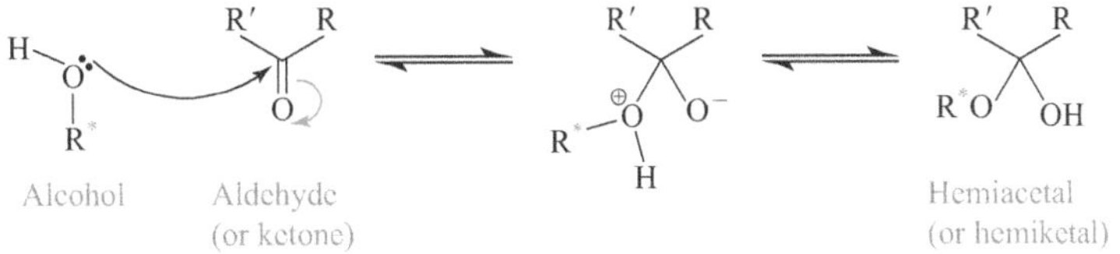

Alcohol   Aldehyde   Hemiacetal
          (or ketone)   (or hemiketal)

Some stability is gained if a cyclic structure can be produced. This can happen if a molecule has an OH group at one end and an aldehyde group at the other. *This means that there can be reactions between different groups within the same molecule.*

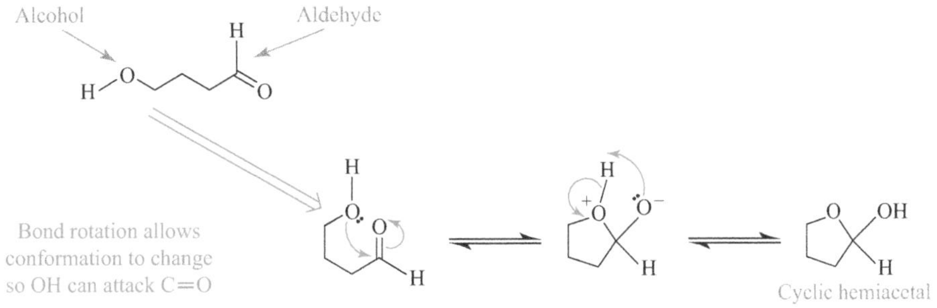

Alcohol   Aldehyde

Bond rotation allows
conformation to change
so OH can attack C=O

                                                              Cyclic hemiacetal

263

This hemiacetal formation happens spontaneously in the simple sugars (monosaccharides – see Chapter 7), for example ribose and glucose, ribose being an *aldopentose* and glucose being an *aldohexose*. These sugars, although often written in a linear form, as on the left below, actually exist almost exclusively in ring structures. This introduces an extra chiral centre and for each sugar there are now *two* ring structures, one with the OH at C-1 above the ring, known as the β form, and one with the OH at C-1 below the ring, known as the α form.

Open chain representations of ribose

β-ribose

* This C is the C-1 position

α-ribose

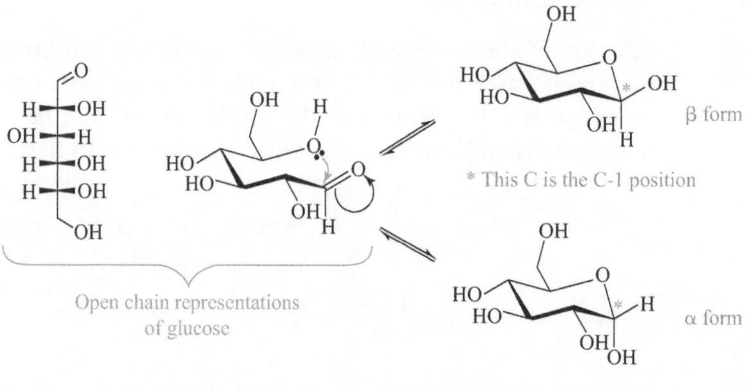

Open chain representations of glucose

β form

* This C is the C-1 position

α form

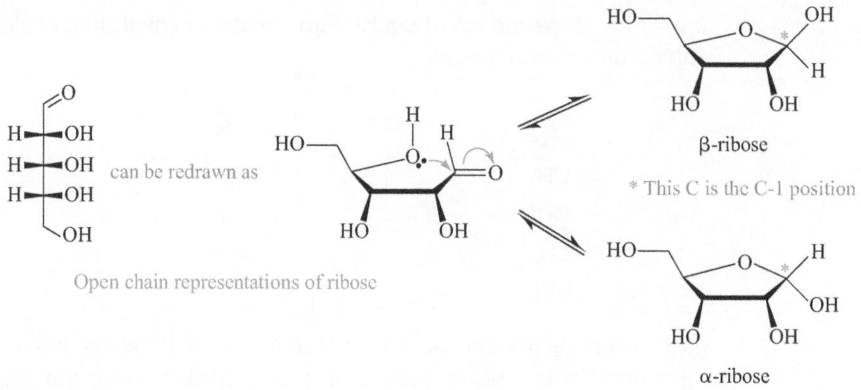

The two forms of glucose indicated above, distinguished by the positioning of the OH on the C-1 carbon, are called *anomers*. In the solid state each form can be isolated independently of the other. If either form is dissolved in water, then an equilibrium state is established where both forms exist (through the open chain form), with 66% β-glucose and 34% α-glucose being present at equilibrium. This process of interconversion between β and α forms is called *mutarotation*.

For some monosaccharides, the open chain structure contains a ketone group. For example, fructose is called a ketohexose. The mechanism for ring formation in fructose is analogous to that for glucose though; because the keto group is attacked, the main cyclic form has a five-membered ring.

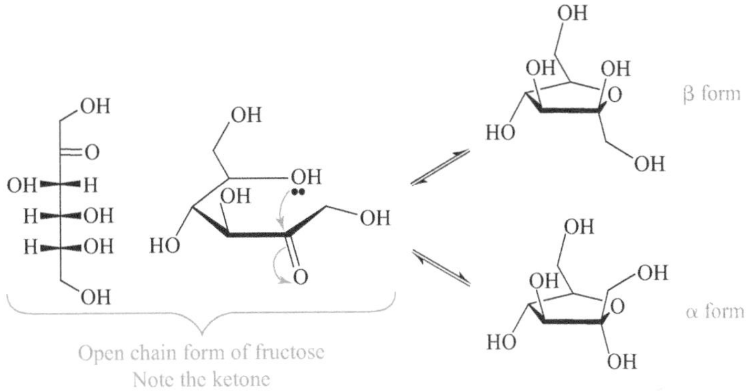

Open chain form of fructose
Note the ketone

β form

α form

The process of hemiacetal and hemiketal formation can be accelerated if acid is present. The H$^+$ that is used to make the intermediate *oxonium ion* is regenerated at the end of the reaction; *it is acting as a catalyst*.

The acid-catalysed reaction is represented thus:

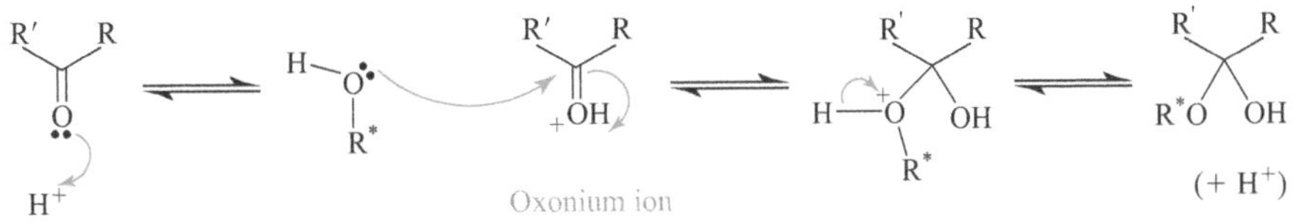

Oxonium ion

8.10.4 Acetal (ketal) formation

If the reaction of an aldehyde (or a ketone) with an alcohol (+ acid catalyst) proceeds in an alcohol as the solvent, rather than in water, then an acetal (or ketal) is formed. This arises because the hemiacetal (hemiketal) formed initially will react further:

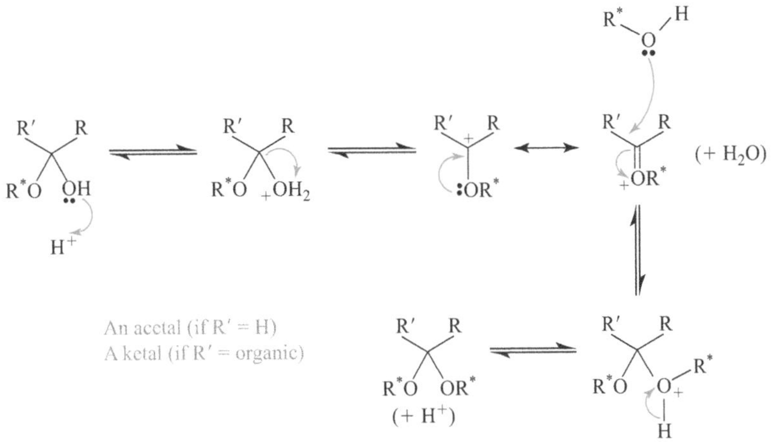

An acetal (if R′ = H)
A ketal (if R′ = organic)

Acetal links hold the constituent sugars together in polysaccharides.

● β-maltose, for example, consists of two glucose units joined from the C-1 of one unit (the left hand side in the picture) to the C-4 of the other, right hand side, unit, via oxygen. The left hand sugar has the acetal structure and this type of link is known as a glycosidic link:

265

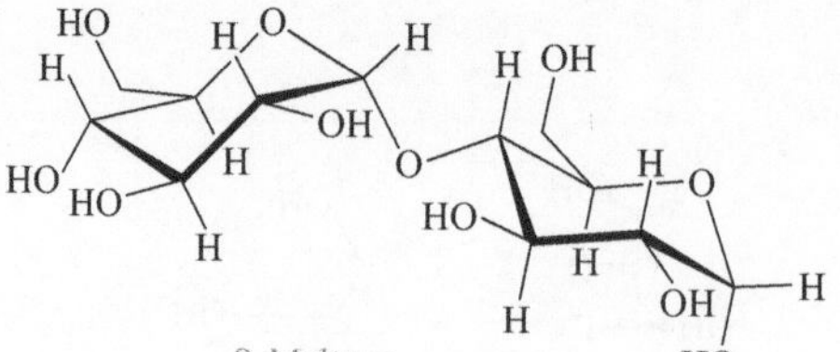

β-Maltose

● Sucrose consists of a glucose linked to a fructose:

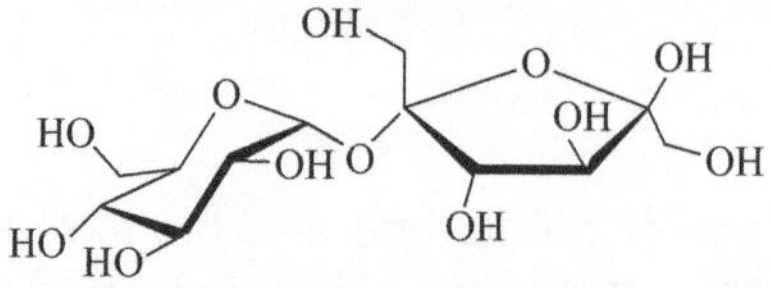

8.10.5 Formation of Schiff's bases and imines

Amines can attack aldehydes and ketones to form imines. The reaction proceeds because the amine has a lone pair of electrons on the nitrogen, enabling it to act as a nucleophile. The overall mechanism of imine formation is:

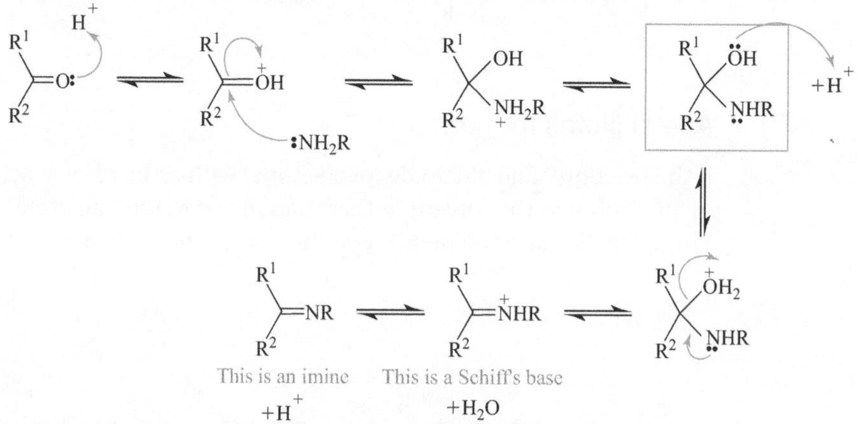

This is an imine   This is a Schiff's base
$+H^{+}$ $+H_2O$

There are a few key points about this mechanism:

● The nucleophilic addition part of the process stops with the structure highlighted in the orange box.

● The second part of the mechanism is an elimination (of water). Where two molecules react together and a molecule of water (or an alcohol) is eliminated. The reaction is known as a *condensation* reaction.

● The process is acid-catalysed. The carbonyl is made more susceptible to nucleophilic attack by the addition of H^{+} to the oxygen, and the acid is regenerated at the end.

This type of mechanism, involving the formation of a Schiff's base, forms part of the glycolysis pathway, specifically where fructose-1,6-diphosphate is then cleaved by the enzyme aldolase into glyceraldehyde-3-phosphate and dihydroxyacetone phosphate. The aldolase contains an amine group which forms a Schiff's base with the carbonyl group of the fructose-1,6-diphosphate. (This will be elaborated on in Section 12.3.4.)

8.10.6 Oxidation of aldehydes and ketones

The oxidation of these compounds is, arguably, the area where their respective chemistries differ the most.

Aldehydes

Aldehydes are relatively easy to oxidise, with the overall process summarised as:

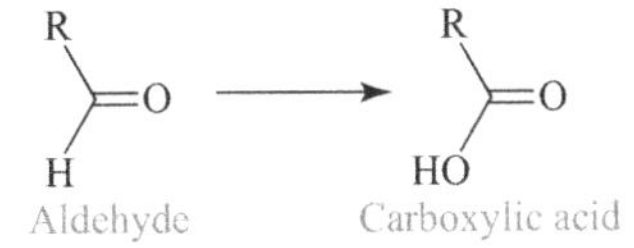

Aldehyde → Carboxylic acid

The key points are:

- The product has the same number of carbon atoms as the starting aldehyde.
- The oxidation can be carried out under gentle conditions (for example, Ag_2O is reduced to Ag, the silver mirror test).

Ketones

Ketones are much more difficult to oxidise, because the ketone has to be split into two parts, and a C—C bond needs to be broken. For example:

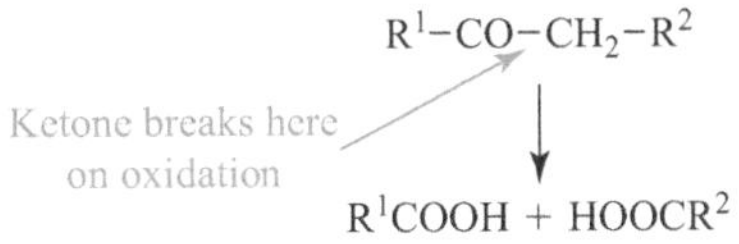

$$R^1-CO-CH_2-R^2$$

Ketone breaks here on oxidation

$$R^1COOH + HOOCR^2$$

The noteworthy points here are:

- The oxidation of ketones is difficult to carry out, with vigorous conditions needed.
- The final product contains fewer carbons than the original ketone – degradation.

8.11 Carboxylic acid derivatives

An important group of compounds are the so-called carboxylic acid derivatives. Examples are esters, amides and acid chlorides. Of these, esters and amides play an important part in biological systems.

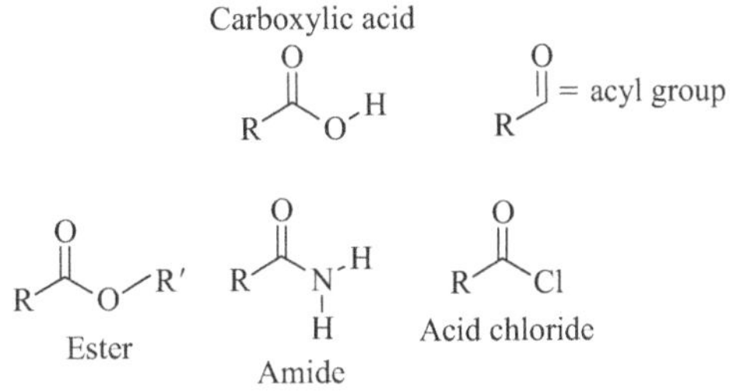

Carboxylic acid

= acyl group

Ester   Amide   Acid chloride

8.11.1 Esters

Esters are one of the more commonplace functional groups in nature, but they are also extensively synthesised as well. Most naturally occurring fats and oils are esters of glycerol. Esters are also an integral part of many pheromones.

Esters are prepared by reacting organic acids with alcohols and, conversely, the hydrolysis of esters results in their break up to acids plus alcohols. There are two ways in which esters can be hydrolysed:

- Acid-catalysed hydrolysis

- Base-induced hydrolysis.

Acid-catalysed hydrolysis of esters

The mechanism for this process is reversible:

Ester

R and R′ are organic

Protonation activates carbonyl meaning even a relatively weak nucleophile such as H_2O can attack CO

+HOR′ + H⁺

H⁺ regenerated. Acts as catalyst

The reversible nature of this reaction means that it is possible to make an ester by reaction of an acid and an alcohol in the presence of acid. The equilibrium can be 'driven' in either direction, using the ideas described by Le Chatelier's principle:

- Removing one component, for example separating the alcohol from the ester plus much water. If the alcohol is methanol, for example, this could be achieved by boiling it off. This would drive the reaction to the right.

- Using a large excess of one component, for example the reaction of an acid plus alcohol with the alcohol as solvent. This would push the reaction to the left.

Base (:OH⁻)-induced hydrolysis of esters

The use of 1 molar equivalent of base (HO:⁻) ensures that the carboxylic acid product is irreversibly converted to its anion. This is because the carboxylate anion is resonance-stabilised, so that the equilibrium moves to the right until all of the ester has been hydrolysed. Note that the hydroxide facilitates the equilibrium steps, but it is consumed, and not regenerated.

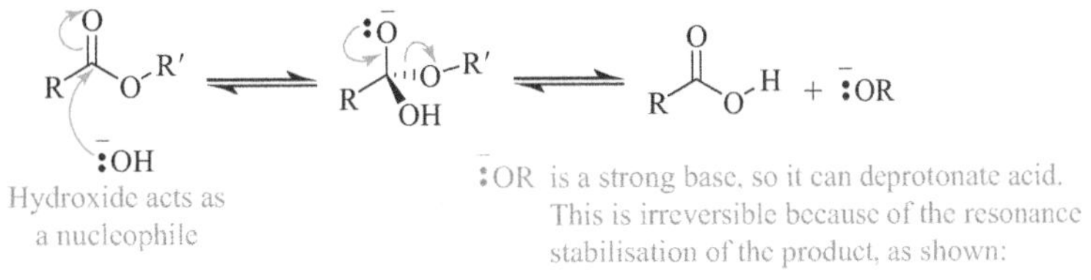

:OH
Hydroxide acts as
a nucleophile

:OR is a strong base, so it can deprotonate acid.
This is irreversible because of the resonance
stabilisation of the product, as shown:

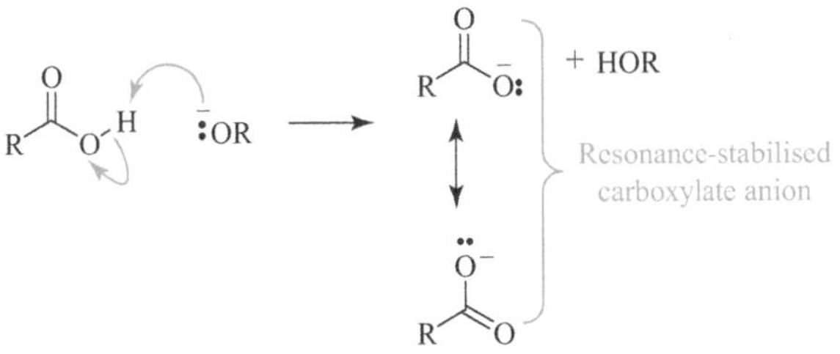

+ HOR

Resonance-stabilised
carboxylate anion

8.11.2 Amides

Amides are the important functional grouping in amino acids, and hence proteins. In biology the C—N bond in amides is usually referred to as a peptide
link. Outside biology they are also important, with materials such as nylon and
Kevlar being polyamides. Structurally their shape is governed by their delocalised
nature.

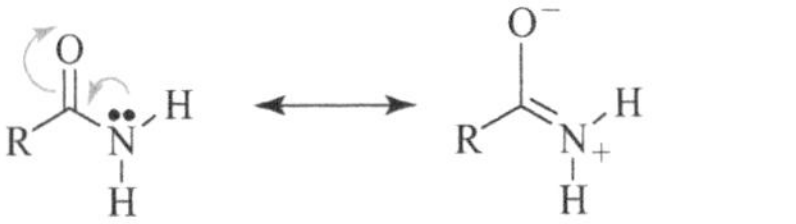

Resonance stabilisation means CN bond has significant
double bond character. Planar group as a result.

The planar nature of the amide grouping has important consequences for the structure of proteins. The mechanisms associated with the reactions of amides are similar
to those associated with esters. They do, for example, undergo both base and acid
hydrolysis, though under more vigorous conditions (usually heating) than esters
because of the stronger CN bond. In nature, for example, the splitting of proteins is
facilitated by enzymes such as cysteine protease or serine protease, for which the key
step is the hydrolysis of the amide.

The mechanism for the acid-catalysed hydrolysis of an amide is:

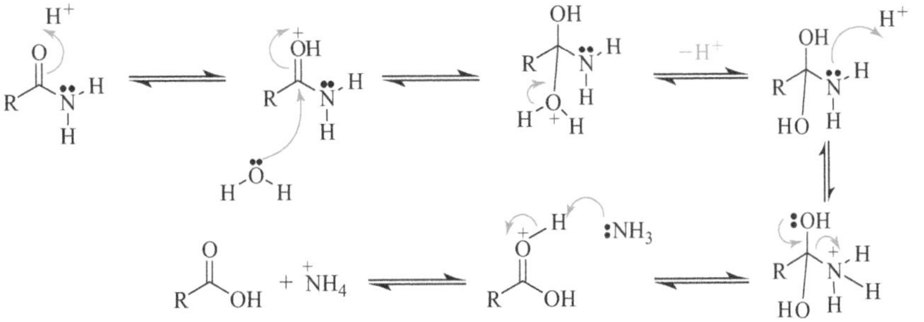

In the enzyme-catalysed splitting of proteins, amino acid residues on the enzyme act as the source of acid

8.12 Enolisation and enolisation reactions

Although the carbonyl group is often subject to nucleophilic attack at the carbonyl carbon, there can be other potentially reactive sites in the molecule. For example, aldehydes and ketones undergo *structural isomerisation* in solution where there is a hydrogen atom on the α carbon to the carbonyl, in which the hydrogen atom relocates to the oxygen of the carbonyl, with the simultaneous relocation of the double bond between the O-bonded carbon and the α carbon. The name given to this type of isomerism is *tautomerism*:

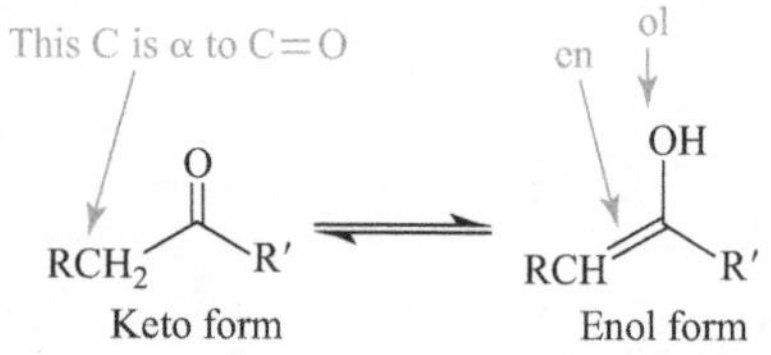

The equilibrium mixture usually has the enol form as a very minor component, but the reactivity of the small equilibrium concentration of enol does affect the overall chemical behaviour of carbonyl compounds. The equilibrium concentration of enol is lower for ketones than for aldehydes.

8.12.1 Enols as carbon nucleophiles

Because enols have the capacity for delocalisation of electrons, from the O to the α-C, this means that the C has the capacity to behave as a nucleophile. This is very important in the chemistry of these species.

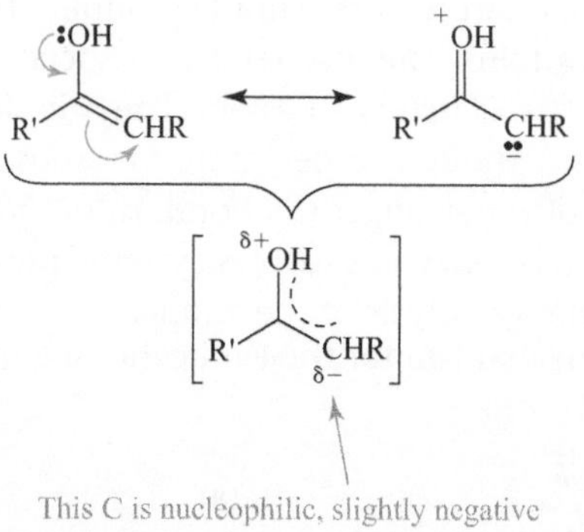

This C is nucleophilic, slightly negative

Under neutral conditions the rate of conversion of the pure keto form into the enol form, a process known as enolisation, is slow, but it is greatly accelerated by operating under base conditions.

8.12.2 Base-catalysed enolisation

To illustrate this, consider the case of propanone (acetone) in the presence of base.

Here the base has removed an H atom from the carbon in an α position to the C=O bond, and the O has then picked up a hydrogen from the solvent, thereby regenerating base.

8.13 Reactions resulting from enolisation

8.13.1 The aldol reaction

The aldol reaction can be regarded as the reaction of an aldehyde (or of a ketone) with itself. It is catalysed by base, and requires a hydrogen atom bonded to the carbon atom *next* to the carbonyl group; in other words an α hydrogen.

From the mechanism shown there are three clear stages.

1. The formation of a resonance-stabilised enolate anion from the first aldehyde (molecule A).
2. The enolate anion (which is a nucleophile) attacks the carbon of the C=O of the second aldehyde (molecule B).
3. Water protonates the alkoxide ion formed in step 2 to give the *aldol* – a β hydroxyaldehyde.

In the product, the $C=O$ from molecule A is retained but the $C=O$ from molecule B is reduced. The same reaction occurs with ketones (as indicated in the mechanism, substituting the R indicated in brackets for the aldehydic H joined to the $C=O$).

Sometimes the reaction goes further by the *elimination* of water from the aldol to form an *enone*. Whether this happens or not depends on the temperature, reactant concentrations and the structure of the aldehyde/ketone. If acetone, $(CH_3)_2C=O$, is used for example, then the aldol that is formed can itself be deprotonated by the base used in the reaction:

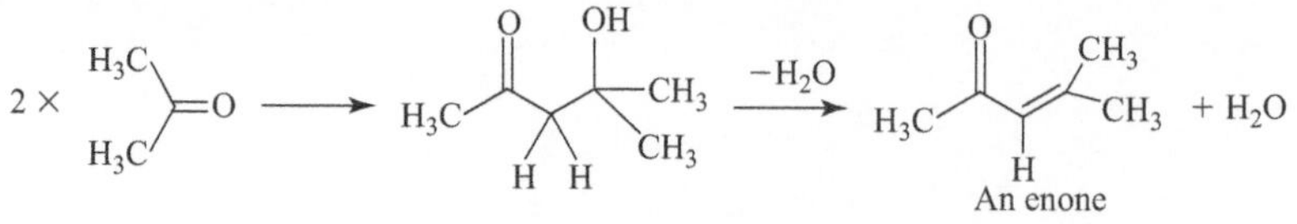

8.13.2 Crossed aldol reactions/condensations

If two different aldehydes (or ketones), represented by X and Y, are present then there are four possible aldol products, which could be represented in the following shorthand manner:

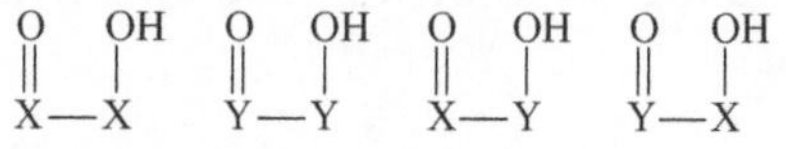

The balance of these can be controlled. One way would involve choosing Y so that it cannot form an enolate anion (using something that does not have a hydrogen on the α carbon would fall into this category, for example benzaldehyde, C_6H_5CHO). Alternatively, by choosing Y so that it is more reactive than X (for example, aldehydes are more reactive than ketones) means that X does not react with itself.

The reaction between propanone (acetone) and benzaldehyde is given by:

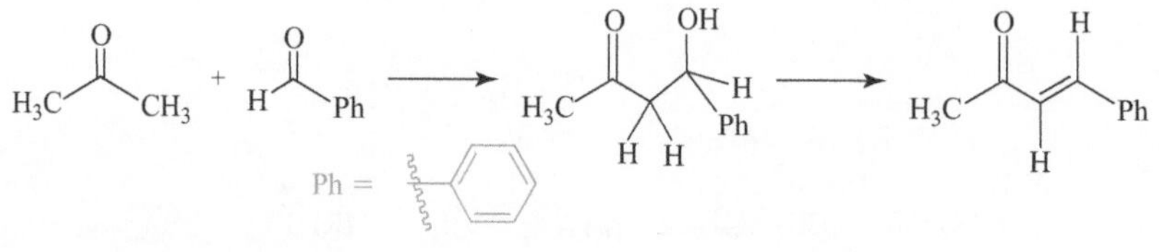

8.13.3 Claisen condensations

The Claisen condensation is the reaction between two esters, and it is possible to summarise a few key points:

- Ethyl esters are normally used.
- They are initiated by base. It is essential to use the alkoxide with the same $-O-$ alkyl group as present in the ester (hydroxide would convert the ester to a carboxylic acid) and the corresponding alcohol is used as the solvent (for example, sodium ethoxide in ethanol).
- It requires a molar equivalent of alkoxide.
- It requires *two* hydrogen atoms bonded to the carbon atom *next* to the carbonyl group; that is two α hydrogen atoms.
- The reaction starts in a way similar to the aldol reaction.

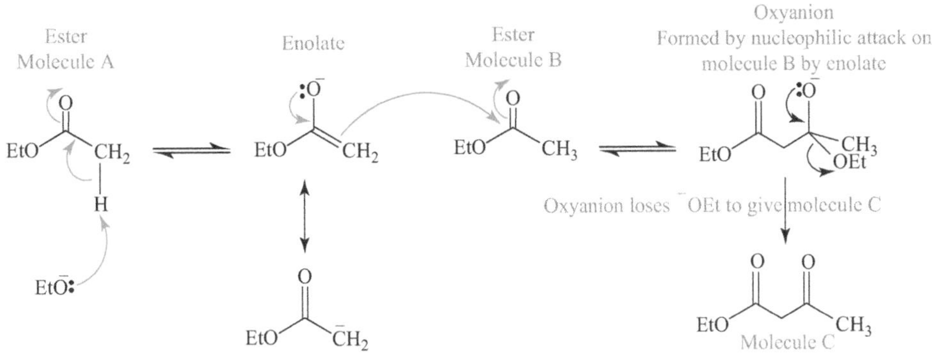

Potentially the OEt⁻ anion could act as a nucleophile and set up the reverse reaction. It does not do so because it acts as a base, and induces further enolisation as shown below. This explains the need for two ∝ hydrogen atoms.

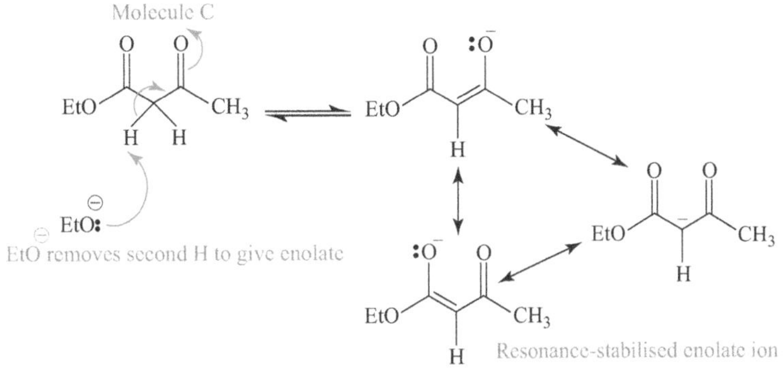

Molecule C can be isolated only if the reaction mixture is treated with an acid (such as aqueous HCl) to re-protonate the resonance-stabilised anion. Molecule C is known as ethyl acetoacetate, which is one of a large group of compounds known as 1, 3-dicarbonyl compounds. These compounds possess a number of interesting properties:

- They very readily give resonance-stabilised anions, which can be used to react with aldehydes/ketones in 'crossed aldol' reactions.
- They will also react with simple alkyl halides replacing the halogen in a nucleophilic substitution reaction.
- They exist as a mixture of two molecules in equilibrium with each other.

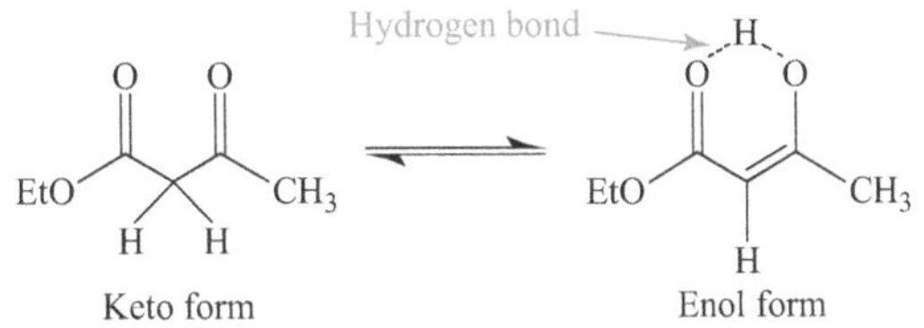

The enol form is more stable than that from a simple ketone because:

- The two double bonds are *conjugated*.
- The OH can form a *hydrogen bond* to the carbonyl group.

Where two isomers are in equilibrium the molecule is said to show *tautomerism*, and the two isomers are called *tautomers*. This occurs when the two isomers have very similar energies and the energy barrier to interconversion is low.

8.14 Reaction mechanisms in biological reactions: synthesis of steroids

The *in vivo* (means in the organism) biosynthesis of steroids involves, in its early stages, Claisen condensations and aldol reactions, and the latter parts of the process involve electrophilic addition reactions. In the very first stages, three acetyl coenzyme A (CoA is used as an abbreviation for coenzyme A) molecules react to give the six-carbon compound mevalonic acid, in which reactions analogous to the Claisen condensation and an aldol reaction occur, as shown (note that the enzyme catalyst acts as a source of key reagents):

The six-carbon molecule * then undergoes enzymic hydrolysis of the thioester at one end (shown as the left hand end) to give the carboxylic acid. This is followed by enzymic reduction with NADPH (NADH with a phosphate attached at the C-2 position on the ribose) of the thioester at the other end, first to give an aldehyde and then to give a primary alcohol, which is mevalonic acid. This last step is analogous to the NADH reduction of pyruvic acid to lactic acid (as referred to in Section 8.10.1).

The next stage of the process involves *electrophilic addition* in the biosynthesis. The mevalonic acid reacts with ATP to *lose* water and CO_2 (elimination) and *gain* a diphosphate unit to form isopentenyl pyrophosphate (IPP). *IPP is the alkene on which addition will take place.*

This molecule undergoes an isomerisation reaction to form dimethylallyl pyrophosphate (DMAP).

The two molecules IPP and DMAP have very similar energies and the activation energy for interconversion is very low, hence they exist in equilibrium and both are always present. *This second molecule (DMAP) is the electrophile.*

DMAP can be thought of as the biological equivalent of HBr in the simple alkene reaction.

● The —OPP plays the part of the Br⁻.

● The *carbon* to which the OPP is joined plays the part of the H⁺ of the HBr.

This is summarised in the reaction scheme below:

The key components here are the *DMAP* and the *IPP*. The electrophilic addition to an alkene involves these two materials, and is illustrated thus:

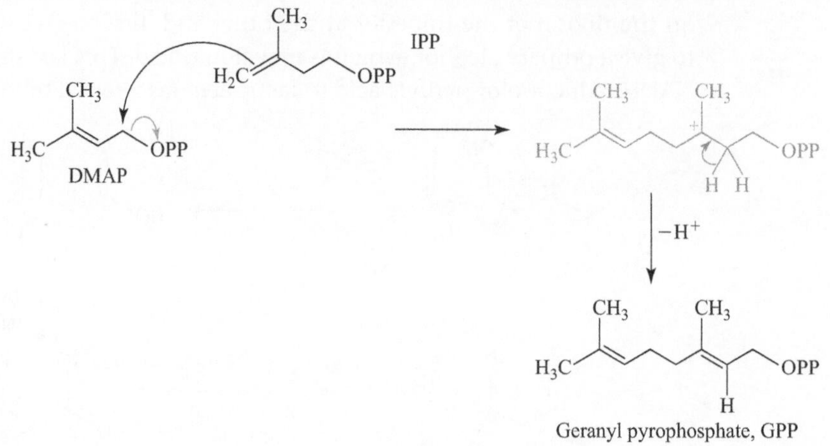

Geranyl pyrophosphate, GPP

In this process it is noteworthy how the ten-carbon (carbocation) unit produced, identified in grey, *loses a proton* (H^+) instead of adding the pyrophosphate ion at the carbon bearing the positive charge (because it gives a lower energy product) to give *geranyl pyrophosphate* (GPP). On examination it is clear that GPP is a longer version of DMAP and as such it reacts with IPP in exactly the same ways as DMAP does, with the result that a 15-carbon molecule known as farnesyl pyrophosphate (FPP) is formed.

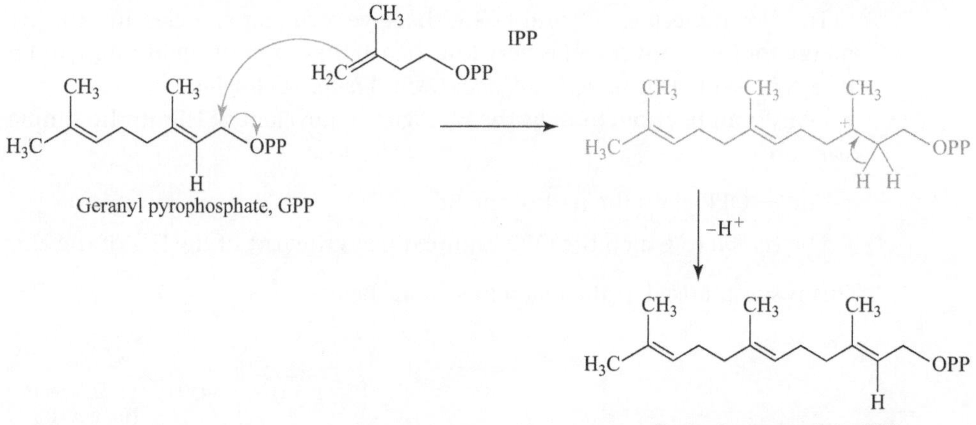

Farnesyl pyrophosphate, FPP

In this case the 15-carbon carbocation, identified in grey, loses H^+ to form the FPP.

The process of steroid synthesis gets more complex hereafter, but, in summary, two molecules of FPP are linked together at the pyrophosphate ends to form a 30-carbon hydrocarbon molecule called squalene. This undergoes an internal cyclisation process to give a molecule with four rings, which is called lanosterol. Finally the lanosterol loses three carbon atoms to give cholesterol.

8.15 | # Summary of mechanisms of carbonyl reactions under different conditions

8.15.1 Acid present

Step 1 Carbonyl oxygen is protonated – 'activates' carbonyl for nucleophilic attack,
Step 2 Nucleophile attacks carbonyl carbon.

Examples: acetal formation (X = R, H); acid catalysed hydrolysis of esters (X = OR); esterification of acids (X = OH)

8.15.2 No acid present

Step 1 Nucleophile attacks carbon – results in formation of a tetrahedral oxyanion
Step 2 This will go either one of two ways:

(a) Oxyanion grabs a proton from solvent – occurs where X = R or H

(b) Oxyanion expels X to form a new carbonyl compound – occurs where X = OR or OH or another leaving group (in carbonyl reactions OR and OH can act as leaving groups).

etc.   etc.
See borohydride reduction, reaction   See base-induced hydrolysis of esters,
with cyanide, aldol reaction   Claisen condensation

In some cases the compound formed at this stage loses water

277

8.15.3 Enolisation

If a strong base (such as :OR$^-$) is present *and* the carbonyl compound has an α-hydrogen, *then* an enolate anion will be formed. This is a nucleophile that can attack the C of the C=O in another molecule:

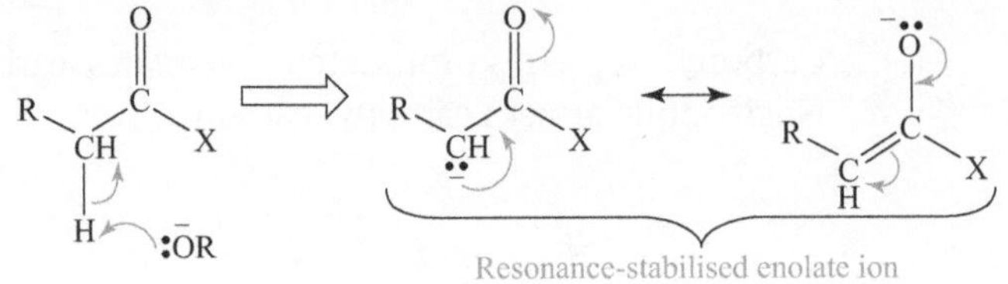

Resonance-stabilised enolate ion

Box 8.4

Acidity and basicity: organic bases

Organic bases usually accept protons (H$^+$), and amines of various kinds are the most common organic bases found in nature. The strength of the base is governed by the degree of availability of the lone pair of electrons that will accept the proton, so anything that inhibits the freedom of donation of the lone pair will reduce the base strength.

To illustrate, compare the data for methylamine (CH$_3$NH$_2$) and aniline (C$_6$H$_5$NH$_2$). The usual way in which this is done is to compare the pK_a values of the conjugate acids of the bases. *The stronger the base, the greater the pK_a of the conjugate acid.*

For methylammonium (the conjugate acid of methylamine), pK_a = 10.62.

For anilinium (the conjugate acid of aniline), pK_a = 4.6.

Aniline is a much weaker base than methylamine because the lone pair associated with the nitrogen is delocalised through the aromatic ring, meaning it is not as available to attract H$^+$. This is shown:

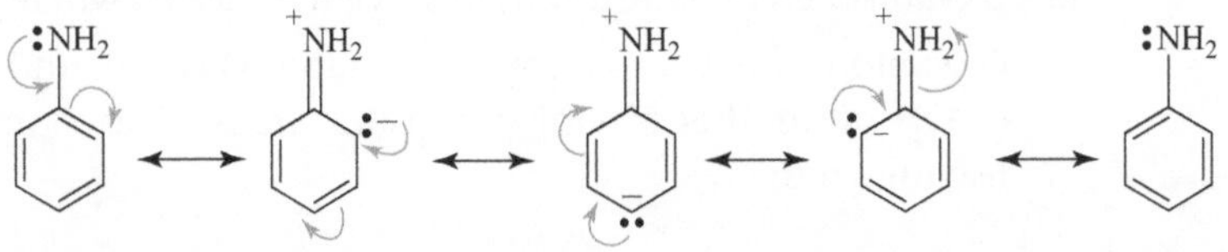

In the above case, resonance delocalisation resulted in the base being weaker. The reverse is found, however, in the case of guanidine, for which the pK_a is 13.6. This is because the resulting guanidinium ion is resonance-stabilised, and this effect outweighs the effect of delocalisation of the lone pair associated with the unprotonated species, as indicated below:

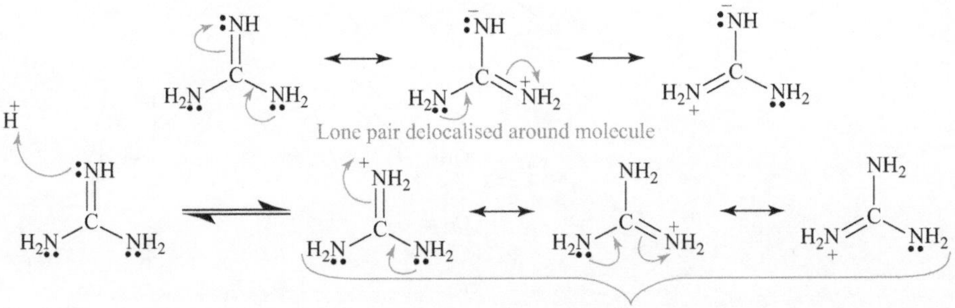

Lone pair delocalised around molecule

Positive charge delocalised around molecule
Three identical structures, so strongly stabilised

Box 8.5

```
Acidity and basicity: organic acids
```

Both ethanoic acid and ethanol possess an OH group, and yet the former is significantly more acidic. The reason for this lies in the nature of the resultant anions. The ethanoate anion, which results from removal of H^+ (called deprotonation) from ethanoic acid, is resonance-stabilised, whereas the removal of H^+ from ethanol results in the ethoxide ion, which does not have any potential for resonance stabilisation. This is shown thus:

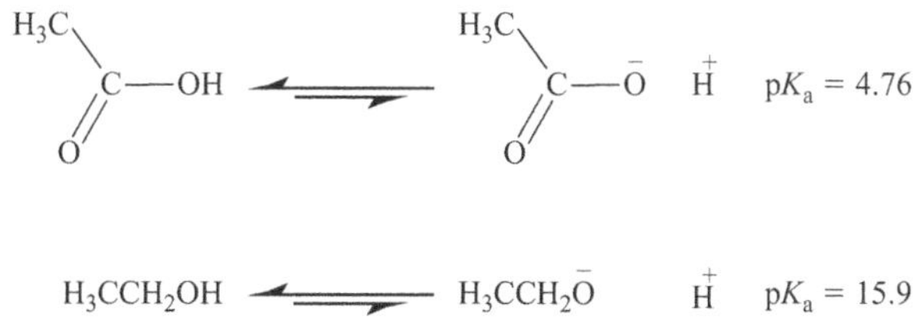

The pK_a values reflect the significant difference in acidity that result from the stabilisation of the ethanoate anion.

Even among alcohols themselves there can be significant increases in acidity as a result of resonance stabilisation of anions resulting from deprotonation. Consider, for example, the situation of phenol, C_6H_5OH, often called carbolic acid. This has a pK_a value of 9.95, due to the resonance stabilisation of the phenolate anion, as shown:

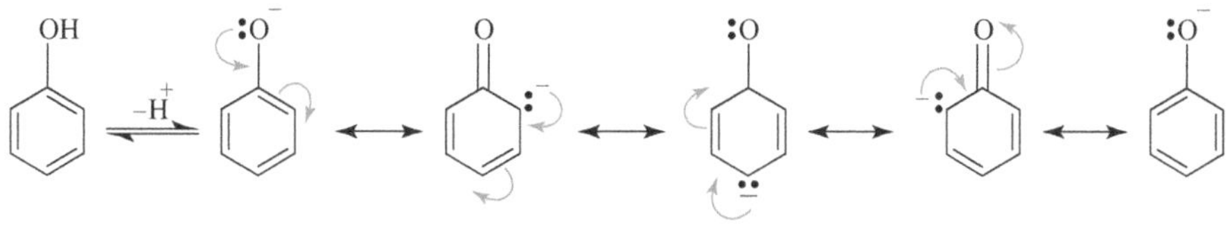

Inductive effects

Recall from section 8.5.2 that alkyl groups tend to 'push' electrons compared with, for example, hydrogen. If the pK_a values of ethanoic acid and methanoic acid are compared then this becomes apparent.

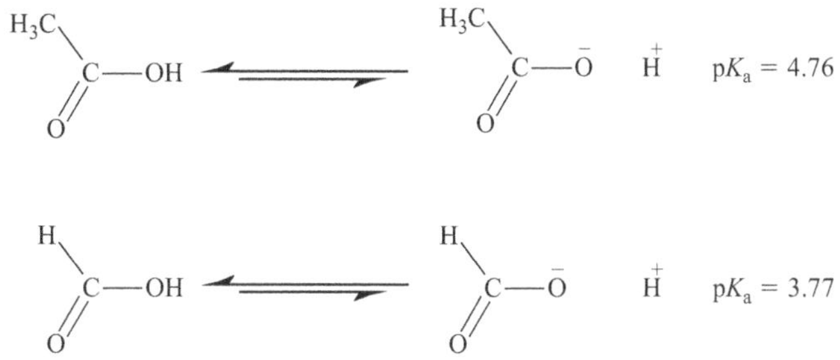

Questions

8.1 Draw full structural diagrams showing the stereochemistry of the following and add $\delta+$ and $\delta-$ to show the bond polarisation, if any:

(a) CH_3-Br
(b) $CH_3CH=CH_2$
(c) CH_3O-H
(d) $(CH_3)_2C=O$
(e) RCO_2H

8.2 (a) State what is meant by the phrase 'two canonical or resonance structures representing a molecule'.

(b) For each of the following four species:
(i) draw additional resonance structures, showing how each of the structures may be converted to the next using 'curly' arrows to denote electron movement
(ii) draw the resonance hybrid structure using dotted lines to indicate the delocalised bonds.

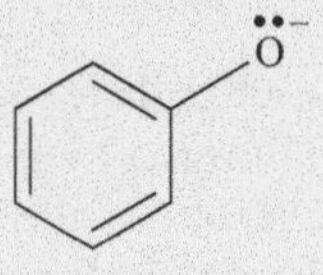

(I)

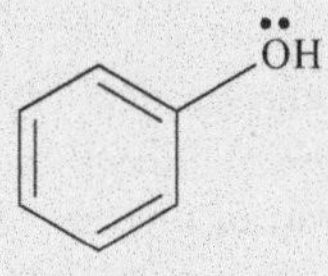

(II)

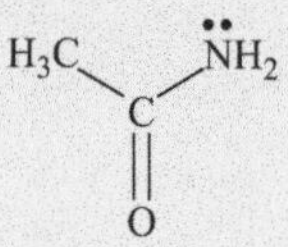

(III)

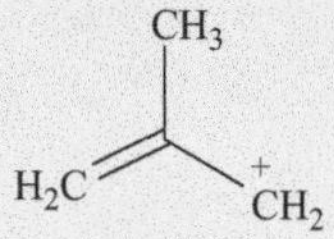

(IV)

(c) Use the information from (I) to explain why phenol, C_6H_5OH, is a much stronger acid than aliphatic alcohols such as ethanol.

(d) Discuss how the result for (III) is important in understanding protein structure.

8.3 Define the terms *electrophile* and *nucleophile* and classify the following as electrophiles or nucleophiles:

(a) H_3O^+
(b) CH_3O^-
(c) CH_3NH_2
(d) BF_3

8.4 Add a 'curly arrow' to the molecule on the left hand side of the diagram below to indicate the movement of the electrons which result in the formation of the ions on the right hand side.

$$H-Cl \longrightarrow \overset{+}{H} + \overset{\cdot\cdot}{\underset{\cdot\cdot}{Cl}}{}^-$$

Describe why the movement of both electrons to one of the joined atoms results in only a single positive charge on the cation and a single negative charge on the anion.

8.5 Predict which will be the stronger acid in the following pairs:

  (a) acetic acid and formic acid
  (b) acetic acid and 2, 2-dichloroacetic acid
  (c) cyclohexanol and phenol
  (d) 4-nitrophenol and phenol

8.6 Account for the following observations, giving the mechanisms in each case:

  (a) Reaction of HBr with hex-3-ene gives two products but the corresponding reaction of hex-1-ene gives only one product.
  (b) Propyne reacts with two molar equivalents of HBr but only one molar equivalent of water in the presence of sulfuric acid.

8.7 The reaction of 1-phenylbutene with hydrogen bromide gives only one product. Write down the structures of the two possible intermediate ions, and explain why one has a lower energy than the other and hence deduce the structure of the product.

8.8 (a) Describe what each element of the terms S_N1 and S_N2 represent.
  (b) Illustrate the mechanisms for S_N1 and S_N2 respectively, for a nucleophile attacking an alkyl halide. Use curly arrows to highlight the movement of electrons.

8.9 (a) Describe what each element of the terms E1 and E2 represents.
  (b) Illustrate the key features of the mechanisms for E1 and E2 respectively, and comment on how they compare with the S_N1 and S_N2 mechanisms. Use curly arrows to highlight the movement of electrons.

8.10 For each of the following alkyl chlorides, complete the equation showing the possible nucleophilic substitution and elimination products. For reactions (b) and (c), predict whether unimolecular (S_N1/E_1), or bimolecular (S_N2/E_2) reaction mechanisms will be dominant.

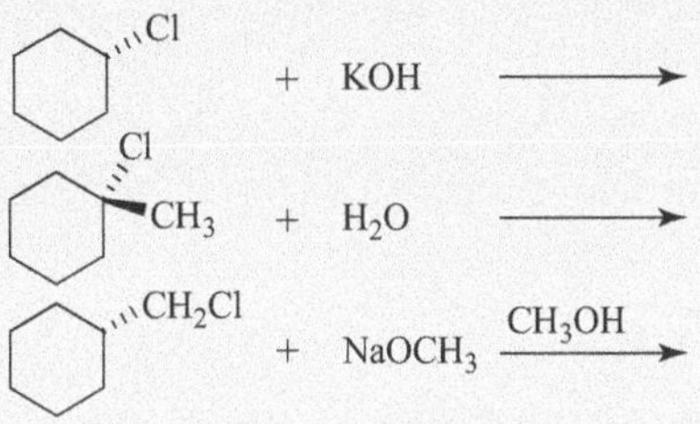

8.11 Write out the general mechanism showing the formation of the intermediate anion when a nucleophile reacts with a ketone.

8.12 Write out the products and the mechanism of their formation for the following reactions:

  (a) Butanone and aqueous sodium cyanide
  (b) Ethyl methyl ketone and hydroxylamine
  (c) Propanal and methanol in the presence of a trace of HCl.

8.13 (a) In describing the mechanism of a chemical reaction, state what a full-headed curved (curly) arrow represents.

(b) Give the structures of the products expected when butanoic acid is reacted with the following reagents:

(i) CH_3CH_2OH / H^+

(ii) dilute NaOH

(c) Write a mechanism for the reactions below, giving in each case the structures of the intermediates formed on the route from the starting materials to the products.

(i)

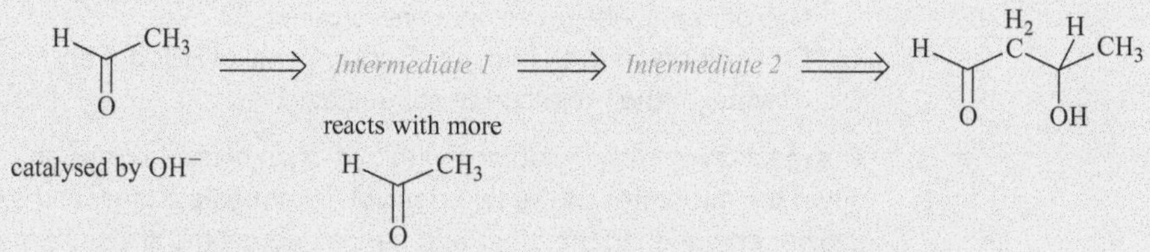

Intermediate 1 ⟹ Intermediate 2 ⟹

catalysed by OH⁻

reacts with more

(ii)

Intermediate 3 ⟹

+NaOH

8.14 *E*-Ethyl cinnamate ($Ph.CH{=}CH.CO_2Et$) is a natural flavouring material which can be prepared in the laboratory using the reaction outlined below. Show the mechanism for this reaction. What type of reaction is taking place?

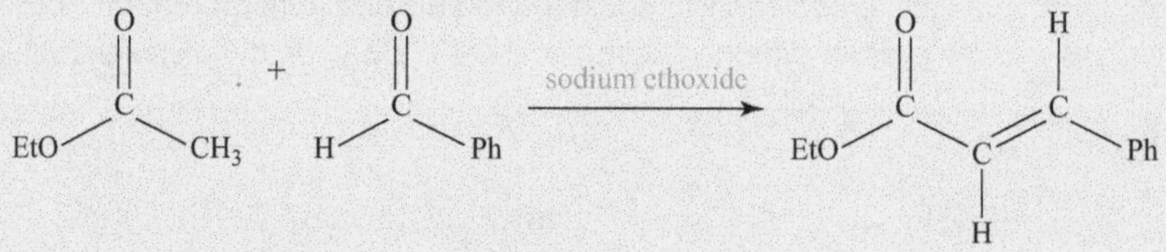

sodium ethoxide

8.15 Identify the functional groups in the molecule below:

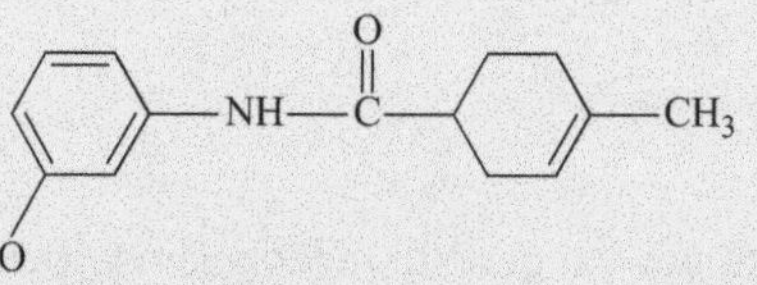

Hence state whether it is likely to react with each of the following reagents, stating your reasons.

(a) Cold dilute aqueous NaOH.
(b) Cold dilute aqueous HCl.
(c) Anhydrous HBr.
(d) Aqueous HCl with heat.

9 Chemical kinetics and enzymes

Learning outcomes

By the end of this chapter students should be able to:

- define and explain the following terms: rate, rate constant, rate law, order of reaction;
- describe the approaches used to study reaction rates;
- use the Arrhenius equation to explain the temperature dependence of rates;
- derive overall rate law for simple reactions using assumptions of rate-determining step, rapid equilibration and the steady-state approximation;
- understand how the steady-state approach is used to derive the Michaelis–Menten model for enzyme kinetics, and how this model is used to analyse enzyme-catalysed reactions.

9.1   Introduction

Chapter 4 contained an overview of how the principles of thermodynamics provided information on whether a reaction should proceed (ΔG) and how the enthalpy change (ΔH) and entropy change (ΔS) contribute to reactions. However, these values do not give information on the *rate* or the details of the *mechanism* of the reaction. Whilst ΔG may be negative, and hence the reaction is technically spontaneous, a reaction may proceed very slowly or not at all if the energy of activation, an 'energy hurdle' between the reactants and products, is large (see Figure 9.1).

Studies of the rates of both chemical and biological processes have been used to establish the mechanisms of many processes including, in the biological sphere, the mechanisms of enzyme behaviour. As a starting point for using kinetics in biological processes, it is helpful to establish the principles of chemical kinetics in a more general chemical sense.

Kinetics is the study of the rates of chemical processes. The information gained in this way can then be used to help to understand the processes by which reactions progress (the mechanisms of reactions). At a fundamental level kinetics is about understanding what happens to molecules in a chemical reaction, specifically the reactive interaction of two reagent molecules. An understanding of these processes could then be used to predict the outcome and rate of reactions.

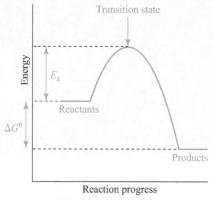

E_a = activation energy

ΔG^θ = free energy difference between reactants and products

Figure 9.1 **Reaction profile diagram.**

9.2   Rates, rate laws and rate constants

Before looking in detail at the principles of kinetics, it is necessary to become familiar with the terminology used to describe the rates of chemical reactions.

9.2.1 Rate of reaction

The rate of a reaction basically means the speed at which a chemical reaction occurs. The speed of a reaction is determined by seeing how the concentrations of the reactants or products vary during given intervals of time. It is generally assumed that the volume of the reaction vessel is kept constant and there is no addition to or removal from the reaction vessel of any of the reactants and products.

Take the hypothetical reaction,

$$R \quad \rightarrow \quad P$$

As the reaction proceeds, the concentration of the reactant (given by [R]) decreases with time while the concentration of the product (given by [P]) increases (see Figure 9.2).

If the concentration of the reactant, [R], is monitored at different times, say t_1 and t_2, then the rate of the reaction is given by the gradient of the line (shown in grey) linking the two points, which is represented by:

$$\text{Rate} = \frac{[R]_2 - [R]_1}{t_2 - t_1}$$

where $[R]_2$ is the concentration of reactant R at time t_2, while $[R]_1$ is the concentration R at a time t_1.

This can be rewritten as,

$$\text{Rate} = -\frac{\Delta[R]}{\Delta t} \qquad\qquad [9.1]$$

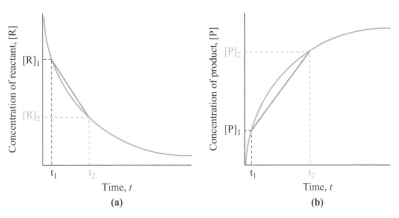

Figure 9.2 Plots of [R] vs time (9.2a) or [P] vs time (9.2b), showing averaged rate.

where $\Delta[R]$ means the change in the concentration of the reactant R and Δt means the change in time, $t_2 - t_1$.

Sometimes, depending on experimental need, the concentration of the product, given by [P], is monitored. In this case the rate of the reaction is given by:

$$\text{Rate} = \frac{[P]_2 - [P]_1}{t_2 - t_1}$$

This can be rewritten as:

$$\text{Rate} = \frac{\Delta[P]}{\Delta t} \qquad [9.2]$$

Equations [9.1] and [9.2] represent the average rate of a reaction over a given, measured, period of time.

In summary, for the reaction

$$R \quad \rightarrow \quad P$$

the rate of the reaction in the terms of appearance and disappearance of the product and reactant is given as

$$r = \text{Rate of disappearance of R}$$
$$r = \text{Rate of appearance of P}$$
$$r = -\frac{\Delta[R]}{\Delta t} = \frac{\Delta[P]}{\Delta t} \qquad [9.3]$$

A negative sign precedes the term in the equation of the rate of reaction when it is expressed in terms of concentrations of the reactants because the average rate of a reaction is a positive quantity, but the value of $\Delta[R]$ will be negative, because the reactant concentration is decreasing.

The average rate of a reaction does not represent the actual rate of a reaction at any instant of time, and it is clear from Figure 9.2 that the rate of reaction actually decreases with time. This is because one of the factors that generally influences the rate of reaction is the concentration of the reactant(s).

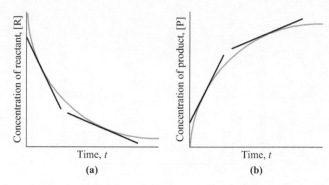

Figure 9.3 Plots of [R] vs time (9.3a) or [P] vs time (9.3b), showing instantaneous rate. The tangents to the curve show rates at different points of reactions.

In order to fully understand the relationship between reaction rate and concentration it is necessary to work with the *instantaneous rate of reaction*.

Instantaneous rate of reaction

Figure 9.3 shows examples of how rate can vary with *infinitely small* changes in time for a reactant and a product respectively. In this case the rate at a given point of the reaction is calculated from the slope of the tangent to the curve.

- Figure 9.3a shows how reactant concentration varies with time.
- Figure 9.3b shows how product concentration varies with time.
- The slopes of the tangents to the curves indicate how the rate of the reaction varies with time. A steeper slope indicates a greater rate.
- The slope of the tangents in Figure 9.3a is negative.
- The slope of the tangents in Figure 9.3b is positive.

As can be seen in Figure 9.3a, the concentration of the reactant decreases with time, but the rate of the loss of the reactant also decreases, as indicated by the slope of the tangent being lower at a larger value of time, t. In contrast, from Figure 9.3b, although the concentration of the product increases with time, the rate of gain of the product decreases, again as indicated by the slope of the tangent being lower at a larger value of time, t.

Thus, reinforcing the points made in equations [9.1] to [9.3], the sign can be positive or negative, and:

- a positive sign means that the concentration is increasing with time; in other words, it relates to the formation of a product;
- a negative sign means that the concentration is falling with time; in other words, it relates to usage of a reactant.

In order to establish both an overarching view of what is happening during a process and the details of how this process is progressing it is necessary to be able to get a mathematical understanding of what is happening. To that end, the rate of a chemical reaction is thought of mathematically as the *derivative* of concentration *with respect to time* (see Appendix 1 for an introduction to the mathematics and Box 9.1 for a brief run through the key mathematical symbols).

Box 9.1

Comments on the mathematical symbols

From Figure 9.2, the average rate of change of formation of product over a period of time is given by:

$$\text{Rate} = \frac{\Delta[P]}{\Delta t}$$

The symbol Δ is used in mathematics and in sciences to denote the change of any variable quantity.

When discussing changes over infinitesimally small periods, then the symbol Δ is replaced by d. Therefore, the instantaneous rate of change for product formation would be given by

$$\text{Rate} = \frac{d[P]}{dt}$$

The instantaneous rate represented thus is the slope of the tangent at a given time t. (Refer to Appendix 1 for more details on the background to this.)

In order to develop this approach, consider a general reaction for converting one mole of a reactant into two moles of products as summed up by:

$$R \quad \rightarrow \quad 2P$$

Mathematically, the relationship between concentration changes and time for this reaction is given by:

$$\text{Rate} = -\frac{d[R]}{dt} = \frac{1}{2}\frac{d[P]}{dt}$$

Looking a bit more closely:

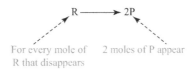

R ⟶ 2P

For every mole of R that disappears   2 moles of P appear

This means that P is appearing at a faster rate than R is disappearing – in fact twice as fast! In mathematical format this is shown as:

$$\text{Rate} = -\frac{d[R]}{dt} = \frac{1}{2}\frac{d[P]}{dt}$$

Rate of disappearance of R = Half the rate of appearance of P

This can also be written as:

$$\text{Rate} = -2\frac{d[R]}{dt} = \frac{d[P]}{dt}$$

$2 \times$ rate of disappearance of R = Rate of appearance of P

If this approach is extended to the general reaction

$$aA + bB \quad \longrightarrow \quad cC + dD$$

The rates of disappearance of A and B, and the rates of appearance of C and D, are related as shown:

$$\text{Rate} = -\frac{1}{a}\frac{d[A]}{dt} = -\frac{1}{b}\frac{d[B]}{dt} = \frac{1}{c}\frac{d[C]}{dt} = \frac{1}{d}\frac{d[D]}{dt}$$

[9.4]

Therefore

- Rates vary with time.
- Rates depend on concentration.

9.2.2 Rates and concentration

Experimentally it is found that rates depend on the concentrations of the species involved in the reaction. The relationship that exists between the rate and these concentrations is represented mathematically in the form of an equation called a *rate law*.

Some rate laws are very simple and some are very complicated. They usually express the rate of a reaction in terms of the molar concentrations of the reactants raised to a power (in some cases product concentrations also appear in rate law). Rate laws are determined experimentally, *and it is important to note that they cannot be predicted from the stoichiometry of the reaction.*

Let us consider the reaction between two species, A and B, such that:

$$aA + bB \longrightarrow P\,(\text{product})$$

The rate law for this is represented by:

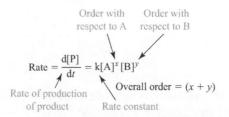

Because the rate of a reaction is defined in terms of the change in concentration of a species with time, then the units are *concentration/time*, usually $mol\,L^{-1}\,s^{-1}$. A rate law in this form is referred to as a differential rate law. In Appendix 1, which covers the mathematical principles used in this text, we see that this stems from the idea that the term $d[P]/dt$ represents the rate of change of concentration, and this is catered for using the mathematical process called differentiation.

To summarise a few key points:

- If, in the example above, the index x was 1, then the reaction would be said to be first order with respect to reagent A.
- Similarly if the index $y = 1$, then the reaction would be first order in B.
- If both $x = 1$ and $y = 1$ then, in addition to being first order with respect to each of reagents A and B, the overall order of the reaction would be second order.

- If the index x was 0, then the reaction would be zero order with respect to A (in other words the rate would not depend on the concentration of A).
- Reaction order must be determined experimentally; it cannot be predicted from the reaction stoichiometry.

The rate constant, k, is a term that defines the speed of a reaction. It is independent of concentration of reactants but (as will be shown in Section 9.3) it is dependent on temperature. Box 9.2 contains a worked example which shows how rate depends on reagent concentration.

Box 9.2

Extraction of rate laws and rate constants from experimental data

Taking the general reaction $A + B \rightarrow C$, the following data were collected.

Experiment	[A] $(mol\,L^{-1})$	[B] $(mol\,L^{-1})$	$\dfrac{d[C]}{dt}$ = Initial rate of formation of C $(mol\,L^{-1}\,s^{-1})$
1	0.2	0.2	8×10^{-4}
2	0.2	0.6	7.2×10^{-3}
3	0.6	0.6	2.16×10^{-2}

From these data it is relatively straightforward to deduce the rate law and the value of the rate constant. The first stage is to establish the rate law, which can be done by initially comparing the results from different experiments. The object is to see how concentration influences rate, so:

- Comparing experiments 1 and 2, when [A] is kept the same, but [B] is trebled, the rate is nine times faster as a result, so it means that rate increase is proportional to $[B]^2$. It is, therefore, second order in B.
- Comparing the data from experiments 2 and 3, [B] is kept the same, but [A] is trebled. The rate is three times faster, so it is proportional to [A]; it is first order in A.

The rate law is, therefore:
$$\frac{d[C]}{dt} \text{ (rate of formation of C)} = k\,[A][B]^2$$

Using data from experiment 1
$$8 \times 10^{-4} = k \times 0.2 \times 0.2^2$$

So, $$k = 8 \times 10^{-4}/0.2 \times 0.2^2 = 0.1 \text{ mol}^{-2}\text{L}^2\,\text{s}^{-1}$$

Example 9.1 The decomposition of N_2O_5

$$2N_2O_5 \longrightarrow 4NO_2 + O_2$$

Note that as no power is expressly shown, the order is 1

$$-\frac{1}{2}\frac{d[N_2O_5]}{dt} = k[N_2O_5]$$

The reaction is first order—no relation to stoichiometry

Example 9.2 The decomposition of ethanal

$$CH_3CHO \longrightarrow CH_4 + CO$$

The power is a non-integer

$$-\frac{d[CH_3CHO]}{dt} = k[CH_3CHO]^{\frac{3}{2}}$$

The reaction is 3/2 order–reaction order does not need to be integral

Example 9.3 The iodination of ethene

$$C_2H_4 + I_2 \longrightarrow C_2H_4I_2$$

$$-\frac{d[C_2H_4]}{dt} = k[C_2H_4][I_2]^{\frac{3}{2}}$$

The reaction is

- first order in ethene;
- 3/2 order in iodine, and
- 5/2 order overall.

Sometimes, however, the rate law does not readily permit the order of a reaction to be clearly determined. For example, the rate law governing many enzymatic reactions is given by the Michaelis–Menten relationship (which will be analysed in more detail later in the chapter).

Example 9.4 Enzyme catalysis

For an enzyme catalysing the transformation of a substrate to a product, represented by the overall process:

$$E + S \quad \rightarrow \quad E + P$$

The rate law takes the form:

$$Rate = \frac{d[P]}{dt} = \frac{k_{cat}[E]_0[S]}{K_M + [S]}$$

In the above it is impossible to state either an overall order or even an order with respect to [S].

9.2.3 Units of the rate constant

The units are determined by the order of the reaction. Using the generalised example:

$$A \quad \rightarrow \quad Product, P$$

Initially consider a first order rate law, for example:

$$\frac{d[P]}{dt} = k[A] \text{ (or, in words, Rate} = k \text{ times concentration)}$$

This can be rearranged in order to have the rate constant on one side of the equation, and all of the other terms on the other side of the equation:

$$k = \frac{\dfrac{d[P]}{dt}}{[A]} \text{ (or, in words, k} = \text{rate divided by concentration)}$$

The units of rate, as shown above, are $mol\,L^{-1}\,s^{-1}$, and those of concentration are $mol\,L^{-1}$, so the units of the rate constant, k, will be:

$$\text{units of k} = \frac{mol\,L^{-1}s^{-1}}{mol\,L^{-1}} = s^{-1}$$

Therefore, for a first order process the units of rate constant k are s^{-1}.

For a second order rate law, for example:

$$\frac{d[P]}{dt} = k[A]^2 \text{ (or, in words, rate} = k \text{ times concentration squared)}$$

This can be rearranged in order to have the rate constant on one side of the equation, and all of the other terms on the other side of the equation:

$$k = \frac{\dfrac{d[P]}{dt}}{[A]^2} \left(\text{or, in words, k} = \frac{\text{rate}}{\text{concentration}^2} \right)$$

Again we insert the units of rate, $mol\,L^{-1}\,s^{-1}$, and those of concentration, $mol\,L^{-1}$, so the units of rate constant, k, will be:

$$\text{units of k} = \frac{mol\,L^{-1}s^{-1}}{(mol\,L^{-1})(mol\,L^{-1})} = L\,mol^{-1}s^{-1}.$$

Therefore, for a second order process the units of rate constant k are $L\,mol^{-1}\,s^{-1}$.

Units of rate constant – the general case

Extending the approach to a general case, for a reaction of order n, such as:

$$\frac{d[P]}{dt} = k[A]^n$$

The units of k derive from $k = \text{rate/concentration}^n$. This is represented by:

$$\text{units of k} = \frac{mol\,L^{-1}s^{-1}}{(mol\,L^{-1})^n} = L^{(n-1)}\,mol^{-(n-1)}s^{-1}.$$

The units of rate constants for the various orders are summarised in Table 9.1.

Table 9.1 **The units of k for different reaction orders**

Reaction order	Rate law	Units of k
Zero	$-d[A]/dt = k$	$mol\,L^{-1}\,s^{-1}$
First	$-d[A]/dt = k[A]$	s^{-1}
Second	$-d[A]/dt = k[A]^2$	$L\,mol^{-1}s^{-1}$
n	$-d[A]/dt = k[A]^n$	$L^{(n-1)}mol^{-(n-1)}s^{-1}$

9.2.4 Determination of rate laws and rate constants

The method of initial rates

The study of reaction rates is geared towards determining the mechanism by which a reaction progresses, and many of the reactions under study may have very complex mechanisms, possessing several steps, some of which may be equilibria. Also, as reactions progress it is quite possible for the products to become involved.

At the very start of a reaction, however, it can be only reactants which influence the reaction rate, and this idea forms the basis of the *initial rates method*, which is a very commonly used approach. To illustrate this, consider the generalised single species reaction

$$A \quad \rightarrow \quad product, P$$

for which the rate law can be represented by:

$$\frac{d[P]}{dt} = rate = k[A]^n$$

The aim is to determine the values of k and of n, and this can be achieved by taking logarithms of this equation, as indicated:

$$\log (rate) = \log (k[A]^n)$$

Using the rules of logarithms provided in Appendix 1, then this can be reshown as:

$$\log (rate) = \log k + \log [A]^n$$

which is the same as:

$$\log (rate) = \log k + n \log [A]$$

This is in the form of the straight line relationship $y = mx + c$, where

- log (rate) replaces the y variable;
- log [A] replaces the x variable;
- n replaces the gradient m (slope); and.
- log k is the constant, c.

It was demonstrated in Figure 9.3 that the rate of a reaction decreases as the reaction progresses, as determined by measuring the concentration of one of the reactants, or the concentration of a product at different times during the reaction. Therefore, in order to carry out the method of initial rates, the following procedures are adopted:

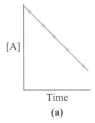

Figure 9.4a **The variation of concentration with time for a zero order reaction.**

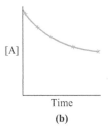

Figure 9.4b **The variation of concentration with time for a non-zero order reaction.**

- Measure how reactant or product concentration varies as reaction progresses for different initial reactant concentrations (it depends which is easiest to measure). If the plot of concentration against time is linear (Figure 9.4a) then the reaction is zero order (the rate does not vary with concentration). In most cases, however, the plot of concentration vs. time will be curved (as in Figure 9.4b), and it will be necessary to repeat the process using a number of initial concentrations.

- Determine the initial rate for the reaction resulting from each initial reactant concentration by measuring the gradient of the tangent drawn to the curve at the origin. The method for doing so is shown in Figure 9.5.

- Tabulate the values of initial rate against initial concentration of A, then plot. This will give a straight line.

- Read off the values of order (from the gradient), and log k (from the intercept).

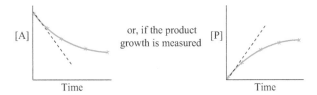

Figure 9.5 **Determination of initial rate for a non-zero order reaction.**

Example 9.5 The hydrolysis of ATP

The enzyme-catalysed hydrolysis of ATP to produce ADP was studied by the method of initial rates, and Table 9.2 lists the initial values of ATP concentration, the resulting initial rates of ADP production, and their respective logarithms (to base 10).

Table 9.2

$[ATP]_o$	$d[ADP]/dt$	$\log [ATP]_o$	$\log (rate)$
0.01	83 000	−2.00000	4.919078
0.005	40 000	−2.30103	4.60206
0.001	8 150	−3.00000	3.911158
0.0005	4 150	−3.30103	3.618048
0.0001	820	−4.00000	2.913814

The slope of the graph in Figure 9.6 is 1, so the reaction is first order with respect to [ATP], and the intercept on the y axis gives a value of $\log k = 6.91$. Therefore $k = 10^{6.91} = 8.1 \times 10^6\ s^{-1}$.

Isolation method

In situations where more than one reactant is present it is possible to set the experiment up to allow one of the components to be studied on its own. This is achieved by ensuring that the concentrations of all reactants except one are present in large excess. For example, taking the general reaction:

$$a\text{A} + b\text{B} \longrightarrow \text{P}(\text{product})$$

for which the rate law can be represented by:

$$\text{rate} = \frac{d[\text{P}]}{dt} = k[\text{A}]^x[\text{B}]^y$$

If B was present in vast excess, then it would be possible to measure the dependence of the rate on [A] with [B] remaining effectively constant (the concentration of B at any time during the reaction will not be very dissimilar to the starting concentration), and hence to find the order with respect to A. Similarly, if A was present in vast excess, then it would be possible to measure the dependence of the rate on [B] with [A] effectively constant, and hence to find the order with respect to B. The isolation method can be combined with the initial rate method.

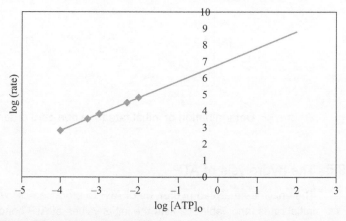

Figure 9.6 **Plot of log (rate) vs log [ATP]$_o$.**

Integrated rate laws

In some instances it is desirable to monitor the progress of a reaction rather than just consider it during the initial stages. In order to do this it would be very useful to be able to use experimental data that are relatively easily measured, such that it will be possible to relate [A] with the time t, and thereby use measured experimental data to see whether the reaction obeys this rate equation, and obtain a value for the reaction order. This can be achieved by integrating the rate equation, and seeing how the results of this depend on the value of n. *(The mathematical background to this is presented more completely in Appendix 1.)*

Using the generalised single species reaction

$$A \quad \rightarrow \quad \text{Product, P}$$

The rate law can be represented as,

$$\text{Rate} = \frac{-d[A]}{dt} = k[A]^n$$

This is a differential rate equation, in which, as has been shown already, k is a constant called the rate constant and n is the order of the reaction.

Example 9.6 The zero order case

For a zero order system, where $n = 0$,

$$-d[A]/dt = k$$

or if this is rearranged to separate the variables [A] and t, then:

$$-d[A] = kdt$$

This is integrated thus:

$$\int d[A] = \int -k \cdot dt$$

So integration gives,

$$[A] = -kt + c$$

If the start of the reaction is said to be at time $t = 0$, then the value of the concentration at the time can be called $[A]_0$ (or, put another way, $[A]_0$ means the value of [A] at time 0). As $t = 0$, then the value of $-kt = 0$, so:

$$c = [A]_0$$

This means that the integrated equation can be rewritten, putting this value in for c:

$$[A] = -kt + [A]_0 \tag{9.6}$$

This is an integrated rate law.

In this relationship, the time t is a variable that is in the control of the experimentalist, and the concentration [A] is the variable that depends on t. Both rate constant, k, and

295

the initial concentration $[A]_0$ are constants, so the integrated rate law is an equation of a straight line:

$$[A] = -kt + [A]_0$$

Therefore if a plot of [A] vs t gives a straight line, then the reaction is zero order.

Example 9.7 The first order case

Looking now at the first order system, where $n = 1$,

$$-\frac{d[A]}{dt} = k[A]$$

Again this can be rearranged to put all of the [A]s on one side and all of the ts on the other:

$$-\frac{1}{[A]} \cdot d[A] = k \cdot dt$$

Integration of this (see Appendix 1 for more detail) gives:

$$\ln [A] = -kt + c$$

As in the zero order case, at time $t = 0$, the value of $[A] = [A]_0$, then

$$c = \ln [A]_0$$

So, replacing c by $\ln [A]_0$ and rearranging gives:

$$\ln [A] = \ln [A]_0 - kt \tag{9.7}$$

Again, in this relationship, the time t is a variable that is in the control of the experimentalist, and the concentration [A] is the variable that depends on t. Both rate constant k and the initial concentration $[A]_0$ are constants, so the integrated rate law is an equation of a straight line, where the value of y is given by the natural logarithm of [A] and the value of x is given by t:

$$\ln [A] = -kt + \ln [A]_0$$

Therefore if a plot of ln [A] vs t gives a straight line, then the reaction is first order.

The relationship between reaction order and rate laws is summarised in Table 9.3.

Table 9.3 **Table showing relationship between reaction order and rate equations**

Reaction order	Differential form	Integrated form	Plot used to determine order
Zero	$-d[A]/dt = k$	$[A] = -kt + [A]_0$	$[A]$ vs t
First	$-d[A]/dt = k[A]$	$\ln [A] = \ln [A]_0 - kt$	$\ln [A]$ vs t
Second	$-d[A]/dt = k[A]^2$	$1/[A] = 1/[A]_0 + kt$	$1/[A]$ vs t

Table 9.4   **Kinetic data for decomposition of dinitrogen pentoxide**

Time, t (s)	[N₂O₅]	ln [N₂O₅]	1/[N₂O₅]
0	0.1000	−2.303	10.00
50	0.0707	−2.645	14.14
150	0.0351	−3.349	28.49
300	0.0125	−4.382	80.00
450	0.0043	−5.449	232.56

Example 9.8   What is the reaction order for the degradation of N_2O_5?

The data in Table 9.4 were obtained from an experiment following the degradation of dinitrogen pentoxide, which has the balanced overall equation:

$$2N_2O_5 \text{ (g)} \quad \rightarrow \quad 4NO_2 \text{ (g)} \quad + \quad O_2 \text{ (g)}$$

In order to determine the reaction order, plots of the time against each of $[N_2O_5]$, ln $[N_2O_5]$ and $1/[N_2O_5]$ respectively were obtained (Figure 9.7). The order of the reaction was shown to be first order, because only the plot of ln $[N_2O_5]$ vs t was a straight line.

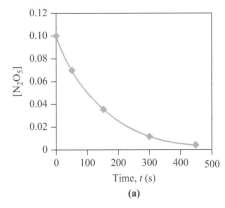

(a)

Figure 9.7a **Plot of $[N_2O_5]$ vs t.**

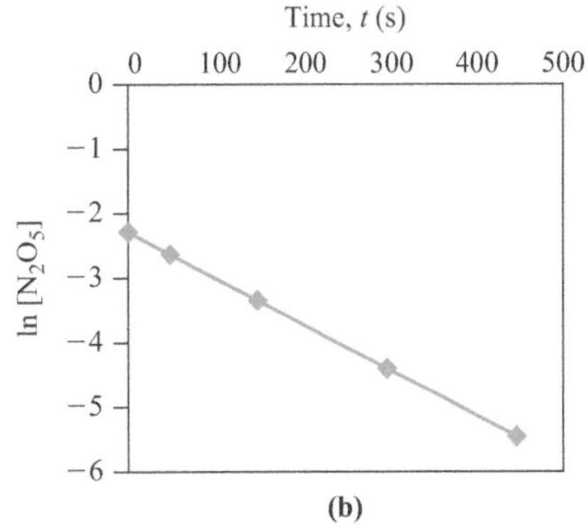

Figure 9.7b **Plot of ln $[N_2O_5]$ vs t.**

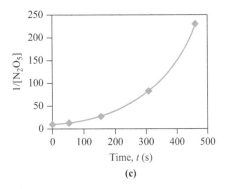

(c)

Figure 9.7c **Plot of 1/ $[N_2O_5]$ vs t.**

Half-life, $t_{1/2}$

The concept of half-life was briefly referred to during the description of radioactive decay (Chapter 1). In the context of studying reaction kinetics, the half-life of a reaction is the time it takes for one-half of the original amount of material to react. The half-life can, therefore, provide a relatively straightforward gauge to measuring reaction rates.

The magnitude of the half-life depends on the reaction order.

For a zero order reaction

Consider the simplest type of reaction:

$$A \quad \rightarrow \quad \text{product, P}$$

Starting with the integrated rate law:

$$[A] = -kt + [A]_0$$

According to the definition, the half-life is the time taken for the reactant concentration to halve. In other words, when $t = t_{1/2}$, then $[A] = [A]_0/2$. Putting this into the integrated rate equation gives:

$$\frac{[A]_0}{2} = -kt_{1/2} + [A]_0$$

This can then rearrange to give:

$$kt_{1/2} = [A]_0 - \frac{[A]_0}{2}$$

or

$$kt_{1/2} = \frac{[A]_0}{2}$$

In order to define the value of $t_{1/2}$ only, then a final rearrangement gives:

$$t_{1/2} = \frac{[A]_0}{2k} \tag{9.8}$$

This means that the magnitude of the half-life for a zero order reaction depends on the starting concentration and on the value of rate constant, k.

For a first order reaction

As above, starting with the integrated rate law:

$$\ln [A] = -kt + \ln [A]_0$$

Again, when $t = t_{1/2}$, then $[A] = [A]_0/2$. Substituting this into the integrated rate equation gives:

$$\ln \frac{[A]_0}{2} = -kt_{1/2} + \ln [A]_0$$

This rearranges to give:

$$kt_{1/2} = \ln [A]_0 - \ln \frac{[A]_0}{2}$$

$$= \ln \frac{[A]_0}{\dfrac{[A_0]}{2}} \quad \text{(from } \ln x - \ln y = \ln x/y; \text{ see rules of logarithms in Appendix 1)}$$

$$= \ln 2$$

Therefore,

$$t_{1/2} = \frac{\ln 2}{k} = \frac{0.693}{k} \tag{9.9}$$

Note that, for a first order reaction, half-life does not depend on the starting concentration.

Table 9.5 summarises the relationships between half-life and reaction order. Only in the case of first order reactions does the half-life have no dependence on the starting concentration. To illustrate these points, Figure 9.8 shows graphical representations of half-life for both first order and second order reactions.

Table 9.5 Half-life values for different reaction orders

Reaction order	Half-life
0	$t_{1/2} = \dfrac{[A]_0}{2k}$
1	$t_{1/2} = \dfrac{\ln 2}{k} = \dfrac{0.693}{k}$
2	$t_{1/2} = \dfrac{1}{k[A]_0}$
n	$t_{1/2} = \dfrac{1}{(n-1)k} \left[\dfrac{2^{n-1} - 1}{[A]_0^{n-1}} \right]$

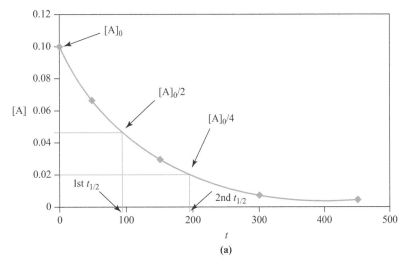

(a)

Figure 9.8a Half-life for a first order reaction.

299

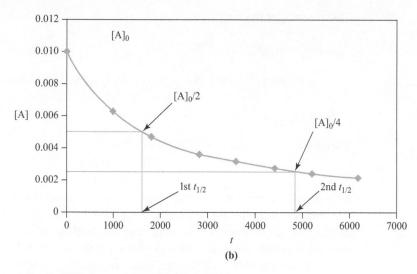

Figure 9.8b Half-life for a second order reaction.

<table>
</table>

9.3 Temperature dependence of reaction rates and rate constants

Reactions can only occur between molecules when the molecules come into contact with one another (with the right amount of energy), and this forms the basis of a theoretical approach to understanding reaction kinetics called *collision theory*. An analysis of this falls outside the remit of this text, but it is reasonable to surmise that anything that will increase both the frequency of molecular collisions, and the energy at which they occur, will increase the rate of the reactions. Increasing the temperature is one way by which these objectives can be achieved.

In general it is found that reaction rates, and hence rate constants, vary (usually increase) with increasing temperature, meaning that it is important to state the temperature at which a rate constant has been determined. The variation in the rate of a reaction with change in temperature is governed by the Arrhenius equation:

$$k = Ae^{-\frac{E_a}{RT}} \qquad\qquad [9.10]$$

where:

- k = the rate constant for the reaction at a temperature of T (K);
- E_a = the activation energy for the process (J mol^{-1});
- A = the frequency factor (a constant for a given reaction). The experimental value of this factor incorporates information about the number of collisions, and how many of the collisions have the molecules aligned correctly to react. It has the same units as the rate constant k;
- R is the gas constant (=8.31 J mol^{-1} K^{-1}).

On taking natural logarithms, the Arrhenius equation can be rewritten in the straight line form:

$$\ln k = \ln A - \frac{E_a}{R}\frac{1}{T}$$

$$\underset{y}{\uparrow}\quad\underset{c}{\uparrow}\quad\underset{m}{\uparrow}\ \underset{x}{\uparrow}$$

[9.11]

The activation energy (E_a) for a given reaction can therefore be determined by measuring the reaction rate at a series of temperatures. From the logarithmic equation above, a plot of ln k against $1/T$ will give a straight line of gradient equal to $-E_a/R$.

Example 9.9 To illustrate this, consider the reaction involving the oxidation of iodide by peroxodisulfate, yielding iodine and sulfate, which is summarised by the equation:

$$2I^- + S_2O_8^{2-} \rightarrow I_2 + 2SO_4^{2-}$$

The reaction follows second order kinetics, the rate law for it being:

$$\text{Rate} = k[I^-][S_2O_8^{2-}]$$

The rate constant for this reaction was determined at four temperatures, yielding data which are summarised in Table 9.6. From the data in Table 9.6, the plot of ln k vs $1/T$ is shown in Figure 9.9. It is a clear straight line, the gradient of which corresponds to $-E_a/R$. The slope is -6521, and this therefore yields a value of activation energy of 54.22 kJ mol^{-1}.

Table 9.6 **Effect of temperature on rate constant for the reaction between iodine and thiosulfate**

Temperature (K)	$1/T$ (K^{-1})	k ($mol^{-1}Ls^{-1}$) ($\times10^{-3}$)	ln k
275.1	3.64×10^{-3}	0.481	-7.640
285.3	3.51×10^{-3}	1.084	-6.827
294.9	3.39×10^{-3}	2.203	-6.118
300.3	3.33×10^{-3}	3.819	-5.568

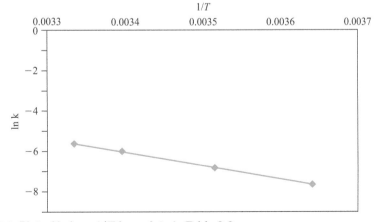

Figure 9.9 **Plot of ln k vs $1/T$ from data in Table 9.6.**

301

Why might the rate of a reaction decrease with increasing temperature?

The main reason for the rate of a reaction to decrease as temperature increases would be if one of the important components of the reaction were unstable at elevated temperatures. For example, rates of reactions involving enzymes will often decrease sharply as temperatures increase, because the enzyme structures 'unravel' at raised temperatures (a process called denaturing; see Section 7.4.7).

9.4 Reaction mechanisms

Arguably, the principal objective of the study of kinetics of reactions is to elucidate the detailed process by which the reaction takes place. Such detailed analyses are referred to as *reaction mechanisms*. To illustrate this, consider the two reactions below.

Dimerisation of butadiene

The gas phase dimerisation of butadiene to form cyclooctadiene is a second order process, as shown:

$$2C_4H_6 \, (g) \rightarrow C_8H_{12} \, (g)$$

For this,

$$\text{Rate} = -\frac{d[C_4H_6]}{dt} = k[C_4H_6]^2$$

This reaction appears to follow the intuitive pathway, whereby two gaseous molecules of butadiene collide to react and form the dimer.

Reaction of nitrogen dioxide with carbon monoxide

The stoichiometrically balanced equation for this reaction is:

$$NO_2 \, (g) \; + \; CO \, (g) \quad \rightarrow \quad NO \, (g) \; + \; CO_2 \, (g)$$

For this reaction the experimentally determined rate law is:

$$\text{Rate} \; = \; k[NO_2]^2$$

This means that the mechanism of the reaction must be more complex than just a straightforward collision between an NO_2 molecule and a CO molecule.

A mechanism that has been proposed for this reaction involves two steps, shown as steps 1 and 2 below:

$$\text{Step 1} \; NO_2 \, (g) + NO_2 \, (g) \quad \rightarrow \quad NO_3 \, (g) + NO \, (g)$$

$$\text{Step 2} \; NO_3 \, (g) + CO \, (g) \quad \rightarrow \quad NO_2 \, (g) + CO_2 \, (g)$$

Both step 1 and step 2 are referred to as *elementary steps*, where an *elementary step* is a reaction whose rate law can be written from its *molecularity*. In other words, it is always possible to write down the rate law for an *elementary reaction* just by inspecting the stoichiometric equation, as such reactions are believed to take place in a single encounter between the species involved in the elementary reaction.

Molecularity is defined as the number of species that must collide to produce the reaction indicated by that step.

- A reaction involving one molecule is called a unimolecular step.
- A reaction involving the collision of two species is called a bimolecular step.
- A reaction involving the collision of three species is called a termolecular step.

The rate law for an elementary step follows directly from the molecularity of that step.
   Therefore, for steps 1 and 2 above, the respective rate laws are:

$$\text{Step 1} \quad \text{Rate} = k_1[NO_2]^2$$

$$\text{Step 2} \quad \text{Rate} = k_2[NO_3][CO]$$

In reactions comprising more than one step, it is common for one of the steps to be significantly slower than the others. Such a step is known as a *rate-determining step*. The overall rate of the reaction is controlled by the rate-determining step. In this reaction, step 1 is the rate-determining (slowest) step.
   If steps 1 and 2 are added together, and any species balanced/cancelled where they occur on both sides:

Step 1: $\quad NO_2(g) + NO_2(g) \quad \rightarrow \quad NO_3(g) + NO(g)$

Step 2: $\quad NO_3(g) + CO(g) \quad \rightarrow \quad NO_2(g) + CO_2(g)$

Sum:

$$NO_2(g) + \cancel{NO_2(g)} + \cancel{NO_3(g)} + CO(g) \longrightarrow \cancel{NO_3(g)} + NO(g) + \cancel{NO_2(g)} + CO_2(g)$$

$$\longrightarrow NO_2(g) + CO(g) \quad \longrightarrow \quad NO(g) + CO_2(g)$$

The mechanism proposed is, therefore, consistent with the experimental rate data.
   Box 9.3 gives an example of a relatively simple reaction (the gas phase reaction between hydrogen + iodine) for which more than one mechanism can be used to account for the observed results.

Box 9.3

Relating experimental data to proposed mechanisms

The gas phase reaction between hydrogen and iodine proceeds according to the stoichiometrically balanced equation:

$$H_2(g) + I_2(g) \rightleftharpoons 2HI(g)$$

The rate law for this reaction is:

$$\text{Rate} = k[H_2][I_2]$$

This experimental observation can be accounted for in two different ways however.

Mechanism 1

This can be accounted for by assuming the reaction follows a 'classic' bimolecular elementary reaction pathway:

$$H_2(g) + I_2(g) \rightleftharpoons 2HI(g)$$

(Continued)

In other words, the stoichiometric reaction would be elementary, which would lead to the rate law:

$$\text{Rate} = k[H_2][I_2]$$

This is not the only rationale however.

Mechanism 2

The data could also be accounted for by assuming a two-step mechanism, for which the first step is a rapid equilibrium and the second step is the slower, rate-determining, step.

Step 1 $\quad I_2 \underset{k_1}{\overset{k_{-1}}{\rightleftharpoons}} I + I \,(\text{rapid})$

Step 2 $\quad H_2 + I + I \xrightarrow{k_2} 2HI\,(\text{slow, rate-determining step})$

The first, rapid, step establishes a 'dynamic equilibrium' (also called a 'pre-equilibrium'). In other words:

$$k_1, k_{-1} \gg k_2$$

Because the second step is the rate-determining one, then the overall rate can be given by:

$$\text{Rate} = k_2[H_2][I]^2$$

This is, however, not the ideal way of referring to the rate, as the expression incorporates atomic iodine, which is an intermediate species, so its concentration is not readily quantified. This can be overcome, however, by recognising that k_2 is so slow that it does not upset the pre-equilibrium. Therefore we can write:

$$K_1 = \frac{k_1}{k_{-1}} = \frac{[I]^2}{[I_2]}$$

On rearranging it is found that

$$[I]^2 = \frac{k_1}{k_{-1}}[I_2] = K_1[I_2]$$

Substituting this into the rate equation gives, therefore:

$$\text{Rate} = k_2 K_1 [H_2][I_2]$$

Both k_2 and K_1 are constants, so they can be replaced by a constant k, and hence

$$\text{Rate} = k[H_2][I_2]$$

This is the same answer as mechanism 1.
More than one mechanism may account for simple experimental results. More complex experiments may be needed to establish which is more likely to be correct.

9.4.1 Deducing reaction mechanisms

Summarising the approach used above, the rate law is always determined firstly, followed by the two rules:

- The sum of the elementary steps must give the overall balanced equation for the reaction.

- The mechanism must agree with the experimentally determined rate law and a possible reaction mechanism is constructed.

Theoretical calculations can be used to provide support for the suggested mechanism but it cannot be proven absolutely.

9.4.2 A more comprehensive look at complex reaction mechanisms

In many mechanisms the first step involves a fast equilibrium reaction, followed by a slower process. Enzyme kinetics are a very important example of this, and will be examined in more detail shortly.

Equilibrium kinetics

Firstly, however, consider the situation of an equilibrium process itself.

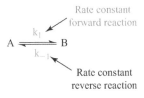

Setting up an ICE table (see Section 5.2.1):

	A	B
Initial	a	0
Change	$(-x)$	x
Equilibrium	$a-x$	x

The equilibrium constant for this is $K = \dfrac{x}{(a-x)}$.

Assuming the process is elementary, then the rate of the forward and reverse reactions at equilibrium are, respectively:

$$\text{Rate (forward)} = k_1[A] = k_1(a - x)$$

$$\text{Rate (reverse)} = k_{-1}[B] = k_{-1}x$$

But, at equilibrium, the rate of the forward reaction must equal the rate of the reverse reaction (otherwise the concentrations of the species would deviate from the equilibrium concentrations), so it is possible to re-write this as:

$$k_1(a - x) = k_{-1}x$$

This can be rearranged to give:

$$\frac{k_1}{k_{-1}} = \frac{x}{(a - x)}$$

(which is the same as the value for K shown above). Therefore, at equilibrium:

$$\frac{k_1}{k_{-1}} = K$$

305

Sequential reactions and the steady-state hypothesis

Consider the situation where there are two steps in a mechanism, one following the other. We will assume both steps are first order.

$$A \xrightarrow{k_1} B$$

$$B \xrightarrow{k_2} C$$

When considering the mechanism of a reaction, one of the steps is usually the slowest, rate-determining, step. Considering the two steps above, let us consider the option of either step 1 or step 2 being rate-determining.

- *Scenario 1: Step 1 is the faster,* and this can be represented in terms of rate constants by $k_1 \gg k_2$. In cases like this, the intermediate product B builds up quickly then slowly converts to the final product C. This can be represented graphically in Figure 9.10. The rate at which the product C is formed depends on the rate of step 2; in other words, step 2 is the *rate-determining step.*

- *Scenario 2: Step 2 is the faster,* which can be represented in terms of rate constants by $k_1 \ll k_2$. In cases like this, the intermediate product B does not get the chance to build up significantly before converting to final product C, and the production of C in fact is closely aligned with the loss of A. Graphically this is represented in Figure 9.11. In this case the *rate-determining step* is step 1.

The steady-state approximation

In scenario 2 above, one very notable feature is that the concentration of the intermediate B does not change much once the reaction is under way. It can therefore be assumed that the concentration of B is constant, which in mathematical terms can be represented by the term:

$$\frac{d[B]}{dt} = 0$$

This is a key concept, and is used in the analysis of enzyme mechanisms.

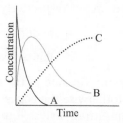

Figure 9.10 Variation of concentrations in sequential reactions when $k_1 \gg k_2$.

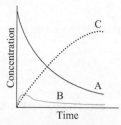

Figure 9.11 Variation of concentrations in sequential reactions when $k_1 \ll k_2$.

Kinetics of enzyme-catalysed reactions

9.5.1 Catalysts and catalysis

A catalyst is a substance that increases the rate of a reaction but does not alter the position of a reaction equilibrium. A catalyst may play an active role in the mechanism of a reaction but it is regenerated and recovered unaltered at the end of the reaction. It works by lowering the activation energy of the reaction compared with the uncatalysed reaction, as represented in Figure 9.12.

9.5.2 Enzymes as catalysts

Enzymes are proteins that act as nature's catalysts. They are very efficient, providing large rate enhancements, with high degrees of selectivity and specificity, and are usually present in relatively small quantities compared with the amount of other species. There are six classes of enzymes, that are named on the basis of their general function, summarised in Table 9.7.

Table 9.7 **Enzyme classification**

Enzyme class	Function
Oxidoreductases	Catalyse redox reactions, where one substrate is oxidised, another reduced
Transferases	Catalyse transfer of a group from one substrate to another
Hydrolases	Catalyse hydrolysis of a substrate
Lyases	Catalyse the formation of double bonds by removing groups from a substrate, or catalyse the addition of chemical groups to double bonds
Isomerases	Catalyse isomerisation reactions
Ligases	Catalyse the joining of two molecules with accompanying hydrolysis of the diphosphate bond in ATP

The word substrate means the material on which the enzyme acts.

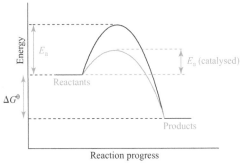

E_a = activation energy

ΔG^0 = free energy difference between reactants and products

Figure 9.12 **Comparison of reaction profiles of a catalysed and an uncatalysed reaction.**

307

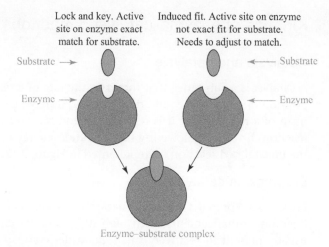

Figure 9.13 **Lock and key vs induced fit.**

There are many enzymes that fall within each category listed in Table 9.7. In many cases the enzyme is specific to one reaction only. Two proposals that were developed to rationalise this are known as the '*lock and key model*' and the '*induced fit model*', which are represented diagrammatically in Figure 9.13. The enzyme has an active site which has a complementary shape to the substrate (the substrate can fit in easily at the active site of the enzyme), and the active site will provide functional groups that will participate in the reaction where the substrate is converted to product.

In the lock and key model, the active site and the substrate have such complementary 3D structure matches that they dock perfectly without major atomic rearrangements. By contrast, the induced fit model proposes that the binding of substrate induces a conformational change in the active site. The substrate fits well in the active site after the conformational change. Experimental evidence favours the induced fit model.

9.5.3 Single substrate enzyme reactions

In terms of a reaction scheme, the enzyme-catalysed conversion of a substrate to a product is summarised thus:

$$E + S \underset{k_{-1}}{\overset{k_1}{\rightleftharpoons}} ES \xrightarrow{k_2} E + P$$

Enzyme   Substrate   Product

This is often easier to picture, however, if considered as two separate elementary steps:

$$E + S \underset{k_{-1}}{\overset{k_1}{\rightleftharpoons}} ES \qquad \text{Step 1}$$

$$ES \xrightarrow{k_2} E + P \qquad \text{Step 2}$$

The establishment of the intermediate, ES, is a rapid process, and the rate-determining (slow) step (step 2) is the conversion of the intermediate to product.

This mathematical model for this mechanism was first proposed by the American biochemist Leonor Michaelis and the Canadian physician Maud Menten in 1913. Nowadays it is generally the steady-state approach (SSA) that is used in the analysis of these reactions, where the assumption is made that the concentration of the intermediate, [ES], is effectively constant during the course of the reaction, with ES both going forward to product formation, but also undergoing the reverse process to E + S.

Applying the SSA to ES means that we assume that:

$$\frac{d[ES]}{dt} = 0$$

Consider, however, the elementary reactions taking place in steps 1 and 2. The rate of production of ES will be given by combining the forward reaction of step 1 with the reverse reaction of step 1 and the reaction involved in step 2. Mathematically this is given by:

Disappearance of ES, step1
Going back to E + S

$$\frac{d[ES]}{dt} = k_1[E][S] - k_{-1}[ES] - k_2[ES] = 0 \qquad [9.12]$$

Formation of ES, step 1

Disappearance of ES, step 2
Forming products E + P

Now it is useful to recognise that [E] is very small compared with [S], so, if the original amount of enzyme is given as $[E]_0$, then during the course of the reaction:

$$[E]_0 = [E] + [ES] \longleftarrow \text{Intermediate}$$

Original enzyme concentration   Unused enzyme

(because the stoichiometry states that for every molecule of E used, one molecule of ES is formed). Therefore,

$$[E] = [E]_0 - [ES]$$

Substituting for [E] in equation [9.12] gives:

$$k_1([E]_0 - [ES])[S] - k_{-1}[ES] - k_2[ES] = 0$$

This can be rearranged to give:

$$[ES] = \frac{[E]_0[S]}{\left(\dfrac{k_{-1} + k_2}{k_1}\right) + [S]}$$

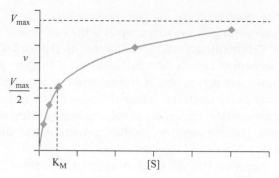

Figure 9.14 **Plot of initial rate vs substrate concentration for an enzyme reaction.**

Because step 2 is the rate-determining step of the mechanism, then that overall rate is given by:

$$\text{Rate} = k_2[\text{ES}]$$

This is a problem, however, because it is given in terms of the concentration of an intermediate, which is difficult to monitor experimentally. This is overcome by substituting in for the value of [ES] in the equation above, thus:

$$\text{Rate} = \frac{k_2[\text{E}]_0[\text{S}]}{\left(\dfrac{k_{-1} + k_2}{k_1}\right) + [\text{S}]}$$

This equation, referred to as the Michaelis–Menten equation, is usually represented in the form:

$$\text{Rate} = v = \frac{V_{\max}[\text{S}]}{K_M + [\text{S}]} \tag{9.13}$$

where $V_{\max} = k_2[\text{E}]_0$ and $K_M = (k_{-1} + k_2)/k_1$.

The Michaelis–Menten equation

The equation contains two variables, namely [S] and rate, V, which are derived from experiments. A typical result of plotting v against [S] for an enzyme reaction is shown in Figure 9.14.

The plot is labelled to indicate how to extract the terms $V_{\max}$ and K_M, where:

- $V_{\max}$ is the maximum rate that can be achieved for a given enzyme concentration. It corresponds to the rate at which the curve plateaus out.

- K_M, called the Michaelis constant, corresponds to the substrate concentration found at a rate with a value of half of $V_{\max}$.

A large value of K_M implies that the enzyme and the substrate do not have a strong affinity between them, and so a large concentration of substrate is needed to achieve half of $V_{\max}$. By contrast, a small value of K_M implies that there is a strong affinity and so a lower substrate concentration is needed to reach half of $V_{\max}$.

Box 9.4

Proving K_M = [S] at half of maximum rate

Starting with

$$v = \frac{V_{max}[S]}{K_M + [S]}$$

then at half of the maximum rate we can say $v = V_{max}/2$. This means that the Michaelis–Menten equation can be rewritten as:

$$\frac{V_{max}}{2} = \frac{V_{max}[S]}{K_M + [S]}$$

This rearranges to:

$$\frac{V_{max}}{2} K_M + \frac{V_{max}}{2}[S] = V_{max}[S]$$

Multiplying through by 2 gives:

$$V_{max} K_M + V_{max}[S] = 2 V_{max}[S]$$

Then cancelling out V_{max}, giving:

$$K_M + [S] = 2[S]$$

so

$$K_M = [S]$$

Box 9.5

Why is the Michaelis-Menten plot curved?

Looking at the plot shown in Figure 9.13, at low values of substrate concentration, [S], the rate increases quickly with increasing [S], whereas once [S] is high the rate stays more or less constant as [S] is varied. This behaviour can be understood by a closer examination of the Michaelis–Menten equation itself. Starting with:

$$\text{Rate} = v = \frac{V_{max}[S]}{K_M + [S]}$$

Low [S]

Consider the situation where the [S] is low. At very low values of [S], the approximation can be made that $[S] \ll K_M$ (the symbol $\ll$ means very much less than). In such cases it is reasonable to make the approximation that:

$$K_M + [S] \approx K_M$$

means approximately equal to

Therefore, when [S] is very low, it is valid to rewrite the Michaelis–Menten equation in the approximate form:

$$v = \frac{V_{max}[S]}{K_M}$$

(Continued)

As both K_M and V_{max} are constants for a given enzyme concentration, then V_{max}/K_M is also constant, so:

$$v = k\,[S]$$

This is a first order rate equation, the reaction being first order in [S] and first order overall. *Therefore, at low values of [S], the kinetics are first order.*

High [S]

Considering the situation where the [S] is high, then at very high values of [S], the approximation can be made that [S] >> K_M (the symbol >> means very much greater than). In such cases it is reasonable to make the approximation that:

$$K_M + [S] \approx [S]$$

Therefore, when [S] is very high, it is valid to rewrite the Michaelis–Menten equation in the approximate form:

$$v = \frac{V_{max}\,[S]}{[S]}$$

This equation reduces down, by cancelling out the [S], to

$$v = V_{max}$$

This is a zero order rate equation, the reaction showing no dependence on [S] at high values of [S].

Why does this happen?

- At low [S] values, there is plenty of enzyme relative to substrate, so the substrate has ready access to enzyme active sites, and the rate depends on [S].
- At high [S], all of the enzyme active sites will be in use, and so the rate cannot be changed by adding more substrate. The enzyme is said to be saturated, and the rate has no dependence on [S].

9.5.4 Analysis of enzyme kinetic data

The nature of the type of curve shown in Figure 9.13 means that extraction of the constants K_M and V_{max} is not a straightforward exercise – deciding where V_{max} is positioned can be difficult. It is possible to rearrange the Michaelis–Menten equation into different forms, which enable the experimental parameters v and [S] to be plotted in the straight line format, $y = mx + c$.

A couple of the rearranged forms are the Lineweaver–Burk and Eadie–Hofstee equations.

The Lineweaver–Burk equation

The Michaelis–Menten equation can be rearranged to produce the relationship known as the Lineweaver–Burk equation. This allows for a plot of $1/v$ against $1/[S]$ to yield a straight line graph, as indicated in equation 9.14:

$$\frac{1}{v} = \left(\frac{K_m}{V_{max}} \times \frac{1}{[S]}\right) + \frac{1}{V_{max}}$$

$$\underset{y}{\uparrow} \qquad \underset{m}{\uparrow} \quad \underset{x}{\uparrow} \qquad \underset{c}{\uparrow}$$

[9.14]

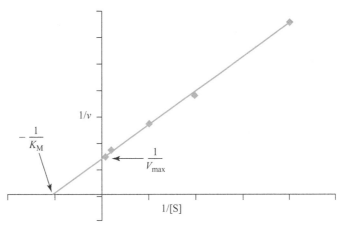

Figure 9.15 Lineweaver–Burk plot for kinetic study of an enzyme-catalysed reaction.

Figure 9.15 shows the plot obtained using the data from the experiment used to produce Figure 9.14. From this the gradient of the line provides the value of K_m/V_{max}, and the intercept with the y axis gives $1/V_{max}$.

It is, however, quite a straightforward process, to show that extrapolating the straight line to cross the x axis, as shown in Figure 9.15, results in the intercept with the x axis being equal to $-1/K_M$.

The Eadie–Hofstee equation

A different way of rearranging the Michaelis–Menten equation results in the Eadie–Hofstee relationship, which is also a straight line relationship.

$$v = -K_M \frac{v}{[S]} + V_{max} \qquad [9.15]$$

In this case, a plot of v against $v/[S]$ yields a gradient of $-K_M$ and a y intercept of V_{max}, as indicated in Figure 9.16.

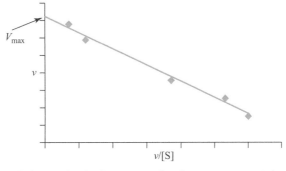

Figure 9.16 Eadie–Hofstee plot for kinetic study of an enzyme-catalysed reaction.

9.6   Enzyme inhibition

Enzyme-catalysed reactions can be inhibited (slowed down) by molecules that interfere with the formation of product. Many drugs function by inhibiting enzymes. Antibiotics, for example, are enzyme inhibitors. On a more everyday basis, the well-known analgesic aspirin also acts as an enzyme inhibitor. Indeed, both the positive effects of aspirin (the relief of pain, fever and inflammation) and the negative effects (principally irritation of the stomach) arise as a consequence of its role as an enzyme inhibitor. The enzymes involved are known as cyclooxygenases, specifically two enzymes called COX-1 and COX-2. These enzymes catalyse a key step in the synthesis of prostaglandins. Prostaglandins are hormones that carry local messages to neighbouring cells.

- COX-2 is the enzyme that is responsible for the prostaglandins involved in inflammation, pain and fever. By inhibiting the function of this enzyme, aspirin can reduce each of these three symptoms.

- COX-1 produces prostaglandins that are needed to synthesise protective gastric mucus in the stomach, for proper blood flow in the kidneys, and it also makes a prostaglandin responsible for platelet cell functioning. The inhibition of this enzyme therefore has a negative effect on the stomach and kidneys but does have a beneficial effect on the circulatory system, hence its use in treatment of patients with heart problems.

As well as enzyme inhibitors providing effective medication, many poisons, both for humans and in the animal world, act by inhibiting enzymes. For example, insecticides often function this way. This means that the chemical design of enzyme inhibitors is one of the major objectives of the pharmaceutical and agrochemical industries.

9.6.1   Mechanisms of inhibition

There are a number of different ways in which inhibitors can act to influence the enzyme–substrate binding, and these will be explored briefly.

Competitive inhibition

With competitive inhibition, the inhibitor provides an alternative option from the substrate for binding to the active site of the enzyme, thereby inhibiting the binding of substrate. In terms of a reaction mechanism, competitive inhibition can be represented thus (E represents enzyme, S represents substrate, I represents inhibitor and P represents product):

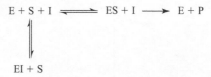

Because the inhibitor is providing an alternative (to substrate) for the enzyme's active site to bind with, competitive inhibition can be reversed by diluting the

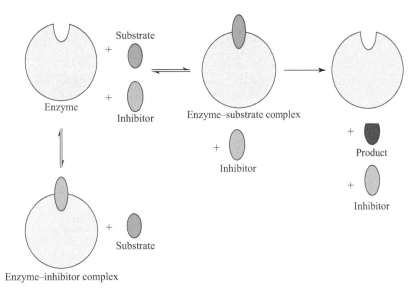

Figure 9.17 **Schematic representation of competitive inhibition.**

inhibitor or adding a large excess of substrate. A simplified pictorial representation is shown in Figure 9.17.

Uncompetitive inhibition

In this form of inhibition (which is not the same as non-competitive inhibition) the inhibitor binds to a site away from the active site, but only if the enzyme–substrate complex is already formed. Mechanistically this can be represented as:

$$E + S + I \rightleftharpoons ES + I \longrightarrow E + P + I$$
$$\updownarrow$$
$$ESI$$

In this case, the amount of enzyme–substrate complex available to form product is lowered when compared with an uninhibited reaction. A simplified pictorial representation is shown in Figure 9.18.

Non-competitive inhibition

Here the inhibitor binds to a site other than the active site and this alters the shape of the active site of the enzyme. The mechanistic representation of this is:

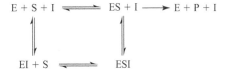

$$E + S + I \rightleftharpoons ES + I \longrightarrow E + P + I$$
$$\updownarrow \qquad\qquad \updownarrow$$
$$EI + S \rightleftharpoons ESI$$

In this mechanism only in the ES complex can the reaction proceed to yield product. Pictorially it can be represented as in Figure 9.19.

315

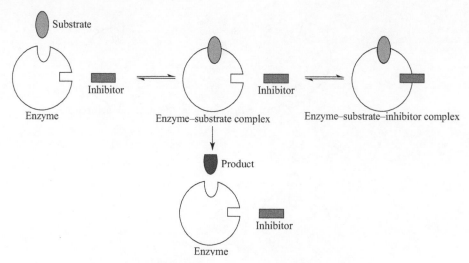

Figure 9.18 **Schematic representation of uncompetitive inhibition.**

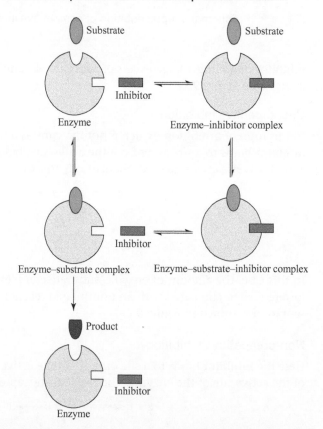

Figure 9.19 **Schematic representation of non-competitive inhibition.**

Distinguishing between the different forms of inhibition

The different types of inhibition result in different effects on the constants K_M and V_{max}, and these can easily be distinguished by looking at the effects on the linear

316

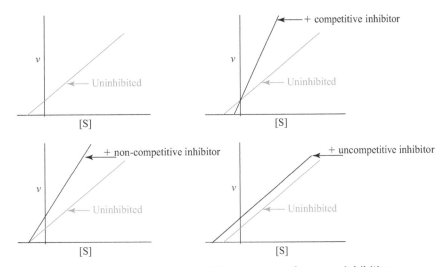

Figure 9.20 **Lineweaver–Burk plots for different types of enzyme inhibition.**

plots (such as the Lineweaver–Burk) when compared with the uninhibited reaction. Figure 9.20 shows how the Lineweaver–Burk plot varies for the different types of inhibition. The outcomes that result from looking at these plots are summarised in Table 9.8.

Table 9.8 **Behaviour of different inhibitor types**

Type of inhibitor	Enzyme binding site	Effect
Competitive	Inhibition at the catalytic site, competes with substrate for binding.	V_{max} is unchanged; K_m is increased
Non-competitive	Binds E or ES complex away from active site. Substrate binding unaltered, but ESI complex cannot form products.	K_m unaltered; V_{max} is decreased proportionately to inhibitor concentration
Uncompetitive	Binds only to ES complexes away from active site. Enzyme structure altered by substrate binding, which makes inhibitor-binding site possible.	V_{max} decreased; K_m decreased

Questions

9.1 The following reaction occurs in the gas phase:

$$A + 2B + 3C \rightarrow 2D$$

The rate law for the reaction was found to be:

$$dD/dt = k[B][C]^2$$

(a) State what happens to the rate of formation of D if

(i) [A] is doubled

(ii) [B] is doubled

(iii) [C] is doubled

(iv) the volume of the container is halved.

(b) State how the rate of formation of D is related to the rates of disappearance of A, B and C.

9.2 The following data were obtained for the reaction $A + B \Rightarrow C$:

Experiment	[A] $(mol\,L^{-1})$	[B] $(mol\,L^{-1})$	Initial rate of formation of C $(mol\,L^{-1}\,s^{-1})$
1	0.2	0.2	4×10^{-4}
2	0.6	0.2	3.6×10^{-3}
3	0.6	0.6	1.08×10^{-2}

Determine the rate law and the value of the rate constant.

9.3 A substance decomposes according to first order kinetics with a half-life of 500 s. Calculate how long, to the nearest second, it takes for 90% of it to decompose.

9.4 The following results were obtained from an experiment looking at the hydrolysis of a sugar in aqueous solution.

Time/minutes	0	30	60	90	120	150
[sugar]/$mol\,L^{-1}$	1.000	0.897	0.807	0.726	0.656	0.588

Determine the reaction order (hint: use integrated rate equations), and hence calculate the rate constant for the hydrolysis reaction.

9.5 In an experiment looking at the hydrolysis of cane sugar it was found that, after one hour, the sugar was hydrolysed to the extent of 25%. Assuming the hydrolysis is first order, calculate the time it takes for the cane sugar to become 50% hydrolysed.

9.6 It is often quoted that, near room temperature, reaction rates double if the temperature is increased by 10°C. Calculate the activation energy for such a reaction.

9.7 Write down the Michaelis–Menten equation. Show how, at low substrate concentrations, it can be modified to be a first order rate expression, whereas at high substrate concentrations it reduces to a zero order rate equation.

9.8 For an enzyme-catalysed reaction that follows the Michaelis–Menten equation:

(a) Sketch the plot for rate, v, against substrate concentration, [S], then

(i) Indicate on the plot where the Michaelis constant K_M occurs.

(ii) On the same plot show where V_{max} is found.

(b) What do the Michaelis constant, K_M, and V_{max} represent?

9.9 The enzyme aspartate-β-decarboxylase catalyses the reaction where L-aspartate is converted to L-alanine:

$$\text{L-aspartate} \quad \rightarrow \quad \text{L-alanine} + CO_2$$

The initial rate for different concentrations of L-aspartate can be followed by monitoring the volume of CO_2 produced. The results of a series of such experiments are tabulated below.

[L-aspartate]	Initial rate
20	14
30	19
40	23
50	28
100	42
200	53

([L-aspartate] in μmol L^{-1}, initial rate in μmol CO_2 per minute per mg enzyme)

(a) The Lineweaver–Burk relationship is given by:

$$\frac{1}{v} = \left(\frac{K_m}{V_{max}} \times \frac{1}{[S]} \right) + \frac{1}{V_{max}}$$

Show how it follows that the intercept of the straight line graph with the x axis provides a measure of $-1/K_M$.

(b) Using the Lineweaver–Burk relationship, calculate the value of K_M.

10 Bioenergetics and bioelectrochemistry

Learning outcomes

By the end of this chapter students should be able to:

- understand basic terms and definitions associated with bioenergetics;
- be familiar with the fundamental equations for calculating thermodynamics of biosystems;
- understand these equations, and relate them to principles of thermodynamics and equilibria;
- apply the equations as tools for calculating the energetics of biological systems.

Introduction

Bioenergetics is the study of the biological driving force for reactions, and follows on from the principles of thermodynamics and equilibria that were described in Chapters 4–6. These chapters introduced the terms ΔG, ΔH and ΔS, which help to describe whether a reaction will 'go', how much energy is needed, and the equilibrium constant K, which defines how far a reaction 'goes'.

In biology many processes involve the movement of charged species, for example:

- The transport of ions across membranes.
- The transfer of electrons between active sites in large molecules. These active sites typically contain redox active inorganic ions at which electrons can be accepted/ donated. See Box 10.1 for an introduction to redox processes.

As an example, in the mitochondrial electron transfer chain (which provides the means by which electrons are removed from NADH and, ultimately, transferred to oxygen, giving H_2O), the general redox process:

$$Fe^{3+} + e^- \rightleftharpoons Fe^{2+}$$

occurs several times, but the energetics associated with the redox processes each time are different. These differences occur because the Fe ions are swathed in biomolecular material, and the different natures of these various materials gives rise to changes in the microenvironment of the ion, which therefore changes the energetics (redox potential) of the process. The energy released each time is optimised to minimise the waste.

Box 10.1

Redox reactions

A chemical reaction for which there is a change in the oxidation numbers of the atoms is called an oxidation–reduction reaction. These reactions are more generally known as redox reactions, this term being derived from **red**uction–**ox**idation reactions.

The definitions are:

- *Oxidation*: loss of electrons

$$M \quad \rightarrow \quad M^{n+} + ne^-$$

- *Reduction*: gain of electrons

$$M^{n+} + ne^- \quad \rightarrow \quad M$$

It is often useful to remember the acronym OILRIG (*Oxidation Is Loss, Reduction Is Gain*).

Example

Displacement reactions, such as that where zinc metal displaces silver from a silver (I) nitrate solution. This can be considered in terms of the two ion-electron half equations below:

$$Zn(s) \quad \rightarrow \quad Zn^{2+}(aq) + 2e^- \text{ (oxidation)}$$
$$Ag^+(aq) + e^- \quad \rightarrow \quad Ag(s) \quad \text{(reduction)}$$

In this:

- Zinc metal is the reducing agent. It donates electrons and it is oxidised.
- The silver ion is the oxidising agent. It accepts electrons and is reduced.

The equation must balance, meaning that the number of electrons lost in the oxidation must be the same as the number of electrons gained in the reduction. Therefore, to bring the half equations together as an equation representing the full redox process, the second equation is multiplied by 2.

Then it is possible to add the two half-equations to get the redox equation:

$$Zn(s) \rightarrow Zn^{2+}(aq) + 2e^-$$
$$\underline{2Ag^+(aq) + 2e^- \rightarrow 2Ag(s)}$$
$$Zn(s) + 2Ag^+(aq) + 2e^- \rightarrow 2Ag(s) + Zn^{2+}(aq) + 2e^-$$

This, therefore, simplifies to:

$$Zn(s) + 2Ag^+(aq) \rightarrow 2Ag(s) + Zn^{2+}(aq)$$

Note that the nitrate ions from the silver (I) nitrate solution do not appear in the redox equation. This is because they are 'spectator ions', existing unchanged on both sides of the reactant/product equation.

In summary, for a reduction or an oxidation half reaction, that is either of:

$$M \rightarrow M^{n+} + ne^-$$

or

$$M^{n+} + ne^- \rightarrow M$$

The M^{n+} can be regarded as an oxidising agent, and the M as a reducing agent. Together they are called a *redox couple*.

This chapter is not specifically about the details of the mitochondrial electron transport chain, but it is about developing an understanding of the energetics behind the redox processes that underpin such chains. The energetics associated with redox centres can be monitored by measuring the electrode potentials. The first stage is to become familiar with the concepts behind these electron transfer processes, and with the terminologies that will be used. This requires an understanding of some key principles of electrochemistry.

10.2   Electrochemical cells

Consider a relatively simple redox reaction in solution; for example, what happens when zinc, Zn, is added to a copper sulfate, $CuSO_4$, solution, as illustrated in Figure 10.1. The Zn dissolves and Cu is precipitated (the sulfate anion, SO_4^{2-}, stays in solution). The reaction is spontaneous, so the free energy change, ΔG, is negative.

The reaction can be broken into two parts:

$$Zn(s) \rightarrow Zn^{2+}(aq) + 2e^- \quad \text{Oxidation}$$

$$Cu^{2+}(aq) + 2e^- \rightarrow Cu(s) \quad \text{Reduction}$$

Carrying out this reaction in a beaker does not give any insight into the energetics (thermodynamics) of the reaction, but if *metal electrodes* are used then it will be possible to make electrons move around an external circuit, from which it will be possible to measure the work done (in other words give the insight into the energetics). This is achieved by creating an electrochemical cell (Figure 10.2).

$$Zn\,(s) + Cu^{2+}\,(aq) \longrightarrow Cu\,(s) + Zn^{2+}\,(aq)$$

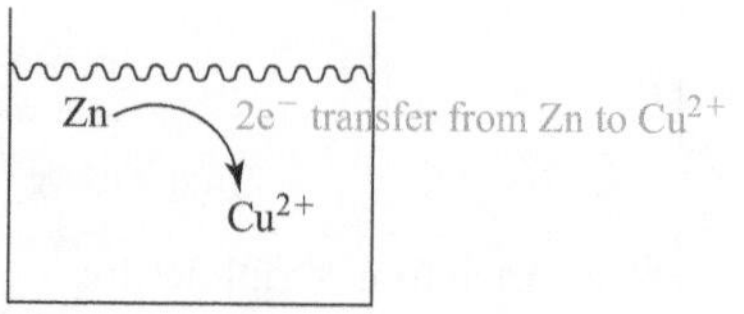

Zn $2e^-$ transfer from Zn to Cu^{2+}

Cu^{2+}

Figure 10.1 Addition of zinc powder to copper sulfate.

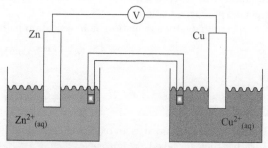

Figure 10.2 A zinc/copper electrochemical cell.

How does this work?

- The reduction reaction in the left hand beaker, where zinc is dipped into a solution of zinc ions, will be:

$$Zn^{2+}(aq) + 2e^- \rightleftharpoons Zn(s)$$

- The reduction reaction in the right hand beaker, where copper is dipped into a solution of copper ions, will be:

$$Cu^{2+}(aq) + 2e^- \rightleftharpoons Cu(s)$$

- Referring back to the chemical reaction illustrated in Figure 10.1, where zinc dissolved in a solution of copper ions, it can be inferred that the formation of zinc ions from zinc metal occurs more readily than does the formation of copper ions from copper metal, so it follows that the equilibrium in the left hand beaker of the electrochemical cell will be further to the left than the equilibrium in the right hand beaker.

- Therefore there are more electrons available from the Zn electrode than from the Cu electrode, and the Zn electrode therefore has a more negative potential than does the copper one.

- Connecting the two electrodes by a wire allows electrons to flow from the left hand to the right hand side along the wire.

- A *salt bridge* is added to prevent the amount of negative charge building up in the right hand beaker and similarly prevent the positive charge building up in the left hand beaker as electrons move across the wire. The salt bridge is a glass tube filled with saturated aqueous KCl. The tube has one end dipping into each beaker, and it is plugged with glass wool, which is permeable to ions. This allows for ions to flow into solution to neutralise charge and maintain current (it completes the circuit).

- It is possible to measure the thermodynamic driving force for the reaction, and this driving force is the voltage (also called the potential difference, or electromotive force, EMF). This is done by inserting a voltmeter, the symbol for this being:

Terminology

- The arrangement in Figure 10.2 is an example of *an electrochemical cell*.

- The voltage is called the *cell voltage*.

- Each beaker with its associated electrodes is called a *half-cell*, so the Zn/Zn^{2+} system is the left hand half-cell, and the Cu/Cu^{2+} system is the right hand half-cell.

Because the electrons flow in a specific direction, then current flow has direction. This means that the cell voltage has a *sign* as well as *magnitude*.

In order to ensure consistency between reported data, the convention used is that the cell voltage is the reduction potential of the right hand electrode with respect to that of the left hand electrode. Mathematically this is shown as:

$$E_{cell} = E_{RH} - E_{LH} \qquad [10.1]$$

where E_{cell} is the cell voltage (potential difference) and E_{LH} and E_{RH} represent the respective electrode potentials.

10.2.1 Cells and cell nomenclature

An abbreviated way of representing cells is used in order to avoid drawing them out in full. This takes the form:

LH electrode | LH solution || RH solution | RH electrode

where

- the symbol | represents a phase boundary (for example, between a solid and a liquid);
- the symbol || represents the salt bridge;
- the external circuitry (wires, voltmeter etc.) is assumed.

The cell shown in Figure 10.2 is therefore represented by:

$$Zn(s) \mid Zn^{2+}(aq) \parallel Cu^{2+}(aq) \mid Cu(s)$$

10.2.2 Types of half-cell

In principle any redox couple can be made into a half-cell reaction. For example, from the zinc/copper system described in Figure 10.2, the half-cells are:

$$
\left[
\begin{array}{ll}
\text{Notation adopted if half-cell is a:} & \\
\text{Left hand half-cell} & \text{Right hand half-cell:}
\end{array}
\right]
$$

	Left hand half-cell	Right hand half-cell:
$Cu^{2+}(aq) + 2e^- \rightleftharpoons Cu(s)$	$Cu(s) \mid Cu^{2+}(aq)$	$Cu^{2+}(aq) \mid Cu(s)$
$Zn^{2+}(aq) + 2e^- \rightleftharpoons Zn(s)$	$Zn(s) \mid Zn^{2+}(aq)$	$Zn^{2+}(aq) \mid Zn(s)$

Half-cells can comprise other physical states too. For example, the hydrogen electrode (represented in Figure 10.3) is an important example:

	Left hand half-cell notation	Right hand half-cell notation
$H^+(aq) + e^- \rightleftharpoons {}^1/_2 H_2(g)$	$Pt, H_2(g) \mid H^+(aq)$	$H^+(aq) \mid H_2(g), Pt$

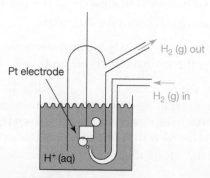

Figure 10.3 **The hydrogen electrode.**

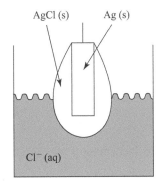

Figure 10.4 **The Ag/AgCl electrode.**

In a hydrogen electrode, the H_2 is adsorbed onto the Pt, with the role of the Pt being to act as an inert source (or sink) of electrons. The platinum acts as an electrode on which the reaction involving reduction of H^+ occurs.

By contrast, sometimes it is possible to use metals in contact with one of its insoluble salts. Important examples of electrodes of this type are the silver/silver chloride (Ag/AgCl) electrode (Figure 10.4) and the calomel electrode (Figure 10.5).

The reaction taking place at the Ag/AgCl electrode in Figure 10.4 is summarised thus:

Right hand half-cell notation

$$AgCl(s) + e^- \rightarrow Ag(s) + Cl^-(aq) \qquad Cl^-(aq)\,|\,AgCl(s), Ag(s)$$

The reaction taking place at the calomel electrode in Figure 10.5 is summarised thus:

Right hand half-cell notation

$$Hg_2Cl_2(s) + 2e^- \rightarrow 2Hg(l) + 2Cl^-(aq) \qquad Cl^-(aq)\,|\,Hg_2Cl_2(s), Hg(l)$$

(The Hg_2Cl_2 and Hg are in the same phase as a paste.)

Both the Ag/AgCl and the calomel electrodes are often used as *reference electrodes* (the reasons for this will be described later).

Biological systems often incorporate redox processes involving two different oxidation states of same (soluble) species, and an important example which was mentioned in Section 10.1 is the Fe^{3+}/Fe^{2+} redox pair:

Right hand half-cell notation

$$Fe^{3+}(aq) + e^- \rightarrow Fe^{2+}(aq) \qquad Fe^{3+}(aq), Fe^{2+}(aq)\,|\,Pt$$

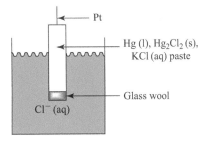

Figure 10.5 **The calomel electrode.**

325

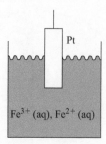

Figure 10.6 Platinum as part of an electrochemical half-cell.

As with the hydrogen electrode, Pt is used in these as a source, or sink, of electrons, as illustrated in Figure 10.6.

10.2.3 Measurement of cell voltage

The voltage of an electrochemical cell is determined by measuring the difference between the reduction potentials of the two half-cells involved. Many cell voltages have been determined, and because measured cell voltages will change with con-centration of redox active ions and with the pressure of redox active gases (which is accounted for by using the Nernst equation; see Section 10.2.6), it is important that results from measurements be specified relative to defined *standard conditions*.

These standard conditions are:

- All activities = 1 (approximately equal to all concentrations of soluble species at 1 mol L^{-1} and all pressures of gases at 1 atmosphere).
- All substances are pure.
- Pt electrodes, when present, are inert.
- A specified temperature (if the temperature is unstated, assume 298 K).

Measured standard cell voltages are then denoted E^θ_{cell}.

For the system described earlier (Section 10.2.1), namely the cell

$$Zn(s)\,|\,Zn^{2+}(aq)\,\|\,Cu^{2+}(aq)\,|\,Cu\,(s)$$

the cell voltage (often referred to as cell potential, though it is a potential difference) is $E^\theta_{cell} = +1.10V$.

Although it is possible to measure cell voltages, it is not possible to measure in-dividual half-cell potentials. In other words, it is only possible to determine relative values. The half-cell *chosen* to be the reference to which others are compared is the *standard hydrogen electrode* (often abbreviated as SHE):

Left hand
half-cell notation

$$H^+(aq) + e^- \rightleftharpoons {}^1/_2H_2(g) \qquad Pt, H_2(g)\,|\,H^+(aq)$$

Recall equation 10.1, which states that the accepted convention for cell potentials is:

$$E^\theta_{cell} = E^\theta_{RH} - E^\theta_{LH}$$

where:

- E^{θ}_{cell} is the cell voltage.
- E^{θ}_{RH} is the right hand half-cell reduction potential,
- E^{θ}_{LH} the left hand half-cell reduction potential.

The left hand electrode is taken as the reference point, the standard hydrogen electrode. Then by using the relationship:

$$E^{\theta}_{cell} = E^{\theta}_{RH} - E^{\theta}_{LH}$$

and recalling the definition,

$$E^{\theta}_{LH} = E^{\theta}(H^+, H_2) = 0$$

Then

$$E^{\theta}_{cell} = E^{\theta}_{RH} - 0$$

The resulting standard cell voltage measured is called the *standard reduction potential* or, sometimes, the *standard electrode potential* of the right hand half-cell, denoted E^{θ}. Table 10.1 lists some of the more common standard electrode potentials.

Table 10.1 **Some standard reduction potentials**

Notation	Reaction	E^{θ}_{298} / V
$E^{\theta}(Na^+, Na)$	Na^+ (aq) $+ e^- \rightarrow Na$ (s)	-2.71
$E^{\theta}(Mg^{2+}, Mg)$	Mg^{2+} (aq) $+ 2e^- \rightarrow Mg$ (s)	-2.37
$E^{\theta}(Zn^{2+}, Zn)$	Zn^{2+} (aq) $+ 2e^- \rightarrow Zn$ (s)	-0.76
$E^{\theta}(Ni^{2+}, Ni)$	Ni^{2+} (aq) $+ 2e^- \rightarrow Ni$ (s)	-0.25
$E^{\theta}(H^+, H_2)$	$2H^+$ (aq) $+ 2e^- \rightarrow H_2$ (g)	0.00
$E^{\theta}(Cu^{2+}, Cu^+)$	Cu^{2+} (aq) $+ e^- \rightarrow Cu^+$ (aq)	$+0.15$
$E^{\theta}(Cu^{2+}, Cu)$	Cu^{2+} (aq) $+ 2e^- \rightarrow Cu$ (s)	$+0.34$
$E^{\theta}(I_2, I^-)$	I_2 (aq) $+ 2e^- \rightarrow 2I^-$ (aq)	$+0.54$
$E^{\theta}(Fe^{3+}, Fe^{2+})$	Fe^{3+} (aq) $+ e^- \rightarrow Fe^{2+}$ (aq)	$+0.77$
$E^{\theta}(Ag^+, Ag)$	Ag^+ (aq) $+ e^- \rightarrow Ag$ (s)	$+0.80$

If two half-cell potentials are known relative to that of the reference standard hydrogen electrode, then it is easy to see that from these values E^{θ}_{cell} can be calculated for *any* two half-cells. Returning to the zinc/copper cell already used, it is straightforward to calculate E^{θ}_{cell}.

Starting from the cell diagram:

$$Zn(s) \,|\, Zn^{2+}(aq) \,\|\, Cu^{2+}(aq) \,|\, Cu(s)$$

The cell potential for this can be determined using:

$$E^{\theta}_{cell} = E^{\theta}_{RH} - E^{\theta}_{LH}$$

$$= E^{\theta}_{298}(Cu^{2+},Cu) - E^{\theta}_{298}(Zn^{2+},Zn)$$

$$= +0.34 - (-0.76)$$

$$= +1.10V$$

Pictorially the numerical value of this difference can be represented by regarding the potentials as points on a line:

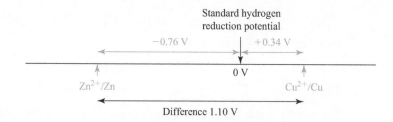

So, knowledge of standard reduction potentials enables the calculation of cell voltages.

It is known that for the zinc/copper system the spontaneous reaction involves the processes:

$$Zn \rightarrow Zn^{2+} + 2e^-$$

$$Cu^{2+} + 2e^- \rightarrow Cu$$

Therefore,

- when E^{θ} is more negative, oxidation occurs,
- when E^{θ} is more positive, reduction occurs.

10.2.4 Free energy relationship

The difference in potential between two electrodes is, therefore, a measure of the tendency of the cell reaction to occur. As discussed in Chapters 4 and 5, the tendency for a reaction to occur can be measured in terms of the change in Gibbs free energy, ΔG, for the reaction.

Combining the two ideas, it can be shown that for any electrochemical cell reaction the standard cell voltage, E^{θ}_{cell}, is related to the standard free energy change, ΔG^{θ}, for the reaction by:

$$\Delta G^{\theta} = -nFE^{\theta}_{cell} \qquad [10.2]$$

where:

- n is the number of electrons transferred in the half-cell reactions;
- F is Faraday's constant (the faraday) $= 96\,487\,C\,mol^{-1}$.

This means that:

- When E^{θ}_{cell} is positive, ΔG^{θ} is negative, the reaction is spontaneous.
- When E^{θ}_{cell} is negative, ΔG^{θ} is positive, the reverse reaction is spontaneous.
- When $E^{\theta} = 0$ 'flat battery', there is no overall voltage to drive the reaction, so the cell is at equilibrium as $\Delta G^{\theta} = 0$.

10.2.5 Determination of the reaction taking place in a cell

By following simple rules it is possible to establish the reaction taking place in a cell. A key feature is to remember the convention used when determining cell voltages. The following examples describe how this can be achieved.

Example 10.1 The $Zn(s)|Zn^{2+}(aq)\|Cu^{2+}(aq)|Cu(s)$ cell

Step 1 Write down *both* reactions as reductions

$$RH \quad Cu^{2+} + 2e^- \rightarrow Cu \quad E^\theta = +0.34\,V$$
$$LH \quad Zn^{2+} + 2e^- \rightarrow Zn \quad E^\theta = -0.76\,V$$

Step 2 Take RH – LH (the right hand reaction minus the left hand reaction) as with the E^θ_{cell} calculation

RH $Cu^{2+} + 2e^- \rightarrow Cu$	$E^\theta = +0.34$ V
LH $Zn^{2+} + 2e^- \rightarrow Zn$	$E^\theta = -0.76$ V
RH−LH $Cu^{2+} + 2e^- - Zn^{2+} - 2e^- \rightarrow Cu - Zn$	$E^\theta_{cell} = +1.10$ V

Cancel out the electrons and rearrange the atoms/ions to make everything positive (in other words, to make chemical sense):

$$Cu^{2+} + Zn \rightarrow Zn^{2+} + Cu$$

This is the cell reaction.

Step 3 Calculation of the standard free energy change, ΔG^θ, for the cell reaction:

$$\Delta G^\theta = -nFE^\theta_{cell}$$
$$= -2 \times 96\,487 \times (+1.10) \ (n = 2 \text{ in both half-cell reactions})$$
$$= -212,300\,J\,mol^{-1} \ (-212.3\,kJ\,mol^{-1})$$

This is the standard free energy change, ΔG^θ, for the cell reaction *as written*. It confirms that zinc will displace copper ions in a solution.

What if the cell was considered the other way round? If the cell were represented the other way around, that is:

$$Cu(s)|Cu^{2+}(aq)\,|\,|Zn^{2+}(aq)|Zn(s)$$

Step 1 Writing down both reactions as reductions

RH $Zn^{2+} + 2e^- \rightarrow Zn$	$E^\theta = -0.76$ V
LH $Cu^{2+} + 2e^- \rightarrow Cu$	$E^\theta = +0.34$ V

Step 2 Take RH−LH

RH $Zn^{2+} + 2e^- \rightarrow Zn$	$E^\theta = -0.76$ V
LH $Cu^{2+} + 2e^- \rightarrow Cu$	$E^\theta = +0.34$ V
RH−LH $Zn^{2+} + 2e^- - Cu^{2+} - 2e^- \rightarrow Zn - Cu$	$E^\theta_{cell} = -1.10$ V

Rearrange to make everything positive:

$$Zn^{2+} + Cu \rightarrow Cu^{2+} + Zn$$

329

This is the cell reaction.

Step 3 Calculation of the standard free energy change, ΔG^{θ}, for the cell reaction:

$$\Delta G^{\theta} = -nFE^{\theta}_{cell}$$
$$= -2 \times 96\,487 \times (-1.10)$$
$$= +212\,300\,J\,mol^{-1}\,(+212.3\,kJ\,mol^{-1})$$

In summary, the cell being used in the opposite sense gives the cell reaction

$$Cu + Zn^{2+} \rightarrow Zn + Cu^{2+}$$

This is the reverse reaction, and has $\Delta G^{\theta} = +212.3\,kJ\,mol^{-1}$, which reflects a non-spontaneous reaction.

Example 10.2 If the electrons do not balance in half-cell equations

Consider the cell:

$$Na\,|\,Na^{+}_{(aq)}\,||\,Zn^{2+}_{(aq)}\,|\,Zn$$

Step 1 Writing the cells as reductions

$$RH \quad Zn^{2+} + 2e^{-} \rightarrow Zn$$
$$LH \quad Na^{+} + e^{-} \rightarrow Na$$

Step 2 Ensure electrons will cancel by making number of electrons the same, then subtract; that is:

$$RH \quad Zn^{2+} + 2e^{-} \rightarrow Zn$$
$$2 \times LH \quad 2Na^{+} + 2e^{-} \rightarrow 2Na$$
$$\overline{RH-2LH \quad Zn^{2+} + 2\cancel{e^{-}} - 2Na^{+} - 2\cancel{e^{-}} \rightarrow Zn - 2Na}$$

Rearrange to make everything positive:

$$Zn^{2+} + 2Na \rightarrow 2Na^{+} + Zn$$

But E^{θ}_{cell} is still $E^{\theta}_{RH} - E^{\theta}_{LH} = -0.76 - (-2.71) = +1.95\,V$. (This is because volts are joules per coulomb; in other words the number of joules per coulomb of charge regardless of the number of electrons involved in the process!)

Step 3 Calculation of standard free energy change for the cell reaction

$$\Delta G^{\theta} = -2 \times 96\,487 \times (+1.95)$$
$$= -376\,300\,J\,mol^{-1} = -376.3\,kJ\,mol^{-1}$$

As the overall reaction involves the transfer of two electrons, $n = 2$.

10.2.6 Effect of concentration

What if standard conditions are not met? In reality experimental conditions such as concentration and/or pressure differ from the standard values, and so it is vitally important to be able to determine cell potentials in non-standard situations. This has the effect that potentials are changed on electrodes. The extent by which

Box 10.2

```
How to calculate new reduction potentials from those already known
```

For example, if the potentials for the two reductions below are:

$$Fe^{2+}(aq) + 2e^- \quad \rightarrow \quad Fe(s) \qquad E^\theta = -0.44 \text{ V}$$

and

$$Fe^{3+}(aq) + e^- \quad \rightarrow \quad Fe^{2+}(aq) \qquad E^\theta = +0.77 \text{ V}$$

then it should be possible to calculate E^θ for the process:

$$Fe^{3+}(aq) + 3e^- \rightarrow \quad Fe(s)$$

This cannot, however, be achieved by just adding the two potentials together, because standard reduction potentials represent *intensive state functions* of the reactions. (*By intensive state function we mean that the property in question, in this case the reduction potential, is the same irrespective of how much material is involved.*)

Therefore it is necessary to use the relationship $\Delta G^\theta = -nFE^\theta$ to convert the potentials to free energy changes. So, for the reactions in question:

$$Fe^{2+}(aq) + 2e^- \quad \rightarrow \quad Fe(s) \qquad \Delta G^\theta = -2 \times 96\,487 \times -0.44 = 84\,908.6 \text{ J mol}^{-1}$$

$$Fe^{3+}(aq) + e^- \quad \rightarrow \quad Fe^{2+}(aq) \quad \Delta G^\theta = -1 \times 96\,487 \times 0.77 = -74\,295.0 \text{ J mol}^{-1}$$

Adding the equations (and their corresponding free energy changes) gives

$$Fe^{3+}(aq) + 3e^- \rightarrow \quad Fe(s)$$
$$\Delta G^\theta = 84\,908.6 + (-74\,295.0) = 10\,613.6 \text{ J mol}^{-1}$$

The free energy change can then be converted to a potential by using the relationship

$$\Delta G^\theta = -nFE^\theta$$
$$\text{So, } 10\,613.6 = -3 \times 96\,487 \times E^\theta$$

This means the three-electron reduction:

$$Fe^{3+}(aq) + 3e^- \rightarrow \quad Fe(s)$$

has a standard reduction potential of $E^\theta = -0.037\,\text{V}$.

observed voltages deviate from standard voltages can be calculated by using the Nernst equation.

The Nernst equation for an electrochemical cell

The derivation of the general Nernst equation makes use of the relationship between the Gibbs free energy change, ΔG, and the voltage, E, of an electrochemical system.

For a redox reaction:

$$aA + bB \rightleftharpoons cC + dD$$

the reaction quotient Q for the cell reaction is given by

$$Q = \frac{[C]^c[D]^d}{[A]^a[B]^b}$$

Chapter 5 provides a description of reaction quotients. (*As described in Chapter 5, it would technically be more correct to use activity in place of concentration, but we will stick with concentrations here.*)

In Chapter 5 it was also demonstrated that:

$$\Delta G = \Delta G^\theta + RT \ln Q$$

and the relationship between free energy change and cell potential was shown (in Section 10.2.4.) as:

$$\Delta G^\theta = -nFE^\theta_{cell}$$

E^θ_{cell} can be used in comparable ways to E^θ_{cell}; for example, $\Delta G = -nFE_{cell}$ to calculate the free energy change *away from* standard conditions.

Therefore, substituting the $-nFE$ terms for the ΔG terms gives:

$$-nFE_{cell} = -nFE^\theta_{cell} + RT \ln Q$$

Thus, it is relatively straightforward to rearrange this to give the relationship:

$$E_{cell} = E^\theta_{cell} - \frac{RT}{nF} \ln Q \qquad [10.3]$$

This is the Nernst equation, and in this equation:

- E_{cell} is the reduction potential under the non-standard conditions;
- E^θ_{cell} is the standard reduction potential for the cell reaction;
- Q is the reaction quotient:
- T is temperature (in kelvin);
- R is the gas constant, 8.314 J K^{-1} mol^{-1};
- F is the Faraday constant, 96 487 C mol^{-1};
- n is the number of electrons involved in the redox process.

Example 10.3 The $Zn \mid Zn^{2+} \parallel Cu^{2+} \mid Cu$ cell

Step 1 Determine the cell reaction as before:

$$Zn + Cu^{2+} \rightarrow Zn^{2+} + Cu$$

Step 2 Use the Nernst equation for cell, which is:

$$E_{cell} = E^\theta_{cell} - \frac{RT}{nF} \ln Q$$

The expression for Q is given by:

$$Q = \frac{a_{Cu} a_{Zn^{2+}}}{a_{Zn} a_{Cu^{2+}}} \approx \frac{[Cu][Zn^{2+}]}{[Zn][Cu^{2+}]}$$

As both Cu and Zn are pure elements, their activity terms = 1, so Q is simplified as shown:

$$Q = \frac{[Zn^{2+}]}{[Cu^{2+}]}$$

This is then introduced into the Nernst equation, to give:

$$E_{cell} = E_{cell}^{\theta} - \frac{RT}{2F} \ln \left\{ \frac{[Zn^{2+}]}{[Cu^{2+}]} \right\}$$

(Note $nF = 2F$, as $n = 2$ for both half-cell reactions.)

The Nernst equation for a half-cell

Consider a half-cell reaction:

$$x\, Ox + n\, e^- \rightarrow y\, Red$$

The Nernst equation for this, using equation 10.3 from above, would be:

$$E = E^{\theta} - \frac{RT}{nF} \ln \left\{ \frac{[Red]^y}{[Ox]^x} \right\}$$

(*Note*: sometimes this is represented in textbooks by

$$E = E^{\theta} - \frac{RT}{nF} \ln \left\{ \frac{[Ox]^x}{[Red]^y} \right\}$$

This is the same as above and is rearranged using the rule of logarithms $\ln x = -\ln 1/x$)

Example 10.4 To establish the Nernst equation for a Ag⁺ | Ag half cell

The reaction is summarised as:

$$Ag^+ + e^- \rightarrow Ag$$

and the Nernst equation that results is:

$$E_{Ag^+/Ag} = E_{Ag^+/Ag}^{\theta} + \frac{RT}{F} \ln \frac{[Ag]}{[Ag^+]}$$

Ag^+ is the oxidised form, and Ag metal is the reduced form.

As seen before, the activity term for Ag can be taken as 1, as it is a pure element in a standard state. The Nernst equation for this half-cell is therefore simplified to:

$$E_{Ag^+/Ag} = E_{Ag^+/Ag}^{\theta} - \frac{RT}{F} \ln \frac{1}{[Ag^+]}$$

This rearranges to:

$$E_{Ag^+/Ag} = E_{Ag^+/Ag}^{\theta} + \frac{RT}{F} \ln [Ag^+]$$

Example 10.5 Devising the Nernst equation for a calomel electrode

The half-cell reaction for a calomel electrode is summarised as:

$$Hg_2Cl_2(s) + 2e^- \rightarrow 2Hg(l) + 2Cl^-(aq)$$

For this, the Nernst equation can be seen to be:

$$E_{cal} = E_{cal}^{\theta} - \frac{RT}{2F} \ln \frac{[Hg]^2[Cl^-]^2}{[Hg_2Cl_2]}$$

Because Hg and Hg_2Cl_2 both exist as a paste and are regarded as being solids, their activity terms $= 1$, and the process is a two-electron process. This gives:

$$E_{cal} = E_{cal}^{\theta} - \frac{RT}{2F} \ln [Cl^-]^2$$

Then, recalling the rule of logarithms stating that $\ln x^2 = 2 \ln x$, this gives:

$$E_{cal} = E_{cal}^{\theta} - \frac{RT}{F} \ln [Cl^-]$$

10.3   Sensors and reference electrodes

It is possible to make cells that measure the concentration of particular types of ion. Such cells are called *ion sensors*.

10.3.1   The silver electrode

In order to make a cell of this type a combination electrode is used. This can be demonstrated by looking at the example of a sensor used to find concentrations of silver, Ag^+, ion. In this case the Ag^+ sensor consists of a probe consisting of two half-cells fused together, as illustrated in Figure 10.7.

The ion sensor is dipped into a test solution of the ions being studied, and E_{cell} is measured.

In the case shown in Figure 10.7, the cell is:

Ag (s), AgCl(s) | Cl⁻ (aq, 0.1 M) ⦙ Ag⁺(aq, unknown concentration) | Ag (s)

The test solution is the Ag^+ solution of unknown concentration.

In cells of this type the salt bridge has been replaced with a porous glass frit (symbolised above by ⦙), which is easier to use in such combination electrodes *but* there is a voltage associated with the frit, albeit a constant. The left hand electrode, which was referred to during the description of types of half-cells earlier, and illustrated in Figure 10.7, is a reference electrode and produces a constant potential. This can be understood by considering the reaction taking place, and the conditions under which it is operating.

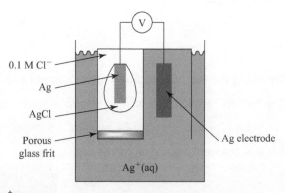

Figure 10.7 An Ag⁺ sensor.

The reaction at the half-cell is summarised by:

$$AgCl(s) + e^- \rightarrow Ag(s) + Cl^-(aq)$$

There are large amounts of AgCl (s), Ag (s) and Cl^- (aq) (the latter has a high concentration at 0.1M).

Small changes in amounts of reagents (for example Cl^-) will not affect the position of the equilibrium, which leads to a constant energy of electrons on the electrode and hence a constant potential. Because this electrode can be perturbed (for example, passing a small current) and still maintain the same potential, this is called a *reference electrode*.

Therefore, the cell voltage is given by:

$$E_{cell} = E_{RH} - E_{LH} + constant$$

(the constant is a constant voltage due to the glass frit, as referred to earlier), where:

- E_{LH} is a constant (for the silver chloride the reference electrode, given by $E^\theta_{Ag/AgCl^-}$);
- E_{RH} is given by the Nernst equation for the half-cell:

$$E_{Ag^+/Ag} = E^\theta_{Ag^+/Ag} + \frac{RT}{F} \ln \left[Ag^+ \right]$$

Looking at E_{cell} in more detail,

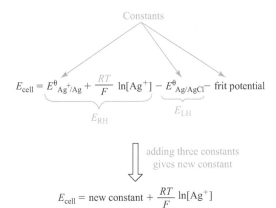

Thus, the new constant incorporates $E^\theta_{Ag^+/Ag}$, the glass frit voltage and $E^\theta_{AgCl/Ag}$ (the reference). Therefore, by dipping the sensor into a solution containing an unknown concentration of a solution of Ag^+, it is possible to use changes in E_{cell} to measure changes in $[Ag^+]$.

10.3.2 The calomel electrode

Recalling the half-cell reaction taking place for this half-cell:

$$Hg_2Cl_2(s) + 2e^- \rightarrow 2Hg(l) + 2Cl^-(aq)$$

Large amounts of Cl^- are present (usually a saturated solution is used) as well as the solids and liquids in paste form, and so the calomel electrode is also used as a reference electrode.

10.3.3 Detecting pH

If it is possible to use electrode sensors to detect ion concentrations, then one potentially important consequence of this would be the ability to measure the concentration of hydrogen ions in solution, and hence to be able to determine the pH of a solution. In this case, the combination electrode shown in Figure 10.8 could be constructed. As noted with the silver electrode, the left hand (reference) electrode and the glass frit operate to a constant voltage.

The reaction taking place at the right hand electrode is summarised as:

$$H^+(aq) + e^- \rightarrow {}^1\!/_2 H_2\ (g)$$

The potential of this half-cell is given by the Nernst equation:

$$E_{H^+/H_2} = E^\theta_{H^+/H_2} + \frac{RT}{F} \ln \frac{[H^+]}{P_{H_2}^{1/2}}$$

In this: $E^\theta_{H^+/H_2} = 0$ by definition.

As with the Ag^+ sensor described earlier, the overall cell reaction can be summarised by:

$$E_{cell} = E_{RH} - E_{LH} + \text{constant}$$

where:

- $E_{RH} = E_{H^+/H_2}$;
- E_{LH} is constant (because the AgCl/Ag electrode is a reference electrode);
- the glass frit potential is constant;
- P_{H_2} is kept constant (for example at 1 atm, but not necessarily this value).

So, on working through the calculations:

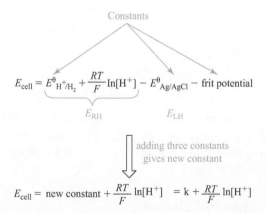

The equation established above, namely $E_{cell} = k + \dfrac{RT}{F} \ln[H^+]$, can be transformed into something much more useful by making use of the relationship between logarithms to base 10 and natural logarithms; that is:

$$\ln x = 2.303 \log_{10} x$$

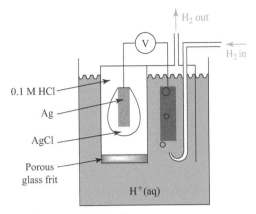

Figure 10.8 **An electrode for measuring pH.**

(See Appendix 1 for a more complete mathematics explanation.)
  This gives:

$$E_{cell} = k + \frac{2.303RT}{F} \log_{10}[H^+]$$

Since pH $= -\log_{10}[H^+]$, then this can be rewritten as:

$$E_{cell} = k - \frac{2.303RT}{F} pH \qquad [10.4]$$

This is a pH sensor, but it is not used in practice, because there are a variety of practical problems associated with this design; for example:

- H_2 is explosive.
- H_2 would need to be stored in a bulky cylinder.
- It is difficult to maintain a constant pressure of hydrogen gas; that is, to keep P_{H_2} constant.
- The Pt electrode is easily 'poisoned' (by, for example, proteins or other biomolecules).
- Pt is very expensive.
- Other redox reagents react on Pt (it is unselective).

A more practical pH sensor

Because of the practical difficulties associated with the construction of a pH sensor using a standard hydrogen electrode, an alternative, more portable, option makes use of an electrode incorporating a *glass membrane*. This preferentially allows H^+ into the membrane because the silica, SiO_2, in glass has protonatable sites. Thus,

$$SiO^- + H^+ \rightleftharpoons SiOH$$

$$Si-O-Si + H^+ \rightleftharpoons \underset{\underset{H^+}{|}}{Si-O-Si}$$

337

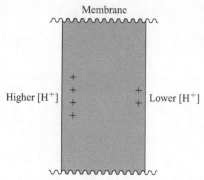

Figure 10.9 Schematic representation of different H^+ concentrations on either side of a glass membrane.

When solutions containing different $[H^+]$ are put on either side of a glass membrane (Figure 10.9) then:

- a spontaneous potential difference develops across the membrane;
- in the diagram there is a more positive potential on the left hand side of the membrane than on the right hand side of the membrane, which acts to drive H^+ through the membrane and thereby to equalise concentrations (a more positive charge gives a higher potential – just like for electrons);
- it only works if the membrane is permselective to H^+ (which means it lets H^+ through, but not other ions);
- the Nernst equation can be used:

$$E_{\mathrm{mem}} = \frac{RT}{zF} \ln \frac{[H^+]_{\mathrm{LH}}}{[H^+]_{\mathrm{RH}}}$$

(E_{mem} is the potential difference between the left and the right hand sides, and z is the charge on the ion , which has a value of $+1$ for H^+). Measuring the voltage would therefore give $[H^+]$.

The glass pH electrode

A typical design for a pH electrode is given in Figure 10.10.

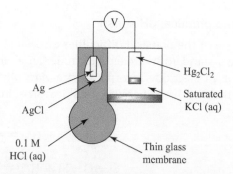

Figure 10.10 A practical design for a pH sensor.

When this combined probe is dipped into solution a cell is produced, which is summarised as:

$$\text{Ag (s), AgCl (s)} \mid \text{HCl (aq, 0.1 M)} \mid \text{glass} \mid H^+ \text{ (aq, test solution)}$$
$$Cl^- \text{ (aq, saturated)} \mid Hg_2Cl_2 \text{ (s), Hg (l)}$$

The system demonstrated in Figure 10.10 has several key points:

- Both the left hand and the right hand electrodes are reference electrodes and give a constant potential.
- The porous frit also gives a constant voltage.
- The voltmeter measures all the potential differences (voltages) between connections in the cell.

Any changes in voltage found are due to changes in glass membrane voltage, which is given by:

$$E_{mem} = \frac{RT}{F} \ln \frac{\left[H^+\right]_{in}}{\left[H^+\right]_{test}}$$

where $[H^+]_{in}$ and $[H^+]_{test}$ are the inside and outside of the bulb respectively.

Because $[H^+]_{in}$ = 0.1M, which is a constant, then it is possible to make use of the relationship $\ln x/y = \ln x - \ln y$, and the membrane voltage simplifies to:

$$E_{mem} = \text{constant} - \frac{RT}{F} \ln \left[H^+\right]_{test} \quad \text{(where constant} = \ln [H^+]_{in} \text{ term)}$$

This can be then incorporated into an equation to represent the voltage for the entire cell:

$$E_{cell} = \text{const} - \frac{RT}{F} \ln \left[H^+\right]_{test}$$

In this the value of const is a new constant which now includes the reference electrode potentials, the glass frit voltage and the $\ln [H^+]_{in}$ term.

As before, this can be rewritten in the form:

$$E_{cell} = \text{constant} + \frac{2.303RT}{F} \text{pH}$$

Therefore, glass electrodes can be used to measure the pH of test solutions.

The equation is often adapted such that:

- the new constant is often given the symbol E*;
- the $2.303RT/F$ term = 0.0591V at 298 K, often given symbol k, and is called the Nernst slope, as the equation is in the form $y = mx + c$, with E_{cell} corresponding to y and pH corresponding to x.

This means that the equation is often shown as:

$$E_{cell} = E^* + k\,\text{pH} \quad\quad\quad [10.5]$$

The constant is determined by a process of calibration, in which the value of E_{cell} is measured using a buffer of known pH and the equation is used to determine E* and hence to calculate k.

The measured value of E_{cell} (with E* and k) can be used to determine pH.

10.4   Biological relevance

10.4.1   Biochemical/biological standard state

As discussed in Box 4.10 in Chapter 4, the definition of biological standard state had the same definition as the chemical standard state *except* that $[H^+] = 10^{-7} \text{mol l}^{-1}$. This definition is extended to electrochemical cells. The superscript symbol used for biological standard states is $^\oplus$; for example, $E^\oplus{}_{cell}$ and $\Delta G^\oplus$.

When is this important?

The significance of this is illustrated by looking at the following examples.

Example 10.6   If H^+ is *not* involved in the reaction under study, for example

$$Fe^{3+} + e^- \rightleftharpoons Fe^{2+}$$

then

$$E^\theta \equiv E^\oplus$$

Example 10.7   If H^+ is involved in the reaction under investigation then it is necessary to use the Nernst equation. For example:

$$O_2 + 4H^+ + 4e^- \rightleftharpoons 2H_2O$$

This half-cell reaction involves a four-electron transfer, which gives:

$$E_{(O_2,H^+,H_2O)} = E^\theta{}_{(O_2,H^+,H_2O)} + \frac{RT}{4F} \ln \frac{P_{O_2}[H^+]^4}{[H_2O]}$$

(H_2O is a pure liquid, so $[H_2O] = 1$.)

In order to establish the relationship between E^θ and $E^\oplus$, it is necessary to make the following substitutions into the equation:

- $E_{(O_2,H^+,H_2O)} = E^\oplus{}_{(O_2,H^+,H_2O)}$
- $P_{O_2} = 1$ atm
- $[H^+] = 10^{-7} \text{mol}^{-1}$

This then gives: $E^\oplus{}_{(O_2,H^+,H_2O)} = E^\theta{}_{(O_2,H^+,H_2O)} + \dfrac{RT}{4F} \ln (10^{-7})^4$

$$= +1.234 + \frac{8.3143 \times 298}{4 \times 96487} \ln (10^{-7})^4$$

$$= +0.820 \text{V}$$

10.4.2   Biological membranes

Membranes are extremely important in biosystems, because they separate the interior of cells from their external environment. They do selectively allow some ionic and organic materials to permeate through them, thereby controlling how such substances move into and out of cells.

Cell membranes are formed from lipid bilayers (as shown in Chapter 7):

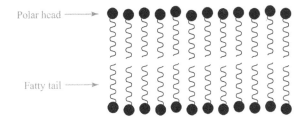

The idea of selective hydrogen ion transport across a glass membrane was described in the context of the operation of the pH meter. Selective transport across cell membranes is also important and (as will be described in a bit more detail in Chapter 11) embedded membrane-bound proteins act as ion channels (for example for ions such as K^+, Na^+, H^+).

The ions are 'pumped' across membranes, and these processes:

- can be used to store/release energy;

- can, when coupled to other reactions, be used to drive unfavourable reactions.

Cells often maintain different ion concentrations inside and outside the cell, and these differences result in a membrane voltage. Charged ions travel across the membrane, meaning that work is done/energy is released.

10.4.3 The thermodynamics of membrane transport

The transport of ions across the cell membrane (Figure 10.11) will be accompanied by a change in free energy:

$$M^{z+}(aq,out) \xrightarrow{\Delta G} M^{z+}(aq,in)$$

('out' for M^{z+} outside cell, 'in' for M^{z+} inside cell).

If the material being transported across the membrane was uncharged (if $z = 0$), then the free energy change, which can be represented by ΔG_r, could be given by:

$$\Delta G_r = \Delta G^\theta + RT \ln Q$$

In this system the reaction quotient, Q, is given by:

$$Q = \frac{\left[M^{z+}\right]_{in}}{\left[M^{z+}\right]_{out}}$$

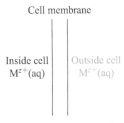

Figure 10.11 **Substance of either side of a cell membrane wall.**

341

This means that the free energy change is:

$$\Delta G_r = \Delta G^\theta + RT \ln \frac{[M^{z+}]_{in}}{[M^{z+}]_{out}}$$

Because the process under study here is simply a transport process (remember, $z = 0$), involving no bond breaking, and given that, by definition, for the standard free energy change $[M^{z+}]_{in} = [M^{z+}]_{out}$ (essentially $= 1.00$ mol L^{-1}) then under standard conditions the system would be at equilibrium, meaning $\Delta G^\theta = 0$, and so the equation for the non-equilibrium situation simplifies to:

$$\Delta G_r = RT \ln \frac{[M^{z+}]_{in}}{[M^{z+}]_{out}} \tag{10.6}$$

If, however, z is not equal to zero, then ions cross the cell membrane. If there is a voltage $\Delta\phi = \phi_{in} - \phi_{out}$ across the membrane where:

- ϕ_{in} is the potential of the inside surface on the membrane and
- ϕ_{out} is the potential of the outside surface,

then there will be a free energy change associated with this potential, given by:

$$\Delta G_{ion} = zF\Delta\phi \tag{10.7}$$

where:

- ΔG_{ion} is the work done;
- zF is the charge on one mole of ions;
- $\Delta\phi$ is potential difference between the inside and the outside of the membrane.

The total free energy change is therefore the sum of [10.6] and [10.7]:

$$\Delta G = \Delta G_r + \Delta G_{ion} = RT \ln \frac{[M^{z+}]_{in}}{[M^{z+}]_{out}} + zF\Delta\phi \tag{10.8}$$

This is a fundamental equation. (Note that if the ions being transported are cations, z is positive. For anions, z is negative.)

Example 10.8 Consider a system at equilibrium:

$$M^{z+}(aq, out) \rightleftharpoons M^{z+}(aq, in)$$

Because the system is at equilibrium, then the free energy change, $\Delta G = 0$. Putting this into [10.9] gives

$$0 = RT \ln \frac{[M^{z+}]_{in}}{[M^{z+}]_{out}} + zF\Delta\phi$$

This can be rearranged to put $\Delta\phi$ on one side of the equation:

$$\Delta\phi = -\frac{RT}{zF} \ln \frac{[M^{z+}]_{in}}{[M^{z+}]_{out}}$$

Because $\ln 1/x = -\ln x$, this can be also written as:

$$\Delta\phi = \frac{RT}{zF} \ln \frac{[M^{z+}]_{out}}{[M^{z+}]_{in}}$$

This means that, at equilibrium,

- when $\Delta\phi$ is negative, $[M^{z+}]_{in} > [M^{z+}]_{out}$
- when $\Delta\phi$ is positive, $[M^{z+}]_{in} < [M^{z+}]_{out}$

The significance of this is that cells can control the internal equilibrium ion concentrations by setting up the appropriate membrane potential (by using other reactions).

Example 10.9 Calculate the internal cellular concentrations of K^+ and Cl^- in equilibrium with 1 mM KCl outside when a membrane potential of $\Delta\phi = -100$ mV is established at 37°C.

Recall that temperatures must be used in kelvin, so

$$T = 273 + 37 = 310 \text{ K}$$

Also recall that 1 mM means 0.001 mol L^{-1}.

The calculation of internal K^+:

$$\Delta\phi = \frac{RT}{F} \ln\left(\frac{0.001}{[K^+]_{in}}\right)$$

(Since $z = +1$ for K^+).

Remember, 100 mV = 0.1 V, so putting in all of the values, the equation can be rearranged to:

$$\ln\left(\frac{0.001}{[K^+]_{in}}\right) = \frac{\Delta\phi \times F}{RT}$$

$$= \frac{-0.1 \times 96487}{8.3143 \times 310} = -3.744$$

From the rules of logarithms, as $\ln x = y$, then $x = e^y$, so:

$$\frac{0.001}{[K^+]_{in}} = e^{-3.744} = 0.02367$$

and

$$[K^+]_{in} = \frac{0.001}{0.02367} = 0.042 \text{ M}$$

The calculation of internal Cl^-:

$$\Delta\phi = \frac{-RT}{F} \ln\left(\frac{0.001}{[Cl^-]_{in}}\right)$$

Note the change of sign compared with the calculation for K^+ because $z = -1$.

Rearranging gives:

$$\frac{0.001}{[Cl^-]_{in}} = e^{+3.774} = 42.27$$

This then leads to:

$$[Cl^-]_{in} = \frac{0.001}{42.27} = 2.37 \times 10^{-5} \text{ M}$$

Example 10.10 Hydrogen ions

These are simply a 'special case' of the general principles outlined above, but specifically referring to hydrogen ions, as summarised:

$$H^+ (aq,out) \rightleftharpoons H^+ (aq,in)$$

The situation is analogous to that of K^+, thus:

$$\Delta\phi = \frac{RT}{F} \ln \left(\frac{[H^+]_{out}}{[H^+]_{in}} \right) \quad \text{(since } z \text{ again equals} +1)$$

For hydrogen ion concentrations it is more useful to work with logarithms to base 10, in order that pH terms can be used, therefore:

$$\Delta\phi = 2.303 \frac{RT}{F} \log_{10} \frac{[H^+]_{out}}{[H^+]_{in}}$$

$$= -2.303 \frac{RT}{F} \log_{10}[H^+]_{in} + 2.303 \frac{RT}{F} \log_{10}[H^+]_{out}$$

This simplifies to:

$$\Delta\phi = 2.303 \frac{RT}{F} (pH_{in} - pH_{out}) \qquad [10.9]$$

10.4.4 Proton motive force

Looking at ion transport away from equilibrium:

$$M^{z+} (aq,out) \rightarrow M^{z+} (aq,in)$$

As seen earlier, in equation [10.8], the free energy change for this process is given by:

$$\Delta G = RT \ln \left(\frac{[M^{z+}]_{in}}{[M^{z+}]_{out}} \right) + zF\Delta\phi$$

For protons, the non-equilibrium situation is represented by:

$$H^+ (aq,out) \rightarrow H^+ (aq,in)$$

and the free energy change for this is given by:

$$\Delta G_{H^+} = 2.303RT(pH_{out} - pH_{in}) + F\Delta\phi$$

(See Box 10.3 to see where this comes from.)

The free energy change ΔG_{H^+} is the thermodynamic driving force for proton transport, and once calculated it can be shown as the term *proton motive force*, Δp_m (a term which is equivalent to EMF for electrons), where:

$$\Delta G_{H^+} = -F\Delta p_m$$

(compare with $\Delta G = -nFE_{cell}$, with $n = 1$ for H^+)

Putting in the expression for ΔG_{H^+} from above, and rearranging gives:

$$\Delta p_m = \frac{-\Delta G_{H^+}}{F} = 2.303 \frac{RT}{F} (pH_{out} - pH_{in}) + \Delta\phi$$

Box 10.3

Thermodynamic driving force for proton transport

Start with equation 10.8, which is:

$$\Delta G = RT \ln \left(\frac{[M^{z+}]_{in}}{[M^{z+}]_{out}} \right) + zF\Delta\phi$$

Substitute in the details for protons, for which $z = +1$, gives,

$$\Delta G_{H^+} = RT \ln \left(\frac{[H^+]_{in}}{[H^+]_{out}} \right) + F\Delta\phi$$

Using the rules of logarithms this rearranges to:

$$\Delta G_{H^+} = RT \ln[H^+]_{in} - RT \ln[H^+]_{out} + F\Delta\phi$$

It is possible to redraft this into logarithms to base 10 format, to facilitate the use of pH. Thus,

$$\Delta G_{H^+} = 2.303 \, RT \log_{10}[H^+]_{in} - 2.303 \, RT \log_{10}[H^+]_{out} + F\Delta\phi$$

Recalling that pH $= -\log_{10}[H^+]$, this can therefore be represented as:

$$\Delta G_{H^+} = -2.303 \, RT \, pH_{in} + 2.303 \, RT \, pH_{out} + F\Delta\phi$$

$$= 2.303 \, RT \, pH_{out} - 2.303 \, RT \, pH_{in} + F\Delta\phi$$

$$= 2.303 \, RT \, (pH_{out} - pH_{in}) + F\Delta\phi$$

(often $pH_{out} - pH_{in}$ is represented by ΔpH)

So
$$\Delta p_m = 2.303 \frac{RT}{F} \Delta pH + \Delta\phi \qquad [10.10]$$

(Δp_m has units of volts, V). In words therefore, proton motive force combines the concentration and voltage effects of a proton gradient, and these equations show that transfer of H^+ across a membrane produces a free energy change.

Therefore energy can be given in/taken out by *either* redox processes *or* moving ions (for example, H^+) across membranes in biological systems and (as will be described in Chapter 11,) biological systems use proteins embedded in cell membranes to *selectively* transport H^+ (or indeed there are channels that are selective for movement of other ions).

In practice therefore:

- Energy given out by redox electron transfer and/or chemical reactions (which have a negative ΔG value) can be used to drive protons across cell membranes in a non-spontaneous direction, thereby providing an 'energy store'.

- The stored proton gradient then drives ATP synthesis, with the negative ΔG from the H^+ travelling back across the membrane fuelling this non-spontaneous reaction.

- In order to work out how many protons need to be pumped for each ATP it is necessary to calculate the free energy change for the proton transfer processes. As the free energy change for the synthesis of ATP is known, then by calculating the ΔG for proton transfer it is possible to establish the theoretical minimum number of protons required to be pumped (though this neglects any energy losses).

Questions

The following table contains data relevant to the questions

Half-cell couple	E^{θ}_{298}/V	$E^{\oplus}_{298}/V$
Zn^{2+}, Zn	−0.763	
Mg^{2+}, Mg	−2.370	
Al^{3+}, Al	−1.660	
Fe^{2+}, Fe	−0.441	
Cu^{2+}, Cu	+0.337	
Ag^+, Ag	+0.799	
AgCl, Cl^-, Ag	+0.222	
Cl_2, Cl^-	+1.360	
O_2, H^+, H_2O		+0.820
NAD^+, H^+, NADH		−0.315
FAD, H^+, $FADH_2$ (in flavo-proteins)		0.000

10.1 Calculate the standard cell potentials, E^{θ}_{cell} or $E^{\oplus}_{cell}$ as appropriate, for the following cells at 298 K:

(a) Fe|Fe^{2+} (aq, 1 M) || Cu^{2+} (aq, 1 M)| Cu

(b) Pt, H_2 (g, 1 atm)| H^+ (aq, 1 M) || Zn^{2+} (aq, 1 M) | Zn

(c) Ag | Ag^+ (aq, 1 M) || Cl^- (aq, 1 M) | AgCl, Ag

(d) Pt | NAD^+ (aq, 1 M), H^+ (pH 7), NADH (aq, 1 M) || H_2O, H^+ (pH 7), O_2 (g, 1 atm) | Pt

10.2 Write down the half-cell reduction reactions and derive the overall cell reactions for the cells illustrated in (a) – (d) from question 10.1. Use the calculated values for the standard cell potential to calculate the standard free energy change.

10.3 Bacteria, fungi and plants reduce nitrate to ammonium in order to assimilate nitrogen into their biomass. This process occurs through two stages, namely the reduction of nitrate to nitrite, followed by the reduction of nitrite to ammonium. The equations for the two stages, and their corresponding reduction potentials are:

$$NO_3^- + 2H^+ + 2e^- \rightarrow NO_2^- + H_2O \quad E^{\theta}=0.42 \text{ V}$$

and

$$NO_2^- + 8H^+ + 6e^- \rightarrow NH_4^+ + 2H_2O \quad E^{\theta}=0.48 \text{ V}$$

Determine the standard reduction potential for the overall process, which is:

$$NO_3^- + 10H^+ + 8e^- \rightarrow NH_4^+ + 3H_2O$$

10.4 (a) Why is the cell

Pt,H$_2$ (g, 1atm) | H$^+$ (aq, test solution) || KCl (aq, saturated) | Hg$_2$Cl$_2$, Hg
not a practical cell for the measurement of pH in a biological system?

(b) Outline the practical advantages of the commonly used glass electrode measurement system.

(c) A glass electrode pH sensor is used to determine the pH of a biological solution at 37°C. The measured cell voltage of this pH sensor system, when calibrated by immersion in a buffer solution at pH 4.00, is +0.344 V. The potential observed when the glass electrode is immersed in a biological solution of unknown pH is +0.422 V.

The measured cell voltage, E_{cell}, of the glass sensor varies with pH according to the equation:

$$E_{cell} = E^* + k\, pH$$

where E* is a constant, independent of the value of pH, and k $= 2.303\, RT/F$.

Use these data to determine a value of E*, and hence calculate the pH of the biological solution.

10.5 Use the Nernst equation to calculate the standard reduction potential, E^θ for the NAD$^+$, H$^+$, NADH half-cell, given that the value of $E^\oplus$ for this half-cell is -0.320 V.

10.6 Consider the following process:

$$\text{NADH} + \text{FAD} + \text{H}^+ \rightleftharpoons \text{FADH}_2 + \text{NAD}^+$$

(a) Calculate the standard cell potential, $E^\oplus_{cell}$ for the process (which represents the reduction of FAD, in flavoproteins, by NADH) at 298 K.

(b) Hence calculate the standard free energy change, $\Delta G^\oplus$.

(c) Is this spontaneous? Can NADH be used to reduce FAD?

10.7 The pH gradient across the membrane of a cell was found to be 0.91 units with the outside more acidic. The membrane potential was measured as 75 mV with the inside of the cell negative relative to the outside. Assume the temperature is 37°C.

(a) Determine the thermodynamic driving force for the transport of protons from the outside of the cell to the inside.

(b) Calculate the proton motive force for this system under these conditions.

(c) What membrane potential would result in this pH difference corresponding to an equilibrium between protons on the inside and outside of the cell?

11 The role of elements other than carbon

Learning outcomes

By the end of this chapter students should be able to:

- recognise that elements other than C, H, N and O play a vital role in biological systems;

- understand the critical roles of phosphates and phosphate esters as a form of energy currency and as linking units in biopolymers;

- describe the nature of interactions between ligands and metals in terms of Lewis acid/base chemistry;

- describe how the crystal field theory accounts for some of the physical and chemical properties of transition metal coordination compounds;

- describe examples of s-block coordination complexes and account for why metals in group 2 have a more extensive coordination chemistry than those in group 1;

- understand how the coordination chemistry of metals plays a fundamental role in biological chemistry.

11.1 Introduction

This chapter will focus on the interface between biology and the area of chemistry known as inorganic chemistry. It is difficult to give a precise definition, but inorganic chemistry is commonly thought to be the branch of chemistry concerned with the properties and behaviour of inorganic compounds (usually regarded as compounds of elements other than carbon). By contrast organic chemistry is the study of organic compounds, which are carbon-based compounds, usually containing C—H bonds. There is a lot of overlap between the inorganic and organic sub-disciplines, however, primarily in the area of organometallic chemistry.

Whilst the title could, in principle, apply to the chemistry of any element other than carbon, specifically this chapter will concentrate on describing some of the basic chemistry, along with a brief look at the biological roles of:

- phosphorus, in the form of phosphate esters;
- main group metals (groups 1 and 2);
- transition metals.

Phosphorus and phosphate esters

Phosphorus is a group 15 element, lying immediately under nitrogen in the periodic table. It has an atomic number of 15 and its main isotope is ^{31}P. It has the full electron configuration $1s^2, 2s^2, 2p^6, 3s^2, 3p^3$. Considering the electron configuration of phosphorus, coupled with its mid-range electronegativity, it is not surprising that the most common oxidation states found are –3, +3 and +5.

- P is –3 in, for example, Ca_3P_2, which has the electron configuration $1s^2, 2s^2, 2p^6, 3s^2, 3p^6$ (the same as argon).
- P is +3 in, for example, PF_3, which has the electron configuration $1s^2, 2s^2, 2p^6, 3s^2$ (it has reached a filled subshell).
- P is +5 in, for example, PF_5, which has the electron configuration $1s^2, 2s^2, 2p^6$ (the same as neon).

Elemental phosphorus is too reactive to be found in nature, but it occurs as phosphate minerals, where it is in the +5 oxidation state. Phosphate minerals contain the PO_4^{3-} group, generally known as phosphate or as orthophosphate, and this is the anion that derives from phosphoric acid.

11.2.1 Phosphoric acid and phosphate esters

Phosphate esters are a group of *organophosphates* that are prepared from phosphoric acid and alcohols.

This is analogous to the formation of an organic ester from the reaction between a carboxylic acid and an alcohol. Where it differs, however, is that the phosphoric acid has three OH groups (it is a triprotic acid), and hence there is the capacity for the formation of monoesters, diesters or triesters.

It is also possible, however, for molecules of phosphoric acid to condense together to form *polyphosphoric acids*, which can in turn form esters. The loss of water means that these compounds can be thought of as anhydrides.

Phosphoanhydride bond

Phosphoric acid   Phosphoric acid   Diphosphoric acid   Water
                                          (often called
                                          pyrophosphoric acid)

Likewise it is possible for three (or more) phosphoric acid molecules to condense together to form longer polyphosphoric acid 'chains'. It is usual for these chains to be represented in a planar form, thus:

Phosphoric acid   Diphosphoric acid   Triphosphoric acid

11.2.2 Relevance to biology

Phosphate esters play important roles in biological systems:

- Phosphodiesters make up the structural 'backbones' of the nucleic acids RNA and DNA (see Figure 11.1).

- The hydrolysis of ATP to ADP results in the triphosphate ester cleaving to ADP + 'inorganic' phosphate, labelled P_i. The bond broken is often referred to as a 'high energy bond' (See Boxes 11.1 and 11.2).

- Phosphate esters are key intermediates in carbohydrate metabolism; for example, looking at the first stage of glycolysis.

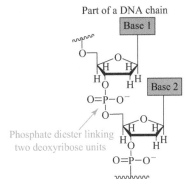

Part of a DNA chain

Phosphate diester linking
two deoxyribose units

Figure 11.1 The phosphodiester linkage as found in DNA and RNA.

Box 11.1

What is ATP?

ATP is an important molecule because of its role as a form of 'energy currency'. The diagram below illustrates its composition:

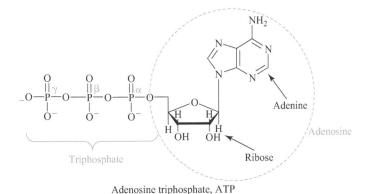

Adenosine triphosphate, ATP

The molecule comprises three components:

- the base, adenine;
- the sugar, ribose;
- the (tri)phosphate.

Box 11.2

Why is the hydrolysis of ATP such an effective energy-giving process?

There are three principal reasons.

Resonance stabilisation of products

ADP and P_i have a greater degree of resonance stabilisation than ATP alone, and indeed phosphate itself has a number of resonance forms.

(Continued)

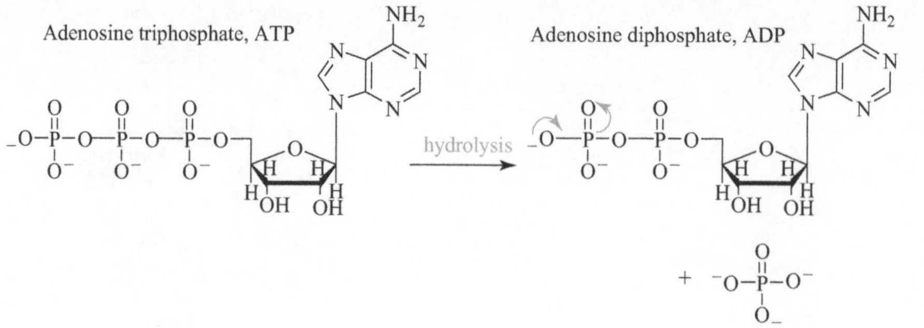

Adenosine triphosphate, ATP

Adenosine diphosphate, ADP

hydrolysis

P_i, inorganic phosphate

P_i, inorganic phosphate

Electrostatic repulsion in ATP

Like charges repel one another when in close proximity, and at pH = 7 ATP carries four negative charges.

Stabilisation due to hydration

Water can interact more effectively (through hydrogen bonds) to ADP and Pi than it can to the phosphoanhydride part of ATP. The phosphoanhydride is so called because of the mode of preparation of these polyphosphates:

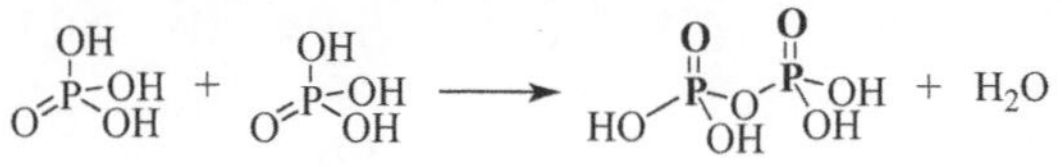

The joining together of two phosphates with the loss of water means that the product can be referred to as an anhydride, and the section of the molecule in bold is the anhydride group. This approach extends to longer chain polyphosphates.

All of these factors contribute to the relative stability of the product (ADP + P_i) compared with ATP.

11.3 Metals in the chemistry of biology

There are two main classes of metals found in biological systems, namely bulk metals and trace metals

11.3.1 Bulk metals

These are the main group elements sodium, Na; potassium, K; magnesium, Mg; and calcium, Ca. These elements exist in ionic form, and their roles are:

- Na^+, K^+ are charge carriers and are involved in maintaining osmotic balance in cells.

- Mg^{2+} forms complexes with ATP and DNA and is found in some enzymes, such as isomerases and hydrolases.
- Ca^{2+} is involved in cell signalling, structures (such as bone), and as a charge carrier.

11.3.2 Trace (catalytic) metals

These metals are principally members of the transition elements, such as V, Mn, Fe, Co, Ni, Cu, Zn and Mo (though zinc, Zn, is sometimes not classified as a transition element). Transition elements can usually display variable valency, which facilitates their roles, some examples of which are summarised in the following list:

- Vanadium, V, is used in nitrogen (N_2) fixation, and found in some oxidases and peroxidases.
- Molybdenum, Mo, is also involved in N_2 fixation and in O atom transfer.
- Manganese, Mn, is an important component of the oxygen evolving complex (known as a water splitting enzyme) which is involved in the photo-oxidation of water during photosynthesis. It is also found in some peroxidases.
- Iron, Fe, has a number of functions such as electron transport; O_2 transport, storage and activation, and it is also important for N_2 fixation.
- Cobalt, Co, is found in vitamin B12, and as such is involved in alkyl group transfer.
- Nickel, Ni, is found in some hydrogenases (along with iron). Hydrogenases facilitate the establishment of the equilibrium process $H_2 \rightleftharpoons 2H^+ + 2e^-$.
- Copper, Cu, has roles in proteins facilitating electron transfer, and in O_2 transport and activation.
- Zn acts as a Lewis acid, enabling it to catalyse hydrolysis reactions in biology. It also has simple roles in stabilising protein structures.

11.4 Transition metals and their role in biological systems

How does biology use these metals for catalysis? A key component of the chemistry of the transition metals is their ability to form *coordination complexes*; they possess an extensive coordination chemistry. The principles of coordination chemistry lie in the idea that metal ions can act as *Lewis acids* (see Box 6.5), where Lewis acids are defined as electron pair acceptors. Metal ions have the capability therefore to interact (through electrostatic attraction) with electron pair donors (which are called *Lewis bases*). These electron pair donors are referred to as *ligands* (see Box 11.3).

In this context a catalyst uses a carefully designed coordination sphere such that catalytic properties are optimised for their function, and the term coordination sphere means both ligation (how the ligand binds to the metal) and geometry around the metal ion.

Box 11.3

Ligands

Some key points about ligands are summarised:

- Usually a ligand can have an independent existence.
- A ligand always has at least one lone pair of electrons available to donate to the metal.
- Ligands can be neutral or charged.

There are a number of different types of ligands: unidentate, bidentate, tridentate . . . multidentate. The term *dentate* (from the Greek word for bite) refers to the number of coordinating atoms the ligand possesses. Unidentate ligands coordinate through one atom, bidentate ligands coordinate through two atoms, tridentate through three atoms, etc.
Some simple examples:

- Monodentate: H_2O, OH^-, NH_3, Cl^-. All of them coordinate through a single atom.
- Bidentate: acetylacetonato, ethylenediamine.

Acetylacetonato
Anionic ligand, delocalised
Coordinates through both O atoms

Ethylenediamine
Neutral ligand
Coordinates through
both N atoms

11.4.1 Introduction to ligands in biological systems

In the wider study of transition metal chemistry all sorts of conditions and all sorts of ligands, such as phosphines and halogens, can be used. But biological systems are usually under conditions such as pH=7, ambient temperature and pressure, and as such ligands are usually provided by amino acids and prosthetic groups (*some proteins have parts which do not derive from amino acids, and these are referred to as prosthetic groups*), usually N, O, or S donors.

The amino acids which bind to metal ions are typically:

Histidine (His) Aspartic acid (Asp) Glutamic acid (Glu)

Phenol OH Thiol Thioether

Tyrosine (Tyr) Cysteine (Cys) Methionine (Met)

One way that is commonly used to describe metal complexes is to consider which atom is acting as an electron donor to the metal ion.

- **N donor.** Histidine has imidazole as a side chain, and either nitrogen of the imidazole can bind to a metal ion, because a tautomeric equilibrium exists as shown.

Imidazole
can coordinate through N atoms

- **O donors.** Aspartic acid and glutamic acid have carboxylate side chains, while tyrosine has a phenolic side chain.
- **S donors.** Cysteine has a thiol side chain, and methionine has a thioether side chain.

Other ligands can be provided by prosthetic groups, such as porphyrin (in myoglobin), corrin (in vitamin B12) and chlorophyll. Biology therefore provides a relatively limited ligand variation, but nevertheless it can produce a wide range of geometries not seen in simple inorganic systems. These may be forced onto the metal by the 'fold' energy of the protein. This allows metalloproteins to form unique sites. More information on porphyrins and similar species is given in Box 11.5.

11.4.2 Introduction to transition metals

In order to understand the behaviour of the transition metals in biological environments it is essential to understand a little about the principles underpinning their chemistry.

Electronic structures of the transition metals

In Section 2.1.4, the aufbau principle was described, which stated that electrons will occupy orbitals of the lowest energy first, and it was shown that the order of relative energy (from low to high) is:

1s 2s 2p 3s 3p 4s 3d 4p 5s 4d 5p 6s 4f 5d 6p 7s 5f 6d 7p

Box 11.4

Hard and soft acids and bases

It has been observed that certain types of metals interact more readily with particular atoms in ligands, and these preferences are referred to as the hardness or softness of the metal ion and ligands involved. The terms derive from a wider concept relating to acids and bases, in which **hard acids** and **hard bases** tend to have the following features:

- small atomic or ionic radius;
- high oxidation state;
- low polarisability;
- high electronegativity (for hard bases).

By contrast, for soft acids and bases, the following features are common:

- large atomic or ionic radius;
- low or zero oxidation state;
- high polarisability;
- low electronegativity.

Therefore, in the context of metal ions acting as Lewis acids and ligand donors behaving as Lewis bases, it relates to the polarisability of the electrons in the metal ion and the ligand atoms, and can be regarded as correlating with the extent of ionic vs covalent character in the bonding. Hard metal ions bond with a large degree of ionic character to hard ligands, while soft metal ions bond with a greater degree of covalent character to soft ligands. Metal ions and ligands are classified as hard, soft or borderline. The hard metals are typically highly charged, smaller and come from the upper left side of the periodic table. Soft metals typically carry a lower charge and come from the lower right side of the periodic table. Some metals/ligands exhibit more borderline behaviour.

Hard and soft acids and bases in bioinorganic chemistry

Type	Metal ion	Coordinating ligand atom
Hard	Na^+, K^+, Mg^{2+}, Ca^{2+}, Mn^{2+}, Fe^{3+}, Al^{3+}, Co^{3+}	Oxygen ligands such as H_2O, HO^-, hydroxyl groups, carboxylates, phenolates, carbonyls. Nitrogen ligands such as amines.
Borderline	Fe^{2+}, Cu^{2+}, Ni^{2+}, Zn^{2+}, Pb^{2+}	Imidazole (a nitrogen ligand).
Soft	Cu^+, Ag^+, Cd^{2+}, Hg^{2+}	Sulfur ligands such as RS^-, R_2S. Carbon ligands such as CN^-.

It is important to note that these ideas reflect general trends, but there are many exceptions; for example, Fe^{3+}– sulfur clusters are well established within bioinorganic systems.

The transition elements, sometimes referred to as d block elements, have valence electrons which fall in the $(n+1)$s and the nd orbitals.

To demonstrate what this means, the first row of the transition elements, which accommodates the elements between scandium, Sc, and zinc, Zn, has the abbreviated electronic structure [Ar] $3d^n$ $4s^2$, as shown in Table 11.1. There are two exceptions to this generalisation, namely the metals chromium and copper for which, in both cases, the s orbital has just one electron.

Table 11.1   **Electron configurations for valence electrons of the first row transition elements**

Name of element (Element symbol)	Electron configuration
Scandium (Sc)	$3d^1\ 4s^2$
Vanadium (V)	$3d^2\ 4s^2$
Titanium (Ti)	$3d^3\ 4s^2$
Chromium (Cr)	$3d^5\ 4s^1$
Manganese (Mn)	$3d^5\ 4s^2$
Iron (Fe)	$3d^6\ 4s^2$
Cobalt (Co)	$3d^7\ 4s^2$
Nickel (Ni)	$3d^8\ 4s^2$
Copper (Cu)	$3d^{10}\ 4s^1$
Zinc (Zn)	$3d^{10}\ 4s^2$

The second row of the transition elements, which accommodates the elements between yttrium, Y, and cadmium, Cd, has the abbreviated electronic structure [Kr] $4d^n 5s^2$ (or [Kr] $4d^n\ 5s^1$). For example molybdenum, Mo, has the structure [Kr] $4d^5 5s^1$, whereas yttrium has the configuration [Kr] $4d^1\ 5s^2$.

Elements in the third row of the transition elements, which accommodates the elements between lanthanum, La, and mercury, Hg, do not play a major role in biological systems.

Whilst in the elemental state the $(n+1)$s orbitals of transition metals fill up before their n d orbitals. *In complexes, the $(n+1)s$ orbital becomes higher in energy than the* n *d orbitals*, therefore the n d ones are filled first. From this it follows that transition metal ions are typically classified by their d electron configuration, since these are the orbitals that play the most significant role in their chemistry. Even though the neutral electron configuration of many transition metals includes a filled or partially filled s orbital (as seen earlier, the 4s orbital for first row transition metals), those electrons are the first lost in the formation of metal ions. Thus the remaining valence electrons are the d electrons.

For example:

- Fe has the configuration [Ar] $3d^6\ 4s^2$; Fe^{2+} has the configuration [Ar] $3d^6$ – referred to as d^6.
- Co is [Ar] $3d^7\ 4s^2$; Co^{3+} is [Ar] $3d^6$ – referred to as d^6.
- Ni is [Ar] $3d^8\ 4s^2$; Ni^{2+} is [Ar] $3d^8$ – referred to as d^8.

Why does this matter?

- For ions found when d orbitals are empty or full (for example, Ca^{2+} has empty d orbitals, and in Zn^{2+} there are full d orbitals), then the electron configuration is not important in defining a particular coordination geometry, and the main factors of those metal ion complexes tend to be size.

● When partly filled d orbitals are present, however, then the bonding and the geometry of the coordination shell are influenced by the stability of the d electrons in that geometry.

Geometries of transition metal compounds

There are a range of coordination numbers (where coordination number is the number of donor atoms bonded to the metal) and geometries exhibited by transition metal complexes.

Amongst the most common are:

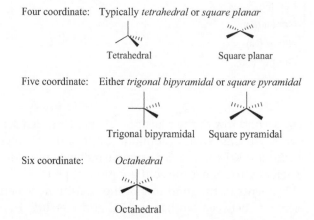

Four coordinate: Typically *tetrahedral* or *square planar*

Tetrahedral   Square planar

Five coordinate: Either *trigonal bipyramidal* or *square pyramidal*

Trigonal bipyramidal   Square pyramidal

Six coordinate: *Octahedral*

Octahedral

A simple approach called crystal field theory, given the abbreviation CFT, is used to describe the bonding found in transition metal complexes, and rationalises many properties exhibited by the complexes.

11.4.3 Crystal field theory

When thinking about the bonding in transition metal complexes there are a few key points to keep in mind:

● The metal orbitals involved are the d orbitals (d_{xy}, d_{yz}, d_{xz}, $d_{x^2-y^2}$, d_{z^2}).

● The pair of electrons in the metal–ligand (M–L) bond comes from the ligand.

● As a result of the M–L bonding, the energy of the five d orbitals is no longer the same.

In specific terms the crystal field theory:

● assumes that the transition metal and ligands are point charges – electron repulsion effects between the metal and the ligand change the energy of the d orbitals.

● the effect of the ligands on the d orbitals of the metal depends on the geometry of the complex.

Figure 11.2 gives a pictorial description of how CFT works for ligands coordinating octahedrally to a metal.

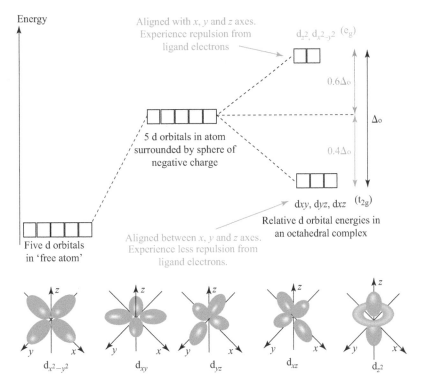

Figure 11.2 **Crystal field splitting for octahedral complexes.**

An isolated metal ion has five d orbitals of equal energy. If the metal ion is put into a symmetric sphere of negative charge then all of the electrons in the d orbitals will be equally repelled by the negative charge, resulting in all five orbitals having their energy increased. If, instead of the negative charge being on a sphere, it is concentrated along the principal axes then this equates to the case in which six electron-rich ligands approach the metal ion along the three axes to form an octahedral complex. In this circumstance, the individual d orbitals are affected differently. The $d_{x^2-y^2}$ and d_{z^2} orbitals (referred to as the e_g orbitals) have electron density on those axes and hence experience a greater repulsion from the negative charges than do the d_{xy}, d_{yz} and d_{xz} orbitals (called the t_{2g} orbitals), which lie between the axes. The e_g orbitals therefore rise to a higher energy than do the t_{2g} orbitals. The energy difference between the two sets of levels is called the *octahedral splitting energy* (Δ_o), with the energy of the e_g orbitals raised by 0.6 Δ_o, and that of the t_{2g} orbitals lowered by 0.4 Δ_o. Specific values of Δ_o are determined by the nature of the ligands involved.

Crystal field theory and colour

Transition metal complexes are often coloured, and it is the absorption of light of correct frequency that promotes electrons from the ground state to an excited state (if that frequency sometimes falls into the visible region of the spectrum then it is responsible for the colour of the metal complex). The frequency of the light absorbed depends on the size of Δ_o.

The size of Δ_o depends on:

- the identity and oxidation state of the metal;
- the nature of the ligand.

The colour seen arises from the wavelengths of light that are transmitted (not absorbed).

Dependence of Δ_o on ligand type

The size of Δ_o is affected by the type of ligand coordinated to the metal, and in particular it depends on the coordinating atom. The list below shows how Δ_o depends on the ligand, with smaller Δ_o values to the left hand side:

$$Br^- < Cl^- < F^- < OH^- < C_2O_4^{2-} < H_2O < NH_3 < NO_2^- < CN^-$$

From the list it is clear that, as a general rule, the dependence of magnitude of Δ_o with coordinating atom follows the pattern:

$$\text{Halide donor} < \text{O donor} < \text{N donor} < \text{C donor}$$

Dependence of Δ_o on metal

- Δ_o increases down a group: thus the order would be $Co < Rh \approx Ir$.
- Δ_o increases as the oxidation number of the transition metal increases. Thus $Fe(II)$ < $Fe(III)$.

How are the orbitals populated by electrons?

This question is answered by drawing on the *aufbau principle* and *Hund's rule of maximum multiplicity* (which were introduced in Chapter 2) and state that lower energy orbitals are filled first, and that if there are *degenerate* orbitals (where they have the same energy), the electrons will fill each singly with parallel spins before spin pairing occurs. Figure 11.3 shows this for the situations where the complex has one, two or three d electrons.

The next question is then what happens when the complex has four or more d electrons? Figure 11.4 illustrates the possibilities for four and for five electrons. The result of this is that, depending on the magnitude of Δ_o, the number of unpaired electrons in the complex can differ. One consequence of unpaired electrons in complexes is that it makes the complexes paramagnetic. This means that it is possible to distinguish between high and low spin forms by measuring a property known as the magnetic susceptibility of complexes.

Crystal field theory and tetrahedral complexes

The disposition of ligands in tetrahedral complexes means that they do not align with any of the d orbitals (see Figure 11.5), but they are closer to the t_{2g} group of orbitals, d_{xz}, d_{yz} and d_{xy}. The consequences of this are:

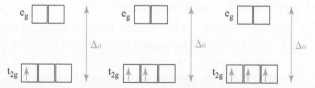

Figure 11.3 The occupation of d orbitals in an octahedrally coordinated complex by one, two and three electrons.

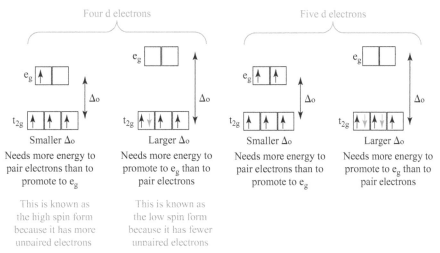

Figure 11.4 **The occupation of d orbitals in an octahedrally coordinated complex by four or five electrons.**

- The t_{2g} orbitals are still degenerate, but are now of higher relative energy than the e_g orbitals, which also remain degenerate.

- The difference in energy between the t_{2g} and the e_g orbitals is much smaller than the gap found in the octahedral complexes (an approximate generalisation is that $\Delta_t = \frac{4}{9}\Delta_o$ (see Figure 11.6). *Tetrahedral complexes are, therefore, nearly always high spin.*

Crystal field theory and square planar complexes

The splitting pattern of the d orbitals found for square planar complexes can be derived from that of octahedral complexes by recognising that removal of the 'z axis' ligands from an octahedral complex results in a square planar complex (as illustrated in Figure 11.7).

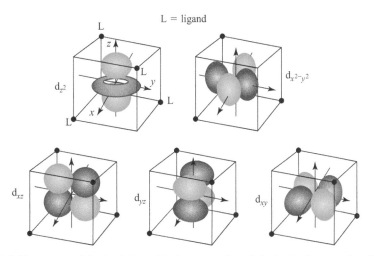

Figure 11.5 **The geometrical relationship between four tetrahedrally coordinating ligands and the five d orbitals.**

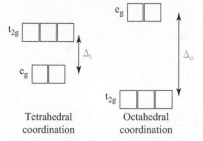

Figure 11.6 Comparison between octahedral and tetrahedral coordination on crystal field splitting.

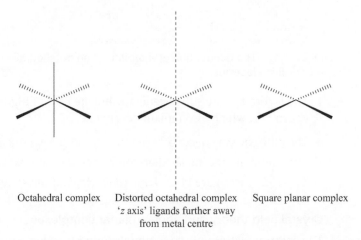

Octahedral complex   Distorted octahedral complex   Square planar complex
'z axis' ligands further away
from metal centre

Figure 11.7 Relationship between octahedral and square planar coordination.

As ligands along the z axis are moving away and, in the case of a square planar complex, disappearing, then there is less electrostatic repulsion on the orbitals directed towards the z axis. This has the effect of lowering the energies of d orbitals with a z axis component relative to those aligned solely along the x and y axes. This is demonstrated in Figure 11.8.

11.4.4 Examples of transition metals in biological systems

Copper in plastocyanin

Plastocyanin (Figure 11.9) is a blue copper protein which cycles between the Cu (I) and Cu (II) states during photosynthesis in plants, and as such is implicated in an electron transfer process. Under normal circumstances, Cu (I) would normally adopt a tetrahedral geometry, and its preferred ligands would be soft ligands such as S. By contrast, Cu (II) is normally square-planar and likes hard ligands with coordinating atoms such as N and O.

In plastocyanin it is believed that the protein fold energy forces the copper ion to accept a distorted geometry (meaning it requires less energy for the copper to adopt

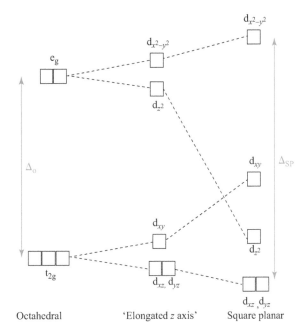

Figure 11.8 Crystal field splitting on progressing from octahedral to square planar coordination.

an unusual structure than is needed to disrupt the protein folding). The copper co-ordination is pseudotetrahedral, with 2 N ligands (histidine, His) and 2 S ligands (methionine, Met, and cysteine, Cys), thus:

$$N(His)$$

$$(His)N \quad Cu \quad S(Cys)$$

$$S(Met)$$

The geometry and distances change little with the change in oxidation state, and this represents an *entatic state* (where entatic state means '*a state of an atom or group which, due to its binding in a protein, has its geometric or electronic condition adapted for function*').

How biology uses metals to deal with oxygen

Oxygen, in the form of diatomic O_2, is a reactive element (it is the second most electronegative element after fluorine), and is only found free in nature because of its generation during photosynthesis. It is believed that approximately 70% of the world's oxygen is produced by marine sources such as green algae and cyanobacteria, with the remainder being produced by land plants. The photosynthetic process is the light-driven splitting of water.

Biological properties of oxygen

Oxygen has many important roles, some of which are as follows:

363

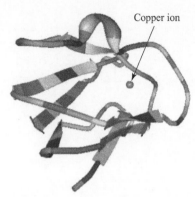

Copper ion

Figure 11.9 **The X-ray crystallographically determined structure of plastocyanin.**

- In oxidative phosphorylation, which involves coupling the process of O_2 reduction with that of ATP synthesis.

- As a metabolite; for example, the insertion of O atoms into organic substrates through the use of oxygenases. Dioxygenases, for example, are nature's way of breaking down aromatic compounds.

- At high partial pressures oxygen behaves as a toxin, resulting in a state known as hyperoxia. The reduction products of oxygen, O_2^- (known as superoxide) and O_2^{2-} (known as peroxide) are toxic. These latter toxins are dealt with using the enzymes known as catalases (to facilitate the breakdown of peroxide, $H_2O_2 \rightarrow H_2O + O_2$) and superoxide dismutases (to break down superoxides, $2O_2^- + 2H^+ \rightarrow H_2O_2 + O_2$).

Oxygen transport

Animals take oxygen into their lungs, where it diffuses through membranes into the blood. It is haemoglobin that takes O_2 from the air in lungs and delivers it to myoglobin in tissues where it is stored. Haemoglobin (abbreviated Hb) is known as an α2β2 tetramer, as it has a quaternary structure comprising four very similar sub-units (see Figure 11.10). The coordination about the iron is shown in Figure 11.11.

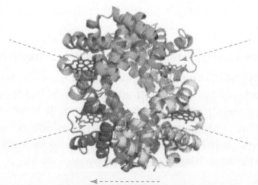

Arrows show where Fe–porphyrin rings are located

Figure 11.10 **The structure of haemoglobin.**

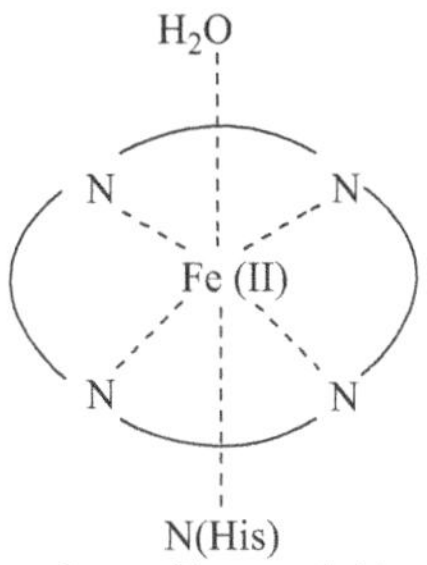

Deoxy form of haemoglobin. Fe is
Fe (II). Fe is slightly below plane
of ring towards His, therefore
called T form (tensed)

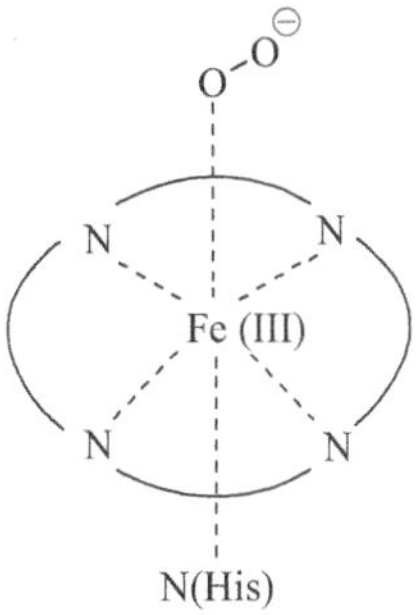

Oxy form of haemoglobin. Fe is
thought to be Fe (III). Fe is more in
plane of ring. Called R form (relaxed)

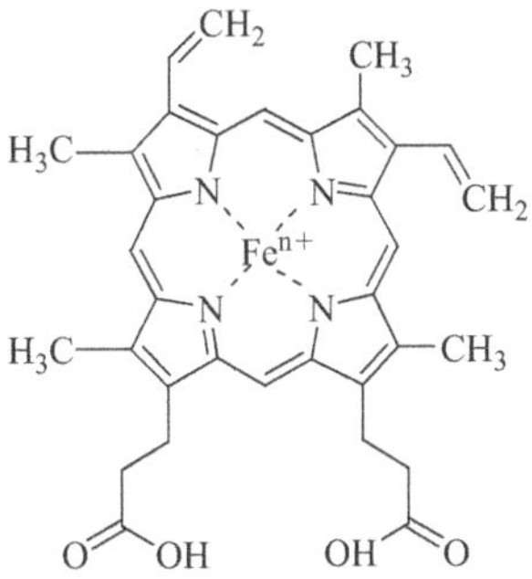

More detailed look at porphyrin ring complexing to iron

Figure 11.11 **Iron in haemoglobin.**

Haemoglobin can bind oxygen in the lungs and release it in oxygen-depleted tissues. This process depends upon the cooperative transition between two states, one low affinity (T) and one high affinity (R), which exist in rapid equilibrium. In the deoxy form of Hb, Fe^{2+} is in its high-spin state and, as such, is paramagnetic because it has four unpaired electrons. This differs from the low spin form which is non-magnetic (called diamagnetic). The electronic configuration possibilities for Fe^{2+} (which has six d electrons) is shown in Figure 11.12. The high spin Fe^{2+} is too large to fit into the space left by the four porphyrin N atoms, resulting in a displacement from the plane of the porphyrin towards the bound (called proximal) His. Hence this is called a 'tensed' situation.

There are differing views about the state of the iron in the oxygenated haemoglobin. One is that, on addition of oxygen, the iron becomes the smaller, low spin Fe^{2+}(which has no unpaired electrons). This smaller ion of iron now moves into the plane of the porphyrin, moving the proximal His, and thereby modifying both the quaternary and tertiary structures of the α and β subunits, in becoming the R form (R = relaxed). These structural changes contribute to a phenomenon called cooperative binding, which means that the binding of oxygen to the iron of haem units becomes easier after it has added to the first one. In haemoglobin there are four haem units and the addition of oxygen to the fourth occurs much more readily than to the first (by about 300 times). The oxy-Hb is diamagnetic because of the low spin nature of the Fe^{2+}.

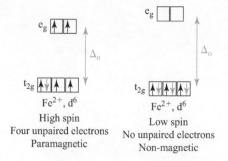

Figure 11.12 **Possible electronic configurations for octahedral d^6 complexes.**

The alternative view is that, on the binding of oxygen, the iron becomes Fe^{3+}, which again is a smaller ion than the high spin Fe^{2+}, and will result in the ion moving into the plane of the porphyrin ring, with the adoption of the R form. The bound oxygen is believed to exist as the superoxide, O_2^- ion. In this model, there is an unpaired electron associated with the Fe (III) ion and there is an unpaired electron associated with the bound superoxide which have to couple in order to provide the overall diamagnetism.

Carbon monoxide

Carbon monoxide poisoning occurs because carbon monoxide binds more strongly to the iron of haemoglobin than does oxygen, and so its presence effectively results in suffocation.

Myoglobin

This is a single subunit protein, related to haemoglobin (it has a very similar subunit structure; Figure 11.13), and it is the oxygen carrier in muscle tissues. It does not exhibit cooperativity of O_2 binding because this is only observed in multi-subunit proteins such as haemoglobin, where a change in structure at one haem is 'felt' by the others.

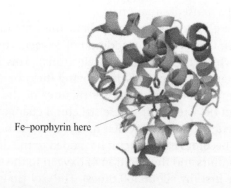

Fe–porphyrin here

Figure 11.13 **Structure of myoglobin.**

Other oxygen carriers

These include haemocyanin, which is found in, for example, molluscs, and haem-erythrin which is found in marine invertebrates. Haemocyanin uses copper to bind to oxygen, whereas haemerythrin uses iron.

Box 11.5

Porphins, porphyrins and chlorins

This is *porphin*, the 'structural base' compound for *porphyrins* which are utilised in nature for many important processes:

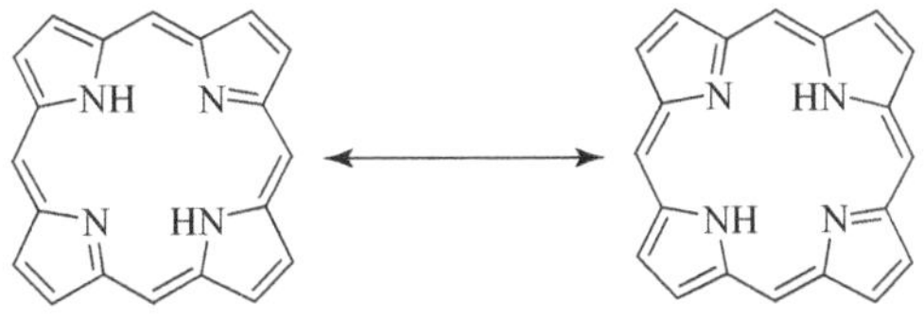

In Section 11.4.4, the interaction between iron, Fe, and the porphyrin within haemoglobin was illustrated, and its role in the transport of oxygen was briefly described. Other proteins, known as cytochromes, contain a porphyrin bound to iron, though in these cases the principal role is electron transfer.

Chlorophyll is the material responsible for the green colouring in plants, and it plays a vital role in the photosynthetic process through which plants convert water, carbon dioxide, and sunlight into carbohydrates. The structure of the form of chlorophyll known as 'chlorophyll *a*' has the apparent features associated with the characteristic porphin structure but with a few differences. A magnesium atom replaces the two nitrogen H atoms at the centre of the ring, and the chlorophyll *a* has an extra ring and other groups attached to the fundamental porphin ring.

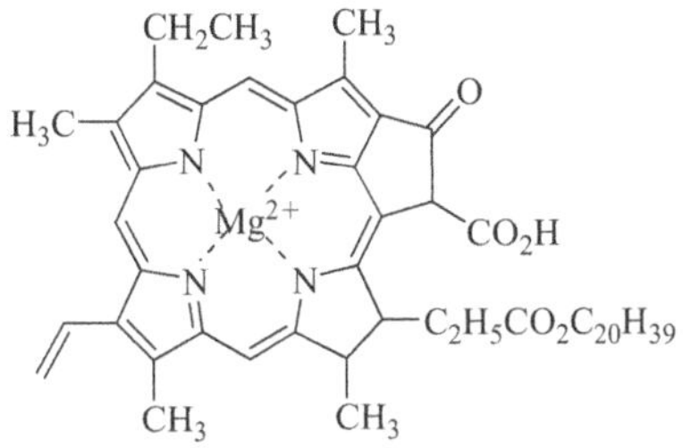

In fact, although the ring system above has global similarities to the porphin system, it is referred to as being based on a ring known as chlorin, the differences between them being:

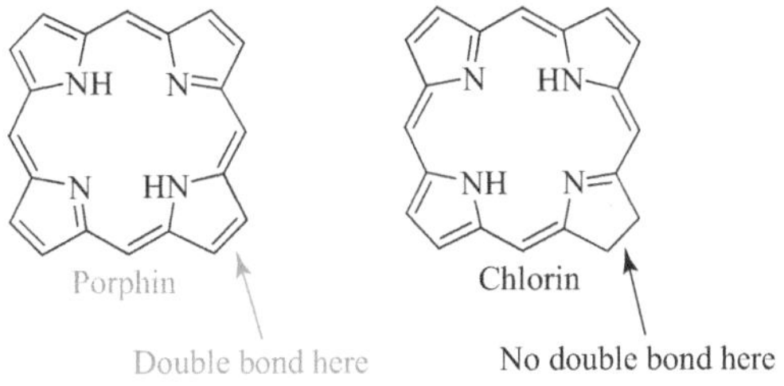

Porphin   Chlorin

Double bond here   No double bond here

367

Table 11.2 **Ionisation energies for the group 1 elements**

Metal	IE 1	IE 2
Li	520	7298
Na	496	4565
K	419	3069
Rb	403	2653
Cs	376	2423

IE 1 and IE 2 are ionisation energy values, and are quoted in kJ mol^{-1}.

11.5   The alkali and alkaline-earth metals

11.5.1 Introduction

The formation of 1+ cations is the dominant feature in the chemistry of the alkali metals (from group 1 of the periodic table), with these cations adopting noble gas configurations. This is reflected in the substantial differences observed between the first and second ionisation energies of these elements, as shown in Table 11.2.

The chemistry of the alkaline-earth metals (from group 2 of the periodic table) is based on their all forming 2+ cations, with these cations adopting noble gas configurations. The second electron to be removed comes from the same shell as the first electron, so it is shielded to some extent from the pull of the nucleus by the inner electrons, and this means that the disparity between the first and second ionisation energy values is not as great as with the alkali metals, as indicated in Table 11.3. Again, however, the removal of an electron from the next shell (in this case the third ionisation energy) is relatively much more difficult. For both alkali and alkaline-earth metals the compounds formed are largely ionic in character (beryllium is an exception).

11.5.2 Solid-state structures

In the solid state, compounds of alkali metals form ionic lattices, in which the cation is electrostatically bound to corresponding anions (known as an ionic bond; Section 2.8.2). As an example, recall the lattice formed by sodium chloride, Na^+Cl^- (common salt), shown in Figure 11.14. In this structure each sodium ion is surrounded by six chloride ions, and each chloride is surrounded by six sodium ions, and as such this type of arrangement is known as 6:6 coordination.

Table 11.3 **Ionisation energies for the group 2 elements**

Metal	IE 1	IE 2	IE 3
Be	899	1757	14849
Mg	738	1450	7730
Ca	590	1146	4941
Sr	549	1064	4207
Ba	503	965	3420

IE 1, IE 2 and IE 3 are ionisation energy values, and are quoted in kJ mol^{-1}.

Table 11.4 **Ionic radii and coordination number for ionic salts**

r^+/r^- (radius ratio)	Structure type	Coordination number around metal
0.225–0.414	ZnS (zinc blende)	4
0.414–0.732	NaCl (rock salt)	6
0.732 upwards	CsCl	8

The ratio of the ionic radii (cation/anion, represented by r^+/r^-) usually defines the structural type found for metal salts (Table 11.4 shows the situation for ionic compounds where the number of cations equals the number of ions, such as the situation found in NaCl). As anions are generally bigger than cations (see Table 11.5), then a smaller cation can fit fewer anions around it.

Figure 11.15 illustrates the three different structure types referred to in Table 11.4, and identifies some common examples for each.

Structures where the cation/anion ratio is not 1:1

So far only systems with cation/anion ratios of 1:1 have been described, and they all fall within the boundaries of cubic unit cells. In fact many other structure types are adopted where the ratio differs from 1:1 (such as $CaCl_2$), and also not all unit cells can be defined within a cubic frame. A detailed discussion does not fall within the scope of this text however.

11.5.3 Coordination chemistry of group 1 and group 2 metals

The transition metals are the group of elements that display the most extensive coordination chemistry (as was illustrated in Section 11.4), but the principles apply to both the group 1 and group 2 metals also. The ions of the metals from group 1, the alkali metals, are singly charged and are also quite large. This means that the charge density of the ions is quite low and as such they are relatively weak Lewis acids. Alkaline-earth metals, which form 2+ ions, have a greater charge density, and hence have a more extensive coordination chemistry (hence they complex with the phosphate oxygen atoms in ATP for example).

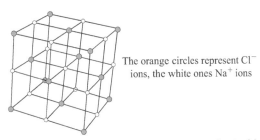

The orange circles represent Cl^- ions, the white ones Na^+ ions

Figure 11.14 **The face-centred cubic lattice structure as adopted by sodium chloride.**

Table 11.5 **Some typical ionic radii for selected cation and anion species**

Cation	Radius (r^+, pm)	Anion	Radius
Li^+	68	F^-	133
Na^+	97	Cl^-	181
K^+	133	Br^-	196
Rb^+	148	I^-	220
Cs^+	169	O^{2-}	140
Mg^{2+}	65	S^{2-}	184
Ca^{2+}	95	N^{3-}	171
Zn^{2+}	69		

The coordination complexes of the metal cations M^+ (for group 1 elements) and M^{2+} (for group 2) involve electrostatic, or ion–dipole, interactions that have no preferred direction of interaction. Therefore the ionic radius of the cation is the principal (but not the only) factor that governs the coordination numbers of the metal in its complexes.

When ionic compounds of alkali, or alkaline-earth, metals are dissolved in water, for example, then the metal ions are surrounded by water molecules, as demonstrated in Figure 11.16. Here the coordination number is six. This is the situation found, for example, for Na^+ and Mg^{2+} ions in aqueous solution.

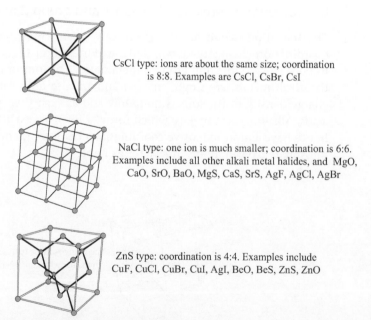

CsCl type: ions are about the same size; coordination is 8:8. Examples are CsCl, CsBr, CsI

NaCl type: one ion is much smaller; coordination is 6:6. Examples include all other alkali metal halides, and MgO, CaO, SrO, BaO, MgS, CaS, SrS, AgF, AgCl, AgBr

ZnS type: coordination is 4:4. Examples include CuF, CuCl, CuBr, CuI, AgI, BeO, BeS, ZnS, ZnO

Figure 11.15 **Some cubic structure types.**

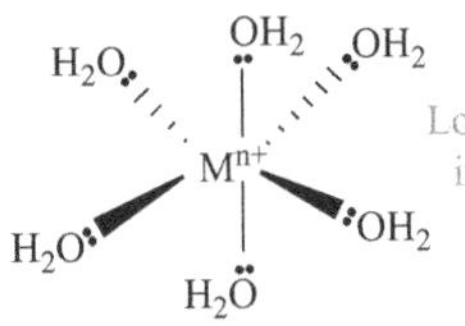

Figure 11.16 **Metal ion surrounded by six water ligands.**

11.5.4 Ions of alkali and alkaline-earth metals in biology

The ions Na^+, K^+, Mg^{2+} and Ca^{2+} are widely found in living organisms, as indicated in Section 11.3, and Table 11.6 illustrates their distribution. It is clear from the table that the ions have quite different intracellular and extracellular concentrations. In Chapter 10 the thermodynamics associated with transport of ions across membranes (for example, cell walls) was described. The method by which such transportation occurs was not commented upon, and under normal circumstances the ions would be unable to travel across the membranes because of the non-polar nature of the lipid 'tails' (see Figure 11.17 for a representation of a lipid bilayer).

It is the ability of the ions to form complexes that facilitates this transport. So far the only ligand that has been described in relation to these main group metals has been water, and in the examples quoted each metal ion was surrounded by six water molecules. They do, however, also coordinate to *multidentate* ligands.

Multidentate ligands

The idea of ligands involving more than one donor atom was introduced in Box 11.3. Porphyrins are examples of such ligands, as are the crown ethers, and Figure 11.18 shows two examples of crown ethers. The important feature here is that the ions are 'enveloped' by the ligand, and as such become soluble in non-polar solvents. It is this kind of coordinating behaviour that allows for transport of ions across membranes, through *ion channels*, which are pores that are found in cell membranes. These pores are proteins, where the hydrophilic (such as oxygen or nitrogen) components point into the pore. Metal ions can be transported across the otherwise impenetrable membranes by forming coordination complexes with these atoms. This is represented in Figure 11.19.

Table 11.6 **Intracellular and extracellular concentrations of important group 1 and 2 ions for mammalian cells**

Metal ion	Typical intracellular concentration (mM)	Typical extracellular concentration (mM)
Na^+	10	140
K^+	140	4
Mg^{2+}	30	1.5
Ca^{2+}	10^{-4}	2.5

The main extracellular anion is chloride, Cl^-; the main intracellular anion is phosphate, PO_4^{3-}.

Box 11.6

`Mg-ATP complexation`

ATP has four negatively charged atoms in neutral solution, and, whilst it can exist as free ions in solution, it can also coordinate to suitable metals. Magnesium has a very strong binding constant with ATP, and in biological systems it is thus the metal that most readily associates with ATP. The nature of this coordination is represented either as:

or as

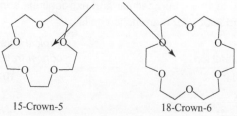

Water is acting as a monodentate ligand, and the ATP is bidentate.

Section of a lipid bilayer
The hydrophobic tails do not allow
for the passage of metal ions

Figure 11.17 Lipid bilayer.

Metals can sit in gaps and will
coordinate to each of the O atoms

15-Crown-5   18-Crown-6

Figure 11.18 Crown ethers as ligands. The first number in the name refers to the ring size and the second to the number of oxygen atoms. Due to ion sizes the 15-crown-5 accommodates the Na$^+$ ion and the 18-crown-6 accommodates the k$^+$ ion.

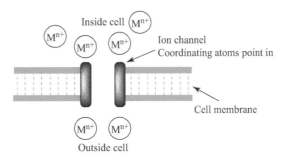

Figure 11.19 **Schematic view of a membrane ion channel.**

Ionophore antibiotics

Ionophore antibiotics have been identified/developed, which act by increasing the permeability of cell membranes to ions by 'sheathing' the ions in a non-polar outer 'coat', thereby allowing the transport of the ions across cell walls, which disturbs the intracellular/extracellular balance of microbial cells. This is represented schematically by Figure 11.20. A typical example of such an antibiotic is valinomycin (Figure 11.21).

The potassium sits inside the valinomycin, and is coordinated to the six ester carbonyl oxygen atoms to form an octahedral arrangement around the metal (as illustrated by the X-ray structure, represented in Figure 11.22, in which the grey coloured atom is potassium and the orange ones are the coordinating oxygen atoms). The outer, organic, part of the complex shields the metal ion from the non-polar alkyl groups of the lipid bilayer.

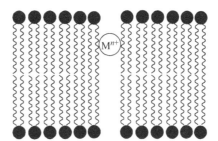

Figure 11.20 **'Sheathed' ion crossing cell wall.**

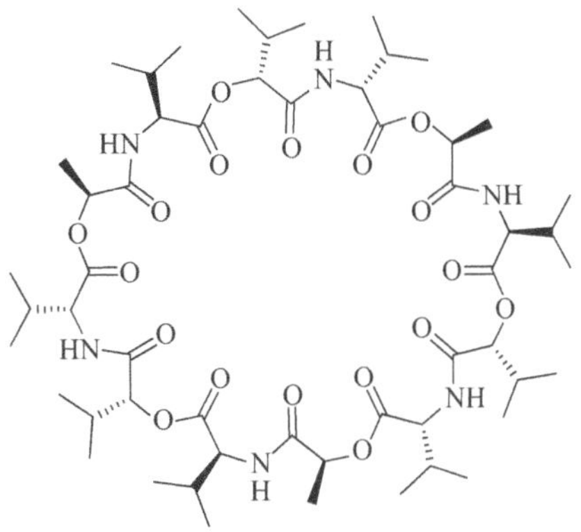

Figure 11.21 **Valinomycin.**

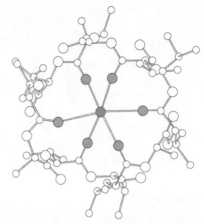

Figure 11.22 **Structure of the potassium–valinomycin complex.**

Questions

11.1 What are the main oxidation states of phosphorus?

11.2 What are phosphate esters, and how do they relate to phosphate?

11.3 Draw a picture of ATP and indicate on this diagram

(a) the positions of any phosphoanhydride linkages;
(b) any phosphate esters.

11.4 Identify the donor atoms in the following ligands by completing their Lewis dot structures.

Which are monodentate, which bidentate, which are tridentate and which, if any, cannot act as a ligand?

(a) $(CH_3)_3N$;
(b) NO_3^-;
(c) $(O_2CCO_2)^{2-}$;
(d) $(H_3NCH_2CH_2NH_3)^{2+}$;
(e) $(SCH_3)^-$;
(f) $CH_3SCH_2CH_2SCH_2CH_2SCH_3$

11.5 Account for the observation that solutions of $Co(H_2O)_6^{2+}$ are a pale pink whereas those of $CoCl_4^{2-}$ are blue.

11.6 Explain what is meant by the term *entatic state,* using the example of plastocyanin to illustrate your answer.

11.7 Which are the most abundant cations and anions

(a) inside animal cells;
(b) outside animal cells?

11.8 Describe the role of ionophores.

12 Metabolism

Learning outcomes

By the end of this chapter students should be able to:

- demonstrate an understanding of some key metabolic pathways, identifying the energy-yielding and energy-requiring reactions;
- describe and appreciate the key reactions at a mechanistic level;
- understand how these biochemical processes are not isolated but interdependent, with key junctions.

12.1   Introduction

Metabolism is the sum of all of the chemical reactions that occur within an organism, and metabolic processes can be thought of as the combination of two different types of process, known as *catabolic* and *anabolic* processes. Thus:

$$Metabolism = catabolism + anabolism$$

- Catabolism. Catabolic processes are involved in the breaking down of relatively complex molecules to simpler ones, and yield energy.
- Anabolism. Anabolic processes involve the net *use* of energy to build relatively complex molecules and structures from simpler ones, hence the use of the term anabolic steroids, which are steroid drugs that gained notoriety in sports, through the illegal use by many sports personnel in building up muscle mass.
- Linking anabolic and catabolic processes. The energy generated by catabolic processes is harnessed by cells to perform anabolic processes, and the metabolic pathways interact in such a way that energy released from the reactions of catabolism can be used to drive the 'energy demanding' reactions of the anabolic pathways.

A knowledge of the chemistry of metabolism is important for several reasons:

- Knowledge of the basic metabolic pathways is fundamental to an understanding of the functions of cells.
- Understanding the details of the reactions involved provides a framework for understanding biochemistry and molecular biology.

- Metabolic diseases such as diabetes mellitus, metabolic syndrome and obesity are major and growing public health problems, and effective treatments can be developed by understanding their natures as comprehensively as possible.

The number of different types of chemical reaction involved in metabolic processes is limited and, as such, much of the chemistry necessary to understand them has been described in earlier chapters. This chapter will put the knowledge gained into a biological context, and in order to demonstrate how these chemical principles apply, the chapter will analyse the details of the chemistry for the following areas:

- the breakdown of sugars to release energy–glycolysis; and
- the release of energy in mitochondria; the tricarboxylic acid (TCA) cycle. This is also known as the citric acid cycle and the Krebs cycle, after Sir Hans Adolf Krebs, who was involved in the elucidation of the process.

Also reference will be made to:

- the synthesis of sugars; gluconeogenesis;
- the synthesis of ATP.

Box 12.1

Some useful definitions relating to terms occurring in this chapter

Cofactor

A substance, such as a metallic ion or a coenzyme, which must be associated with an enzyme for it to function. Cofactors work by changing the shape of an enzyme or by actually participating in the enzymatic reaction.

Coenzyme

A coenzyme is a non-protein organic substance, usually containing a vitamin or mineral, which combines with a specific protein, called an *apoenzyme*, to form an active enzyme system.

Apoenzyme

The protein component of an enzyme, to which the coenzyme attaches to form an active enzyme.

12.2 Glycolysis

12.2.1 Introduction to glycolysis

Glucose is very important in human metabolism, and several pathways are involved with the usage, storage and regeneration of glucose:

- Glycolysis, the degradation of glucose to pyruvate, with the goal of generating energy (ATP). Glycolysis generates some ATP directly, but much more indirectly (through the subsequent oxidation of pyruvate).
- Glycogen synthesis. Glycogen is a polymeric form of glucose.
- Breakdown of glycogen to glucose. This occurs if other supplies of glucose are low.

- Gluconeogenesis, which turns pyruvate derived from amino acids into glucose. It also occurs when other (external) glucose supplies are low. Whilst appearing to be the reverse of glycolysis, components of the pathway cannot be directly related to the reverse of glycolysis (see Section 12.8).

- The hexose monophosphate shunt, where glucose is also degraded, but here the main product is NADPH, which is used as a reducing agent in cell metabolism.

12.2.2 The glycolysis pathway

The overall pathway is summarised in Figure 12.1. Glycolysis is the sequence of reactions that metabolises glucose to pyruvate with an accompanying production of energy (through the production of 'energy-rich' ATP).

In humans, glucose in the blood (obtained, for example, from glycogen or starch or gluconeogenesis) enters most cells by a specific glucose transporter into the cytosol where the enzymes of glycolysis are located.

Overall, glycolysis converts glucose, which is a 6-carbon compound, to two 3-carbon units (pyruvate). The glycolysis pathway as illustrated in Figure 12.1 can be summarised as comprising the following enzyme-catalysed stages (indicated by the orange numbers in Figure 12.1):

1. Phosphorylation of glucose at position 6 by *hexokinase*.
2. Isomerisation of glucose-6-phosphate to fructose-6-phosphate by *glucose-6-phosphate isomerase*.
3. Phosphorylation of fructose-6-phosphate to fructose-1,6-bisphosphate by *phosphofructokinase*.
4. Cleavage of fructose-1,6-bisphosphate by *aldolase*. Two different products, dihydroxyacetone phosphate and glyceraldehyde-3-phosphate, result from this.
5. Isomerisation of dihydroxyacetone phosphate to a second molecule of glyceraldehyde-3-phosphate by *triose phosphate isomerase* (TIM).
6. Conversion of glyceraldehyde-3-phosphate to glycerate-1,3-diphosphate (via dehydrogenation and accompanying phosphorylation) by *glyceraldehyde-3-phosphate dehydrogenase*.
7. Transfer of the 1-phosphate group from glycerate-1,3-diphosphate to ADP (using *phosphoglycerate kinase*), which yields ATP and glycerate-3-phosphate.
8. Isomerisation of glycerate-3-phosphate to glycerate-2-phosphate by *phosphoglycerate mutase*.
9. Dehydration of glycerate-2-phosphate to phosphoenolpyruvate by *enolase*.
10. Transfer of the phosphate group from phosphoenolpyruvate to ADP by *pyruvate kinase*, which gives further ATP, and pyruvate.

The pyruvate produced during the glycolysis pathway is then converted to acetyl coenzyme A, which is a link into a series of other metabolic pathways, such as the TCA cycle, fatty acid synthesis and steroid synthesis.

The different stages of glycolysis will now be looked at from a chemical mechanistic viewpoint. In most cases, for simplicity, the surface of the enzyme acting as the catalyst is represented by a ⌇⌇⌇⌇. The active components of the enzyme will be referred to in the reaction scheme and/or in the text.

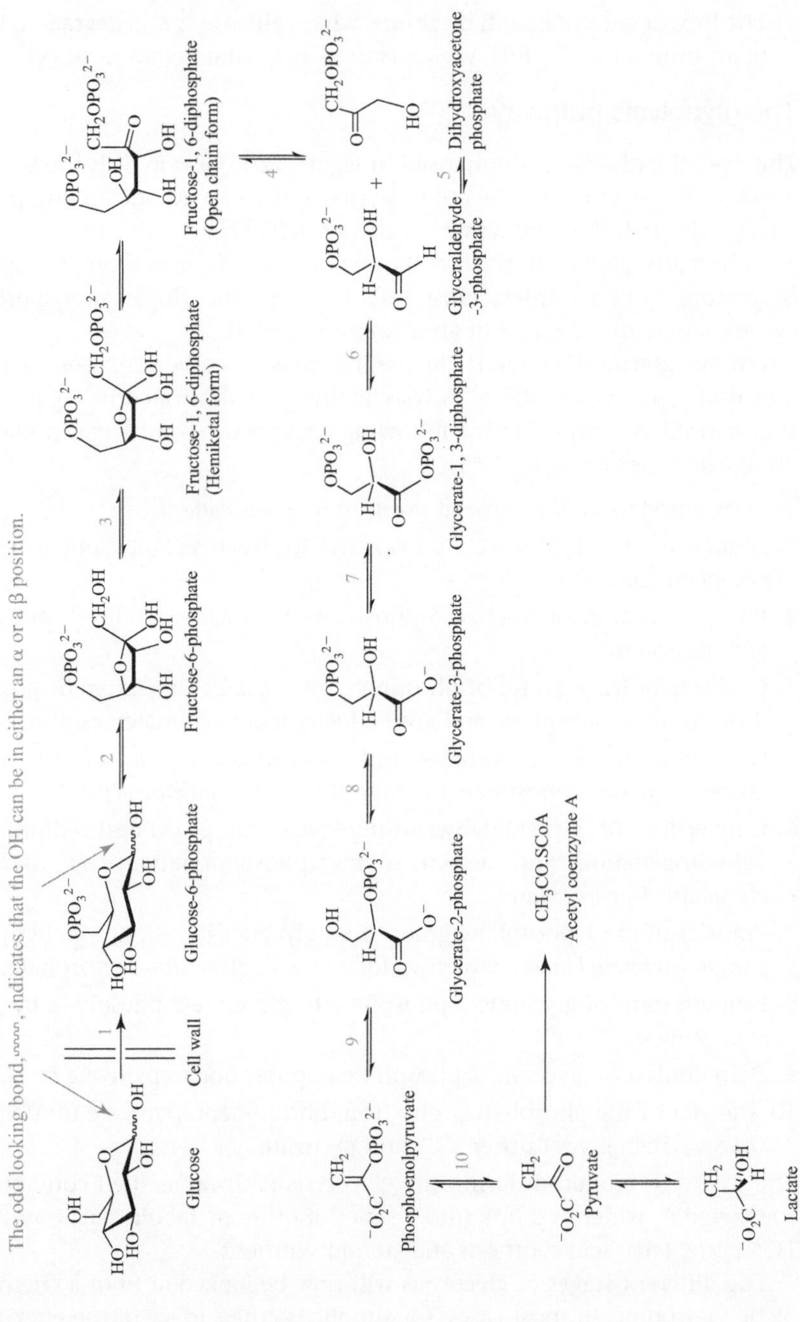

The odd looking bond, ∿∿, indicates that the OH can be in either an α or a β position.

Figure 12.1 Overview of the glycolysis pathway.

Analysis of the mechanism of glycolysis

12.3.1 Glycolysis step 1

Mechanistically this stage represents a nucleophilic attack of an alcohol (the C-6 oxygen) on an anhydride, where the anhydride is the phosphate attached to ATP (specifically the γ P atom), and the mechanism is summarised thus:

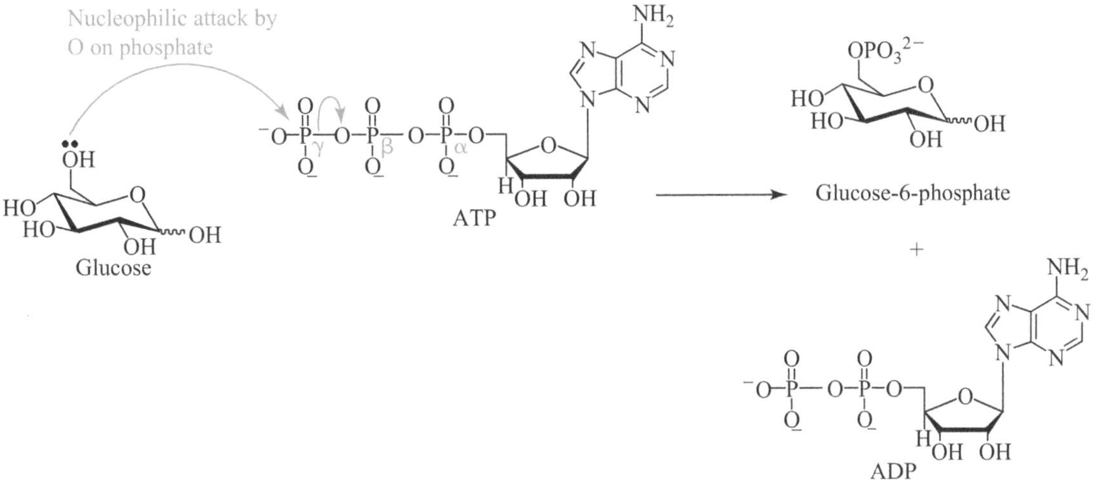

The reaction is catalysed by a hexokinase enzyme in which, after the glucose binds, the enzyme undergoes conformational change. This allows ATP binding and, because the glucose sits in a water-free region, it means that ATP is not hydrolysed thereby allowing it to provide phosphate for the glucose.

This is an example of an induced-fit mechanism. In addition, the product glucose-6-phosphate inhibits the enzyme, thereby keeping its rate of production under control.

12.3.2 Glycolysis step 2

This involves the transformation of glucose-6-phosphate to fructose-6-phosphate, which is an *isomerisation* reaction converting a *pyranose* to a *furanose*. The overall reaction is summarised thus:

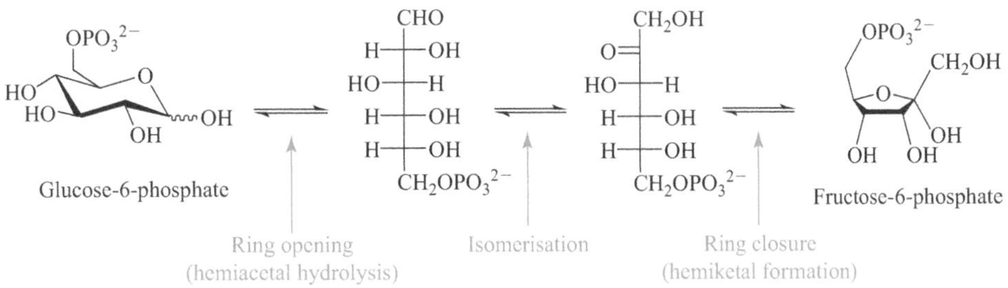

In a more detailed sense this process is catalysed using acid–base chemistry, in which the active site of the enzyme glucose-6-phosphate isomerase acts as sources of acid protons and of base. This is summarised below (for simplicity/clarity of presentation the ring structures are shown as Haworth forms, and the glucose is drawn in the α anomer form):

379

The first step, which involves ring opening, occurs through a base catalyst removing the H of the anomeric OH at position 1 of the glucose-6-phosphate. An acid catalyst protonates the 'leaving group'. The second phase involves a base-catalysed enolisation, and the third stage involves the enol formed being converted to a ketone. The final stage is the ring closure to the five-membered fructose-6-phosphate, initiated by base-catalysed deprotonation of the OH at position 5.

12.3.3 Glycolysis step 3

Fructose-6-phosphate is converted into fructose-1,6-diphosphate by nucleophilic attack of the C-1 alcohol on the γ phosphorus of the phosphate anhydride of ATP, in an analogous mechanism to that found in step 1.

Note that so far the glycolysis pathway has used up two molecules of ATP, and the sugar unit is still a 6-carbon unit.

12.3.4 Glycolysis step 4

In this stage the 6-carbon fructose-1,6-diphosphate is broken into two 3-carbon fragments. This is achieved through a *retro-aldol* reaction (which is a reverse aldol reaction; aldol reactions were described in Chapter 8, and a detailed look at them was given in Section 8.13). The mechanism is represented in Figure 12.2, and the first stage involves the formation of an imine (as referred to in Section 8.10.5).

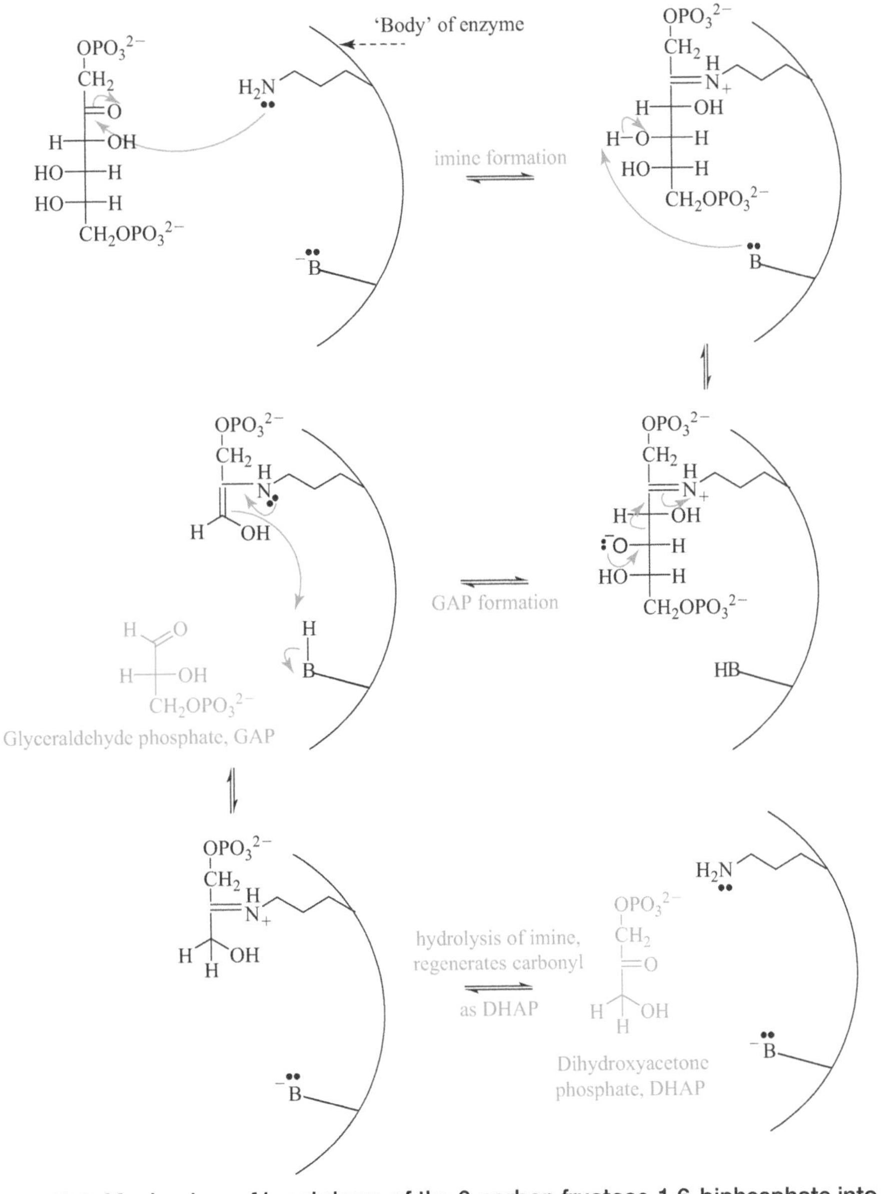

Figure 12.2 **Mechanism of breakdown of the 6-carbon fructose-1,6-biphosphate into two 3-carbon fragments.**

381

The 'glucose' (via fructose-1,6-diphosphate) has now been broken down into two 3-carbon units, namely dihydroxyacetone phosphate, abbreviated to DHAP, and glyceraldehyde-3-phosphate, abbreviated to GAP. Only GAP is further metabolised by the next stage in the glycolytic pathway, which would mean that half of the glucose would be 'wasted' as DHAP. The situation is resolved, however, by converting the 'unwanted' DHAP into the required GAP via an isomerisation reaction, and there is an enzyme that facilitates this conversion of DHAP into GAP, called *triose phosphate isomerase* (usually abbreviated either as TIM or as TPI). This is step 5 of the glycolysis cycle.

12.3.5 Glycolysis step 5

The structures of the TIM from many sources have been determined, and they are all very similar. Analysis of the amino acid sequence showed that two residues are involved in the mechanism of the reaction, these being a histidine (His) and glutamic acid (Glu), and their role in the isomerisation process is to act as acid/base, as indicated in Figure 12.3.

The acid/base chemistry of the conversion of DHAP to GAP is controlled by the enzyme TIM. Potentially the enediol intermediate can decompose, but TIM holds it in position, enabling the preferential conversion to GAP.

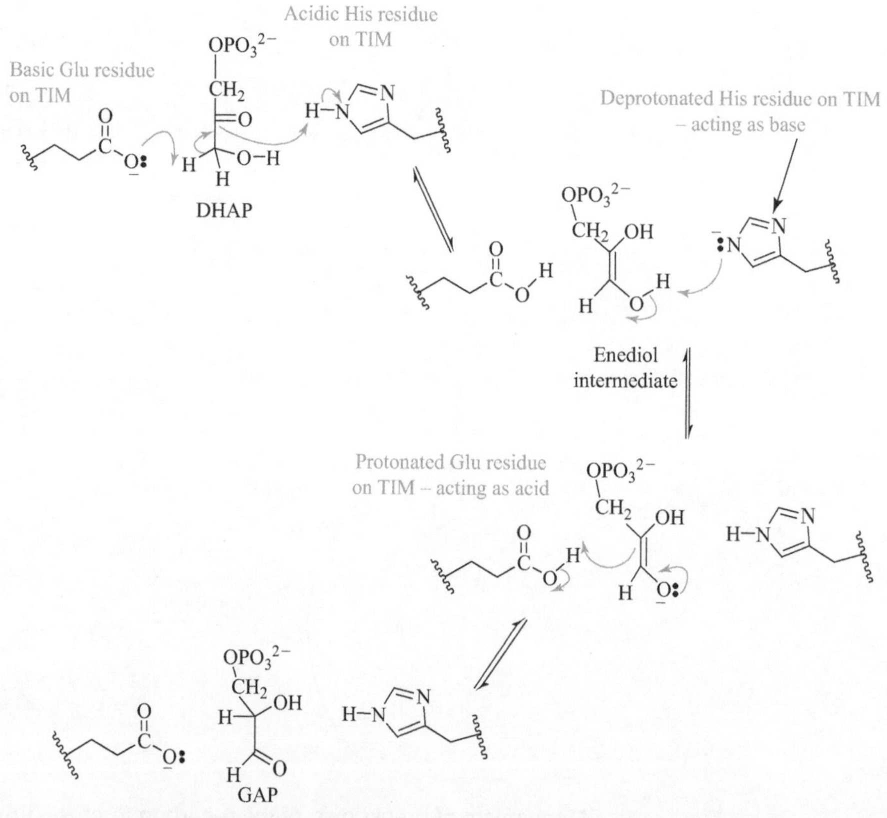

Figure 12.3 **Conversion of DHAP to GAP.**

12.3.6 Glycolysis step 6

The preceding five steps in the glycolysis pathway have transformed one molecule of glucose into two molecules of GAP, during the course of which 2 ATP molecules were used. The second five steps will 'harvest' energy (by generating ATP).

The first part of the 'energy harvesting' stage of glycolysis involves the conversion of GAP to glycerate-1,3-diphosphate, which is also called 1,3-bisphosphoglycerate (abbreviated as 1,3-BPG). Note that the 1,3-BPG formed is a 'high energy' acyl-phosphate (see Box 11.2 for clarification of this idea). The process of converting GAP to 1,3-BPG is catalysed by the enzyme *glyceraldehyde-3-phosphate dehydrogenase* (GAPDH), and the reaction is the sum of two processes:

- The oxidation of the aldehyde to a carboxylic acid by NAD^+.
- The joining of the carboxylic acid and orthophosphate to form the acyl-phosphate product.

The reaction is summarised in Figure 12.4.

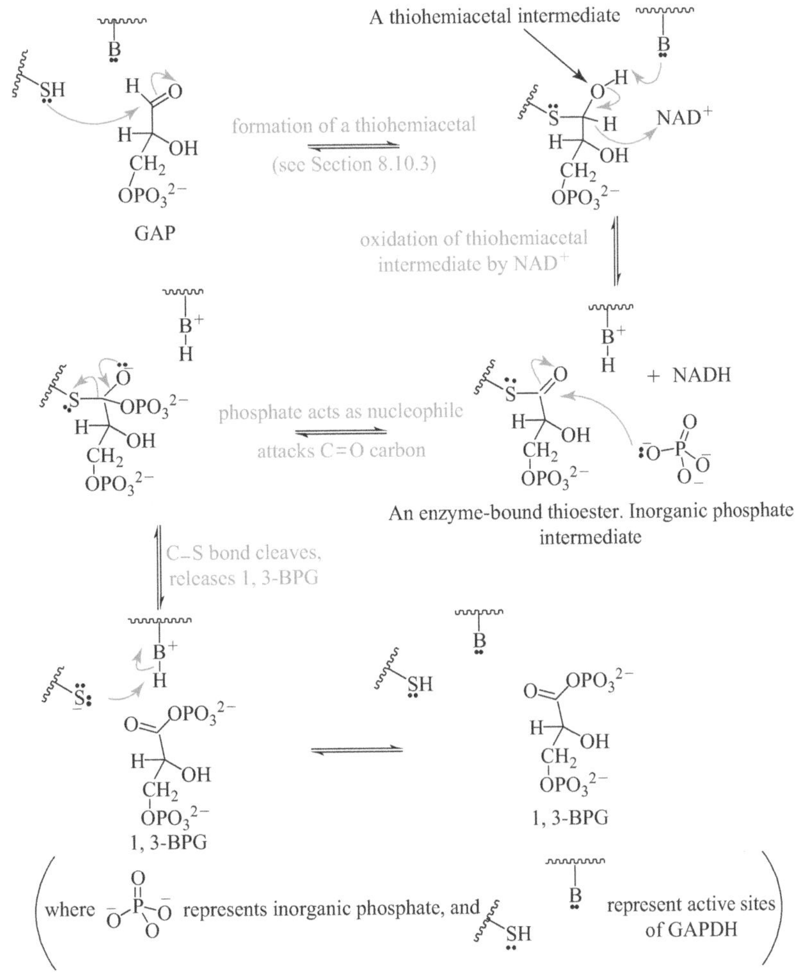

Figure 12.4 **Mechanism of conversion of GAP to 1,3-BPG.**

12.3.7 Glycolysis step 7

This step involves the transfer of the 1-phosphate group from glycerate-1,3-diphosphate to ADP using the enzyme *phosphoglycerate kinase,* which acts as the catalyst (enzyme abbreviated as PGK), giving ATP and glycerate-3-phosphate. The precise details of the mechanism for this at an enzymic level are not certain, but in a chemical sense it can be summarised by the direct attack of ADP on 1,3-BPG, as shown:

1, 3-BPG

ADP

(functions using enzyme PGK)

3-Phosphoglycerate
(or glycerate-3-phosphate)

ATP

The formation of ATP in this reaction is referred to as substrate-level phosphorylation because the phosphate donor, 1,3-BPG, is a substrate with high phosphoryl-transfer potential $\Delta G^{\oplus} = -49.4 \text{ kJmol}^{-1}$).

Remember 2 x 3-carbon units go through PGK for every glucose molecule being metabolised, so 2 x ATP are produced at this step!

12.3.8 Glycolysis step 8

The conversion of 3-phosphoglycerate to 2-phosphoglycerate is catalysed by *phosphoglycerate mutase* (abbreviated as PGM). A *mutase* is an enzyme that catalyses the transfer of a functional group from one position to another *on the same molecule*; in other words it facilitates a rearrangement or intramolecular shift (intramolecular means within the same molecule).

The reaction appears to be a simple transfer process, but has an unusual phosphorylated-enzyme intermediate, in which one histidine unit has a phosphate bound to it, and the oxygen attached to the 2 position of the 3-phosphoglycerate initially binds to this phosphate, hence forming a biphosphate intermediate. The phosphate at the 3-position is then abstracted by another histidine, as indicated in Figure 12.5.

12.3.9 Glycolysis step 9

This is the conversion of the 2-phosphoglycerate to 2-phosphoenolpyruvate (PEP), which involves an E2 elimination of water (a dehydration reaction). The reaction is

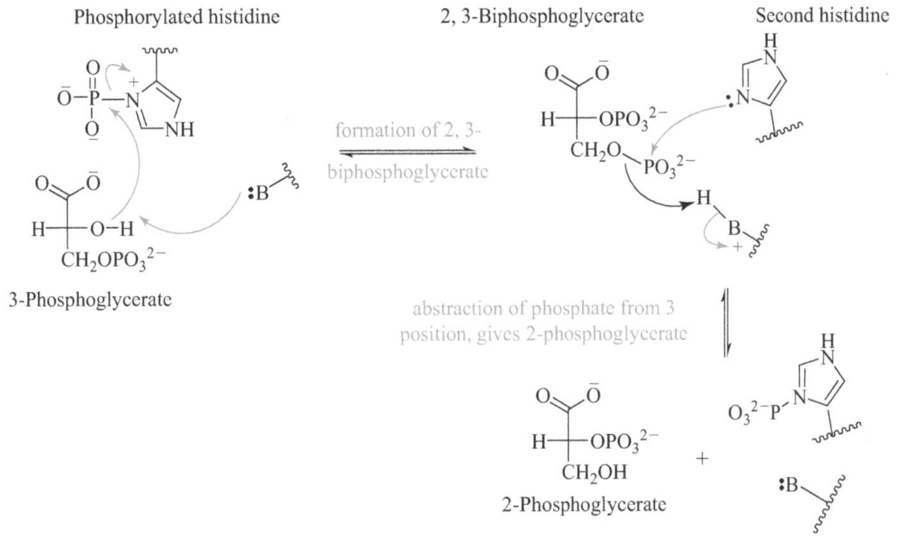

Phosphorylated histidine   2, 3-Biphosphoglycerate   Second histidine

3-Phosphoglycerate

formation of 2, 3-
biphosphoglycerate

abstraction of phosphate from 3
position, gives 2-phosphoglycerate

2-Phosphoglycerate

Figure 12.5 Isomerisation reaction, converting 3-phosphoglycerate to
2-phosphoglycerate.

an acid–base reaction in which a lysine residue associated with the enzyme *enolase*
acts as a base which removes a proton from the carbon at the α position, and the
OH group on the 3 position is protonated by a nearby glutamic acid residue on the
enzyme. The mechanism is summarised thus:

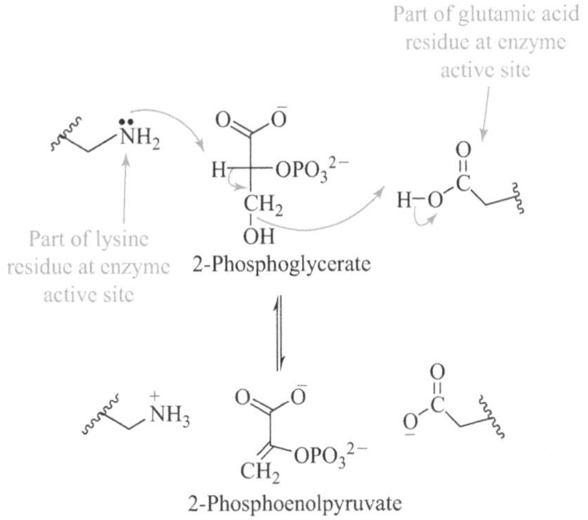

Part of glutamic acid
residue at enzyme
active site

Part of lysine
residue at enzyme
active site

2-Phosphoglycerate

2-Phosphoenolpyruvate

12.3.10 Glycolysis step 10

This is the final step in the glycolysis pathway, and involves transfer of the phosphate on 2-phosphoenolpyruvate to ADP, thereby forming ATP and pyruvate itself.
Based on the initial glucose molecule at the start of the cycle, there will be two ATP
molecules given during this step (similar to step 7). The enzyme that catalyses this
process is *pyruvate kinase*.

This reaction is spontaneous, and PEP has a larger $\Delta G^{\oplus}$ for the hydrolysis of the
phosphate than does ATP.

12.3.11 Summary

The free energy changes for the individual steps of the glycolysis pathway are given in Table 12.1 (and graphically in Figure 12.6) where $\Delta G^{\oplus}$ is the free energy change under biochemical standard state conditions, and ΔG is the free energy change for the concentration conditions that exist in the cell.

Table 12.1 **Free energy changes associated with glycolysis**

Step	$\Delta G^{\oplus}/kJ\,mol^{-1}$	$\Delta G/kJ\,mol^{-1}$
1	−16.7	−33.5
2	1.7	−2.2
3	−14.2	−22.2
4	23.8	−1.3
5	7.5	2.5
6	6.3	2.5
7	−18.5	1.3
8	4.4	0.8
9	7.5	−3.3
10	−31.4	−16.7

The energy profile in Figure 12.6 represents that associated with the free energy change for the concentration conditions that exist in the cell. It is clear from this view that most of the steps operate at, or close to, equilibrium under these conditions, but three steps, namely 1, 3 and 10, very much favour the forward reaction, and hence drive this process forward.

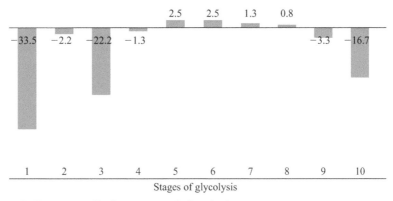

Figure 12.6 **Energy profile for stages of glycolysis.**

It is possible to summarise the ten steps of the complete glycolysis pathway by the equation:

Glucose $+ 2NAD^+ + 2ADP + 2$ phosphate $\rightarrow 2$ pyruvate $+ 2NADH + 2ATP + 2H^+ + 2H_2O$

Whilst glycolysis has broken glucose down into pyruvate, that is not the sole function, and it is also responsible for the production of the high energy molecules ATP and NADH, which are both participants in other metabolic pathways.

12.4 What now? Where does the pyruvate go?

The ultimate fate of pyruvate produced by glycolysis depends on its environment. There are three main routes by which pyruvate can be used:

- In situations where there is an absence of oxygen (called anaerobic conditions), such as in muscle tissue during strenuous exercise, when pyruvate is converted into lactic acid (as the anion lactate).
- In certain bacteria and in yeast, pyruvate is converted into ethanol, via acetaldehyde (ethanal).
- In the presence of an adequate supply of oxygen (called aerobic conditions), it is converted to acetyl coenzyme-A, and this enters into the TCA cycle. This process uses a 3-enzyme system known as the *pyruvate dehydrogenase complex.*

12.4.1 Conversion of pyruvate into lactate

This reaction is catalysed by lactate dehydrogenase (LDH). Lactate is produced by many microorganisms, the process being known as lactic acid fermentation. As mentioned above, it is also formed in muscle tissue during intense exercise. Overall the production of lactate from glucose can be summarised as:

$$\text{Glucose} + 2P_i + 2ADP \longrightarrow 2 \text{ lactate} + 2ATP + 2H_2O + 2H^+$$

The conversion from the pyruvate to the lactate can be regarded as a simple reduction:

387

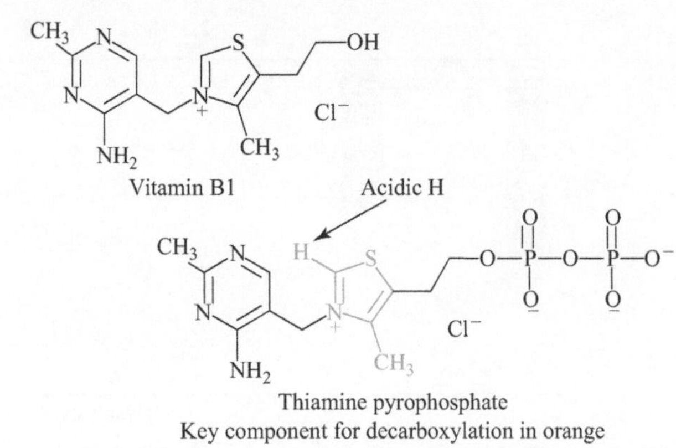

Figure 12.7 Structure of vitamin B1 (thiamine) and its derivative thiamine pyrophosphate.

Lactate from muscle is then carried by the blood to the liver where it can be converted to glucose via the process called gluconeogenesis.

12.4.2 Conversion of pyruvate into ethanol

This is a two-step process in which the pyruvate is firstly converted into acetaldehyde, which is itself then reduced to ethanol. Step 1 is a decarboxylation catalysed by the enzyme pyruvate decarboxylase. This enzyme requires the coenzyme thiamine pyrophosphate (TPP), where TPP is derived from the vitamin thiamine (vitamin B1; see Figure 12.7). It is an important cofactor for many α-decarboxylation reactions (reactions involving the removal of CO_2 from a site in an α-position to a carbonyl group).

The first step of the mechanism, the decarboxylation, has the mechanism shown:

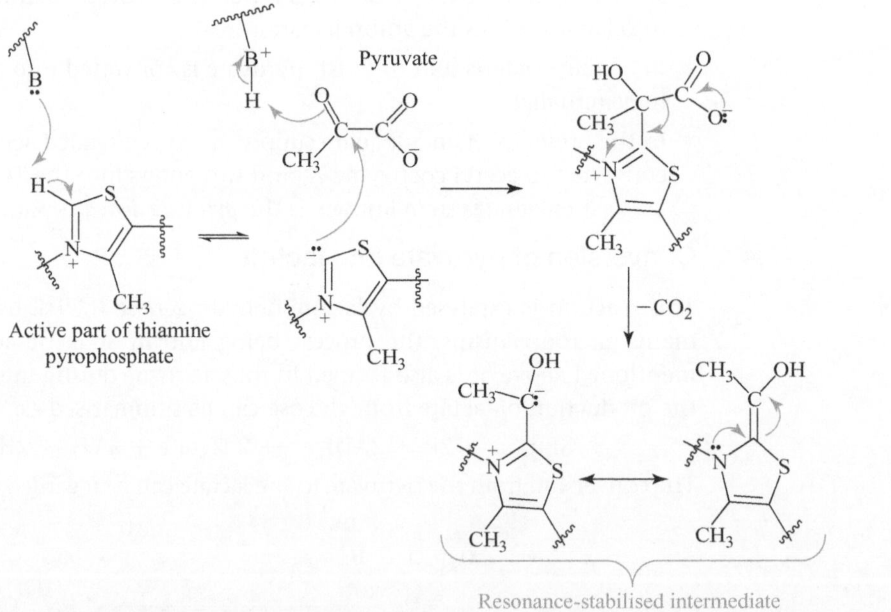

Protonation of this resonance-stabilised intermediate takes the reaction onwards:

The second step, which is the reduction reaction converting ethanal to ethanol, is catalysed by the enzyme alcohol dehydrogenase and is zinc-dependent. Zinc is co-ordinated to the enzyme through two cysteines and a histidine residue, and binds to the oxygen of the aldehyde, essentially activating the carbonyl group to attack by the H$^-$ from NADH.

The human liver performs this reaction as an oxidation, metabolising alcohol, whereas in other circumstances (such as brewer's yeast) glucose is taken all of the way through to alcohol.

389

12.4.3 Conversion of pyruvate into acetyl coenzyme A

This conversion is very important as it forms the link which allows the products from the glycolysis pathway (pyruvate) to be converted into a form where it provides a principal participant in the TCA cycle. The first step of this process involves the decarboxylation of pyruvate. Whilst this process uses the enzyme *pyruvate dehydrogenase* (rather than pyruvate decarboxylase), it nevertheless still requires TPP as a cofactor, and is mechanistically similar to that described in Section 12.4.2, leading to the resonance-stabilised anion shown:

Resonance-stabilised intermediate

At this stage other enzymes, coenzymes and cofactors become important, namely lipoic acid, FAD (flavin adenine dinucleotide (FAD), a *redox* cofactor, can be reduced to FADH$_2$), NAD$^+$ and CoA (coenzyme A).

The lipoic acid is converted to lipoate by linking to a lysine residue of the second enzyme in the complex, *dihydrolipoyl transacetylase*, and the subsequent mechanism to produce acetyl coenzyme A is:

The dihydrolipoate, which is still bound to a lysine residue of the complex, then moves to the active site of the third enzyme of the complex, called *dihydrolipoyl dehydrogenase*. Here FAD oxidises dihydrolipoate back to lipoate, thereby producing $FADH_2$. An NAD^+ cofactor then oxidises $FADH_2$ back to FAD, and is converted to NADH.

The structure of lipoic acid itself is shown below:

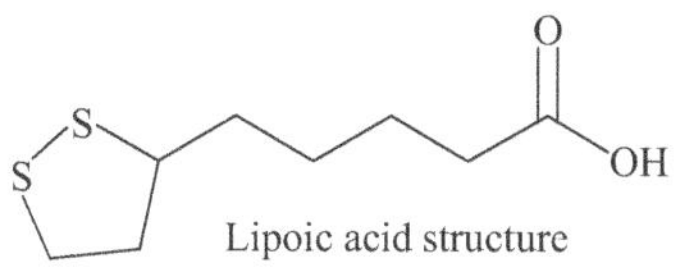

Lipoic acid structure

12.5   The TCA cycle

12.5.1   Introduction and overview

Acetyl CoA (Box 12.2) enters the TCA cycle, which is called a cycle because one of the molecules it starts with, the four-carbon oxaloacetate, is regenerated by the end of the cycle, allowing the process to start again. It begins when acetyl CoA and oxaloacetate interact to form the six-carbon compound citric acid. This citric acid molecule then undergoes a series of eight chemical reactions that strip carbons to produce a new oxaloacetate molecule. The extra carbon atoms are expelled as CO_2 (the TCA cycle is the source of the CO_2 exhaled on breathing).

In the process of breaking up citric acid, energy is produced. It is stored in ATP, NADH and $FADH_2$. The NADH and $FADH_2$ proceed on to the electron transport chain.

Box 12.2

Properties of acyl CoA thioesters

- The CoA part is 'recognised' by enzymes, meaning it binds readily with them.
- The C—S bond of thioesters is much weaker than the corresponding C—O bond of esters, so thioesters are more readily hydrolysed, which can be a driving force for reactions.
- S is much less electronegative than O, so the C—S bond is not very polarised. This means that thioesters are chemically more akin to ketones than to esters. The C=O bond of thioesters is more polarised, and the α-hydrogens are more acidic.

Thioester hydrolysis

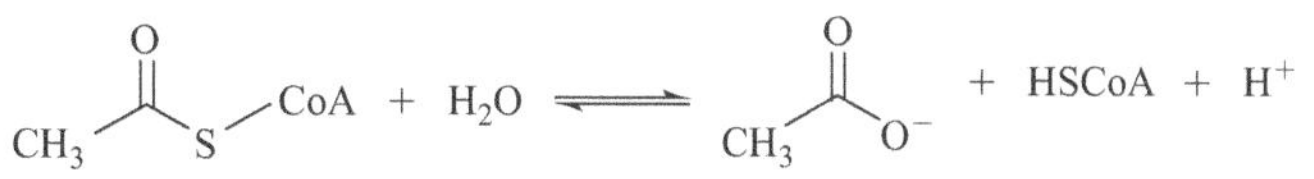

The free energy change, $\Delta G^{\oplus}$ for this process is $-31\,kJ\,mol^{-1}$, which is much more favourable than the comparable ester hydrolysis would be.

Figure 12.8 shows an overview of the key points of the TCA cycle, with the individual steps numbered from 1 to 8. These steps are now described in brief.

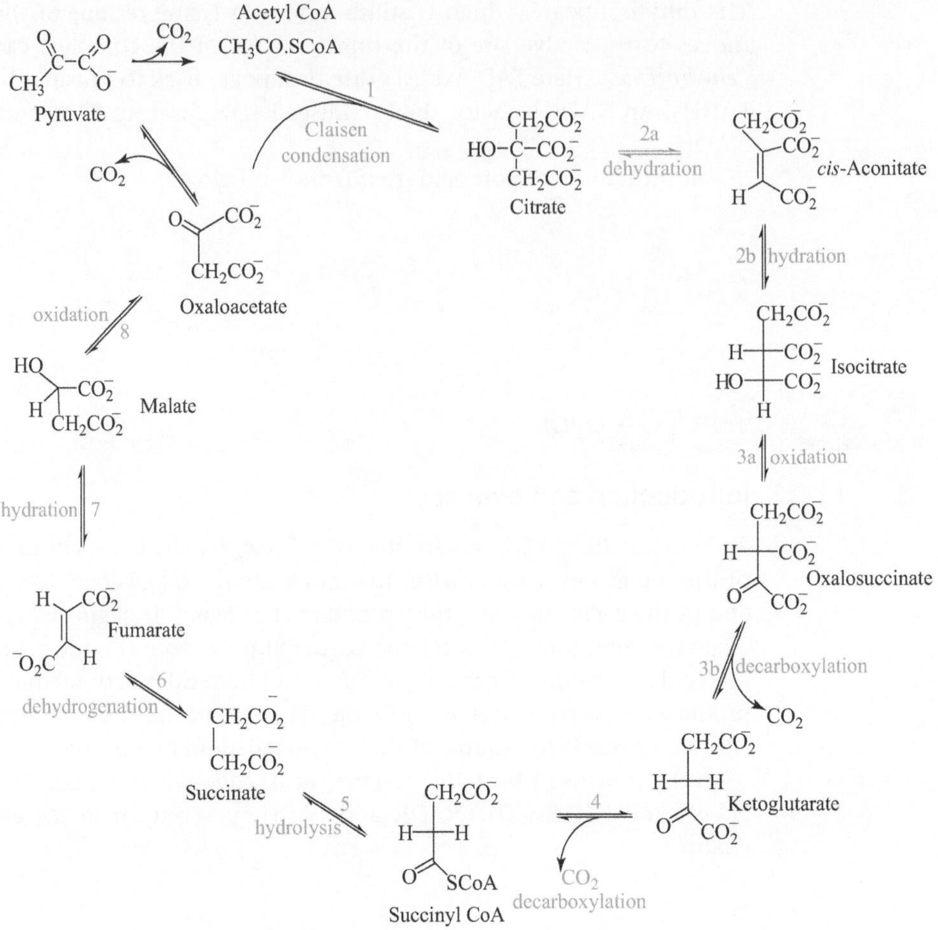

Figure 12.8 **Overview of the TCA cycle.**

1. The acetyl group of acetyl CoA reacts with the carbonyl group of oxaloacetate (via a Claisen condensation), with the corresponding release of coenzyme A in the process. *Citrate synthase* is the enzyme responsible for catalysing this stage.

2. The hydroxyl group of the citrate is moved to an adjacent carbon to give isocitrate, proceeding via cis-aconitate. It involves removal of OH and H, as water (in other words, dehydration) across the two carbons involved. Rehydration occurs in the reverse sense. The enzyme involved for both steps is *aconitase*.

3. The isocitrate is dehydrogenated and decarboxylated by *isocitrate dehydrogenase*, initially giving oxalosuccinate followed by ketoglutarate.

4. *Ketoglutarate dehydrogenase* converts ketoglutarate to succinyl CoA by a mechanism similar to that involved in the decarboxylation of pyruvate.

5. *Succinate thiokinase* facilitates the conversion of succinyl CoA to succinate. Succinyl CoA (like acetyl CoA) has a high value for $\Delta G^{\oplus}$ of hydrolysis (-33.5 kJmol^{-1}). In the citrate synthase reaction described in step 1 above, acetyl CoA hydrolysis was coupled to the synthesis of the 6-carbon citrate from oxaloacetate and acetyl CoA. In this case the succinyl CoA thioester bond cleavage is coupled with the

phosphorylation of ADP to give ATP (or of guanosine diphosphate, GDP, to give GTP, the triphosphate).

6. The CH—CH bond of succinate is dehydrogenated (using *succinate dehydrogenase*) to give fumarate. This reaction needs a cofactor, flavin adenine dinucleotide (FAD), which takes up the hydrogen.

7. Fumarate is hydrated to malate with *fumarase* (also known as *fumarate hydratase*).

8. Oxaloacetate is generated from malate by dehydrogenation using *malate dehydrogenase*. This reaction has regenerated oxaloacetate, which was used in step 1 described above, and has therefore completed the cycle.

The different stages of the TCA cycle will now be looked at from a chemical mechanistic viewpoint.

12.6 Analysis of the mechanism of the TCA cycle

12.6.1 TCA cycle, step 1

The enzyme citrate synthase is responsible for catalysing this step, and it only binds and hydrolyses acetyl coenzyme A when oxaloacetate is already bound; it is therefore an 'induced fit' mechanism. From a chemical perspective, the reaction is highly favourable ($\Delta G^{\oplus} = -31.5$ kJ mol^{-1}), and it is an example of a Claisen condensation, as shown in Figure 12.9.

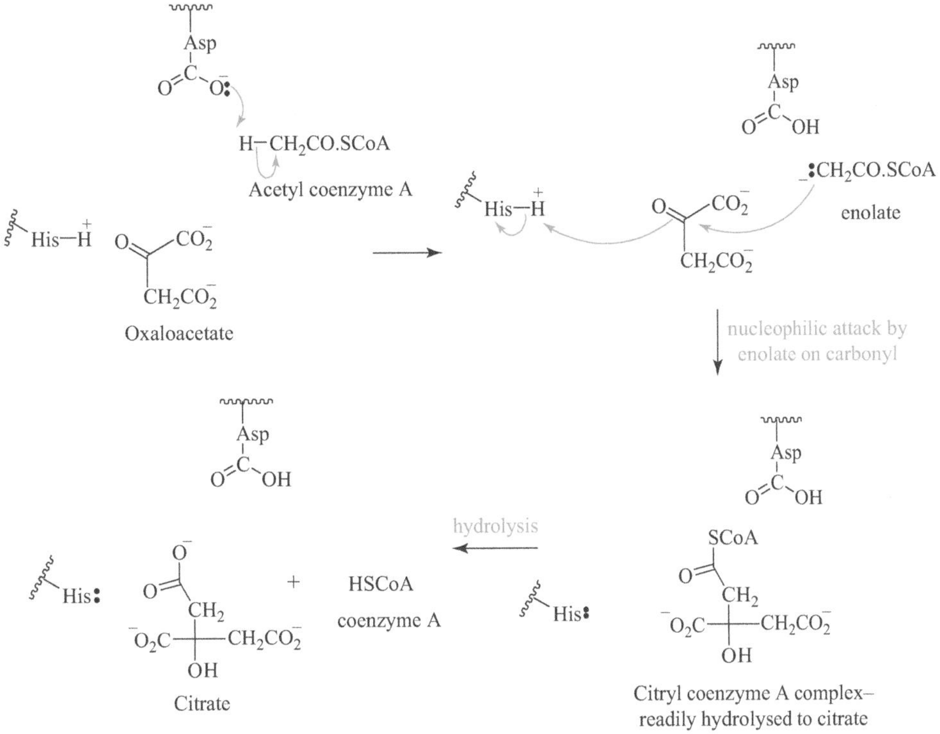

Figure 12.9 **Formation of citrate from oxaloacetate and acetyl coenzyme A.**

393

12.6.2 TCA cycle, step 2

Ultimately citrate is converted to its isomer, isocitrate, using the enzyme *aconitase*. This occurs initially via a dehydration reaction, which gives *cis*-aconitate, followed by a hydration reaction which gives isocitrate.

The OH group in citrate is not a naturally good leaving group, but the aconitase enzyme provides help in the form of an Fe_4S_4 cluster present at the active site. Three of the cluster Fe atoms are bound to the enzyme through three cysteine groups from the protein. This means that one iron of the cube is vacant and the iron atom in this position can coordinate the C-3 carboxyl and hydroxyl groups of citrate (see Figure 12.10). This iron atom thus acts as a Lewis acid, accepting an unshared pair of electrons from the hydroxyl, making it a better leaving group, and the OH can leave as water through protonation by a nearby histidine.

The mechanism is summarised below, with the enzyme providing a base, in the form of a serine side chain, which deprotonates the citrate.

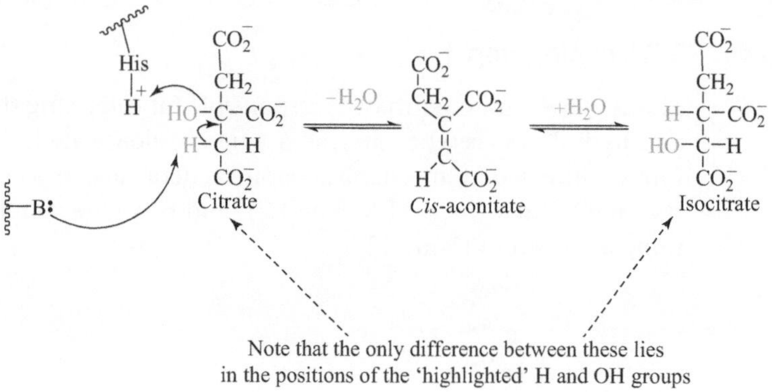

Note that the only difference between these lies
in the positions of the 'highlighted' H and OH groups

The standard free energy change for the overall process of going from citrate to isocitrate is $\Delta G^{\ominus} = 13.3\,\mathrm{kJ\,mol^{-1}}$, which is thermodynamically unfavourable, and equates to less than 10% of isocitrate in the equilibrium mixture. The driving force, however, is removal of the product by the next stages in the cycle, which do provide a thermodynamic boost.

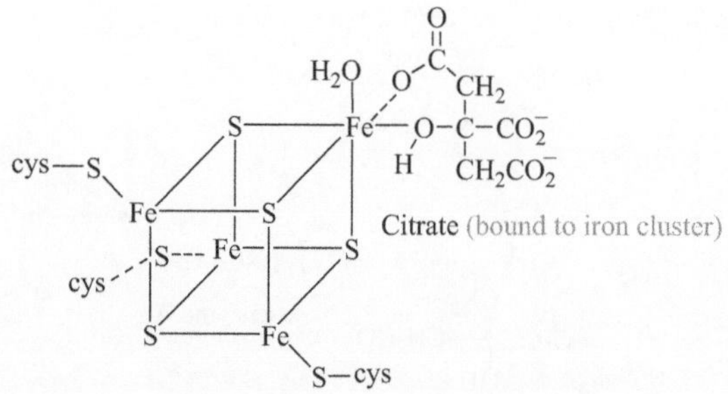

Figure 12.10 Fe₄S₄ cluster associated with aconitase enzyme.

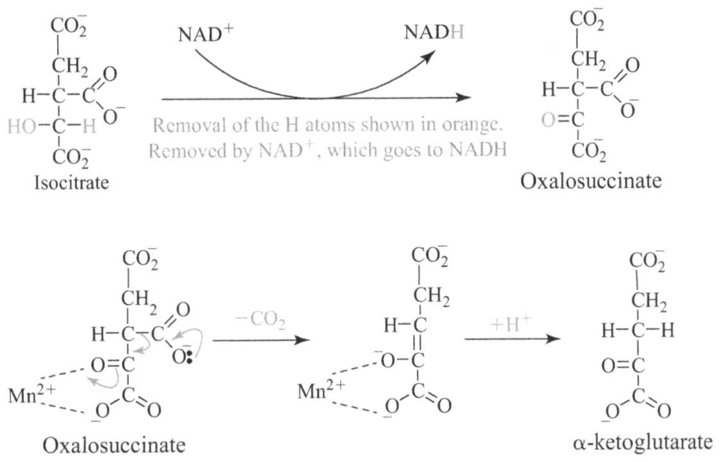

Figure 12.11 **Formation of α-ketoglutarate from isocitrate.**

12.6.3 TCA cycle, step 3

Here isocitrate undergoes oxidative decarboxylation to α-ketoglutarate and the overall process is catalysed by the enzyme isocitrate dehydrogenase. This occurs in two stages:

- the oxidation of isocitrate to oxalosuccinate, which is the oxidation of a secondary alcohol to a ketone, followed by
- a decarboxylation, which results in the α-ketoglutarate.

The overall free energy change that results from the conversion of isocitrate to α-ketoglutarate is $\Delta G^{\oplus} = -21\,\mathrm{kJ\,mol}^{-1}$. The key features of the mechanism are as shown in Figure 12.11.

In the second step, the decarboxylation of oxalosuccinate:

- Bound Mn^{2+} polarises the ketone, increasing the polarisation of the C=O bond. This 'pulls' electrons from the carboxylate, weakening the $C-CO_2$ bond.
- This is an example of a polarised ketone acting as an electron 'sink'.

12.6.4 TCA cycle, step 4

Succinyl CoA is formed by the oxidative decarboxylation of α-ketoglutarate, and the process is catalysed by the α-ketoglutarate dehydrogenase complex.

This reaction also produces NADH, CO_2 and a CoA thioester and is analogous to the pyruvate dehydrogenase (Sections 12.4.2 and 12.4.3) reaction:

$$\text{pyruvate} + \text{CoA} + \text{NAD}^+ \rightarrow \quad \text{acetyl CoA} + CO_2 + \text{NADH}$$

(Both dehydrogenases require thiamine, lipoic acid and FAD cofactors.) The process is summarised:

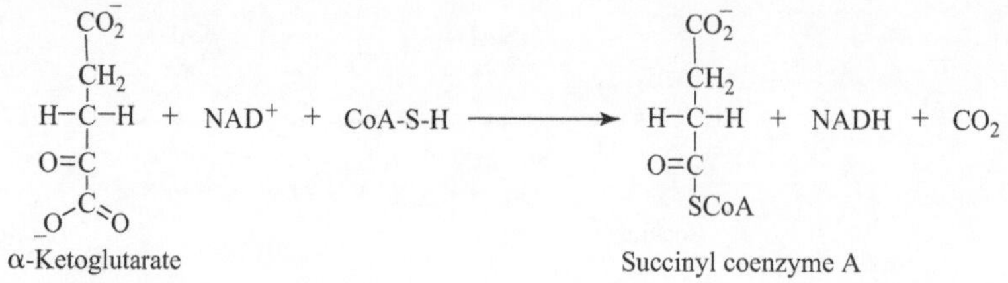

α-Ketoglutarate + NAD$^+$ + CoA-S-H ⟶ Succinyl coenzyme A + NADH + CO$_2$

12.6.5 TCA cycle, step 5

Succinyl CoA is hydrolysed to succinate with the formation of CoA and ATP (or GTP), with the process catalysed by succinyl CoA synthetase. Succinyl CoA (like acetyl CoA, as described in Box 12.2) has a high $\Delta G^{\oplus}$ of hydrolysis ($\Delta G^{\oplus} = -33.5$ kJmol^{-1}).

The energetically favourable hydrolysis of acetyl coenzyme A was an important part of the citrate synthase reaction (Section 12.6.1), in which the acetyl CoA hydrolysis was coupled to the synthesis of the 6-carbon citrate molecule from oxalo acetate and acetyl CoA. In this case, however, the succinyl CoA synthetase thioester bond cleavage is coupled to the phosphorylation of GDP to give GTP (in mammals) or ADP to give ATP (in plants). This is illustrated in Figure 12.12 .

12.6.6 TCA cycle, step 6

Succinate is dehydrogenated to fumarate, with the reaction being catalysed by succinate dehydrogenase. Succinate dehydrogenase links the TCA cycle and the aerobic electron transport chain, and it is the only membrane-bound enzyme in the TCA cycle.

FAD, rather than NAD$^+$, is the electron acceptor, since the reducing power of succinate is not sufficient to reduce NAD$^+$. The mechanism involves hydride abstraction

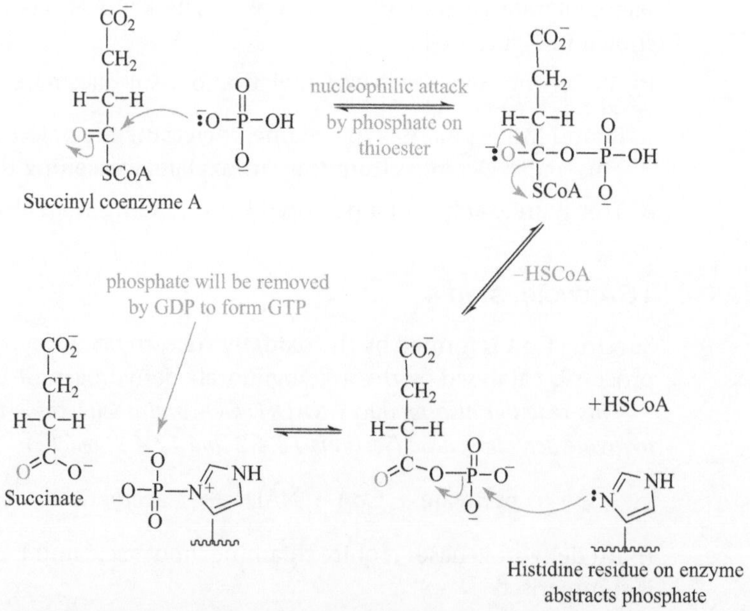

Figure 12.12 Conversion of succinyl coenzyme A to succinate.

but, despite the apparent simplicity of the reaction, it is too complex a mechanism to describe in any detail in this text.

In summary, therefore, the process is:

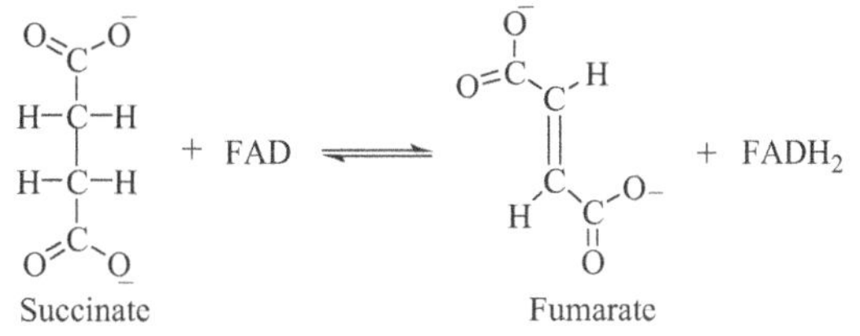

12.6.7 TCA cycle, step 7

Fumarase (also known as fumarate hydratase) catalyses the hydration of fumarate to form malate, and whilst this looks like a simple addition of H_2O to a double bond, because the double bond is conjugated to two $C=O$ groups, it is not very reactive and does not protonate easily. There are two types of fumarase enzyme, and only the mechanism of the type II enzyme is known. It is believed to proceed through a carbanion intermediate, resulting from attack of OH^- on the double bond:

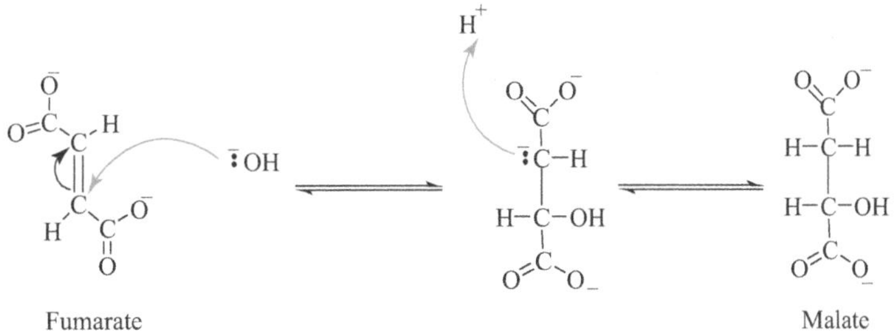

12.6.8 TCA cycle, step 8

The step that completes the TCA cycle is the generation of oxaloacetate from malate, which is catalysed by malate dehydrogenase. The detailed mechanism is unknown, though it is likely to involve a hydride transfer to NAD^+, akin to other dehydrogenases.

The reaction is summarised thus:

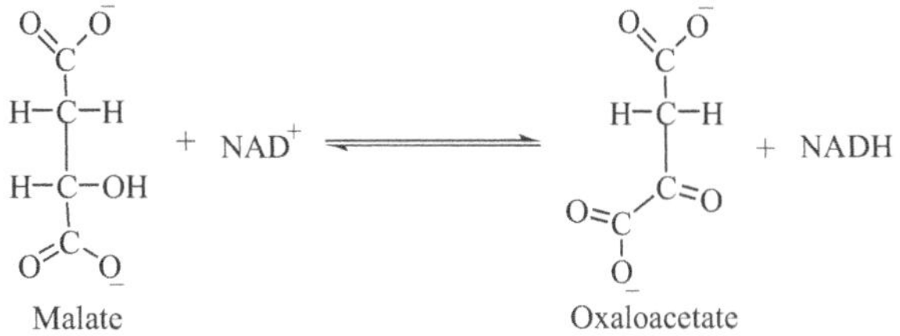

This process is energetically unfavourable, with the $\Delta G^\oplus$ for the reaction being $+29.7\,kJ\,mol^{-1}$. This means that the equilibrium concentration of oxaloacetate is low. It is the removal of the product by citrate synthase, during the production of citrate, that is responsible for driving this reaction forward.

12.7   Summary of outcomes of the glycolysis and TCA cycles

In glycolysis, each glucose produces four ATP molecules and two NADH molecules, but the process uses two ATP molecules. Therefore there is a *net* production of two ATP and two NADH molecules for each glucose molecule taken in. Two pyruvate molecules are produced, which proceed to the TCA cycle. For each glucose molecule that is metabolised at the start of the glycolysis pathway, the TCA cycle produces two ATPs, eight NADHs, and two $FADH_2$s.

Electron transport phosphorylation (sometimes called chemiosmosis) is a theory which links the breaking down of NADH and $FADH_2$ produced above, through pumping protons across mitochondrial membranes, and the energy released is used to produce ATP. It typically produces 32 ATPs.

12.8   Gluconeogenesis

If the only source of glucose in bodies came from ingestion then there would be a risk of blood sugar levels falling too low. The body, therefore, has to have methods of producing glucose and gluconeogenesis is a metabolic pathway that enables the body to regenerate glucose. There are a number of problems associated with this, and a couple of major ones are:

- How to get the carbon to use to build glucose.

- How to overcome the energetics of generating glucose. Recall that the production of pyruvate from glucose (glycolysis) is an energetically favourable process, so the exact reverse of the process would not be likely to happen. In particular there are three steps in the glycolysis pathway (steps 1, 3 and 10) which are irreversible.

Lactate and some amino acids are converted to pyruvate or to other TCA intermediates, and these can act as carbon sources. Progress on to glucose uses the reversible reactions from glycolysis (steps 2 and 4–9), along with reactions that permit the bypassing of those irreversible steps identified in the glycolysis pathway (steps 1, 3 and 10). These points are summarised in the overview illustrated in Figure 12.13. In this diagram:

- the molecules participating in the glycolysis path are in italics;

- the molecules involved in the TCA cycle are in grey;

- the molecules in bold type can be produced by conversion from amino acids;

- the reaction arrows in orange indicate the stages of gluconeogenesis which do not follow the reverse path of glycolysis.

As mentioned above there are three steps in the glycolysis pathway which are irreversible, resulting in its being necessary to find alternatives to the reverse reactions for these steps. The first stage in gluconeogenesis is the conversion of pyruvate into phosphoenolpyruvate (PEP). This is a two-step process in which a carboxylation reaction is catalysed by the enzyme *pyruvate carboxylase* (often abbreviated to PC), and a decarboxylation-phosphorylation reaction is catalysed by the enzyme *phosphoenolpyruvate carboxykinase*. These are the steps labelled with the orange reaction processes 1 and 2 in Figure 12.13.

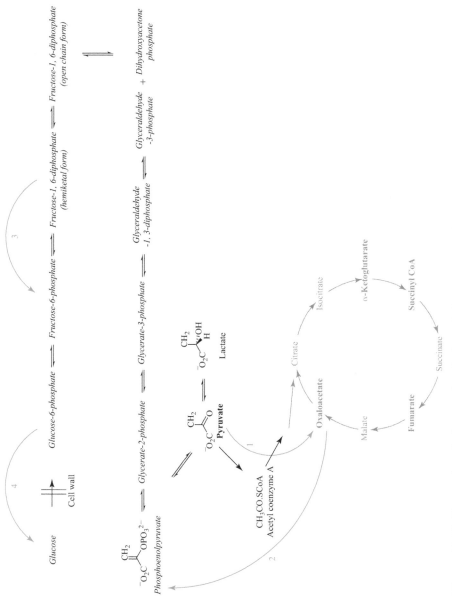

Figure 12.13 **Comparison between glycolysis and gluconeogenesis.**

The first step involves the carboxylation of pyruvate to give oxaloacetate, requiring the enzyme pyruvate carboxylase, and it is dependent on the cofactor biotin, which has the structure:

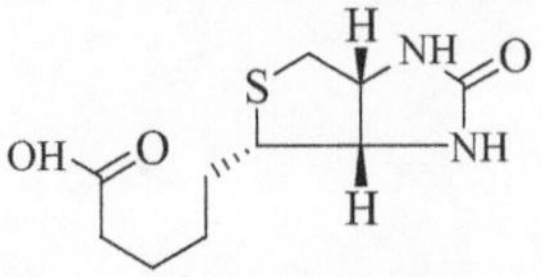

Biotin (which is bound to its carrier protein through the formation of an amide linkage) acts as the transient CO_2 carrier, as indicated below:

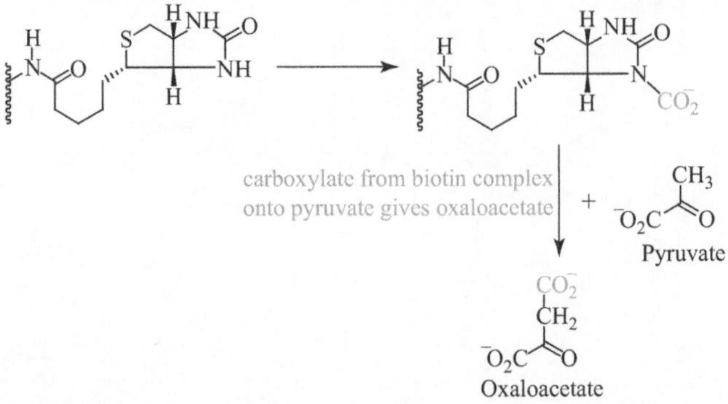

carboxylate from biotin complex onto pyruvate gives oxaloacetate

+ Pyruvate

Oxaloacetate

In the second step, the oxaloacetate is converted to PEP, a process which is catalysed by *phosphoenol pyruvate carboxykinase* (PEPCK), the activity of which is enhanced by Mg^{2+} and Mn^{2+}. *There is a problem however.* Pyruvate carboxylase is a mitochondrial enzyme, whereas the other enzymes in gluconeogenesis are cytoplasmic. Oxaloacetate has to be transported to the cytoplasm but there is no transporter for it. The problem is overcome by firstly reducing oxaloacetate to malate using the enzyme *malate dehydrogenase*. It can be transferred to the cytoplasm, where it is then re-oxidised to oxaloacetate by another malate dehydrogenase.

Decarboxylation of the oxaloacetate produces the enol pyruvate, which is then phosphorylated by guanosine triphosphate (GTP). The phosphorylation of the pyruvate occurs via nucleophilic attack of the γ-phosphoryl group of GTP, by the enolate anion, to form PEP. Overall, $\Delta G^{\oplus} = +0.13\,\text{kJ mol}^{-1}$).

Decarboxylation is the driving force for this reaction:

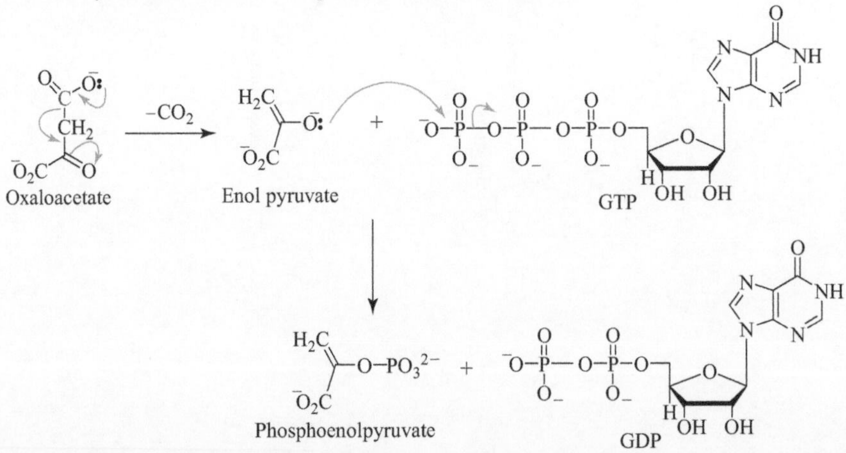

Oxaloacetate

$-CO_2$

Enol pyruvate

+ GTP

Phosphoenolpyruvate

+ GDP

Gluconeogenesis continues as the reverse of glycolysis from PEP to fructose 1,6-bisphosphate (F-1,6-BP). The six enzymes in the glycolysis steps 4–9 inclusive are working close to equilibrium under intracellular conditions, as evidenced by the graph in Figure 12.6; however, two of the last three steps in gluconeogenesis are different from glycolysis (the step involving phosphoglucose isomerase is the same).

The next major difference between gluconeogenesis and glycolysis relates to the following.

Phosphofructokinase is used in glycolysis to catalyse the reaction:

$$\text{fructose-6-phosphate} + \text{ATP} \quad \rightleftharpoons \quad \text{fructose-1,6-diphosphate} + \text{ADP}$$

By contrast, during gluconeogenesis *fructose-1,6-bisphosphatase* catalyses the hydrolysis reaction (step 3):

$$\text{fructose-1,6-diphosphate} + H_2O \quad \rightleftharpoons \quad \text{fructose-6-phosphate} + P_i$$

Finally, *hexokinase* is used in glycolysis to catalyse:

$$\text{glucose} + \text{ATP} \quad \rightleftharpoons \quad \text{glucose-6-phosphate} + \text{ADP}$$

During gluconeogenesis, however, *glucose-6-phosphatase* catalyses the hydrolysis (step 4):

$$\text{glucose-6-phosphate} + H_2O \quad \rightleftharpoons \quad \text{glucose} + P_i$$

12.9 Summary

This chapter has focused on bringing together the ideas described in the earlier parts of the book with the objectives of:

- showing how a relatively small number of organic reaction types account for many of the chemical processes that are found in metabolic pathways;
- demonstrating the important role of biological catalysts (enzymes) in the metabolic pathways;
- demonstrating how the principles of energetics help us to identify those reactions that are responsible for driving metabolic processes.

Questions

12.1 Answer the following questions about the glycolysis pathway.

(a) What is the carbohydrate starting material?

(b) This carbohydrate is broken down ultimately into which 3-carbon entity?

(c) What happens to this 3-carbon entity?

(d) Glycolysis is often said to have two stages, one energy demanding and one energy harvesting. Explain this.

12.2 During gluconeogenesis the synthesis of fructose-1,6-diphosphate is catalysed by the enzyme aldolase, and this reaction is summarised below:

Fructose-1, 6-diphosphate
(open chain form)

(a) Name the two compounds on the left hand side of the equation.
(b) Draw a plausible mechanism for the formation of fructose-1,6-diphosphate in this reaction.
(c) What type of reaction is taking place?
(d) In the cell, fructose-1,6-diphosphate exists mainly as a cyclical hemiketal form. Draw this structure, and show the mechanism by which ring closure takes place.

12.3 Citrate synthase catalyses the reaction between acetyl coenzyme A and oxalo acetate to produce citrate, which is known to proceed via a two-step process, for which citryl coenzyme A is a product.

(a) Draw out the structures of the molecules oxaloacetate, citrate and citryl coenzyme A.
(b) Name the type of reaction occurring between acetyl coenzyme A and oxalo acetate.
(c) Draw the mechanism for this process.

12.4 In brewing, yeast is used in the process of metabolising glucose into alcohol. The final part of this process involves two steps to convert pyruvate to ethanol, via ethanal.

(a) Draw the structures of pyruvate, ethanol and ethanal.
(b) What types of reactions are involved in steps 1 and 2?
(c) Name the enzymes, cofactors and substrates involved in each step.

13 Structural methods

Learning outcomes

By the end of this chapter students should be able to:

- outline the methods of operation of the different techniques;
- recognise the importance of the molecular ion in mass spectrometry as a means of determining relative molecular mass;
- recognise how the presence of isotopes can affect the appearance of mass spectra;
- make use of fragmentation patterns in mass spectra to suggest partial molecular structures;
- define the wavelength/frequency range of radiofrequency, microwave, infrared, visible and ultraviolet radiation;
- relate the absorbance position and intensity of a compound in its UV–visible absorption spectrum to give information about chromophores (conjugation, aromatic character, etc.);
- identify common functional groups with their characteristic IR group frequencies;
- explain the concepts of nuclear spin energy levels, resonance and chemical shift, in nuclear magnetic resonance (NMR) spectroscopy, and define the relationship between chemical shift and resonance frequency;
- understand how to account for the splitting of signals in ^{1}H NMR spectra;
- use results from broadband and distortionless enhancement by polarisation transfer (DEPT) ^{13}C NMR experiments to identify the nature and number of different types of carbon group (methyl, CH_3; methylene, CH_2; methane, CH, or quaternary, C) present in a molecule;
- combine different types of spectrometric/spectroscopic data to determine the structures of simple organic molecules.

13.1   Introduction

In Chapter 1 the study of chemistry was described as dealing with the determination of the composition of substances (elements and molecules), how they react with each other, and the factors that drive these reactions. In order to pursue these studies it is important to have tools which will allow for chemical structures to be determined and for reactions to be followed. In these studies there are some key questions that need to be addressed:

- Which atoms are present?
- How many atoms are present?
- How are the atoms connected?
- Which functional groups are present?
- What is the shape of the molecule?

This chapter will focus on introducing some techniques used by chemists in such structure determinations.

13.2   Mass spectrometry

13.2.1   Background

Mass spectrometry is a technique commonly used in chemistry to help identify unknown compounds. The technique is used primarily to obtain the molecular mass, but can yield further information (e.g. identity of functional groups and the empirical formula) if the identity of the elements present in the compound are known. Figure 13.1 shows a schematic diagram which illustrates the basic instrumentation involved.

The stages of the experimental procedure are itemised as follows:

- The sample is vaporised and introduced into an ionisation chamber.
- The sample is ionised, for example by being bombarded with electrons. The ionisation occurs with enough energy to remove an electron from the sample to form a positive ion. This ion is called the *molecular ion* – or sometimes the *parent ion*.

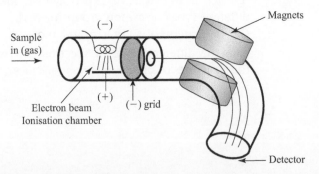

Figure 13.1 Schematic diagram of the essential components of a mass spectrometer.

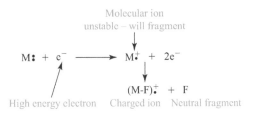

Figure 13.2 Schematic representation of ionisation and fragmentation in a mass spectrometer.

The molecular ion is often given the symbol M^+ or $M^+\cdot$. (The $\cdot$ represents a single unpaired electron in the ion. This was originally part of a pair of electrons, the other half of which is the electron removed during the ionisation process).

- The molecular ions are energetically unstable, and some of them will break up into a series of positively charged ions, and there will be uncharged radicals form also. The process of ionisation to form a molecular ion, and its subsequent breakdown is represented in Figure 13.2.

- The ions accelerate towards a grid and pass through an aperture to form a narrow beam.

- This positively charged beam then travels through a strong magnetic field, which deflects the ions according to their mass-to-charge ratio. Light ions are deflected the most; heavy ions the least. This means that uncharged particles such as radicals do not show up in the results.

- Each ion then arrives at a different place on the detector, and the resulting spectrum is a plot of fractional abundance vs mass-to-charge ratio (m/z; m is mass and z is charge).

13.2.2 Analysis of a mass spectrum

Figure 13.3 shows what would be expected for the mass spectrum of n-heptane.

The molecular ion peak and the base peak

In Figure 13.3, which shows the mass spectrum of heptane, the line produced by the heaviest ion passing through the instrument (at $m/z = 100$) is due to the *molecular ion*. Formally the mass of the molecular ion is the sum of the masses of the most abundant naturally occurring isotopes of the various atoms that make up the molecule.

The tallest line in the spectrum (in this case at $m/z = 43$) is called the *base peak*. This is usually given an arbitrary height of 100, and the height of everything else is measured relative to this. The base peak is the tallest peak because it represents the commonest fragment ion to be formed – either because there are several ways in which it could be produced during fragmentation of the parent ion, or because it is a particularly stable ion.

What about the rest of the spectrum?

Looking again at the mass spectrum of heptane shown in Figure 13.3, the line at $m/z = 71$ arises because of the ion $C_5H_{11}^+$ which is $[CH_3CH_2CH_2CH_2CH_2]^+$, and

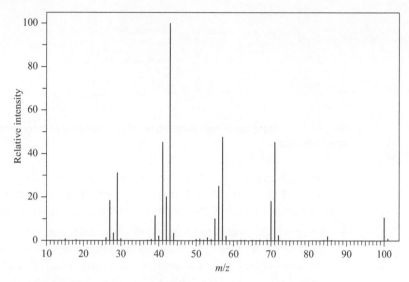

Figure 13.3 **Mass spectrum of n-heptane.**

This, and all of the mass spectra in this chapter, have been produced using data from the Spectral Data Base System for Organic Compounds (SDBS) at the National Institute of Materials and Chemical Research in Japan.

indeed in alkanes mass spectra will have peaks at m/z values corresponding to the ions produced by breaking C—C bonds. In heptane therefore, peaks may occur at m/z values of 15, 29, 43, 57, 71 and 85, which correspond to the ions CH_3^+, $C_2H_5^+$, $C_3H_7^+$, $C_4H_9^+$, $C_5H_{11}^+$ and $C_6H_{13}^+$.

The first of these would be produced by the following fragmentation:

$$[CH_3CH_2CH_2CH_2CH_2CH_2CH_3]^+ \rightarrow [CH_3CH_2CH_2CH_2CH_2CH_2]^+ + CH_3 \cdot$$

The methyl radical, $CH_3 \cdot$, being uncharged, is undetected by the instrument.

The line at $m/z = 71$ can be worked out similarly, and it corresponds to a break producing a 5-carbon ion:

$$[CH_3CH_2CH_2CH_2CH_2CH_2CH_3]^+ \rightarrow [CH_3CH_2CH_2CH_2CH_2]^+ + CH_3CH_2 \cdot$$

There are other lines in the mass spectrum, and these are more difficult to explain. For example there are lines for which the m/z values are 1 or 2 less than one of the lines described by the approach outlined above. These are often due to loss of one or more hydrogen atoms during the fragmentation process.

13.2.3 Isotopes: complicating factors or diagnostic tools?

Table 13.1 shows the relative proportions of naturally occurring isotopes for some key elements. The spectrum of heptane, shown in Figure 13.3, contains a small peak at 1 mass unit above the molecular ion (the molecular ion is at m/z 100) at $m/z = 101$. It is referred to as an $[M + 1]^+$ peak. The reason for this additional peak lies in the fact that naturally occurring carbon has three isotopes of which two are present in significant quantity: namely ^{12}C (almost 98.89%) and ^{13}C (1.11%). The radioactive

Table 13.1 **List of isotope abundances and masses**

Element	Isotope	Natural abundance (%)
Carbon	^{12}C	98.89
Carbon	^{13}C	1.11
Hydrogen	^{1}H	99.98
Hydrogen	^{2}H	0.02
Nitrogen	^{14}N	99.64
Nitrogen	^{15}N	0.36
Oxygen	^{16}O	99.76
Oxygen	^{17}O	0.04
Oxygen	^{18}O	0.20
Fluorine	^{19}F	100
Phosphorus	^{31}P	100
Sulfur	^{32}S	95.02
Sulfur	^{33}S	0.75
Sulfur	^{34}S	4.22
Chlorine	^{35}Cl	75.78
Chlorine	^{37}Cl	24.22
Bromine	^{79}Br	50.54
Bromine	^{81}Br	49.46
Iodine	^{127}I	100

isotope ^{14}C occurs only in trace quantities. The $[M + 1]^+$ peak is due to the presence of ^{13}C.

The size of the $[M + 1]^+$ peak depends on how many carbon atoms are present in the molecule. Consider the three cases of methane, ethane and heptane:

- The mass spectrum of methane, CH_4, would show an $[M + 1]^+$ peak which was 1.11% the intensity of the M^+ peak because statistically the likelihood of a molecule of $^{13}CH_4$ being present is 1.11%.

- The mass spectrum of ethane, CH_3CH_3, would show an $[M + 1]^+$ peak which was (2×1.11)% of the intensity of the M^+ peak because statistically there is a 1.11% chance of a molecule of $^{13}CH_3{}^{12}CH_3$, but there is an equal chance of there being a molecule of $^{12}CH_3{}^{13}CH_3$. Therefore the $[M + 1]^+$ peak will be 2.22% of the size of the M peak.

- The mass spectrum of heptane, $CH_3CH_2CH_2CH_2CH_2CH_2CH_3$, would show an $[M + 1]^+$ peak which was (7×1.11)% of the intensity of the M^+ peak, because there is statistically a 1.11% chance of any of the carbon atoms in the molecule being ^{13}C.

This suggests that the size of the $[M + 1]^+$ peak relative to the M^+ peak can assist in confirming how many carbon atoms are present in a molecule. This is, indeed, true in some cases, but there can be problems if other elements are present. For example, molecules containing nitrogen will have an $[M + 1]^+$ component due to the

407

presence of ^{15}N, which has a natural abundance of 0.36% (^{14}N is 99.64%). Whilst this is relatively less abundant than ^{13}C, it nevertheless is enough to distort any interpretations.

Chlorine and bromine

Chlorine has two main stable isotopes, ^{35}Cl (75.78%) and ^{37}Cl (24.22%), and bromine has two main stable isotopes also, namely ^{79}Br (50.54%) and ^{81}Br (49.46%). These have a profound effect on the appearance of mass spectra of compounds containing these elements.

Figure 13.4 shows the mass spectra of 2-chloropropane and 2-bromopropane. The spectrum of the chloro-compound shows the molecular ion $[C_3H_7{}^{35}Cl]^+$ at

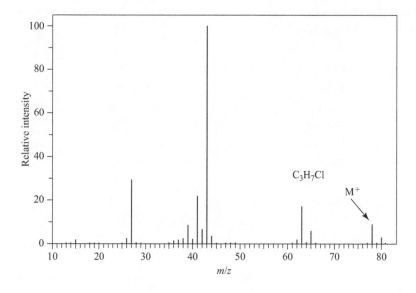

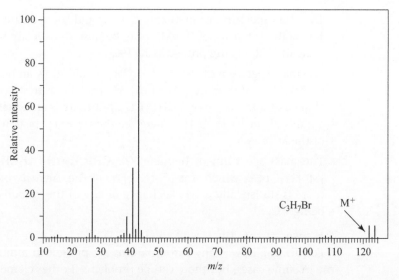

Figure 13.4 **Mass spectra of 2-chloropropane and 2-bromopropane.**

78 amu, with a smaller peak at 80 amu which corresponds to $[C_3H_7{}^{37}Cl]^+$. The spectrum of the bromo-compound shows the molecular ion $[C_3H_7{}^{79}Br]^+$ at 122 amu, with a smaller peak at 124 amu corresponding with $[C_3H_7{}^{81}Br]^+$. In both cases the base peak is at 43 amu, consistent with $[C_3H_7]^+$.

13.2.4 Fragmentation pathways involving functional groups

Carbonyl compounds of various types are important in both mainstream chemical and biological settings, so any technique that adds to the arsenal of analytical methods is valuable. Mass spectrometry can assist in determining the position of a carbonyl within a chain. Consider, for example, butan-2-one. This gives the mass spectrum shown in Figure 13.5.

After the molecule is ionised there are two main routes by which it breaks down, which essentially involve breaking on either side of the C=O, leaving the positively charged fragments at $m/z = 57$ and $m/z = 43$, that are detected. This is indicated below:

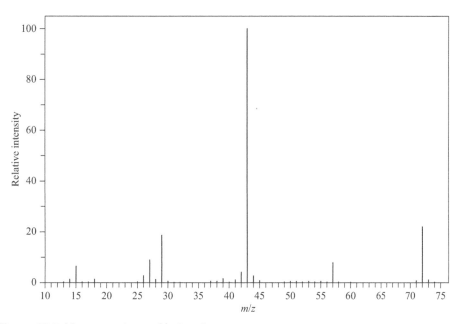

Figure 13.5 **Mass spectrum of butan-2-one.**

409

Table 13.2 **Molecular ions and fragmentation paths for common functional groups**

Functional group	General formula	Molecular ion and fragmentation mechanism
Alkyl halide	RCH$_2$X	
Alkanone	RCH$_2$.CO.CH$_2$R*	
Alkanol	RCHROH	
Amine	R*CH$_2$NHR	
Ether	R*CH$_2$OR	

Table 13.2 summarises the main initial fragmentation mechanisms from a selection of different functional groups. It is not a comprehensive list of mechanisms, but does give an insight into how such molecular ions break down. (Note the use of single headed curly arrows to indicate the movement of single electrons.)

13.2.5 Uses in biology

The principal feature of mass spectrometry involves the measurement of the mass-to-charge (m/z) ratio of ions. In the context of its application to the study of biological systems it was limited by the need to generate ions in the vapour phase. In recent years these problems have been addressed by the development of two techniques that are often used with liquid and solid biological samples, namely electrospray ionisation and matrix-assisted laser desorption/ionisation (MALDI). As a result, mass spectrometry is becoming an essential analytical tool in biological research and can be used to characterise biomolecules such as sugars, proteins and oligonucleotides, gaining insight into their sizes and their compositions.

<table>
<tr><td>13.3</td><td></td></tr>
</table>

13.3 Introduction to electromagnetic radiation

The electromagnetic spectrum (which was described in Chapter 2) is represented in Figure 13.6. Electromagnetic radiation refers to packets of energy (called photons) that are composed of orthogonal oscillating electric and magnetic fields (orthogonal means they are at 90 degrees to each other), as illustrated in Figure 13.7.

The oscillating electric field interacts with electrons in atoms and molecules, with the result that:

● atoms and molecules can absorb energy from electromagnetic radiation;

● atoms and molecules can give out energy in the form of electromagnetic radiation.

Measuring a compound's electromagnetic radiation absorbance pattern is the basis of *spectroscopy*, a very common method in chemistry used to identify and characterise an unknown compound. Because each different type of electromagnetic radiation possesses a different energy, each form of radiation will induce a different effect in matter, as illustrated in Table 13.3.

The remainder of this chapter will provide an introduction to some of these spectroscopic techniques (specifically ultraviolet–visible (UV–vis), infrared (IR) and nuclear magnetic resonance (NMR) spectroscopy) and will finish with a brief description of some of the basic principles of X-ray diffraction.

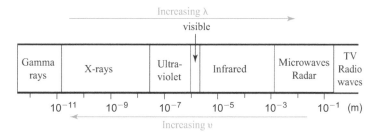

Figure 13.6 The electromagnetic spectrum.

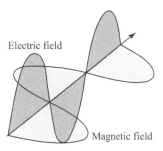

Figure 13.7 Electrical and magnetic components of electromagnetic radiation.

Table 13.3 **Summary of spectroscopic techniques using electromagnetic radiation**

Radiation (E = hυ)	Effect on atoms and molecules	Basis of technique
Radio waves	Transitions between nuclear energy levels	NMR spectroscopy
Microwaves	Transitions between molecular rotational energy levels	Rotational spectroscopy
Infrared (IR)	Transitions between vibrational energy levels	IR spectroscopy
Ultraviolet–visible (UV–vis)	Transitions between electronic energy levels	UV–vis spectroscopy
X-rays	1. Absorbance – knocks out inner-core electrons	X-ray photoelectron spectroscopy
	2. Reflected	X-ray diffraction

13.4 Ultraviolet and visible (UV–vis) spectroscopy

13.4.1 Introduction

Absorption of UV and visible light is associated with electronic transitions in the absorbing molecule. These transitions are usually between a ground state (lowest energy) bonding or lone pair orbital and an unfilled non-bonding or antibonding orbital. The energy difference between orbitals is given by $\Delta E = hc/\lambda$ so the wavelength of the absorption shows the difference in energy between the orbitals.

The highest energies involve s orbitals which absorb in the 120–200 nm range. Absorptions in this range, known as the vacuum UV (since air has to be excluded from the instrument), are difficult to measure and not very informative.

Above 200 nm (2×10^{-7} m), however, the electronic transitions of lone pair, d orbitals, π orbitals and particularly conjugated π systems (see Section 2.9 for an explanation of conjugation), give rise to absorptions which are extremely useful for structure elucidation and quantification of biological molecules.

Absorption

The basic equation which describes the absorption of light at a particular wavelength by a compound of concentration c (moles per litre) in a cell of length l (cm) placed in a beam is known as the Beer–Lambert Law, and is:

$$\log_{10}(I_o/I) = \varepsilon cl$$

where:

- I_o = the intensity of the incident beam;
- I = the intensity of the exiting beam;
- ε is the molar extinction (or absorption) coefficient which has units of $M^{-1}cm^{-1}(mol^{-1}L\,cm^{-1})$ but the units are frequently ignored. The extinction coefficient is characteristic of a specific absorption at a specific wavelength for a specific *chromophore* in the compound.

- The term $\log_{10}(I_0/I)$ is usually abbreviated to A, the absorbance.

Chromophore means light-absorbing group.

13.4.2 Measurement of the spectrum

UV spectra are typically recorded in very dilute solutions in solvents such as ethanol or water which do not absorb above 200 nm. In a double beam spectrophotometer (see Figure 13.8 for a schematic representation of this type of instrument) monochromatic light is split into two identical beams. The reference beam passes through a cell (typically 1 cm long) containing only the solvent while the sample beam passes through a solution of the sample in an identical cell. The spectrophotometer measures the ratio $\log_{10}(I_0/I)$ and expresses it in absorbance units. To record a complete spectrum the wavelength of the light is scanned from 200 to 700 nm. Since glass adsorbs UV light the cells are normally made of quartz but for spectra recorded at wavelength values of greater than 350 nm, glass and plastic cells can be used.

Peaks observed in a typical spectrum are broad. This is because the excitation of electrons is also accompanied by changes in vibrational and rotational states in the molecule.

13.4.3 Using UV–visible spectra for characterising compounds

Compounds with single chromophores (isolated C=C, C=O and lone pairs) absorb energy only in the vacuum UV region (below 200 nm) and cannot be observed using a conventional spectrophotometer. However, compounds with conjugated π-electron systems (for example, C=C—C=C and C=C—C=O; see Chapter 2) undergo $\pi \rightarrow \pi^*$ transitions above 200 nm. Table 13.4 shows the absorption wavelengths and ε values for a range of linear compounds containing conjugated double bonds. Two important trends are apparent here:

- As the number of conjugated bonds increases the electrons are less tightly held and undergo the transition at lower energy (higher wavelength).

- Since there are more electrons undergoing the transition the peaks get bigger (ε increases). A *rule of thumb* is that there is a 30 nm increase in the wavelength (λ_{max}) for every conjugated double bond and that ε increases by 10 000 for every double bond.

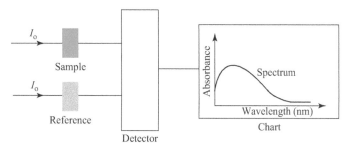

Figure 13.8 **A double beam UV-visible spectrometer.**

413

Table 13.4 **Examples of absorptions caused by $\pi \rightarrow \pi^*$ transitions**

Compound	λ_{max} (nm)	ε (approx)
$H_2C{=}CH_2$	160	10 000
$H_2C{=}CH{-}CH{=}CH_2$	218	20 000
$CH_3{-}CH{=}CH{-}CH{=}O$	222	20 000
$CH_3{-}CH{=}CH{-}CH{=}CH{-}CH_3$	225	20 000
$CH_3{-}CH{=}CH{-}CH{=}CH{-}CH{=}CH{-}CH_3$	255	30 000
$H_2C{=}CH{-}CH{=}CH{-}CH{=}CH{-}CH{=}O$	280	40 000
$H_2C{=}CH{-}CH_2{-}CH{=}CH_2$*	180	20 000

*is a non-conjugated diene.

To illustrate these points, consider the two naturally occurring compounds, β-carotene and lycopene.

β-Carotene

Lycopene

Both are coloured (the former is responsible for the orange colouration in carrots, and the latter for the reddish colour in tomatoes). It is possible to predict, with reasonable accuracy, the expected data for these.

Both carotene and lycopene have 11 conjugated double bonds. On the basis of adding 30 nm to the wavelength λ for each conjugated double bond, the simple way of doing this would be to take the value for ethylene (160 nm) and add on 300 nm for the other ten double bonds. This would give an estimated λ_{max} of 460 nm.

The values found for carotene and lycopene are, respectively, 452 nm and 469 nm. These are in the *visible* light region of the electromagnetic spectrum.

13.4.4 Aromatic compounds

This group of compounds does not behave in a way that would be immediately apparent on the basis of the guidelines given above. For example, benzene has a λ_{max} at 254 nm and an extinction coefficient $\varepsilon = 204$. It would be expected that, if it were to be a conjugated triene, it would have an ε of about 30 000. What is being observed at this frequency is a so-called 'forbidden' band transition which has a

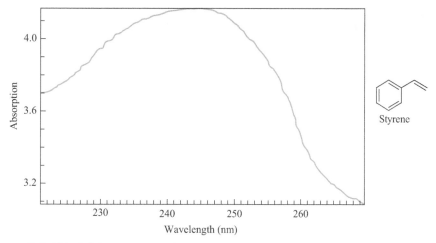

Figure 13.9 UV–visible spectrum of styrene.

much lower extinction coefficient. Monosubstituted benzenes have λ_{max} values between 250 and 290 nm and ε values between 200 and 4000. As the electron density of the substituent increases the λ_{max} and ε increase.

Polycyclic aromatic and heterocyclic compounds can have very complex spectra, the explanation for which is beyond the scope of this book. Figure 13.9 shows the UV–visible spectrum of the molecule styrene, which has an aromatic ring joined to an ethene. The λ_{max} value is about 250 nm, with an ε value of about 14 700.

13.4.5 Using UV–visible spectra for measuring concentrations of biologically important compounds

One of the main uses of UV–visible spectrophotometry in biology is to measure concentrations of materials and to follow changes of concentration with time. For example, suppose it is known that a pure compound has λ_{max} 320 nm with ε 50 000. If in a solution of this compound with unknown concentration (using a 1 cm cell), the measured absorption (A) is 2.2 at 320 nm, then since:

$$A = \varepsilon c l$$

it follows that:

$$c = A/\varepsilon l$$

$$= 2.2/50\,000$$

$$= 4.4 \times 10^{-5}\ \text{mol}\,\text{L}^{-1}$$

One important point to note is that this technique can be used even in very complex mixtures provided there are no other components with interfering absorptions at 320 nm.

13.5 Infrared (IR) spectroscopy

13.5.1 Introduction

Photons in the IR region of the electromagnetic spectrum (roughly from 2×10^{-6} to 25×10^{-6} m) do not have enough energy to promote electrons. Instead they excite vibrations in molecules, causing them to have a larger amplitude. The molecular vibrations occur at fixed frequencies which depend upon how heavy the atoms are and how strong the bonds are that connect the atoms together.

13.5.2 Measurement of the spectrum

IR spectrophotometers operate along similar principles to the UV–visible spectrophotometer described in Section 13.3. There are, however, practical difficulties involved in its construction and the data it produces are expressed in a different format.

Since both glass and quartz absorb IR radiation, then these cannot be used and the sample cells have to be constructed of alkali-metal halide crystals, such as NaCl, KBr or CsCl, which have no absorptions in the normal region of interest. Because most solvents have interfering absorptions, solution IR spectra are normally measured in $CHCl_3$ and CCl_4. Alternatively, pure liquid films or solid dispersions fused into KBr are used.

Units

By convention, IR absorption spectra are displayed 'upside down', with the wavelength expressed in reciprocal wavelength units. Thus an absorption found at a wavelength of 5×10^{-6} m ($=5 \times 10^{-4}$ cm) is expressed as 2000 cm^{-1} (as 2000 = $1/5 \times 10^{-4}$). In practice this is not difficult to handle since the spectrophotometer does the conversion.

13.5.3 Interpretation of IR spectra

By contrast with UV–visible spectra, which produce broad, or very broad, 'lines', the IR spectrum of a molecule is characterised by a large number of relatively narrow absorption bands (peaks) each of which corresponds to an individual bond vibrational frequency. The vast majority of organic compounds have a large variety of C—C and C—H bonds, which exhibit a complex set of vibrations, so IR spectroscopy is not used to identify the carbon skeleton of a molecule, but these regions can provide a fingerprint of the molecule, since no two compounds will have identical absorbances in them.

The major use of IR spectroscopy is to identify functional groups. Stretching vibrations occur in the order of bond strengths, with triple bonds > double bonds > single bonds in general. If the single bond stretch involves the very small hydrogen, it occurs at high frequency (or wavenumber).

Key regions of the spectrum, in which strong peaks may occur, can be identified as follows:

$$3200\text{–}3400\,\text{cm}^{-1} \quad \text{OH stretch}$$

$$3300\text{–}3500\,\text{cm}^{-1} \quad \text{NH stretch}$$

(These may be difficult to distinguish, especially as the positions can vary due to hydrogen bonding.)

$$3267\text{–}3333\,\text{cm}^{-1} \quad \text{CH stretch (alkyne CH)}$$

$$3020\text{–}3080\,\text{cm}^{-1} \quad \text{CH stretch (alkene CH)}$$

$$2850\text{–}2980\,\text{cm}^{-1} \quad \text{CH stretch (alkane CH)}$$

$$2200\text{–}2300\,\text{cm}^{-1} \quad \text{C}\equiv\text{N stretch}$$

(This is diagnostically useful, because not many groups absorb in this region.)

$$1600\text{–}1800\,\text{cm}^{-1} \quad \text{C}=\text{O stretch}$$

The carbonyl region is very important, and different carbonyl-containing functional groups absorb at different, though overlapping, parts of the overall range. Below are listed more detailed carbonyl absorbance regions:

Aldehydes (RCH=O)	$1690\text{–}1740\,\text{cm}^{-1}$
Ketones (R_2C=O)	$1680\text{–}1730\,\text{cm}^{-1}$
Ester (RCO_2R')	$1720\text{–}1750\,\text{cm}^{-1}$
Amide ($RCONR_2$)	$1630\text{–}1670\,\text{cm}^{-1}$

The region between 900 and $1500\,\text{cm}^{-1}$ is usually referred to as the fingerprint region and, as the name implies, it is a very complex area of overlapping peaks. Nitro compounds (RNO_2), however, have two intense peaks in the range 1500–1570 and 1300–1370 cm^{-1}.

If a compound contains two (or more) functional groups then the typical features of each group appear in the spectrum.

IR is now used routinely in studies of protein structure studies where the different C=O stretching frequency values can give valuable information on peptide (—CONH—) bond conformations.

Example 13.1 Two compounds with the molecular formula C_4H_8O are butan-2-one and but-3-en-2-ol, the infrared spectra of which are shown in Figure 13.10. There are many differences between the two, but the key features are indicated by an asterisk,*. In the ketone (spectrum a), a strong 'peak' is seen at $1704\,\text{cm}^{-1}$, consistent with the presence of a C=O bond. In the enol (spectrum b) the significant peak is the strong and very broad signal found at about $3370\,\text{cm}^{-1}$, consistent with the presence of an alcohol.

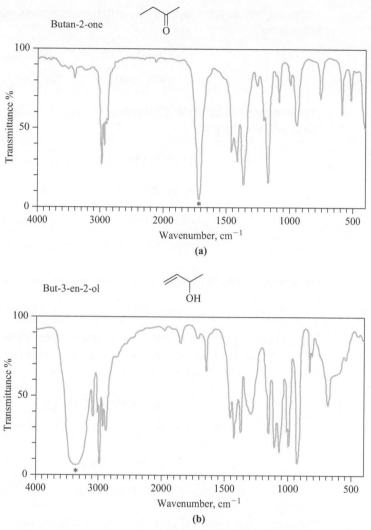

Figure 13.10 IR spectra of (a) butan-2-one and (b) but-3-en-2-ol.

13.6 Nuclear magnetic resonance (NMR) spectroscopy

13.6.1 Introduction and basic principles

The advent of modern high field magnets, coupled with the incredibly rapid development of computer technology, has meant that NMR has become an indispensable tool in chemistry, biology and medicine. The diversity of applications of NMR techniques in the chemical, biological and medicinal sciences are enormous, ranging from the detailed determination of protein, nucleic acid and oligosaccharide structures to three-dimensional imaging of abnormal cells in human organs (magnetic resonance imaging). This section will concentrate on the basic principles of NMR, with the aim of understanding how simple single dimensional NMR spectra of molecules can be obtained, analysed and interpreted to give detailed structural information.

Basic principles

- Many (but not all) nuclear isotopes (e.g. ^1H, ^{13}C, ^{31}P) possess *spin*. Nuclear spin is given the symbol I. As a consequence, nuclei have an associated spin angular momentum (like electrons; see Chapter 2).
- Spin gives magnetic properties to the nuclei; they behave like small magnets.
- In the presence of an external magnetic field (given the symbol B_0, the units of which are tesla, symbol T) only a small number of orientations of these nuclear magnets (with respect to B_0) are allowed. Figure 13.11 shows the situation for a nucleus with $I = \frac{1}{2}$, such as ^1H.
- Nuclei in different orientations have different energies. Those aligned against the applied magnetic field have the higher energy.
- The orientation can be changed by using electromagnetic waves whose energy quanta ($h\nu$) are the same size as the energy difference between the orientations (ΔE).

For a fixed external magnetic field strength (B_0):

- Different nuclear isotopes (e.g. ^1H, ^{13}C, ^{31}P) require very different frequencies (ν).
- Nuclei of the same isotope (e.g. ^1H) at different sites in a molecule require only very slightly different frequencies (ν).

Nuclear spin, *I*

In Chapter 2 the concept of electron spin ($s = 1/2$, $m_s = \pm 1/2$) was introduced. Protons and neutrons likewise have spin, and again for these, $s = 1/2$ and $m_s = \pm 1/2$. Nuclei are made up of protons and neutrons, so it is not unexpected that, in some cases, nuclei will also have a resultant spin. I is the symbol often used to represent the nuclear spin quantum number of the nucleus (S may also be used).

I can have values of 0, 1/2, 1, 3/2, . . . In other words nuclear spin, I, can have either integral and half-integral values, though *any particular nucleus has ONLY ONE value of* I. The nucleus will have a non-zero spin if it has an odd mass number or an odd atomic number, as shown in Table 13.5.

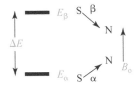

Figure 13.11 The separation of nuclear energy levels as a result of applying an external magnetic field. Note how the nuclear magnets can align either in the general direction of the applied magnetic field (the α orientation) or against it (the β orientation). They differ from bar magnets in this way.

Table 13.5 **Summary of relationship between atomic number, mass number and nuclear spin**

Atomic number	Mass number	Nuclear spin (I)
Even or odd	Odd	1/2, 3/2, 5/2 …
Even	Even	0
Odd	Even	1, 2, 3 …

Many of the biologically most important nuclei have $I = 1/2$, such as ^1H, ^{13}C and ^{31}P. In the absence of any externally applied magnetic field, there is no preferred orientation for a magnetic moment, and hence the energy levels are degenerate. In the presence of an externally applied magnetic field however, a nucleus of spin quantum number I has $2I + 1$ possible orientations when in a magnetic field. For a nucleus of spin $I = 1/2$ placed in a magnetic field this means there are $(2 \times 1/2 + 1) = 2$ orientations, and as each orientation has an associated energy, this means there are two energy states. For field strength of B_0 T, the difference in energy between the two energy levels (aligned with and against the field) is given by:

$$\Delta E = \gamma . B_o . h/2\pi$$

Thus energy level separation, illustrated in Figure 13.12, depends on:

- the gyromagnetic ratio, γ, which is an intrinsic property for each type of nucleus;
- the applied magnetic field, B_o.

Absorption of electromagnetic radiation of frequency v, such that $hv = \Delta E$, causes nuclei to move from one energy level to an adjacent energy level, and NMR experiments are usually referred to in terms of the operating absorption frequency, which is given by:

$$v = \gamma B_o/2\pi$$

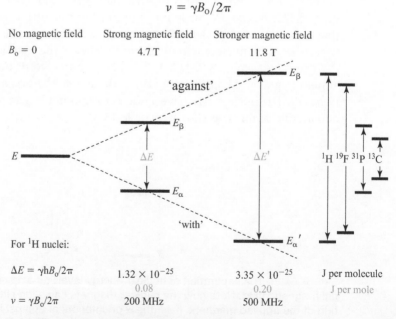

Figure 13.12 The relationships between ΔE, B_o and γ.

Therefore, summarising and developing the key points:

- On increasing the applied magnetic field the energy difference (and the frequency of the observed transition) gets bigger.

- The actual energy difference (ΔE) between the excited ('against') state and the ground ('with') state is extremely small, and even with extremely intense magnetic fields there are a significant proportion of nuclei in the excited state. Therefore, NMR is a relatively insensitive technique, though from the point above, increased field strength leads to increased sensitivity.

- For a given value of B_o the energy gap for different nuclei differs significantly; the diagram shows the relative gaps for ^1H, ^{19}F, ^{31}P and ^{13}C. The proton is, therefore, comparatively sensitive with respect to other nuclei (because the energy gap, ΔE, depends directly on the gyromagnetic ratio as shown in Table 13.6).

Also, importantly, isotopic abundance is proportional to sensitivity, which means that ^{13}C is relatively insensitive, as it forms only approximately 1% of the natural abundance of carbon.

13.6.2 Design of the NMR spectrometer

Essentially the NMR instrument is made up of a magnet, a probehead (containing transmitter and receiver coils), a radiotransmitter, a receiver and a computer (Figure 13.13).

Table 13.6 **Magnetic properties of some nuclei**

Isotope	Natural abundance	Spin (I)	Gyromagnetic ratio, $\gamma (\times 10^{-7} rad\, T^{-1} s^{-1})$
^1H	99.98	1/2	26.752
^2D	0.016	1	4.106
^{10}B	18.83	3	2.875
^{11}B	81.17	3/2	8.583
^{12}C	98.89	0	–
^{13}C	1.11	1/2	6.762
^{14}N	99.63	1	1.933
^{15}N	0.037	1/2	– 2.711
^{16}O	99.76	0	–
^{17}O	0.037	5/2	– 3.627
^{19}F	100	1/2	25.167
^{27}Al	100	5/2	6.971
^{28}Si	92.28	0	–
^{29}Si	4.67	1/2	– 5.314
^{31}P	100	1/2	10.829
^{195}Pt	33.7	1/2	5.752

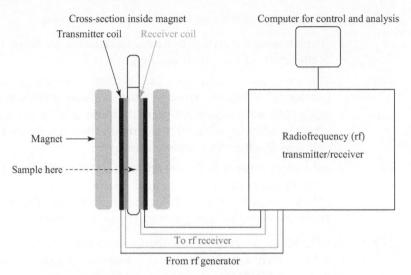

Figure 13.13 **Representation of an NMR spectrometer.**

Summary of the procedure

● The sample under analysis is placed inside the probehead in the magnetic field. A short (in the order of microseconds) pulse of radiofrequency is sent to the sample. This flips the nuclear spins.

● When the transmitter is turned off the nuclei return (decay) to the ground state populations. This is accompanied by emission of radiation of specific frequency values (each frequency corresponding to the energy of a spin transition).

● These are detected by a sensitive receiver coil, amplified by a powerful amplifier. The signal is converted by a mathematical process called Fourier transformation, using a computer, to give a frequency plot.

Since the energies of nuclear spin transitions for different nuclei are quite distinct (remember that $\Delta E = \gamma \cdot B_0 \cdot h / 2\pi$ where γ is different for each nucleus), then the system can be set up to detect transition frequencies from different magnetic nuclei, say ^1H, ^{13}C or ^{31}P, thus generating a proton, carbon, or phosphorus NMR spectrum of a sample.

The main focus from now will be on the most heavily studied nucleus, namely the proton, ^1H.

13.6.3 The ^1H NMR spectrum

For the sample $CH_3.CO.CH_3$ (propanone) the ^1H NMR spectrum that is obtained comprises one line, as shown in Figure 13.14. However, for the sample $CH_3.CO.O.CH_3$ (methyl ethanoate, also called methyl acetate), looking at its spectrum over the same range gives two lines (Figure 13.15).

The rationale for this lies in the fact that, for any nucleus, the resonance frequency is directly proportional to the magnetic field experienced by the nucleus,

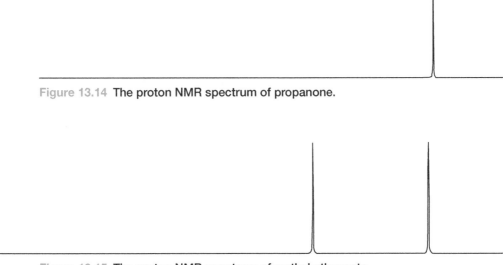

Figure 13.14 **The proton NMR spectrum of propanone.**

Figure 13.15 **The proton NMR spectrum of methyl ethanoate.**

but the field experienced by the nucleus is not necessarily exactly B_o. The electrons within atoms have the property of spin and therefore they also act as mini-magnets. This local magnetic field opposes the external applied field and therefore:

$$B_{local} < B_o$$

The exact frequency depends on the *site of the nucleus in the molecule*. The effect is known as *nuclear shielding*. This is most readily illustrated by referring to the structures of propanone and of methyl ethanoate (Figure 13.16).

In $CH_3.CO.O.CH_3$ the ^1H nuclei on the CH_3 group next to the $C=O$ experience a local magnetic field which can be represented by $(B_o - b_2)$ and the ^1H nuclei on the CH_3 group next to the O experience a different local field, which can be given by $(B_o - b_1)$. The magnitude of B_o is much greater than that of either b_1 or b_2, so the differences are tiny, but they are sufficient to change the energy of the transition and hence the radiofrequency of the signal. Thus two resonances, corresponding to the two magnetically dissimilar environments of the ^1H nuclei in the molecule, are found in the proton spectrum. The ^1H nuclei on the two methyl groups are said to be in *different magnetic environments*.

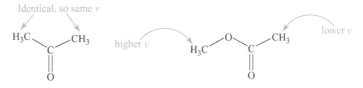

Figure 13.16 **The structures of propanone and of methyl ethanoate.**

423

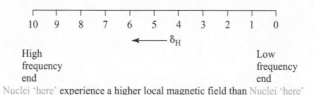

Figure 13.17 **Relation between chemical shift scale and frequency (see Box 13.2 for more detail).**

Because the environments of the two methyl groups in propanone are identical, it means that in the proton NMR spectrum of CH_3OCH_3 there is only one signal (resonance) since all of the ^1H nuclei in the molecule are in similar magnetic environments.

13.6.4 The chemical shift

Because ν_H (see Box 13.1) depends on the spectrometer field (B_o) as well as the site in the molecule, it makes it impossible to directly compare frequency results from spectrometers with different field strengths. (In modern systems, magnetic field strengths vary from 5 to 18 T). This potential problem was overcome, however, by using a *ratio* of two frequencies, one of which, ν_{STD}, is from the protons in a universally accepted 'standard' molecule, the other, ν_H, the frequency of the proton of interest leads to a dimensionless number which is independent of B_o and depends only on the site of the proton in the molecule of interest.

This number is called the *chemical shift*, and the definition of the chemical shift (symbol δ_H for protons) of the nucleus is:

$$\delta_H = \left\{ \frac{\nu_H - \nu_{STD}}{\nu_{STD}} \right\} \times 10^6$$

This is the fractional change in frequency *in parts per million* (ppm) *from* that of the standard *to* that of the nucleus of interest.

The standard chosen for protons (and also ^{13}C and ^{29}Si) is tetramethylsilane, $(CH_3)_4Si$, TMS. It is used because:

- ν_{TMS} is lower than ν for almost all other H signals.
- It gives a single sharp line.
- It is very volatile, and so easily removed from samples.
- It is very unreactive.

Chemical shift (δ_H), like frequency (ν_H) increases from right to left, as represented in Figure 13 17. Therefore nuclei giving signals on the 'left' absorb radiation at a higher absolute frequency, so a resonance at 3.2 ppm is said to be *to high frequency of* a resonance at 3.1 ppm.

Box 13.1

Local diamagnetic shielding – the background

The spectrometer field B_o causes the electron clouds to precess, i.e. the electrons circulate around the nucleus causing a small current which induces a very small magnetic field B_{IND} which is in the *opposite* direction to B_o.

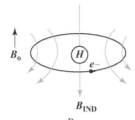

B_{local}	=	B_o	−	B_{IND}
magnetic field		external magnetic		electron-induced field
experienced by the nucleus		field		

The strength of B_{IND} depends on (and is proportional to) B_o. Thus
$$B_{IND} = \sigma_H B_o$$
and so
$$B_{local} = B_o(1 - \sigma_H)$$
Note: σ_H is called the *shielding constant* (it is very small, approximately 10^{-4} to 10^{-6}) and is a measure of how much the spectrometer field is reduced by the local electrons; its magnitude depends upon the electron density.

The resonance frequency $v_H = \gamma B_{local}/2\pi$

i.e. $$v_H = \gamma B_o(1 - \sigma_H)/2\pi$$

Box 13.2

How to convert shift differences to frequency differences and vice versa

Let us compare the results for an instrument operating at 4.7 T (200 MHz) with those for an instrument operating at 11.8 T (500 MHz).

Consider the situation where two signals have a chemical shift separation of x ppm (ppm is parts per million). This means that:

at 1 MHz, 1 ppm is 1 Hz (remember 1 MHz is 10^6 Hz)

so

at 200 MHz, 1 ppm is 200 Hz, hence x ppm is $(x \times 200)$ Hz

and

at 500 MHz , 1 ppm is 500 Hz, hence x ppm is $(x \times 500)$ Hz

So, for two signals separated by 3 ppm, using an instrument operating at 200 MHz, the separation is $(3 \times 200) = 600$ Hz, whereas using an instrument operating at 500 MHz, the separation is $(3 \times 500) = 1500$ Hz.

Comments on ^{1}H chemical shift differences

The chemical shift (ppm) value provides a lot of information about the chemical environment of specific hydrogen sites in the molecule. For example the methyl group hydrogen atoms in $(CH_3)_4C$, $(CH_3)_3N$ and $(CH_3)_2O$ have ^{1}H chemical shifts

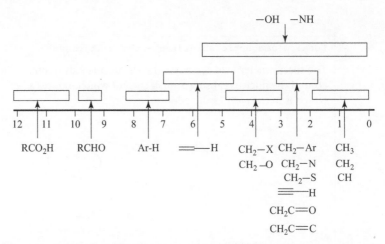

Figure 13.18 Some typical chemical shift ranges for ^{1}H NMR signals in organic molecules.

(δ_H) of 1.0, 3.5 and 4.0 ppm. Figure 13.18 gives a guide to the range of expected chemical shifts for the ^{1}H signals for hydrogen atoms in different chemical environments.

Looking at the shifts identified in Figure 13.18, a number of patterns emerge:

- **Electronegativity of neighbouring atoms.** Consider the difference in chemical shifts for the $C-CH_3$ protons compared with the $O-CH_3$ protons in methyl acetate. Oxygen is more electronegative than carbon, and electronegative neighbours are said to have a deshielding effect (they 'pull away' the electrons that would otherwise shield the nucleus). This has the effect of moving the signals to higher chemical shifts (*to the left*). This effect wears off quickly as the electronegative atom is more remote.

- **The nature of bonds.** Bonding electrons are affected by the applied magnetic field, and this has a resultant impact on shielding. Thus there is a significant effect on the chemical shifts of ^{1}H signals associated with alkenic or aldehyde H atoms, both of which are significantly deshielded compared with signals arising from alkyl hydrogens.

- **Delocalisation.** The chemical shift values of hydrogen nuclei associated with aromatic rings, such as hydrogens on phenyl groups, are also found at relatively high frequency values, indicating that they are also deshielded relative to alkyl hydrogens.

13.6.5 Peak areas – integration

The ^{1}H spectrum of a compound can also reveal information about the proportion of hydrogen nuclei in different environments. Consider, for example, the spectrum of methylbenzene (toluene) shown in Figure 13.19. This consists of a signal (may look broadened) at about 7 ppm, due to the five magnetically similar (but not identical) phenyl hydrogens, and a signal at about 1.5 ppm, arising from three magnetically equivalent hydrogens on a carbon bonded to the ring.

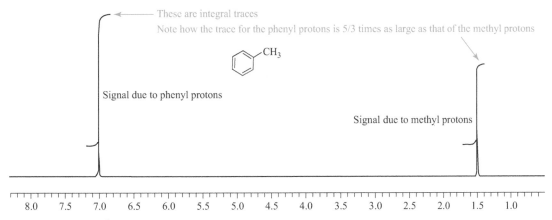

Figure 13.19 The ^{1}H NMR spectrum of toluene.

In addition to the ^{1}H NMR spectrum showing two resonances as described above, by comparing the *areas under each peak* it is possible to determine the *ratio* of the different types of hydrogen in the molecule. Here the peak areas have a ratio of 5:3. Modern NMR instruments display this as an integral and assign a value to the areas.

13.6.6 The solvent

NMR is not restricted to liquid samples, but most spectra of solid compounds are determined in solution. It is difficult to measure ^{1}H NMR spectra of compounds in solvents containing hydrogen atoms, because the solvent is present in such vast excess that the solvent signal dominates the spectrum and makes it difficult to see signals from the solute (though this is less problematic with modern instruments). Spectra are, therefore, normally determined in deuterated solvents such as D_2O or $CDCl_3$, which do not have a ^{1}H NMR spectrum. There is a trace of protonated material in these solvents so, for example, the spectrum of a compound in deuterated chloroform ($CDCl_3$) will show a small peak at 7.25 ppm which results from a small (ca. 0.1%) impurity of $CHCl_3$ in the deuterated chloroform.

13.6.7 Exchangeable hydrogens

Compounds like ethanoic acid (CH_3COOH), methanol (CH_3OH) and methylamine (CH_3NH_2) dissociate relatively easily. The O—H and N—H bonds are relatively weak. When the ^{1}H spectra of these compounds are recorded in $CDCl_3$, signals due to the O—H or N—H protons are detected, but when a small amount of D_2O is introduced into the samples and their spectra re-recorded these ^{1}H signals disappear. The reason is that the O—H (or N—H) protons are exchanged for D, and the resultant deuterated sites are not detectable using ^{1}H NMR.

This is an important application. Exchange with D_2O can be used to determine which protons are directly bonded to O or N. This technique is especially valuable in NMR studies of proteins where the rate of exchange of amide (—CONH—) hydrogens is used to determine whether a particular hydrogen lies on the solvent exposed surface of the protein or is buried in the hydrophobic core (for which exchange occurs very slowly).

13.6.8 Nuclear spin–spin coupling

Frequently more than one line is seen for an NMR signal. Consider the example in Figure 13.20, the ^{1}H NMR spectrum of a molecule involving just two different hydrogen atoms, such as dichloroethanal.

This spectrum comprises two signals, due to H_A and to H_X, but each signal consists of two lines. When a signal is made up of two lines it is called a *doublet*.

Rationale for splitting of signals into multiple lines

As described earlier, in the presence of a magnetic field, nuclei can align either 'with' (↑) or 'against' (↓) the applied magnetic field B_o (they adopt α and β states). This situation applies to both H_A and to H_X.

Consider the effect of this on the signal arising from nucleus H_A

H_A 'senses' the alignment of H_X, and this information is transmitted *through the electrons in the bonds* linking H_A to H_X. Thus, H_A will experience a field B_{local} and also a field (b) due to H_X. Depending on the orientation of the H_X nuclei (either 'with' (↑) or 'against' (↓)), the field experienced by H_A will be either

$$B_{local} + (b) \qquad \text{or} \qquad B_{local} - (b)$$

Because the energy gap between nuclear energy levels is so small, the populations of α and β states in a sample containing many molecules are almost identical.

Thus, in approximately half of the molecules in an NMR sample, H_A will sense H_X in the α state, and in the other half of the molecules in the NMR sample H_A will sense H_X in the β state. Thus, two lines of equal size are observed, as shown in Figure 13.20.

The same argument holds for the signal due to H_X, which also shows two lines of equal intensity. This process is called *spin–spin coupling*, and the separation of each pair of lines is known as the *coupling constant*, J (called J_{AX} in the case described above). Hence the spectrum of dichloroethanal comprises two doublet signals due to H_A and to H_X.

What happens if more protons are present?

If the approach described above is extended to look at molecules containing three hydrogen atoms, it is possible to consider two different scenarios where coupling is found; one where an H couples to two identical Hs, such as 1,1,2-trichloroethane and another, such as bromoethene, where there are three distinct H environments.

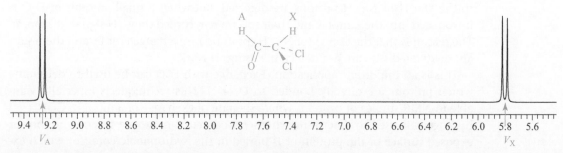

Figure 13.20 The ^{1}H NMR spectrum of dichloroethanal.

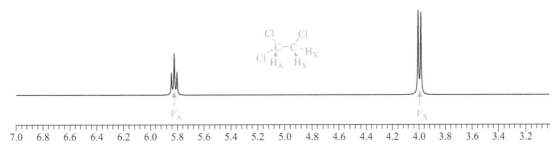

7.0 6.8 6.6 6.4 6.2 6.0 5.8 5.6 5.4 5.2 5.0 4.8 4.6 4.4 4.2 4.0 3.8 3.6 3.4 3.2

Figure 13.21 The ^{1}H NMR spectrum of 1,1,2-trichloroethane.

1,1,2-trichloroethane

Take the spectrum obtained for 1,1,2-trichloroethane firstly. In this molecule the hydrogen atoms on the C-2 carbon (labelled 'X') are both chemically and magnetically identical to each other, giving rise to the signal at 3.96 ppm, whereas the hydrogen on the C-1 position (labelled 'A') differs from them, giving rise to the signal at 5.76 ppm. The spectrum is shown in Figure 13.21.

Extending the approach described for the analysis of the spectrum of dichloroethanal, it is clear that in the above the signal for H_A will be affected by the presence of the two H_X hydrogens.

- Assume that ↑ represents H_X nuclei aligned 'with' the applied magnetic field.
- Assume that ↓ represents H_X nuclei aligned 'against' the applied field.

There are four possible ways that H_A can sense the alignments of the H_X nuclei:

These are identical

Because the two 'antiparallel' options are identical, the signal of H_A contains three lines, of relative intensity 1:2:1.

The signal arising from the two equivalent H_X protons will only interact with the single H_A proton, so will be a doublet. *Magnetically equivalent protons do not show coupling to one another.* Hence there is no coupling between the pair of H_X protons.

Bromoethene

The hydrogen atoms in this molecule are all in chemically (and hence magnetically) distinct environments. In the ^{1}H NMR spectrum, in $CDCl_3$ solution, three signals are found, at 5.84 ppm, 5.97 ppm and 6.44 ppm. Each of these three signals is made up of four lines, as shown in Figure 13.22.

Consider the appearance of the signal due to H_A. It will sense both H_M and H_X. There are four possible ways that H_A can sense the alignments of the H_M and H_X:

$$
\begin{array}{cccc}
\text{M X} & \text{M X} & \text{M X} & \text{M X} \\
\uparrow\uparrow & \uparrow\downarrow & \downarrow\uparrow & \downarrow\downarrow
\end{array}
$$

Each arrangement is equally probable, and this time the two 'antiparallel' arrangements are non-identical, so there are four lines found for H_A (referred to as a *doublet*

429

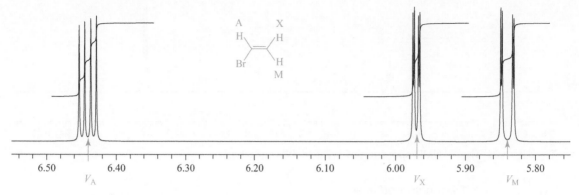

Figure 13.22 The ^{1}H NMR spectrum of bromoethene.

of doublets) which are (approximately) the same height, and hence the NMR signal for H_A would be a four-line multiplet.

By the same reasoning there are four lines for H_M and four lines for H_X.

Some features about coupling constants

- Coupling constants, J, are measured in Hz.
- The magnitude of J is independent of the applied magnetic field B_o.
- In ^{1}H NMR spectra, the magnitude of J depends on the geometrical relationship between the coupled protons.
- The magnitude of J depends on the number of bonds separating the coupled protons. For coupling between protons they are often not seen when the number of bonds separating the nuclei is more than 3.
- Coupling constants are mutual. If the splitting of H_A by H_X is, say, 6 Hz, then the splitting of H_X by H_A will also be 6 Hz.

The N + 1 rule

For a signal from a proton coupling to 'N' other nuclei, if all of the J-values are the same (usually if the nuclei coupling to the proton are equivalent), then $N + 1$ lines are seen, the intensities of which are given by Pascal's triangle, as shown below:

Number of coupled nuclei	Pattern	Name of pattern
$N = 0$	1	A singlet
$N = 1$	1 1	A doublet
$N = 2$	1 2 1	A triplet
$N = 3$	1 3 3 1	A quartet
$N = 4$	1 4 6 4 1	A quintet
$N = 5$	1 5 10 10 5 1 etc.	A sextet

(i.e. if $N = 0$, the proton is uncoupled and gives a single line; if $N = 1$ the proton couples to one other proton, giving a doublet; if $N = 2$, there are two couplings of the same magnitude, and so the signal has three lines of ratios 1:2:1).

Table 13.6   **Typical proton–proton coupling constants**

(free-rotation)	6–7 Hz	12–20 Hz
(cis)	7–12 Hz	0–3 Hz
(trans)	13–18 Hz	
(free rotation)	6–13 Hz	Aromatic nuclei J_{HH} ortho 7–9 Hz meta 2–3 Hz para 0–1 Hz
	0–3 Hz	
	5–8 Hz	

Coupled H nuclei indicated in orange.

Some typical values of J are shown in Table 13.6.

Example 13.2 Determine the structure of C_3H_7Cl from the spectrum shown in Figure 13.23

This is tackled by paying attention to the key data from the spectrum, namely the:

- integrals;
- chemical shifts;
- coupling patterns and constants.

Taking the information in turn:

- Integrals: Consistent with 6:1 ratio for signals at 1.8 ppm and 4.3 ppm respectively.
- Shifts: Signal at 4.3 ppm (the high frequency one) is most deshielded, so will derive from H nearest to electronegative Cl.
- Couplings: Signal at 4.3 ppm is a 6.5 Hz septet. Applying the 'N+1 rule' (Pascal's triangle) it follows that coupling is to six equivalent Hs. Signal at 1.8 ppm is a 6.5 Hz doublet – coupled to one H.

431

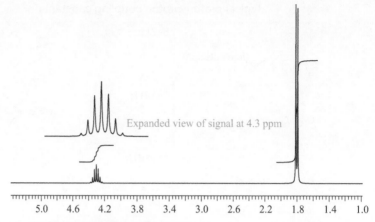

Figure 13.23 The ^{1}H NMR spectrum of C_3H_7Cl.

The formula of C_3H_7Cl has only two possible structures, these being:

(A)   Cl or   **(B)** Cl

Of the two possible structures, there are three types of H in A and two types in B (the $2 \times CH_3$ groups are identical in B). Therefore structure B is the one consistent with the spectrum.

Structure B would exhibit two signals; one due to six Hs associated with two equivalent CH_3 groups, and the other due to the CH proton on the same C as the Cl group. The six Hs associated with two Me groups will couple with the CH proton (and, of course, vice versa). The CH near to the Cl would be deshielded, and be found at higher frequency than the other signal. This is entirely consistent with the spectrum shown in Figure 13.23.

13.6.9 ^{13}C NMR spectroscopy

^{1}H NMR spectroscopy is the most studied form by far, but other nuclei also can reveal a lot of helpful information. Despite problems of sensitivity, one very useful form of NMR spectroscopy is that where ^{13}C is the nucleus being studied.

In a ^{13}C NMR spectrum, each different type of carbon atom in a molecule comes as a separate peak, and its chemical shift (represented by δ_c) is normally found between 0 and 220 ppm (see Figure 13.24; the range is much greater than a proton). There are four different types of carbon atom possible in a compound; a CH_3 (methyl), a CH_2 (methylene), CH (methine) or C (quaternary).

In the normal way of recording the spectrum, the effects of interactions between carbon nuclei and proton nuclei are removed (the interactions are spin–spin coupling, and their removal is therefore known as broadband decoupling), and the

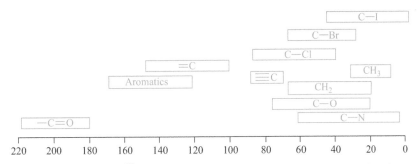

Figure 13.24 Some typical ^{13}C NMR chemical shifts found in organic molecules.

signals obtained in this way are single lines, though quaternary carbon signals are often distinguishable by being smaller than the others. It is, however, possible to adjust certain experimental parameters to perform an experiment called a *DEPT* experiment (DEPT is taken from Distortionless Enhancement by Polarisation Transfer) such that:

- only carbons associated with CH are recorded (using a form of the experiment called a DEPT $\pi/2$ experiment); or
- carbons from CH and CH$_3$ show positive peaks and those associated with CH$_2$ show negative ones (called a DEPT $3\pi/4$ experiment).
- Quaternary carbons do not show up in the DEPT experiments.

Hence, all carbons in a molecule can be distinguished by looking at the standard spectrum and the two different DEPT spectra. To illustrate this, Figure 13.25 shows the standard spectrum and the two different DEPT spectra obtained for

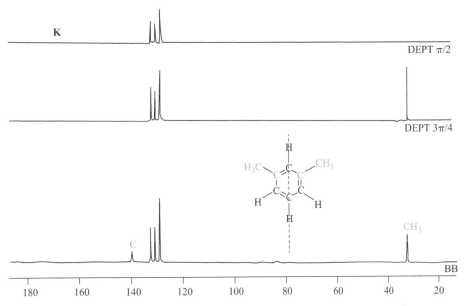

Figure 13.25 The standard carbon NMR spectrum of 1,3 dimethylbenzene (bottom trace), and two DEPT spectra.

433

1,3-dimethylbenzene. This molecule possesses an axis of symmetry, as indicated by the dotted line shown in the figure, and hence there are five different C environments. Three key points arise from analysis of the spectra.

- There are three protonated carbon environments, and the signals for these are found in all three spectra.

- The assignments of the signals at approximately 20 ppm and at 136 ppm approximately are confirmed as being due to the CH_3 and the quaternary carbons respectively. The methyl carbon presents a positive signal in the DEPT $3\pi/4$ experiment, but is absent in the DEPT $\pi/2$ spectrum.

- The quaternary signal is absent from either DEPT spectrum.

13.7   X-ray diffraction

X-ray diffraction is one of the most important tools at a scientist's disposal to positively identify an unknown compound. It is the only method that generates a precise 'picture' of the molecule as its output, *and* not just connectivity information (what order the atoms are connected), but also structural information (the distances and angles between atoms). Substances from the smallest molecules to large proteins and viruses have all been studied by this method (see Figure 13.26 for an example).

X-rays are diffracted by the core electrons associated with atoms. An X-ray crystal structure is, therefore, really a map of the electron density in a crystal. The larger an element, the more core electrons it possesses, and hence the more effective the data obtained.

The basic principle of this method requires that the sample to be analysed must be a crystal (meaning that the molecules are regularly orientated on a three-dimensional lattice, and the repeating unit is called the *unit cell*). When the X-rays scatter off different planes in the crystal the emerging waves will either be in phase

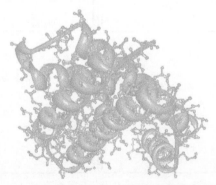

Figure 13.26 The structure of one haem unit in haemoglobin, obtained by X-ray diffraction.

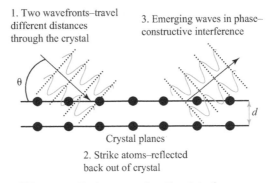

Figure 13.27 **Effect of X-ray scattering: constructive interference.**

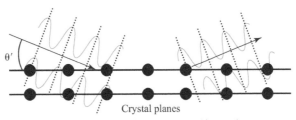

Figure 13.28 **The effect of changing the angle of incidence, resulting in destructive interference.**

(giving constructive interference) or out of phase (giving destructive interference). The stages by which this occurs are summarised in Figures 13.27 and 13.28.

It is possible to determine whether constructive or destructive interference takes place, and the criteria for this are established by Bragg's law:

$$2d \sin\theta = n\lambda$$

where

- d = distance between the diffracting planes,
- θ = angle of incidence
- n = reflection order (usually value of 1)
- λ = wavelength of X-rays.

The resulting data are collected as diffraction patterns (see Figure 13.29), and from the position of the white spots (which are constructive interference points) the dimensions of the unit cell can be determined. The intensity of the spots allows the identity and the positions of the atoms to be determined.

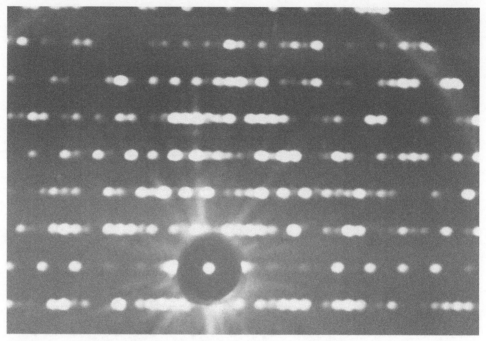

Figure 13.29 **A typical diffraction pattern for a molecular crystal.**

(Pasieka/Science Photo Library)

13.8   Summary of the techniques

- Mass spectrometry gives some information as to how those groups are joined together *and* measures the gram formula mass of the compound (from the molecular ion peak).

- Infrared and UV–vis spectroscopy help to identify the functional groups present.

- NMR spectroscopy is often used to obtain information on the structure of the carbon-hydrogen framework of the molecules. It also can be used to determine/confirm functional group identity. More advanced techniques, beyond the scope of this book, allow for full 3D structural information to be obtained.

- X-ray crystallography gives a picture of a molecule, and therefore its connectivity. It gives precise values of bond lengths and angles, and also provides information about intermolecular interactions (the structure is of a crystal, not isolated molecules). It does, however, require crystalline solid material.

In many current research laboratories, when investigating structural and functional properties of biomacromolecules such as proteins or nucleic acids, combinations of mass spectrometry, NMR spectroscopy and X-ray crystallography are often used in parallel.

Questions

13.1   A liquid compound gives a mass spectrum in which the molecular ion appears as a pair of equal intensity peaks at $m/e = 108$ and $m/z = 110$. Small fragment ion peaks are seen at $m/z = 93$ and 95 (approximately equal intensity), and at $m/z = 79$ and 81 (approximately the same size). Large fragment ions are seen at $m/z = 29$ (base peak), 28, 27 (also large) and 26. Suggest a name for this compound.

13.2   The following mass spectrum was acquired from a compound with a molecular formula of $C_3H_6O_2$.

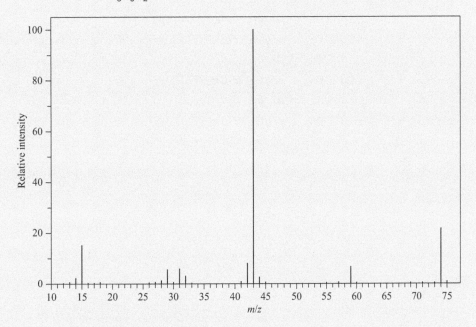

(a)   Identify the molecular ion.
(b)   Identify the base peak.
(c)   Explain the fragmentation peaks and hence suggest a structure for the compound.
(d)   Suggest how other spectroscopic techniques could be used to confirm the assignment.

13.3   Which of the following compounds would give a UV absorption in the 200–400 nm region?

CH$_3$

(I)   (II)   (III)   (IV)

(V)   (VI)

13.4 A solution of 4-methyl-penten-2-one in ethanol shows an absorbance of 0.2 at 236 nm in a cell with a 1 cm path length. Its ε value at 236 nm is 12 600. Calculate the concentration of the solution.

13.5 A student was given a flask containing exactly 250 mL of a solution (called solution A). The student then took 10 mL of this solution and diluted it to 100 mL (call this solution B). They ran a UV–visible spectrum on solution B. The compound has a quoted molar absorbance (ε) value of 6990, and the spectrum gave an absorbance of 0.25. Given that the UV cell has a path length of 1 cm, what is the concentration of solution A?
(Note that this is not a single step calculation.)

13.6 State how you would use IR spectroscopy to distinguish between the following pairs of isomers:

(a) $(CH_3)_3N$ and $CH_3CH_2NHCH_3$

(b) $CH_3CH_2CCH_3$ and $CH_3CH=CHCH_2OH$
   ($\overset{\|}{O}$)

(c) $H_2C=CHOCH_3$ and $CH_3CH_2CH=O$

(d)

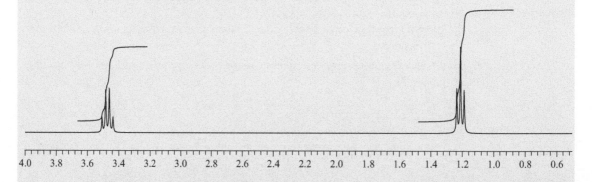

13.7 Which band of electromagnetic radiation is absorbed in NMR spectroscopy?

13.8 Explain the appearance of the ^1H NMR spectrum of diethyl ether, $CH_3.CH_2.O.CH_2.CH_3$, shown below:

4.0 3.8 3.6 3.4 3.2 3.0 2.8 2.6 2.4 2.2 2.0 1.8 1.6 1.4 1.2 1.0 0.8 0.6

13.9 (a) A student is trying to identify an unknown aromatic compound as part of a project. From elemental analysis of the compound, it is known to contain only carbon, hydrogen and bromine.

(i) The molecular ion (M$^+$) in the mass spectrum appears as a pair of equal intensity peaks at 170 and 172 amu (atomic mass units), and the base peak was found at m/z = 91. Assuming that the atomic masses for C, H and Br are 12, 1 and 79/81 amu, respectively, state the molecular formula of the unknown compound.

(ii) Sketch four plausible structures for the unknown compound based on this molecular formula.

(b) The student now runs some spectroscopic experiments to make a definitive identification of the unknown compound in part (a). Account for the following spectroscopic data.

(i) UV–visible spectroscopy gave a peak with a maximum at 254 nm.

(ii) ^{1}H NMR spectroscopy gave resonances at δ = 2.3 ppm (singlet, integral corresponding to 3H) and δ = 7.0 (complex signal pattern, integral = 4H).

(iii) ^{13}C NMR spectroscopy gave five different resonances.

(c) Taking all of the experimental evidence together, propose a structure for the unknown compound.

Appendix 1: Basic mathematical tools for biological chemistry

Introduction

Mathematics helps all scientists to understand the processes that they observe. In the context of biology and chemistry there are several areas where a mathematical look at things can help to illustrate what is happening.

In many cases the mathematical principles needed for these purposes are not, in themselves, very difficult, but it is not always immediately obvious how to apply them to the problems being dealt with. There are certain core skills that will be reviewed throughout the text, but in the context of this text the main focus will be at the mathematical tools needed to understand:

- chemical equilibria (including pH, buffers, etc.);
- energetics (the concepts of heat and energy);
- kinetics and biocatalysis (how fast reactions/biological processes happen).

Rearranging equations

Sometimes mathematical expressions are found in a format that is not easy to handle. In these cases an important skill is to be able to change the format to something a bit more straightforward.

The key fact to remember is that an equation means that what is on one side is equal to what is on the other side. Therefore, if something is done to one side of the equation, the same thing must be done to the other side.

Example A.1 Given the following equation linking y and x:

$$\frac{y}{2} = 3x + 1$$

This would be much more useful if the left hand side were just y. To achieve this, both sides of the equation must be multiplied by 2, so:

$$\frac{y}{2} \times 2 = (3x + 1) \times 2$$

$$y = 6x + 2$$

The same ideas apply even if the situation becomes more complex.

Example A.2

$$\frac{y + 1.3}{2} = 2x + 3$$

Multiply both sides by 2:

$$\frac{y + 1.3}{2} \times 2 = (2x + 3) \times 2$$

Giving

$$y + 1.3 = 4x + 6$$

Now subtract 1.3 from both sides, to give:

$$y = 4x + 4.7$$

Example A.3 Now to look at a 'real' example, namely the rearrangement of the Michaelis–Menten equation into the expression to produce a Lineweaver–Burk plot (done in order to ensure a straight line plot can be used).

The Michaelis–Menten expression is usually shown as:

$$v = \frac{V_{max} \times [S]}{K_m + [S]}$$

(The definitions of these terms are provided in Chapter 9.)

The components of this expression that are measurable during the experiment are [S] and v, and we need to be able to see, in the simplest possible way, how one varies with the other. The Lineweaver–Burk plot is one method of achieving this, and is given by:

$$\left(\frac{K_m}{V_{max}} \times \frac{1}{[S]} \right) + \frac{1}{V_{max}} = \frac{1}{v}$$

To go from the Michaelis–Menten form to the Lineweaver–Burk form, first multiply both sides of the Michaelis–Menten equation by $(K_m + [S])$, so:

$$v = \frac{V_{max} \times [S]}{K_m + [S]}$$

goes to:

$$v (K_m + [S]) = V_{max} \cdot [S]$$

The left hand side is then multiplied out:

$$K_m v + [S]v = V_{max} \cdot [S]$$

Now, divide both sides through by $V_{max} \cdot [S]$

$$\frac{K_m v}{V_{max}[S]} + \frac{v[S]}{V_{max}[S]} = 1$$

Cancel [S]s in $\dfrac{v[S]}{V_{max}[S]}$ to give:

$$\frac{K_m v}{V_{max}[S]} + \frac{v}{V_{max}} = 1$$

Finally, divide both sides by v

$$\left(\frac{K_m}{V_{max}} \times \frac{1}{[S]}\right) + \frac{1}{V_{max}} = \frac{1}{v}$$

Self-test

Rearrange the Michaelis–Menten expression into the Eadie–Hofstee form:

$$v = -K_M \frac{v}{[S]} + V_{max}$$

Equation of a straight line

One of the most important mathematical expressions that you will use, not just in chemistry but in many biology-related subjects, is the equation of a straight line, a useful form of which is:

$$y = mx + c$$

Take the example below:

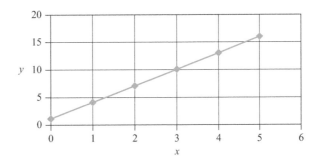

The equation for this line is

$$y = 3x + 1$$

where m = 3 and c = 1.

From the graph we see that

for a value of $x = 2$, then $y = 7$
for a value of $x = 4$, then $y = 13$

Therefore,

- as x changes by 2 units, y changes by 6 units

or, put another way,

- for a change in x of 1 unit, there is a corresponding change in y of 3 units.

The value of 3 is the gradient (slope) of the line, and corresponds to m.

Note also that for a value of $x = 0$, $y = 1$. This is known as the intercept on the y axis, and corresponds to the c part of the expression

$$y = mx + c$$

The value of m, the slope, can be deduced from a graph by making use of the relationship:

$$m = \frac{y_2 - y_1}{x_2 - x_1}$$

where the different x and y coordinates are indicated thus:

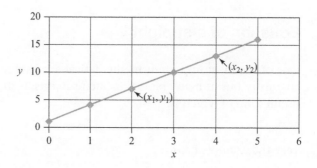

Self-test

Estimate the values of m and c from the graph below:

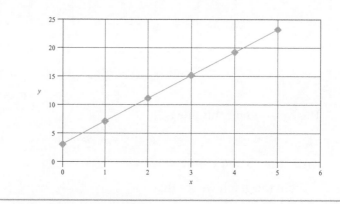

Note: Developing the above definitions:

- Lines with positive gradients slope up, from left to right:

- Lines with negative gradients slope down, from left to right:

- Lines parallel to the x axis have a gradient of zero:

  ———————————————

- Lines parallel to the y axis have an undefined gradient.

- Two lines are said to be *parallel* when they have the same gradient (or when both lines are vertical).

Example A.4   Extracting data from UV–visible spectroscopy

One of the more common experimental techniques used in both chemistry and biology is UV–visible spectroscopy. Results from such experiments make use of the Beer–Lambert law, which tells us that the intensity of a transmitted beam of light, I_t, is related to the concentration of the absorbing species, C, by:

$$\log(I_i/I_t) = C\varepsilon l$$

where

- I_i is the intensity of the incident beam before it enters the sample;
- l is the thickness of the sample;
- ε is the molar absorption coefficient, sometimes known as the extinction coefficient, and which is a property of the sample being looked at.

The expression $\log(I_i/I_t)$ is usually referred to as the absorbance, which is normally given the symbol A, leading to the normal representation of the equation being:

$$A = C\varepsilon l$$

Example A.5   What information can be obtained from the straight line plot $A = C\varepsilon l$?

Work out the gradient and, as the sample length l is the same for each concentration, then it is possible to work out the value of ε. Note:

- A has no units;
- C has units of $M (\equiv mol\,L^{-1})$;
- l has units of cm.

This means that the equation can be rearranged to $\dfrac{A}{C \cdot l} = \varepsilon$. Hence the units of ε are $M^{-1}cm^{-1} (\equiv mol^{-1}\,L\,cm^{-1})$.

ε is the molar extinction coefficient, a measure of how strongly a species absorbs light at a given wavelength.

Example A.6   Some specimen results

Below are some results showing how the absorbance of a coloured iron complex varies with its concentration, along with the plot that results.

[Iron(III)] (mol L⁻¹)	Absorbance
0.000245	1.17
0.0001208	0.602
0.0000805	0.404
0.0000604	0.307

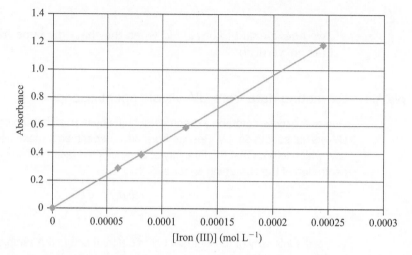

From the graph take values of absorbance at two concentrations, say at $[Fe^{3+}] = 0.0002$ M and at $[Fe^{3+}] = 0.0001$ M. These are 0.98 and 0.49 respectively.

The gradient (slope, m) is therefore:

$$m = \frac{y_2 - y_1}{x_2 - x_1}$$

$$= \frac{0.98 - 0.49}{0.0002 - 0.0001}$$

$$= \frac{0.49}{0.0001}$$

$$= 4.9 \times 10^3$$

In this case the sample length was 1 cm. Therefore:

$$\varepsilon = 4.9 \times 10^3 \, M^{-1} \, cm^{-1}$$

Example A.7 The study of enzyme kinetics

The study of the kinetics (rates of reactions) of enzymes is based on the Michaelis–Menten model of the mechanism of behaviour (as described in more detail in Chapter 9), which is summarised by the equation:

$$v = \frac{V_{max} \times [S]}{K_m + [S]}$$

The definitions of these terms are provided below. In a typical experiment the enzyme concentration is kept constant and the dependence of the initial rate, v, on substrate concentration, [S], is determined. If the initial rate v were plotted against substrate concentration, [S], we would get something like:

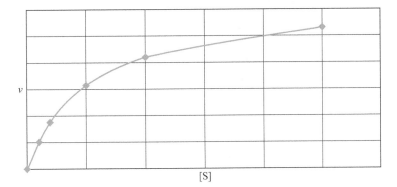

- The initial rate, v, will eventually reach a plateau at high values of [S], and this value is referred to as the maximum rate, V_{max}.
- The value of K_m corresponds to the concentration [S] at which $v = V_{max}/2$.

Determining values of K_m and V_{max} from a plot of the type shown above is difficult, so instead the Michaelis–Menten equation is rearranged to give a straight line graph, from which K_m and V_{max} can be estimated directly.

The Lineweaver–Burk plot, which is:

$$\left(\frac{K_m}{V_{max}} \times \frac{1}{[S]} \right) + \frac{1}{V_{max}} = \frac{1}{v}$$

This provides a plot of the form $y = mx + c$, where:

- $y = 1/v$;
- $x = 1/[S]$;
- m (gradient) $= K_m/V_{max}$ (both K_m and V_{max} are constants, so the value of K_m/V_{max} must be a constant too);
- c (intercept) $= 1/V_{max}$.

Hence a graphical plot of $1/v$ against $1/[S]$ should yield a straight line of gradient (K_m/V_{max}) and an intercept $1/V_{max}$ (a Lineweaver–Burk plot).

What do we plot, and on what axis do we plot it?

Plots of experimental results usually involve plotting one variable against another. The horizontal axis (called the x axis normally) is usually used for the variable that the experimenter has control over, and the vertical axis (usually called the y axis) is usually used for the variable that depends on the value of the other one.

The variable plotted along the x axis is called the *independent variable* and the variable plotted on the y axis is the *dependent variable*.

Therefore, in example 4, where we used the relationship,

$$A = C\varepsilon l$$

the absorbance value, A, depends on the concentration C. Therefore C is plotted on the x axis and A is plotted on the y axis.

In example 5, the situation appears slightly more complex because we are not plotting simple values. The value of $1/v$ does, however, depend on the value of $1/[S]$, because the rate, v, depends on the concentration, [S]. Therefore, $1/[S]$ is plotted on the x axis and $1/v$ on the y axis.

Numbers as powers and the use of logarithms

In science it is often necessary to use very large (or very small) numbers, and these could be extremely unwieldy. There are 'shorthand' ways of presenting numbers to make them easier to represent and to work with.

Consider these three examples:

(a) The number 2 000 000 can also be written as 2×10^6.

(b) Looking at 2 500 000, this can be represented as 2.5×10^6.

(c) The number 0.00075 can be represented as 7.5×10^{-4}.

[*In calculators or in graph plotting programs, example a would usually be represented as 2 E6 or 2 E06 and example c as 7.5 E-4 or 7.5 E-04.*]

- In cases a and b the decimal point was moved from the right hand side of the number by 6 digits, hence the 10^6 notation. *In case a, we would refer to the number as '2 times 10 to the power of 6' or '2 times 10 to the 6'.*

- In case c, where the number is less than 1, the decimal point has moved from left to right. In cases like this the superscript is represented by a negative number. *We would refer to the number as '7.5 times 10 to the minus 4'.* This number could also be represented as $7.5 \times 1/10^4$ or $7.5/10^4$ (although it rarely is).

This type of notation, referred to as *exponential notation*, is common in science, and the powers (indices) do not have to be integers. For example, $10^{1/2}$, which could also be shown as $10^{0.5}$, represents the square root of 10 (see example 5 below). $10^{-1/2}$ (or $10^{-0.5}$) is the same as $1/10^{1/2}$, i.e. it is the reciprocal of the square root of 10.

All numbers can be expressed in this way. For example 2^2 is (2×2), and similarly $2^{1/2}$ is the square root of 2.

General rules

- x^n means raising x to the nth power.
- $x^{m/n} = \sqrt[n]{x^m}$ and means taking the nth root of x to the power m.
- $x^{-n} = 1/x^n$.
- $x^{-m/n} = 1/\sqrt[n]{x^m}$.

Multiplying and dividing exponential numbers

- x^n multiplied by x^m gives x^{n+m}. In other words the indices are added together.
- x^n divided by x^m gives x^{n-m}. In other words the index associated with the number doing the dividing is taken from the index of the number being divided.

Examples

1. $2^2 \times 2^6 = 2^{(2+6)} = 2^8$
2. $10^3 \times 10^9 = 10^{(3+9)} = 10^{12}$

3. $10^8 \div 10^5 = 10^{(8-5)} = 10^3$

4. $5^{6.7} \div 5^6 = 5^{(6.7-6)} = 5^{0.7}$

5. $10^{0.5} \times 10^{0.5} = 10^{(0.5+0.5)} = 10^1 = 10$

6. $(4 \times 10^{-3}) \times (6 \times 10^{-2}) = 4 \times 6 \times 10^{-3} \times 10^{-2} = 24 \times 10^{(-3+-2)} = 24 \times 10^{-5}$
or 2.4×10^{-4}

Some chemical/biological examples using exponential numbers

There are several areas where exponential notation is used in the context of chemistry/ biology; for example in dealing with mole calculations, in dealings with equilibrium constants, etc.

Two examples are presented below and further examples will be used in the subsequent sections and will be specifically highlighted in the section on logarithms.

Example A.8 There are 6.022×10^{23} units (i.e. atoms or molecules or ions) in a mole of a substance. By definition, 1 L of a 1 molar solution contains 6.022×10^{23} units. Therefore, how could we determine the number of molecules in 200 cm^3 of a 0.1 molar solution of an organic acid?

As 1 L of a 1 molar solution contains 6.022×10^{23} molecules, so 200 cm^3 (0.2 L) of a 1 molar solution contains

$$(6.022 \times 10^{23}) \times 0.2 \text{ molecules} = 1.2045 \times 10^{23} \text{ molecules}$$

and therefore 200 cm^3 of a 0.1 molar solution contains

$$(1.2045 \times 10^{23}) \times 0.1 \text{ molecules} = 1.2045 \times 10^{22} \text{ molecules}$$

Example A.9 If 25.0 cm^3 of 0.2 M sodium hydroxide is neutralised by 15 cm^3 of a sulfuric acid solution, what is the concentration of the acid?

$$2NaOH + H_2SO_4 \longrightarrow Na_2SO_4 + 2H_2O$$

From the stoichiometry of the reaction, 1 mole of acid neutralises 2 moles of base, so:

1 L (i.e. 10^3 cm^3) of 0.2 M NaOH contains 0.2 moles

1 cm^3 of 0.2 M NaOH contains $(0.2/1000)$ moles $= 2 \times 10^{-4}$ moles

25 cm^3 of 0.2 M NaOH contains $2 \times 10^{-4} \times 25$ moles $= 5 \times 10^{-3}$ moles of alkali.

As 1 mole of acid reacts with 2 moles of alkali, the number of moles of acid required to neutralise 5×10^{-3} moles of alkali will be 2.5×10^{-3} moles, i.e. 15 cm^3 of acid contain 2.5×10^{-3} moles.

So, 10^3 cm^3 (i.e. 1 dm^3 or 1 L) of acid contain $2.5 \times 10^{-3} \times (10^3/15) \times 0.167$ moles (or 1.67×10^{-1} moles).

Therefore the concentration is 0.167 mol L^{-1} or 0.167 M.

Logarithms

General rules

$$y = a^x \qquad \text{(}y \text{ equals } a \text{ to the power of } x\text{)}$$

then

$$x = \log_a y \qquad \text{(}x \text{ equals the log to the base } a \text{ of } y\text{)}$$

Example A.10

$$100 = 10^2 \qquad \text{(100 equals 10 to the power of 2, also called 10 squared)}$$

so,

$$2 = \log_{10} 100 \qquad \text{(2 equals the log to the base 10 of 100)}$$

Example A.11

$$0.01 = 10^{-2}$$

so,

$$-2 = \log_{10} 0.01 \qquad \text{(}-2 \text{ equals the log to the base 10 of 0.01)}$$

Only powers of 10 have been used so far, but the approach is a general one.

Example A.12

$$25 = 5^2 \qquad \text{(25 equals 5 to the power of 2)}$$

So,

$$2 = \log_5 25 \qquad \text{(2 equals the log to the base 5 of 25)}$$

Example A.13

$$81 = 3^4 \qquad \text{(81 equals 3 to the power of 4)}$$

So,

$$4 = \log_3 81 \qquad \text{(4 equals the log to the base 3 of 81)}$$

Example A.14

$$4 = 2^2$$

so,

$$2 = \log_2 4 \qquad \text{(2 equals the log to the base 2 of 4)}$$

Logarithms to the base 10 are used very often in science. From a chemistry/biology perspective they are particularly valuable in understanding acid/base equilibria and pH. It is quite common for them to be written as log x rather than in the more complete form as $\log_{10} x$.

The other commonly used type of logarithm is the *natural logarithm* or logarithm to the base e. Logarithms to the base e are not usually written as $\log_e$ but are more commonly shown as ln.

What is e?

e is just a number, like π. Whilst π has a value of 3.1415 . . ., e has a similarly infinitely long number of decimal places. The value of e is, in fact, 2.7182. . . . .

The rules of logarithms

There are some rules for handling logarithmic numbers which are common to logarithms to any base. These rules are summarised below, and shown in both $\log_{10}$ and in ln formats.

$$\log x^a = a \log x$$

$$\log xy = \log x + \log y$$

$$\log \frac{x}{y} = \log x - \log y$$

$$\log \frac{1}{x} = \log x^{-1} = -\log x$$

$$\ln x^a = a \ln x$$

$$\ln xy = \ln x + \ln y$$

$$\ln \frac{x}{y} = \ln x - \ln y$$

$$\ln \frac{1}{x} = \ln x^{-1} = -\ln x$$

A bit more on the background to logarithms

Concentrate, for the moment, on base 10.

- $10 = 10^1$, so $\log_{10} 10 = 1$
- $100 = 10^2$, so $\log_{10} 100 = 2$
- $1000 = 10^3$, so $\log_{10} 1000 = 3$

What about numbers between 10 and 100?

What about numbers between 100 and 1000?

- If $\log_{10} 10 = 1$ and $\log_{10} 100 = 2$, then any number between 10 and 100 should have $\log_{10}$ values between 1 and 2.
- If $\log_{10} 100 = 2$ and $\log_{10} 1000 = 3$, then any number between 100 and 1000 should have $\log_{10}$ values between 2 and 3.

Consider 50 and 500

Use a calculator to see what value is obtained on taking log 50 and then log 500. (The log button on your calculator represents log to base 10.) The results are:

$$\log_{10} 50 = 1.699 \text{ (rounded to 3 d.p.) and}$$

$$\log_{10} 500 = 2.699 \text{ (rounded to 3 d.p.)}$$

Note the common factor is the part following the decimal point.

Now consider the number 5

Using a calculator to see what value is obtained on taking $\log_{10} 5$, gives:

$$\log_{10} 5 = 0.699 \text{ (rounded to 3 d.p.)}$$

Putting the bits together, we can see that

$$\underset{1.699}{\downarrow} \quad \underset{0.699}{\downarrow} \quad \underset{1.000}{\downarrow}$$
$$\log_{10}50 = \log_{10}5 + \log_{10}10$$

and

$$\log_{10}500 = \log_{10}5 + \log_{10}100$$
$$\underset{2.699}{\uparrow} \quad \underset{0.699}{\uparrow} \quad \underset{2.000}{\uparrow}$$

So, how does this happen? From the definition of logs, we see that:

$$5 = 10^{0.699}$$
$$50 = 10^{1.699}$$

and

$$500 = 10^{2.699}$$

It is also known that:

$$50 = 5 \times 10^1$$

and

$$500 = 5 \times 10^2$$

So, it follows that,

$$50 = 10^{0.699} \times 10^1$$

and

$$500 = 10^{0.699} \times 10^2$$

Example A.15 Why is log 1 = 0, and equally, ln 1 = 0?

Any number to the power 0 = 1,

Putting this mathematically,

$$x^0 = 1$$

Elaborating on this:

$$\frac{10^2}{10^2} = 1$$

But

$$\frac{10^2}{10^2} = 10^{2-2} = 10^0$$

So, $10^0 = 1$. Therefore, using the rules of logarithms,

$$0 = \log_{10}1$$

Similarly, taking a different number

$$\frac{e^2}{e^2} = 1$$

But

$$\frac{e^2}{e^2} = e^{2-2} = e^0$$

So, $e^0 = 1$. Therefore, using the rules of logarithms,

$$0 = \log_e 1 = \ln 1$$

Example A.16 This can be done using rules given above. Starting with

$$\log_{10} x = y$$

prove that

$$\ln x = 2.303 \log x$$

Answer:

$$\log_{10} x = y$$

so

$$x = 10^y$$

Take logs to the base e:

$$\ln x = \ln 10^y$$

But, using the rules of logarithms this rearranges to:

$$\ln x = y \ln 10$$

Now, $\ln 10 = 2.303$, so this is the same as:

$$\ln x = 2.303y$$

But at the start we saw that $\log_{10} x = y$, so:

$$\ln x = 2.303 \log_{10} x$$

Some chemical/biological examples illustrating the use of logarithms and of numbers using exponential notation

Most calculators have the capability to handle calculations involving logarithmic or exponential notation. There are, however, many different makes of calculators on the market, so it is difficult to make reference to how data should be 'keyed in' to calculators – this will need to be practised.

Below we will look, in a stepwise fashion, at some chemical/biological calculations involving the use of the notations referred to above.

Example A.17 This involves pH calculations, so remember the definition:

$$pH = -\log_{10}[H_3O^+] \quad \text{(could also write } \log_{10}[H_3O^+] = -pH; \text{ it is the same thing)}$$

An acid solution has a pH of 3. What is the hydrogen ion concentration $[H_3O^+]$?

453

Remember the general rule is that if

$$y = a^x \quad (y \text{ equals } a \text{ to the power of } x)$$

then

$$x = \log_a y \quad (x \text{ equals the log to the base } a \text{ of } y)$$

In the case of this question remember that $pH = -\log_{10}[H_3O^+]$, so therefore we could write the equation, using $x = 3$, $a = 10$ and $y = [H_3O^+]$, as

$$pH = 3$$
$$= -\log_{10}[H_3O^+]$$
$$= \log_{10}[H_3O^+]^{-1}$$

Therefore,

$$[H_3O^+]^{-1} = 10^3$$

So,

$$[H_3O^+] = 10^{-3} \, mol \, dm^{-3}$$

Example A.18 An acid solution has a pH of 8.85. What is the hydrogen ion concentration $[H_3O^+]$?
Using the technique adopted in the previous example, we see that

$$[H_3O^+] = 10^{-8.85} \, mol \, dm^{-3}$$

This is fine, but numbers raised to fractional indices are not easy to visualise. Convert $10^{-8.85}$ to a decimal number using a calculator. This should show:

$$10^{-8.85} = 1.413 \times 10^{-9}$$

Therefore,

$$[H_3O^+] = 1.413 \times 10^{-9} \, mol \, dm^{-3}$$

Example A.19 A solution has a hydrogen ion concentration of $2.30 \times 10^{-7} mol \, dm^{-3}$. What is its pH?

1. Enter 2.30×10^{-7} into your calculator.
2. Press the log key (it may say $\log_{10}$ on the key).

   (You may need to press log followed by 2.30×10^{-7}; it depends on the make of the calculator.)

3. You will see a number to many decimal places; to the first three decimal places it is -6.625.
4. Thus $\log_{10}[H_3O^+] = -6.625$.
5. Answer is, therefore, pH $= 6.625$.

Means and medians

A *mean* is an average of a series of numbers. The formal representation of this is:

$$\bar{x} = \frac{\sum x}{n}$$

where

- x represents the individual values of the numbers/measurements;
- $\sum x$ is the sum of these values;
- $\bar{x}$ is their mean value;
- n refers to the number of numbers/measurements used.

In other words, you add all of the numbers (measurements/results, etc.) and divide by the number in the sample. For example, if we have a series of four titration results, 11.64 mL, 11.67 mL, 11.59 mL and 11.66 mL, the mean titre would be:

$$\frac{11.64 + 11.67 + 11.59 + 11.66}{4} = 11.64 \, \text{mL}$$

A *median* is the middle value in a series of numbers. For an odd number of items this is an easy item to identify, thus the median of 1, 2, 4, 7, 10 would be 4, as it is the third number from each end of the list.

For an even number of items the matter is slightly trickier. In this case it is necessary to identify the two middle values and take the average of them. In the case of the four titration results above, namely 11.64 mL, 11.67 mL, 11.59 mL and 11.66 mL, the two middle readings are 11.64 and 11.66, and the average of those is $(11.64 + 11.66)/2$, which is 11.65. Therefore the median reading is 11.65 mL.

Standard deviations

A standard deviation (SD) gives an indication of the spread of a series of numbers, and in the context of data obtained in the laboratory this can be a very useful figure to determine. For many laboratory experiments that you are likely to do, one way of determining the SD for the data that are obtained can be calculated from the expression:

$$SD = \sqrt{\frac{\sum (x - \bar{x})^2}{(n - 1)}}$$

where:

- x is the value of a reading;
- $\bar{x}$ is the mean value from a series of readings.

Now, whilst $(x - \bar{x})$ may be negative, $(x - \bar{x})^2$ will always be positive. The value n refers to the number of measurements used. Therefore, in the example used above, with four titration results, we would get the following:

x	$(x - \bar{x})$	$(x - \bar{x})^2$
11.64	0	0
11.67	0.03	0.0009
11.59	−0.05	0.0025
11.66	0.02	0.0004

Now, $\sum (x - \bar{x})^2 = 0.0038$, and $(n - 1) = 3$, so $SD = \sqrt{(0.0038/3)} = 0.036$

Solving quadratic equations

Quadratic equations usually take the form:

$$ax^2 + bx + c = 0$$

They appear quite often as part of equilibrium calculations, and this will be illustrated in the examples below. Before tackling any examples, let us define how to solve these equations. There are several ways of solving them, but there is a general method for solving an equation like this (to work out what x is). In such equations, x can have two solutions (called roots). The formula that is used is:

$$\frac{-b \pm \sqrt{b^2 - 4ac}}{2a}$$

The use is best illustrated by example.

Example A.20 Solve the equation

$$2x^2 + 4x - 6 = 0$$

In this, $a = 2$, $b = 4$ and $c = -6$.

$$x = \frac{-4 \pm \sqrt{4^2 - [4 \times 2 \times (-6)]}}{2 \times 2}$$

$$= \frac{-4 \pm \sqrt{16 + 48}}{4}$$

$$= \frac{-4 \pm \sqrt{64}}{4}$$

$$= \frac{-4 \pm 8}{4} = \frac{-4 + 8}{4} \text{ or } \frac{-4 - 8}{4}$$

Thus $\qquad x = 1 \text{ or } -3$

Examples of quadratic equations in chemical/biological problems

Example A.21 60.0 g of ethanoic acid (CH_3CO_2H) and 46.0 g of ethanol (CH_3CH_2OH) were reacted at 50 °C until equilibrium was established. By this time 12.0 g water and 58.7 g ethyl acetate ($CH_3CO_2C_2H_5$) were produced.

(a) What is the equilibrium constant for this reaction at this temperature?

(b) What quantities of ester and water would be produced if 90.0 g of ethanoic acid and 92.0 g ethanol were brought to equilibrium at 50 °C?

First, show the equilibrium:

$$CH_3CO_2H + CH_3CH_2OH \rightleftharpoons CH_3CO_2C_2H_5 + H_2O$$

Relative molecular mass	60.0	46.0	88.0	18.0
No. of moles: -				
Initially	1	1	0	0

At equilibrium, there are 58.7 g ethyl acetate, which is (58.7/88.0) moles = 0.666 moles. Similarly there are 12 g water, which is 12.0/18.0 = 0.666 moles.

Therefore, the number of moles of ethanoic acid = 1 − 0.666 = 0.333, and

$$\text{number of moles of ethanol} = 1 - 0.666 = 0.333$$

Therefore as the equilibrium constant

$$K = \frac{[CH_3CO_2C_2H_5][H_2O]}{[CH_3CO_2H][C_2H_5OH]}$$

then

$$K = \frac{0.666 \times 0.666}{0.333 \times 0.333} = 4$$

Now consider the second question. We are starting with 90.0 g ethanoic acid (=1.50 moles), and with 92.0 g ethanol (=2.00 moles). Let us say that these will react together to produce x moles of ethyl acetate and x moles of water.

Tabulating the data:

	CH_3CO_2H	CH_3CH_2OH	$CH_3CO_2C_2H_5$	H_2O
Initial concentration	1.50	2.00	0.00	0.00
Equilibrium concentration	$1.50 - X$	$2.00 - X$	X	X

It is known that

$$K = \frac{[CH_3CO_2C_2H_5][H_2O]}{[CH_3CO_2H][C_2H_5OH]} = 4$$

so, putting in the concentrations from the table gives:

$$\frac{X \times X}{(1.5 - X)(2 - X)} = 4$$

Remembering the techniques for rearranging equations, multiply both sides of the equation by $(1.5 - X)(2 - X)$, to give:

$$X^2 = 4.00(1.50 - X)(2.00 - X)$$
$$= 4.00(3.00 - 3.50X + X^2)$$
$$= 12.00 - 14.00X + 4.00X^2$$

Then take X^2 from both sides which gives:

$$0 = 3.00X^2 - 14.00X + 12.00$$

which is a quadratic equation. So,

$$X = \frac{-(-14) \pm \sqrt{(-14)^2 - (4 \times 3 \times 12)}}{2 \times 3}$$

$$= \frac{14 \pm \sqrt{196 - 144}}{6}$$

$$= \frac{14 \pm \sqrt{52}}{6}$$

$$= \frac{14 + 7.21}{6} \quad \text{or} \quad \frac{14 - 7.21}{6}$$

This leads to:

$$X = 3.54 \text{ or } 1.13$$

In fact the answer cannot be 3.54, because that would mean we would use more moles than we put in! The answer is, therefore, 1.13 moles – we need common sense as well!

Non-linear graphs and differentiation

We saw earlier how the equation of a straight line was

$$y = mx + c$$

In this expression m represents the gradient of the line. The gradient of a straight line can be thought of as showing what the *rate of change of* y *is with respect to* x. In other words, for a change in x of one unit, there is a corresponding change in y of m units. Now look at the graph below, which is the graph representing part of the curve $y = x^3$ (in other words, for a value of $y = 1$, $x^3 = 1$; for $y = 2$, $x^3 = 8$; for $y = 3$, $x^3 = 27 \ldots$ etc.).

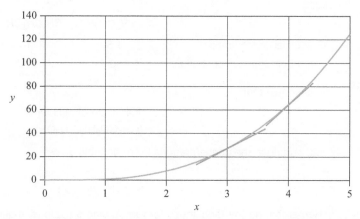

For a non-linear plot like that shown (in other words, a plot that is not a straight line), there is a different value for the gradient at different points on the graph. The picture shows two such points, where x is about 3 and where x is about 4. In the example shown it is clear that the gradient, which is determined by drawing a tangent at a given point on the curve, is steeper at the higher value of x.

There are examples where the opposite situation can hold true; for example if we consider the graph of $y = 1/x$ shown below:

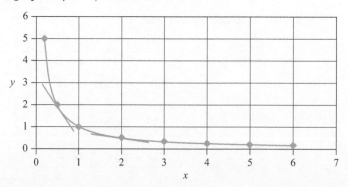

In this case the gradient at $x = 0.5$ is steeper than that at $x = 2$.

In the non-linear relationships where we compare two different variables (usually called y and x) how can we get an idea of rates of change of y with respect to x? *This is achieved through the process known as differentiation.*

The *rate of change of* y *with respect to* x is usually denoted as $\dfrac{dy}{dx}$.

In this representation the $\dfrac{d}{dx}$ part merely means rate of change with respect to x.

Below is a table showing the differentials $\dfrac{dy}{dx}$ for six of the most commonly occurring types of relationships in chemistry and biology.

$y =$	$\dfrac{dy}{dx} =$
constant, c	0
x^n	nx^{n-1}
$\dfrac{1}{x^n} = x^{-n}$	$-nx^{-(n+1)} = -\dfrac{n}{x^{(n+1)}}$
e^x	e^x
e^{ax}	ae^{ax}
$\ln x$	$\dfrac{1}{x}$

If $y = a + b$, where a and b are both functions of x, then a and b are treated independently, i.e.

$$\frac{dy}{dx} = \frac{da}{dx} + \frac{db}{dx}$$

(See example 5 below for an example of this.)

All this seems a bit abstract at the moment, but we will look at some general examples firstly, then focus on examples of the real application of these techniques.

Example

1. $y = x^2$ $(n = 2)$ so $\dfrac{dy}{dx} = 2x$

2. $y = x^3 + 3$ $(n = 3, c = 3)$ so $\dfrac{dy}{dx} = 3x^2$

(c is a constant, and its differential is 0)

3. $y = x^{-3}$ $(=1/x^3)$ so $\dfrac{dy}{dx} = -3x^{-4}$ $(= -3/x^4)$

4. $y = \ln (2x)$ so $\dfrac{dy}{dx} = 1/x$

Remember from the section on logarithms that

$$\ln(xy) = \ln x + \ln y$$

so, $y = \ln 2 + \ln x$

Now, ln 2 is a constant, so its differential is 0, meaning that the differential for the expression as a whole will be the differential for ln x, so: dy/dx = 1/x.

5. $y = e^{2x} + e^{5x}$ so $dy/dx = 2e^{2x} + 5e^{5x}$

(each of the terms is differentiated independently).

Have another look at $y = x^3$

For $y = x^3$

$$\frac{dy}{dx} = 3x^2$$

So, the slope (gradient) of the tangent to the curve at any point will have value of $3x^2$.

- For $x = 3$, slope of tangent is therefore $3(3^2) = 27$
- For $x = 4$, slope of tangent is therefore $3(4^2) = 48$

Visual inspection of the graph shows that this is correct.

Where is differentiation important in a chemical/ biological context?

In the context of this text, the use of differentiation is limited to Chapter 9 on kinetics. Thinking in a general sense, if we have a reaction:

$$A + B \rightarrow C + D$$

then one of the things we may want (or indeed need) to know is how fast does the reaction go. We can think of this either as:

What is the rate of disappearance of A or B with respect to time?

or

What is the rate of appearance of C or D with respect to time?

Let us consider the reaction in terms of, for example, reagent A. We need to measure how much of A is present (usually in terms of its concentration) at different times during the reaction in order to measure its rate of disappearance.

The rate of disappearance of A would be represented by $-\dfrac{d[A]}{dt}$

(The − sign is there to show that [A] is decreasing, and the $\dfrac{d}{dt}$ part means that we are interested in how [A] varies with time t. In other words, compared with

$\dfrac{dy}{dx}$, we have the same approach, with [A] being used instead of y and t being used instead of x.)

In the case of the reaction shown above we could say

$$-\frac{d[A]}{dt} = -\frac{d[B]}{dt} = +\frac{d[C]}{dt} = +\frac{d[D]}{dt}$$

This is because for each mole of A and B being used on the left hand side (represented by the $-$ sign), one mole of each of C and D are produced on the right hand side (indicated by the $+$ sign).

This measures the rate of the reaction.

The process we are describing is known as reaction kinetics. This is an experimental science, and we see *experimentally* how rate varies with concentration of reagents. For example if we were studying the reaction:

$$A \quad \rightarrow \quad products$$

we would measure the concentration of A with time, and hence produce a *rate equation* that looks like:

$$-\frac{d[A]}{dt} = k[A]^{\alpha} \; (k = \text{rate constant})$$

The value of α is referred to as the *order* of the reaction.

Integration

Whenever we have an equation involving $\dfrac{dy}{dx}$, we can use a process called integration to establish a relationship between y and x. *This is therefore simply the reverse of differentiation.*

We have to use some new symbols here, which again are just forms of shorthand. If we say that

$$y = f(x)$$

That means that y is a function of x (or the value of y depends on the value of x). Taking the differential (differentiating $y = f(x)$) produces a new relationship, which is represented as:

$$\frac{dy}{dx} = f'(x)$$

If we are in the position of knowing what $f'(x)$ is, then integration will enable us to determine $f(x)$.

Integration is represented by the symbol $\int$, so the integration of $f'(x)$ would be represented as $\int f'(x).dx$ (the .dx tells us that the integration is being carried out with respect to x).

There are some standard procedures, tabulated below, that can do this (note each has a constant, c, added – see after the table for reason):

f'(x)	∫f'(x)·dx
constant, a	$ax + c$
x	$\dfrac{x^2}{2} + c$
x^n	$\dfrac{x^{(n+1)}}{(n+1)} + c$ (except for $n = -1$)
$\dfrac{1}{x^n} = x^{-n}$	$\dfrac{1}{(n-1)x^{(n-1)}} + c = -\dfrac{x^{-(n-1)}}{(n-1)} + c$
e^x	$e^x + c$
e^{ax}	$\dfrac{e^{ax}}{a} + c$
$\dfrac{1}{x} = x^{-1}$	$(\ln x) + c$

Why do we have a constant, c, added to the end of each integral?

Remember example 2 from the section on differentiation, where it was shown $y = x^2 + 3$ ($n = 2$, $c = 3$) so $dy/dx = 2x$ (c is a constant, and its differential is 0). Similarly, if we had written $y = x^2 + 12$ ($n = 2$, $c = 12$) so $dy/dx = 2x$ (c is a constant, and its differential is 0), i.e. we get the same answer from different equations.

Therefore, if we are re-establishing a relationship between y and x, we need to take account of the possibility that a constant may be present.

At a pictorial level, let us see what this means. The graphs below show the plots for the equations $y = (x^2 + 3)$ and for $y = (x^2 + 12)$ (the dotted and dashed lines respectively). Also shown, on the same graphs, are the tangent lines, indicating the gradients at a value of $x = 3$.

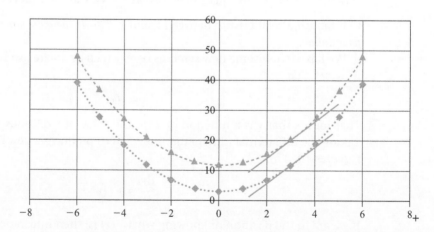

As the graphs show, the gradient at $x = 3$ is the same for both, which would be expected as dy/dx has a value of $2x$ for both.

The consequence of this, however, is that when this term is integrated, whilst we know that it contains an x^2 term, it does not tell us the vertical placement of the curve, and that is why we need the constant c to be introduced.

Integration in the chemical/biological context

The rate equation,

$$\text{Rate} = -d[A]/dt = k[A]^\alpha$$

is a differential rate equation. In this equation:

- k is a constant called the rate constant;
- α is the order of the reaction.

What we would like to be able to do, in order to be able to use experimental data we can easily measure, is to be able to relate [A] and the time t, and thereby use measured experimental data to see whether the reaction obeys this rate equation, and obtain a value for α.

To do this we need to integrate our rate equation, and see how the results of this depend on the value of α.

Example A.22 For a zero order system, where $\alpha = 0$,

$$-d[A]/dt = k$$

or, if we rearrange to separate the variables (which means putting [A] and time, t, on opposite sides of the equation), we get

$$-d[A] = kdt$$

which can be further manipulated by multiplying both sides by −1 to

$$d[A] = -kdt$$

This is integrated thus:

$$\int d[A] = \int k.dt$$

(On the left hand side, the term $\int d[A]$ can be re-written as $\int 1 \cdot d[A]$, so it is the integral of a constant.) So integration gives

$$[A] = -kt + c$$

If we say, at the start of the reaction (when $t = 0$), that $[A] = [A]_0$, where $[A]_0$ means the value of [A] at time 0, then we can say that:

$$c = [A]_0$$

so, this means that we can rewrite the integrated equation putting this value in for c:

$$[A] = -kt + [A]_0$$

Now note this is in the form $y = mx + c$, *with [A] corresponding to y and t corresponding to x.*

$[A]_0$ is a constant, and will be the intercept, and the gradient m is −k. *Therefore if a plot of [A] vs t gives a straight line, then the reaction is zero order.*

Example A.23 Now let's look at the first order system, where $\alpha = 1$,

$$-\frac{d[A]}{dt} = k[A]$$

Again this can be rearranged to put all of the [A]s on one side and all of the ts on the other.

$$-\frac{1}{[A]} \cdot d[A] = k \cdot dt$$

We can multiply through by -1 to give:

$$\frac{1}{[A]} \cdot d[A] = -k \cdot dt$$

This will now be integrated:

$$\int \frac{1}{[A]} \cdot d[A] = \int -k \cdot dt$$

Integration of this (see the list above) gives us

$$\ln [A] = -kt + c$$

As in example A.22, if we say at time $t = 0$, that $[A] = [A]_0$, then

$$c = \ln [A]_0$$

So, we can replace c by $\ln [A]_0$ and rearrange this to give us

$$\ln [A] = \ln [A]_0 - kt$$

Now note this is in the form $y = mx + c$, *with* $\ln [A]$ *corresponding to* y *and* t *corresponding to* x.

In $[A]_0$ is a constant, and will be the intercept, and the gradient m *is* $-k$. Therefore if a plot of $\ln [A]$ vs t gives a straight line, then the reaction is first order.

Self-test

Repeat this for a second order reaction. The answer is in Chapter 9.

Appendix 2: Answers to end of chapter questions

Chapter 1

1.1 (d)

1.2 (c)

1.3

	Number of protons	Number of neutrons	Number of electrons
$^{10}_{5}B$	5	5	5
$^{11}_{5}B$	5	6	5

$^{10}_{5}B$ and $^{11}_{5}B$ are isotopes

1.4 (a) $^{32}_{15}P \rightarrow ^{32}_{16}S + e^- + \bar{\nu}$

The element is sulfur

(b) Every 14 days the mass halves, so after 112 days the mass of phosphorus remaining will be 1/256 of its original value

1.5 54.23%

1.6 (a) 57.1% (b) 72.7% (c) 67.3%

(d) 34.8% (e) 41.3% (f) 57.7%

1.7 (a) 1.204×10^{24}

(b) 3.74×10^{24}

1.8 (a) $2KClO_3 \rightarrow 2KCl + 3O_2$

(b) $Pb(OH)_2 + 2HCl \rightarrow 2H_2O + PbCl_2$

(c) $2AlBr_3 + 3K_2SO_4 \rightarrow 6KBr + Al_2(SO_4)_3$

(d) $C_6H_{14} + \dfrac{19}{2}O_2 \rightarrow 6\,CO_2 + 7H_2O$

1.9 (a) $N_2(g) + 3H_2(g) \rightarrow 2NH_3(g)$

(b) 18.75 moles

(c) 6.25 moles, 175 g

(d) 12.5 moles or 212.5 g

1.10 Dissolve 20 g of NaOH in water, and make the volume up to 500 cm^3 in a volumetric flask

1.11 $0.1\,mol\,L^{-1}$; contains 0.01 moles

1.12 $0.034\,mol\,L^{-1}$

1.13 26.24 mL or 0.02624 L

Chapter 2

2.1 (a) P: $1s^2\,2s^2\,2p^6\,3s^2\,3p^3$ [Ne] $3s^2\,3p^3$

(b) K: $1s^2\,2s^2\,2p^6\,3s^2\,3p^6\,4s^1$ [Ar] $4s^1$

(c) Se: $1s^2\,2s^2\,2p^6\,3s^2\,3p^6\,4s^2\,3d^{10}\,4p^4$ [Ar] $4s^2\,3d^{10}\,4p^4$

(d) I: $1s^2\,2s^2\,2p^6\,3s^2\,3p^6\,4s^2\,3d^{10}\,4p^6\,5s^2\,4d^{10}\,5p^5$ [Kr] $5s^2\,4d^{10}\,5p^5$

2.2 (a) Mg^{2+}, (b) O^{2-}, (c) P^{3+}, (d) Cs^+

2.3 (a) isoelectronic,

(b) isoelectronic,

(c) not isoelectronic,

(d) not isoelectronic,

(e) isoelectronic

2.4 Only (b) is possible

2.5 (a) Na^+ (b) S^{2-} (c) Cl^-

2.6 Ionisation energy tends to *increase* from left to right across a period and to *decrease* down a group, so: (a) He > Ar > Kr, (b) Te > Sb > Sn, (c) Ca > K > Rb, (d) Xe > I > Cs

2.7 (a) Compound formed is K_2O

(b) Compound formed is $MgCl_2$

2.8 (a) Bond length: S—F < S—Cl < S—Br

Bond strength: S—Br < S—Cl < S—F

(b) Bond length: C≡O < C=O < C—O

Bond strength: C—O < C=O < C≡O

(c) Bond length: S—F < Si—O < Si—C

Bond strength: Si—C < Si—O < Si—F

2.9 (a) $O^{\delta-}$—$H^{\delta+}$ (b) $F^{\delta-}$—$N^{\delta+}$ (c) $I^{\delta+}$—$Cl^{\delta-}$

2.10 (a) H—C < H—N < H—O
(b) Cl—Cl < Br—Cl < Cl—F

2.11 (a) ionic, (b) covalent, (c) metallic, (d) polar covalent, (e) polar covalent, (f) polar covalent, (g) metallic, (h) ionic, (i) covalent

2.12 (a) tetrahedral, (b) linear, (c) tetrahedral, (d) bent, (e) bent, (f) t-shaped, (g) linear

2.13 (a) Resonance structures

(b)

(c)

2.14 (a) H—C—N angle < 109.5°; H—N—H about 107°; C and N are both sp^3; all σ-bonds; note that N is (close to) tetrahedral with lone pair

(b) H—C—O angle > 120°, H—C—H <120°; C and O sp^2 hybridised; σ and π-bonds between C=O; O lone pairs in plane of molecule

(c) H—C=O angle>120°; H—O—C ~105°; C=O both sp^2 (planar); other O is sp^3 (tetrahedral); C=O one σ one π bond, all others σ

2.15

This O atom is sp^2

This C atom is sp^2

All C's in these rings sp^2

these C atoms are sp^3

*These O and N atoms are sp^3

2.16 (a) Dipolar and van der Waals, (b) ionic lattice, (c) van der Waals only, (d) hydrogen bonding, (e) dipolar and van der Waals, (f) hydrogen bonding

2.17 The difference is the electronegativity of oxygen.

2.18 (a) LiCl is ionic, (b) NH_3 has H bonding; (c) I_2, due to van der Waals; (d) NO due to dipolar interactions

Chapter 3

3.1 A is 3-methylpentane, B is 3,3-dimethylpentane

3.2

4-Ethyl-5-methylnonane

3.3 A is

B is

3.4 A is 5-ethylnon-3-en-6-yne; B is 2-ethylbutene

3.5 $CH_3CH_2CH_2OCH=CH_2$ $CH_3CH=CHCH(OH)CH_3$

Ether Alkene   Alkene Hydroxyl

3.6

Name	Structure
Pentanal	$CH_3{\cdot}CH_2{\cdot}CH_2{\cdot}CH_2{\cdot}CHO$
4-Hydroxybutanal	$HO{\cdot}CH_2{\cdot}CH_2{\cdot}CH_2{\cdot}CHO$
Pent-2-en-4-ynamide	$HC{\equiv}C{\cdot}CH{=}CHCONH_2$
n-Propyl ethanoate	$CH_3{\cdot}CH_2{\cdot}CH_2{\cdot}O{\cdot}CO{\cdot}CH_3$
Pent-3-en-2-ol	$CH_3{\cdot}CH{=}CH{\cdot}CH(OH)CH_3$

The five compounds contain:
- Pentanal has an aldehyde (or alkanal)
- 4-Hydroxybutanal has an aldehyde and an alcohol
- Pent-2-en-4-ynamide contains an amide, an alkene and an alkyne

- *n*-Propyl ethanoate contains an ester
- Pent-3-en-2-ol contains an alkene and an alcohol

3.7 (i)

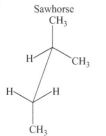

Sawhorse   Newman

(ii)

3.8 (a)

E isomer   *Z* isomer

(I)   (II)

(b)

(I)
$$
\begin{array}{c}
\text{H} \qquad\qquad \text{H} \\
\text{C}==\text{C} \\
\text{Cl} \qquad Z \qquad \text{Br}
\end{array}
$$
or
$$
\begin{array}{c}
\text{Cl} \qquad\qquad \text{H} \\
\text{C}==\text{C} \\
\text{H} \qquad E \qquad \text{Br}
\end{array}
$$

(II) None

(III)   Each double bond can be labelled as *Z* or *E*, so there are 4 isomers.

E,E   *Z,E*

E,Z   *Z,Z*

(IV)

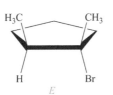

or

E   *Z*

467

3.9

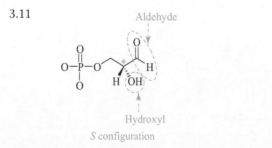

Most stable

Z isomer

E isomer

Identical

(a) and (b)

E isomer can display enantiomerism

Mirror images cannot be superimposed

(c)

3.10

S centre

R centre

(a)

(b)

S centre

R centre

S centre

S centre

(c)

(d)

3.11

Aldehyde

Hydroxyl

S configuration

Chapter 4

4.1 (a) −950 kJ   (b) 484 JK^{-1} mol^{-1}

4.2 $C_{12}H_{22}O_{11}(s) + 12O_2(g) \rightarrow 12CO_2(g) + 12H_2O(1)$

$\Delta H_f^{\theta} = -2225.8$ kJmol^{-1}

4.3 (a) Increase, liquid to vapour

(b) Decrease, water more ordered

(c) Minimal change, gas same on both sides

4.4 (a) $\Delta S^{\theta} = -114.4$ JK^{-1} mol^{-1}

(b) $\Delta H^{\theta} = -73.3$ kJ mol^{-1}

(c) $\Delta G^{\theta} = -38.2$ kJ mol^{-1}, spontaneous

4.5 (a) $\Delta G = 0$ (b) $T_d = \Delta H/\Delta S$ (c) $T_d = 1112$ K

Chapter 5

5.1 (a) [A] falls

(b) [A] stays the same

(c) [A] increases

5.2 (a) [H$_2$] falls

(b) [H$_2$] rises

(c) [H$_2$] rises

(d) [H$_2$] rises

(e) No change

5.3 Equilibrium constant given by:

$$K = \frac{[\alpha\text{-ketoglutarate}][\text{L-alanine}]}{[\text{pyruvate}][\text{L-glutamate}]}$$

Under reaction conditions, $Q = 1.5 \times 10^3$; reverse reaction

5.4 (a) It is product favoured (spontaneous)

(b) Exothermic – raising temperature favours reverse reaction

(c) At 310 K, $\Delta G^{\theta} = -4370.2$ J mol^{-1}; at 298 K, $\Delta G^{\theta} = -4355.2$ Jmol^{-1}

(d) $\Delta H^{\theta} = -3982.7$ J mol^{-1}; $\Delta S^{\theta} = -1.25$ JK^{-1} mol^{-1}.

(e) No effect on equilibrium, except to achieve it more quickly

5.5 (a) −26.2 kJ mol^{-1}

(b) 10.4 JK^{-1}mol^{-1}

(c) 1.4×10^5

The principal driver is the enthalpy

5.6 (a) $K = \dfrac{\chi_U}{[1 - \chi_U]}$; check by putting in values of χ_U and seeing if corresponding K value is correct

(b) At 63°C, $\Delta G^{\theta} = 3658$ J mol^{-1}; at 78°C, $\Delta G^{\theta} = -3697$ J mol^{-1}.

(c) $\Delta S^{\theta} = 490$ JK^{-1} mol^{-1}; $\Delta H^{\theta} = 168$ kJmol^{-1}

(d) Unfolding is entropy driven

Chapter 6

6.1 (a) A proton donor (donates H^+)
  (b) A proton acceptor
  (c) A strong acid is fully dissociated in solution. A strong base will be fully dissociated in aqueous solution
  (d) A weak acid will not be completely dissociated in solution. A weak base will not be completely protonated (Brønsted–Lowry definition)

6.2 $pH = -\log_{10}[H_3O^+]$; $pOH = -\log_{10}[OH^-]$
  (a) $pH = 1.3$
  (b) $pH = 3.1$
  (c) $pH = 2.1$
  (d) $pH = 12.7$

6.3 (a) $K_a = \dfrac{[H_3O^+][A^-]}{[HA]}$
  (b) Henderson–Hasselbalch expression is:
  $$pH = pK_a + \log_{10}\frac{[base]}{[acid]}$$
  (c) $pH = 5.19$

6.4 (a) $K_a = \dfrac{[H_3O^+][CH_3CH_2CH_2COO^-]}{[CH_3CH_2CH_2COOH]}$
  (b) $pH = 3.07$
  (c) $pH = 4.61$
  (d) The solution is acting as a buffer, so addition of a small amount of strong acid has little effect on the pH

6.5 (a) $pH = 2$. This has changed from $pH = 7$ for pure water
  (b) $pH = 9.03$
  (c) The pH changes to 8.98
  (d) The drop in pH in (c) is only 0.05, compared with 5 pH units in (a). The second solution is behaving as a buffer

6.6 (a) $H_3PO_4 + H_2O \rightleftharpoons H_2PO_4^- + H_3O^+$
  $H_2PO_4 + H_2O \rightleftharpoons HPO_4^{2-} + H_3O^+$
  $HPO_4^{2} + H_2O \rightleftharpoons PO_4^{3-} + H_3O^+$
  (b) For each equilibrium, those species in orange are the acids, those in grey are their corresponding conjugate bases.

Chapter 7

7.1 (a) Dotted circles around glycerol components. Remainder are fatty acid components.

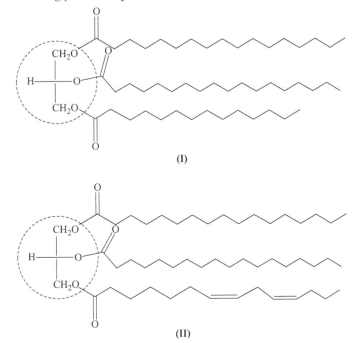

(I)

(II)

(b) Compound 1 is the higher melting one

(c) It is chiral, with the 'middle C' of the glycerol being a chiral centre

(d) It will form micelles

7.2

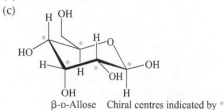

Fischer   Haworth

(a) It is a pyranose

(b) It is an aldose

(c)

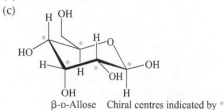

β-D-Allose   Chiral centres indicated by *

(d) It is a D-isomer

(e) β-anomer

7.3 (a) It is a hexose

(b) It is a ketose

(c)

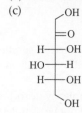

It is a D-isomer

(d)

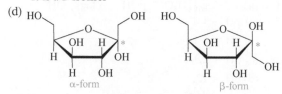

α-form   β-form

Differ in configuration at*

7.4 Amylose has α-1, 4 linkages between glucose monomers whereas cellulose has β-1, 4 linkages. Mammalian enzymes much more readily digest the former compared with the latter

7.5 Amide groups, also called peptide links or peptide bonds

7.6

$$H_3\overset{+}{N}-CH-\overset{O}{\overset{\|}{C}}-\overset{H}{\overset{|}{N}}-CH-\overset{O}{\overset{\|}{C}}-NH-CH-\overset{O}{\overset{\|}{C}}-NH-CH-CO_2^-$$
with substituents CH_3, CH_2-$CONH_2$, CH_2-OH, H

or

$$\bar{O}_2C-CH-NH-\overset{O}{\overset{\|}{C}}-CH-NH-\overset{O}{\overset{\|}{C}}-CH-NH-\overset{O}{\overset{\|}{C}}-CH-\overset{+}{N}H_3$$
with substituents H_3C, H_2C-$CONH_2$, H_2C-OH, H

7.7 Aliphatic/non-polar: Glycine, leucine, isoleucine, proline

Aromatic: Tyrosine

S-containing: Cysteine

Amides: Asparagine, glutamine

7.8 Looking for discussion of β-sheets and α-helices, and the importance of hydrogen bonding

7.9 Description of how the elements of secondary structure interact

7.10 (a) The heterocyclic (purine or pyrimidine) base, the ribose/deoxyribose sugar and the phosphate

(b) Guanine and cytosine are complementary as are adenine and thymine

(c)

(d) DNA contains a deoxyribose sugar along with the heterocyclic bases guanine, cytosine, adenine and thymine. RNA has a ribose sugar and adenine is replaced by uracil

(e) Guanine, cytosine, uracil and thymine

7.11 Nucleotides are phosphorylated nucleosides

7.12 Draw a diagram showing the phosphate diester linkage. Example is:

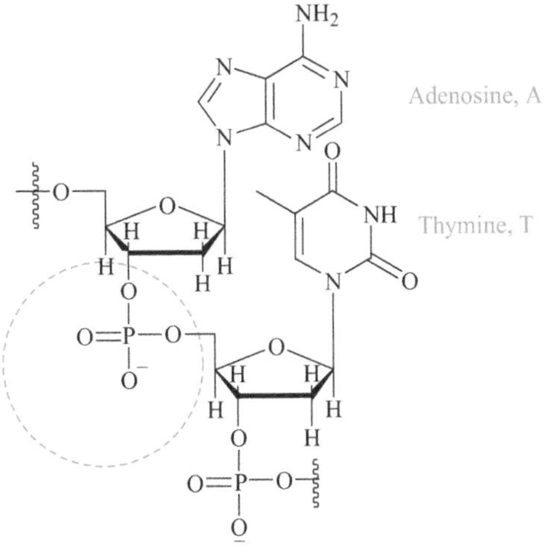

Adenosine, A

Thymine, T

7.13 5′—CGCAGT—3′

Chapter 8

8.1

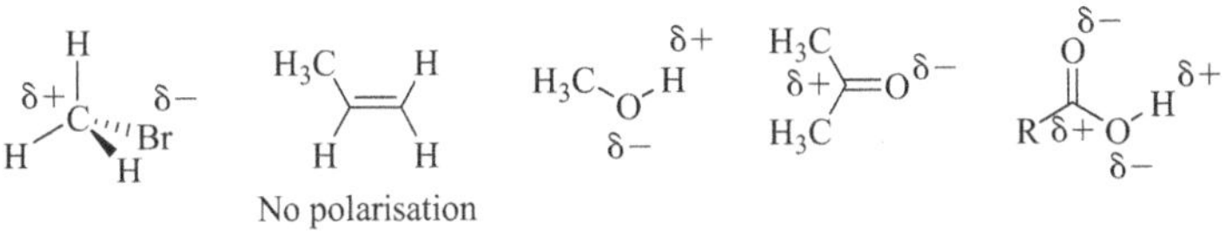

(a)   (b)   (c)   (d)   (e)

8.2 (a) Canonical/resonance structures are approximate 'localised electron bond' structures that can be drawn for molecules possessing delocalised electrons. These structures are *not* true representations of the electronic structure and are the 'average' of all the structures considered the best approximation. The positions of the nuclei are the same in all the resonance structures. Molecules do not 'change' from one resonance structure to another and the structures are not 'in equilibrium' with each other.

For two molecules existing in equilibrium, each has a different structure and there is a continual process of bond breaking and reforming as each transforms to the other.

(b)

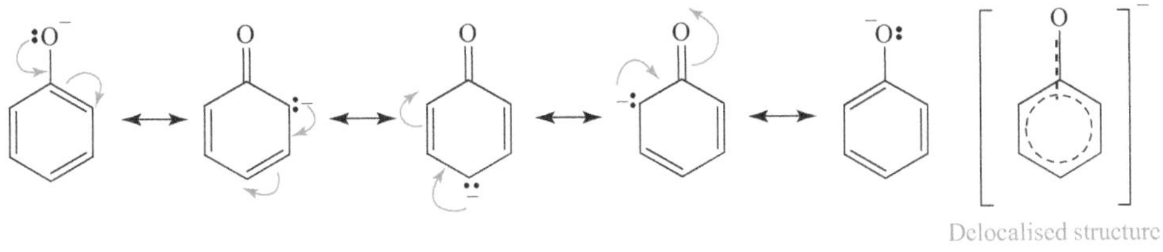

Delocalised structure

(I)

471

Delocalised structure

(II)

(III)

(IV)

(c) The phenolate anion (I) exhibits a lot of resonance stabilisation, as the negative charge is delocalised about the ring. The anion that results from deprotonation of aliphatic alcohols does not experience any resonance stabilisation.

(d) In an amide the C and N bonded together, along with the four atoms directly bonded to them are *coplanar* and the bond angles are 120°. The CN bond has partial double bond character and rotation about this bond is restricted. This therefore affects the structures of polymers that result from linking amides together.

8.3 A molecule or ion that accepts a pair of electrons to make a new covalent bond is called an electrophile; a molecule or ion that donates a pair of electrons to make a new covalent bond is called a nucleophile.

(a) H_3O^+ is an electrophile

(b) CH_3O^- is a nucleophile

(c) CH_3NH_2 is a nucleophile

(d) BF_3 is an electrophile

$$H-Cl \longrightarrow H^+ + \overset{\bullet\bullet}{\underset{\bullet\bullet}{Cl}}{}^-$$

8.4 Since one of the electrons making the bond originally came from the hydrogen, the movement of the pair to the Cl results in removing this electron completely from the H (so it becomes singly

+ charged) and giving it to the Cl which also keeps its original electron. The chlorine gains one NEW electron and therefore gains one negative charge.

8.5 (a) Formic acid (HCOOH)

(b) 2,2-Dichloroacetic acid

(c) Phenol

(d) 4-Nitrophenol

8.6 (a) Addition of HBr to alkenes proceeds via the more stable (lower energy) carbocation. For hept-3-ene, there are two possible intermediate carbocations, both of which are secondary and so they are formed in roughly equal amounts. These yield the products 3- and 4-bromoheptane.

For hept-1-ene, the attack of H^+ at the C1 position gives a secondary carbocation (leading to 2-bromoheptane) which is more stable than the primary carbocation formed by attack of H^+ at the C2 position. So 2-bromoheptane only is formed.

(b) The reaction of HBr with propyne initially yields 2-bromopropene. This then reacts further to give 2,2-dibromopropane. The reaction of propyne with aqueous acid initially forms 2-hydroxypropene, which undergoes rearrangement to acetone (propanone) before further reaction can occur. Acetone does not react with aqueous acid.

8.7 Summarise something like:

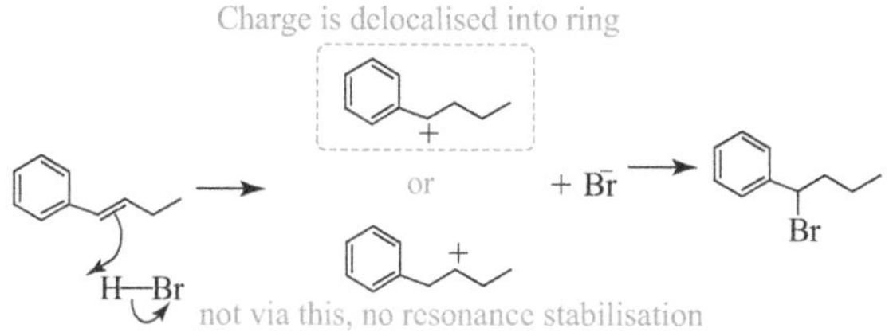

Charge is delocalised into ring
or
+ Br⁻
Br
H—Br
not via this, no resonance stabilisation

8.8 (a) **S**ubstitution **N**ucleophilic (the 1 stands for unimolecular, 2 for bimolecular)

(b) S_N1

Nu⁻
X⁻
R
R¹
R²
Nu

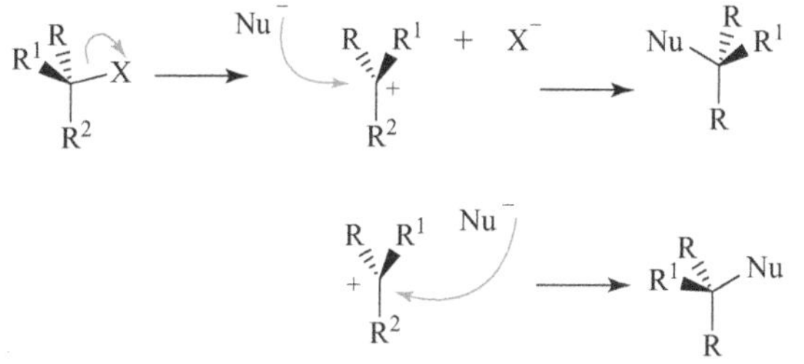

Comment on planar carbocation, and attack by nucleophile equally possible on either side

(c) S_N2

Nu⁻
R
R¹
R²
X
X⁻

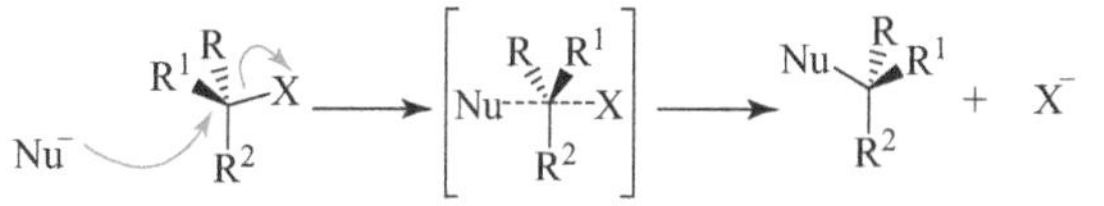

Comment on inversion of configuration for chiral molecules

8.9 (a) E means elimination. E1 is unimolecular elimination, E2 is bimolecular elimination

(b) E1 follows the mechanism:

HO⁻:
H
Br
−:Br⁻
−H₂O
+H₂O
+:Br⁻

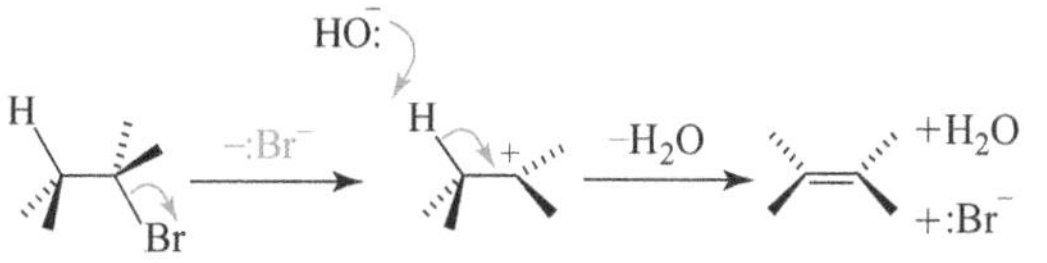

Comment on intermediate formation, similar to S_N1

E2 is illustrated thus:

HO⁻:
H
Br
+H₂O
+:Br⁻

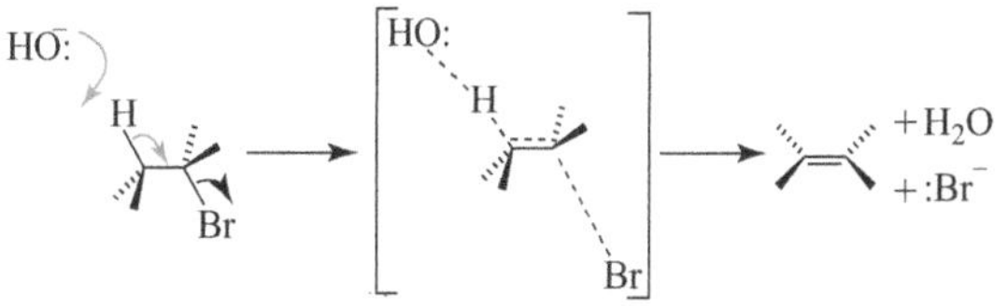

This mechanism proceeds via a transition state

8.10

Cl
+ KOH
OH
+

(a)

Mainly SN1/E1
Cl
CH₃
+ H₂O
OH
CH₃
+
CH₃
+
CH₂

(b)

Mainly SN2/E2
CH₂Cl
+ NaOCH₃
CH₃OH
CH₂OCH₃
+
CH₂

(c)

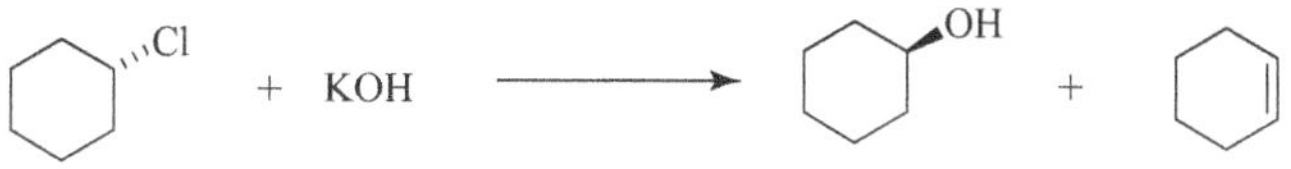

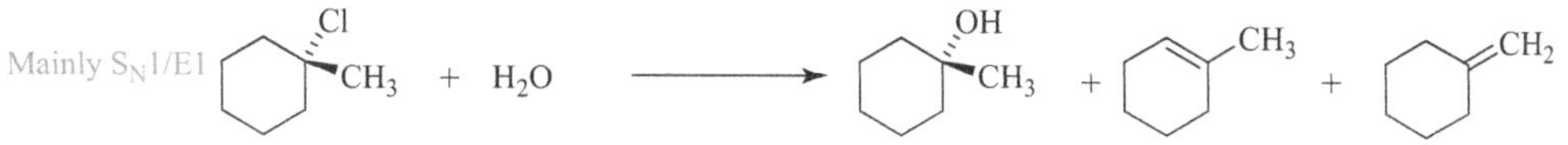

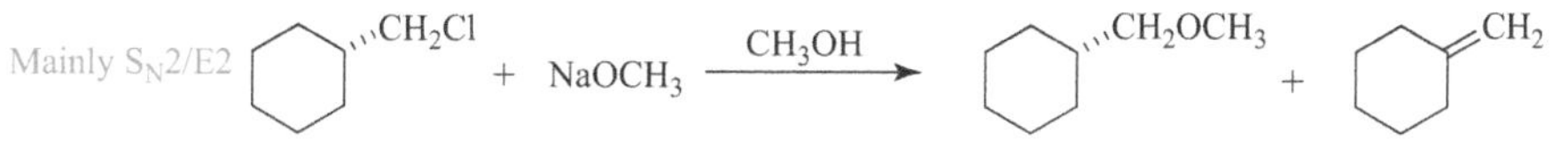

8.11

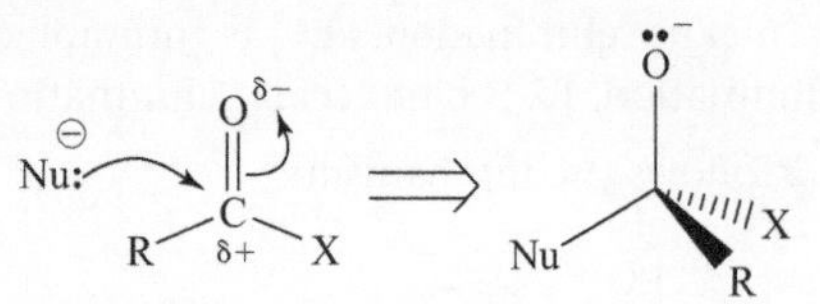

8.12 (a)

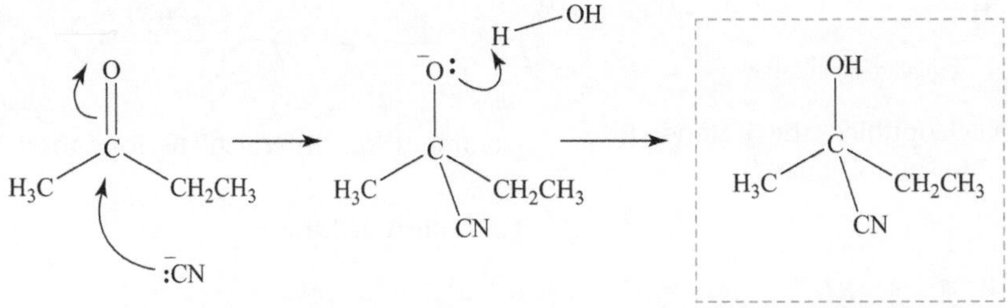

(b)

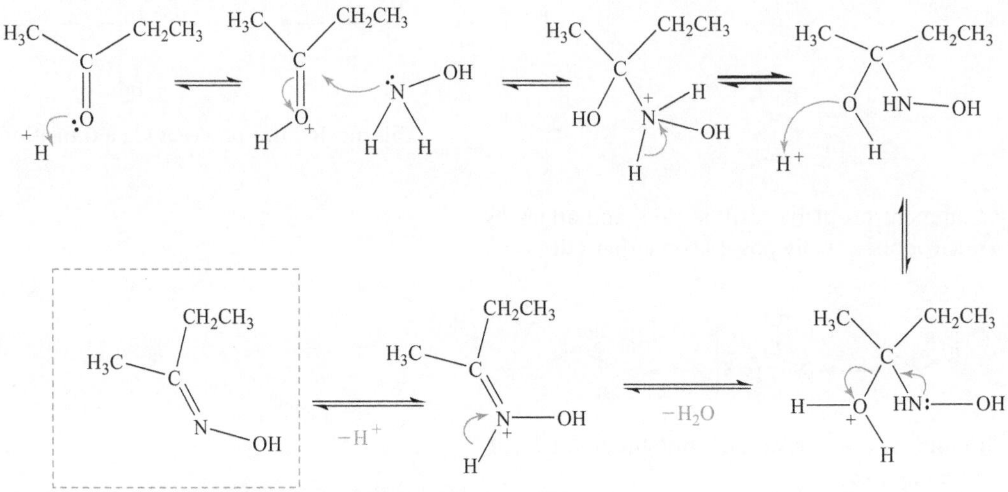

(c) Involves firstly the formation of a hemiacetal then a full acetal:

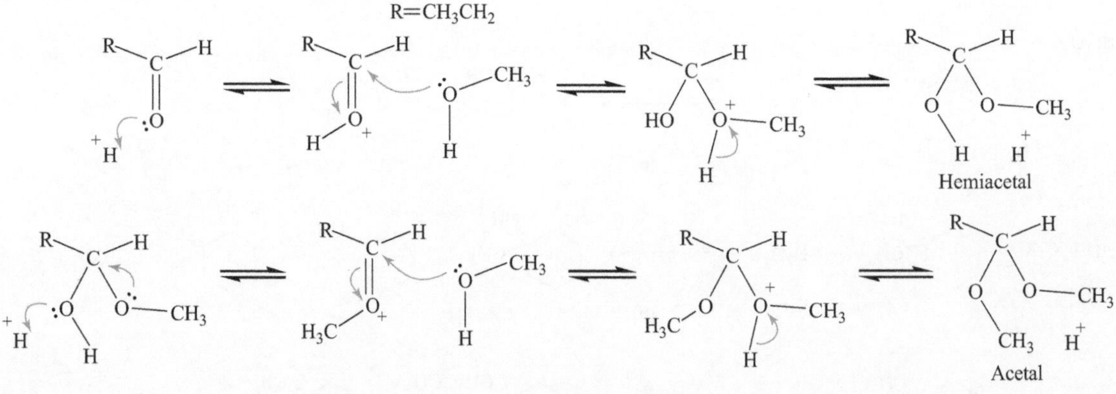

8.13 (a) A full-headed curved (curly) arrow represents the movement of a pair of electrons. The tail shows where they are moving from, the head shows where they are moving to

(b)

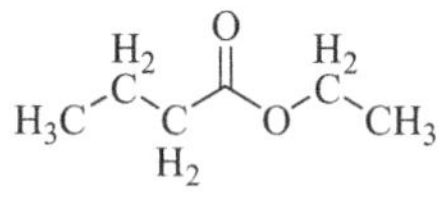

(I)

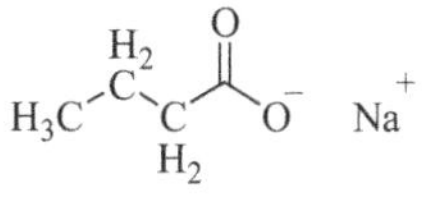

(II)

(c)

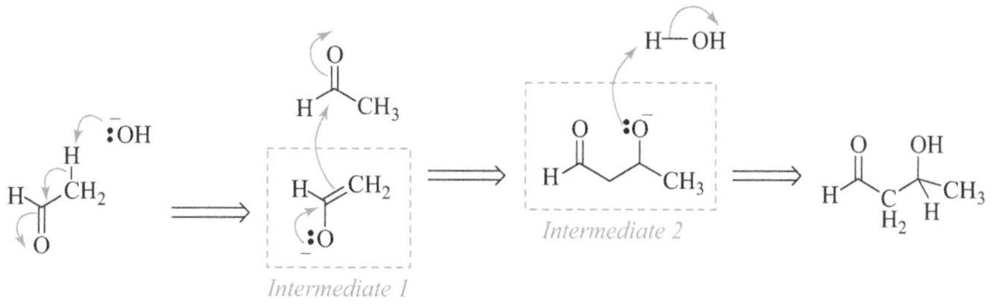
Intermediate 1
Intermediate 2

(I)

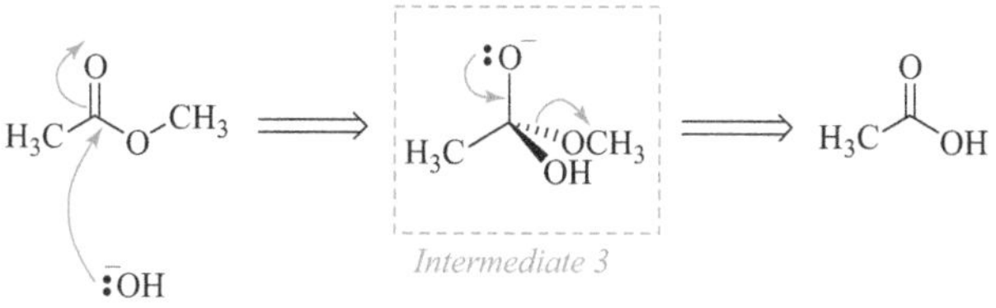
Intermediate 3

(II)

8.14 This is a 'crossed Claisen' reaction

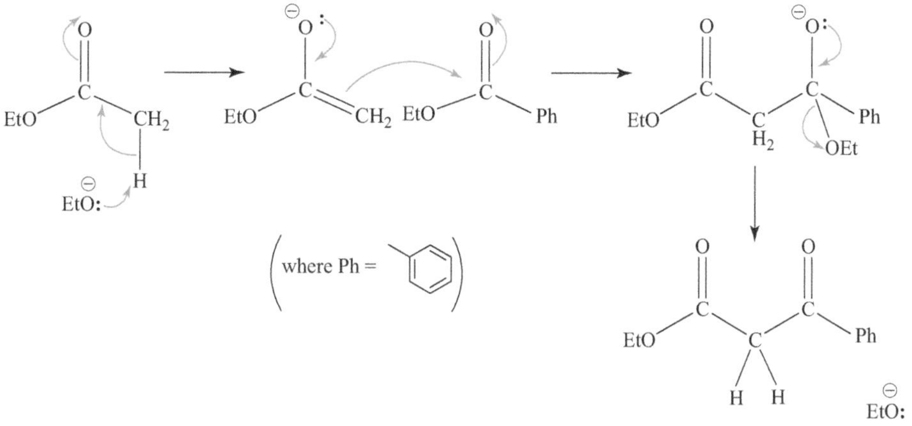
(where Ph =)

8.15 The functional groups in the molecule are circled and named:

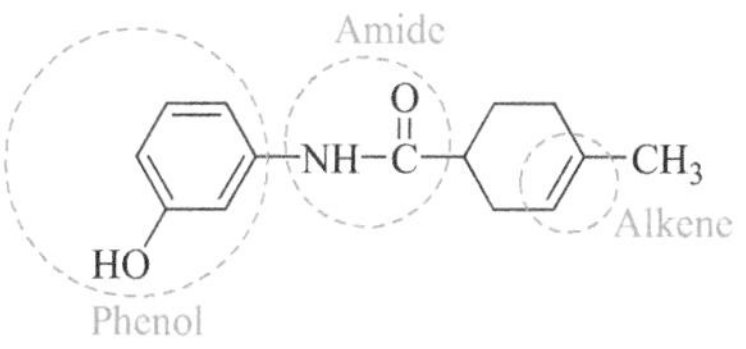
Amide
Alkene
Phenol

(a) Will dissolve a phenol (an acid/base reaction) (heat is needed to hydrolyse amide)

(b) No reaction

(c) Will react with the alkene

(d) Will hydrolyse amide to carboxylic acid

Chapter 9

9.1 (a)

   (i) No change to the rate

   (ii) The rate doubles

   (iii) The rate quadruples

   (iv) The rate increases eightfold

(b) $\dfrac{d[D]}{dt} = -2\dfrac{d[A]}{dt} = -\dfrac{d[B]}{dt} = -\dfrac{2}{3}\dfrac{d[C]}{dt}$

9.2 The rate law is $\dfrac{d[C]}{dt} = k[A]^2[B]$

   The rate constant $k = 5 \times 10^{-2}\,mol^{-2}L^2s^{-1}$

9.3 72 s

9.4 The reaction is first order, and $k = 5.9 \times 10^{-5}s^{-1}$
   or $3.5 \times 10^{-3}\,min^{-1}$

9.5 2.41 hr

9.6 52.9 kJmol^{-1}

9.7 See Box 9.5

9.8 (a)

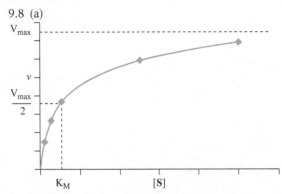

(b) V_{max} is the maximum rate of an enzyme-cata-lysed reaction when the substrate concentration is very high. K_m is the substrate concentration that lets the reaction occur at a rate of $0.5\,V_{max}$, and it is a measure of the affinity that the enzyme has for that substrate. The more tightly bound the enzyme–substrate complex, the smaller the value of K_M.

9.9 (a) $\dfrac{1}{v} = \dfrac{K_M}{V_{max}}\dfrac{1}{[S]} + \dfrac{1}{V_{max}}$

   This is in the form $y = mx + c$, where

   $y = \dfrac{1}{v}$ and $x = \dfrac{1}{[S]}$

   Line crosses x axis when $y = 0$, so:

   $0 = \dfrac{K_M}{V_{max}}\dfrac{1}{[S]} + \dfrac{1}{V_{max}}$

So, $-\dfrac{1}{V_{max}} = \dfrac{K_M}{V_{max}}\dfrac{1}{[S]}$

Hence, $-\dfrac{V_{max}}{V_{max}} = -1 = K_M\dfrac{1}{[S]}$

So, $-\dfrac{1}{K_M} = \dfrac{1}{[S]}$

(b) $K_M = 90.9\,\mu mol\,L^{-1}$ or $9.09 \times 10^{-5}\,mol\,L^{-1}$

Chapter 10

10.1 (a) E^θ cell $= +0.778\,V$

   (b) E^θ cell $= -0.763\,V$

   (c) E^θ cell $= -0.577\,V$

   (d) $E^\oplus$ cell $= +1.140\,V$

10.2 (a) Half cell reactions are:

   Right hand: $Cu^{2+} + 2e^- \rightarrow Cu$

   Left hand: $Fe^{2+} + 2e^- \rightarrow Fe$

   Overall cell reaction:

   $Cu^{2+} + Fe \rightarrow Cu + Fe^{2+}$

   Standard free energy change is
   $\Delta G^\theta = -150\,kJ\,mol^{-1}$

   (b) Half cell reactions are:

   Right hand: $Zn^{2+} + 2e^- \rightarrow Zn$

   Left hand: $2H^{2+} + 2e^- \rightarrow H_2$

   Overall cell reaction:

   $Zn^{2+} + H_2 \rightarrow Zn + 2H^+$

   Standard free energy change is
   $\Delta G^\theta = +147.2\,kJ\,mol^{-1}$

   (c) Half-cell reactions:

   Right hand: $AgCl + e^- \rightarrow Ag + Cl^-$

   Left hand: $Ag^+ + e^- \rightarrow Ag$

   Overall cell reaction:
   $AgCl \rightarrow Ag^+ + Cl^-$

   Standard free energy change is
   $\Delta G^\theta = +55.7\,kJ\,mol^{-1}$

   (d) Half-cell reactions:

   Right hand:
   $O_2 + 4H^+ + 4e^- \rightarrow 2H_2O$

   Left hand:
   $2NAD^+ + 2H^+ + 4e^- \rightarrow 2NADH$

   Overall cell reaction:

   $O_2 + 2H^+ + 2NADH \rightarrow 2H_2O + 2NAD^+$

Standard free energy change,

$\Delta G^{\ominus} = -440.0\,\text{kJ}\,\text{mol}^{-1}$

10.3 0.465 V

10.4 (a) There are several problems, these being:

- H_2 explosive, and requires bulky cylinders;
- difficult to maintain a constant pressure of H_2;
- Pt electrodes poison easily;
- Pt electrodes are non-selective, so other redox processes can happen;
- not easy to make a single pH probe from a cell which uses a salt bridge.

(b) Glass electrodes:

- are selective to H^+;
- do not poison easily;
- can be combined into one probe (glass frit instead of salt bridge);
- do not use gases.

(c) $E^* = 0.098$

pH $= 5.27$

10.5 $E^{\theta}(\text{NAD}^+,\text{H}^+,\text{NADH}) = -0.113\,\text{V}$

10.6 (a) $E^{\oplus}{}_{\text{cell}} = 0.315$ V

(b) $\Delta G^{\oplus} = -60.7\,\text{kJ}\,\text{mol}^{-1}$

(c) Yes, NADH can be used to reduce FAD

10.7 (a) $-9390\,\text{J}\,\text{mol}^{-1}$

(b) 0.097 V

(c) 0.056 V

Chapter 11

11.1 3 and 5

11.2 A group of *organophosphates* that are prepared from phosphoric acid and alcohols.

11.3

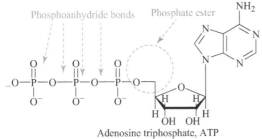

Adenosine triphosphate, ATP

11.4

(a) Donor atom N, monodentate

(b) Bidentate, donor atoms O

(c) Bidentate, donor atoms O

(d) No lone pairs, no ability to coordinate

(e) Donor atom S, monodentate

(f) Three S donor atoms, tridentate

477

11.5 $Co(H_2O)_6^{2+}$ is octahedral, with oxygen as the donor atom of the ligand, whereas $CoCl_4^{2-}$ is tetrahedral, with Cl as the donor atom/ligand. In both cases the cobalt is in the $+2$ oxidation state. The diagram illustrates the situation for each:

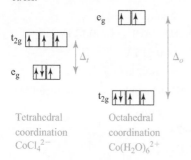

Tetrahedral coordination $CoCl_4^{2-}$

Octahedral coordination $Co(H_2O)_6^{2+}$

$CoCl_4^{2-}$ needs less energy to excite electrons to a higher level than does $Co(H_2O)_6^{2+}$. Hence different colours transmitted.

11.6 An entatic state is 'a state of an atom or group which, due to its binding in a protein, has its geometric or electronic condition adapted for function'. Plastocyanin is a copper protein which cycles between the Cu (I) and Cu (II) states during photosynthesis in plants. Normally Cu (I) would adopt a tetrahedral geometry, binding to soft ligands such as S. Cu (II) is normally square-planar and likes hard ligands with coordinating atoms such as N and O. In plastocyanin the protein fold energy forces the copper ion to accept a distorted geometry and the copper coordination is pseudotetrahedral, with two N ligands (histidine, His) and two S ligands (methionine, Met and cysteine, Cys).

11.7 (a) K^+ and phosphate, PO_4^{3-}

(b) Na^+ and Cl^-

11.8 Ionophores are (organic) compounds that can facilitate the transport of ions across cell membranes. They can bind specific ions, and 'shroud' them, shielding their charge from the nearby environment, thereby allowing them to pass across the hydrophobic membrane interior.

Chapter 12

12.1 (a) Glucose

(b) Pyruvate

(c) It can be converted to acetyl coenzyme A and enters the Krebs cycle (in aerobic respiration). It can also be converted to lactic acid (anaerobic respiration).

(d) The first five steps require a net input of energy into the pathway, in order to break up the 6-carbon glucose into two 3-carbon units. The energy is provided by the hydrolysis of two molecules of ATP to ADP per glucose unit. The second phase of glycolysis produces four molecules of ATP and two molecules of NADH for every glucose molecule used. This is seen as the 'payback' phase, as both ATP and NADH are regarded as energy rich.

12.2 (a) Glyceraldehyde phosphate and dihydroxy-acetone phosphate

(b) Assuming that the provision of reagents such as base and acid are provided by the enzyme, then the mechanism can be summarised thus:

Enolisation of DHAP

Enol attack on GAP

(c) Aldol condensation

(d)

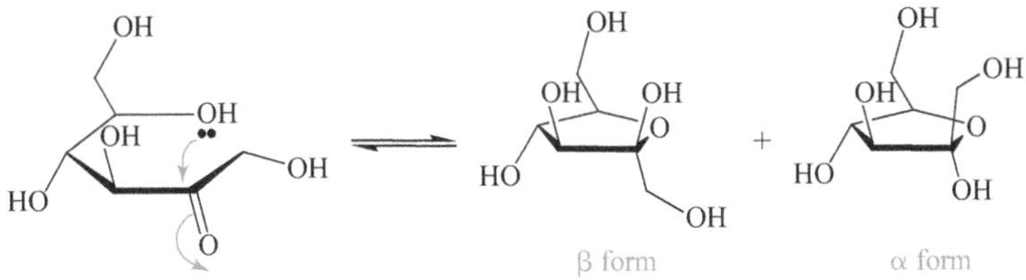
β form
α form

12.3 (a)

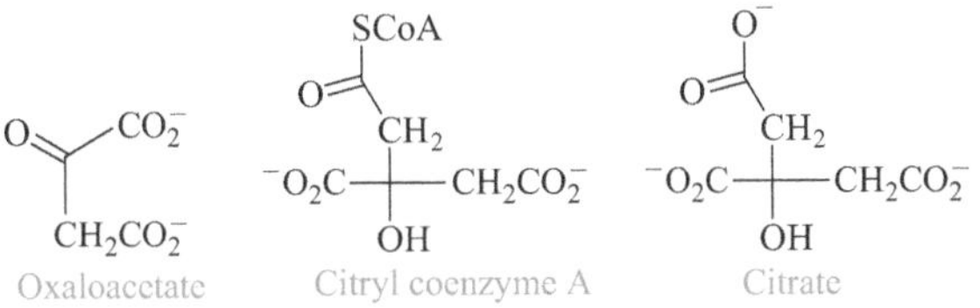
Oxaloacetate
Citryl coenzyme A
Citrate

(b) Claisen condensation.

(c) Assume source of acid/base is the enzyme:

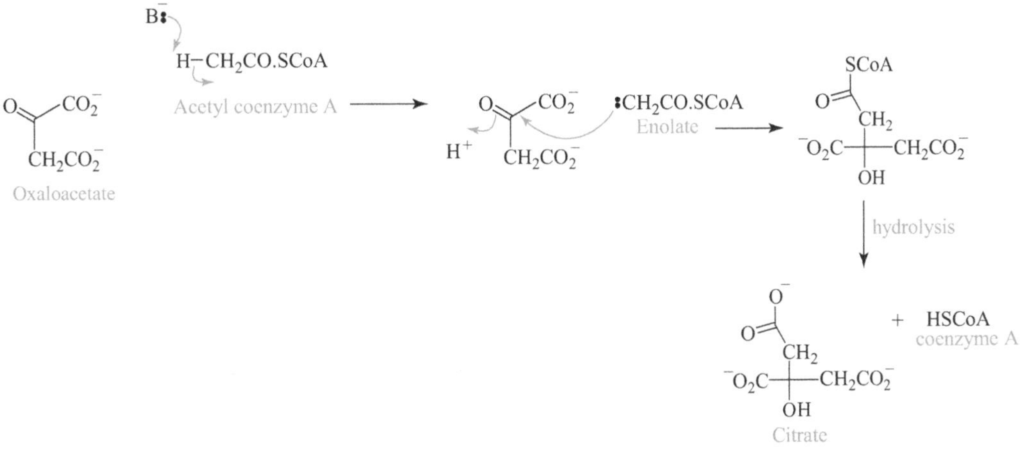
Acetyl coenzyme A
Oxaloacetate
Enolate
hydrolysis
HSCoA
coenzyme A
Citrate

12.4 (a)

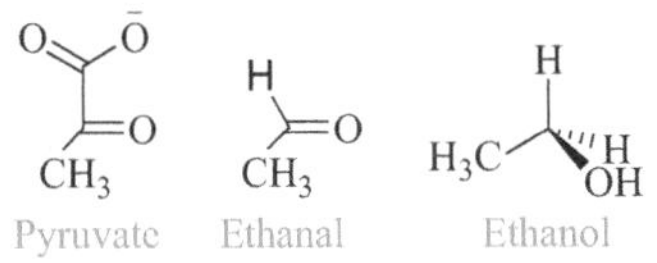
Pyruvate
Ethanal
Ethanol

(b) Step 1 is a decarboxylation and step 2 is a reduction

(c) Step 1 is catalysed by the enzyme pyruvate decarboxylase. This enzyme requires the coenzyme thiamine pyrophosphate (TPP), and TPP is derived from the vitamin thiamine. Step 2 uses the enzyme alcohol dehydrogenase, and the reaction is zinc dependent

Chapter 13

13.1 Ethyl bromide

13.2 (a) $m/z = 74$

(b) $m/z = 43$

(c) $m/z = 59$ suggests loss of CH_3, with detector picking up $[C_2H_3O_2]^+$

$m/z = 43$ suggests loss of CH_3O, detector picking up $[CH_3CO]^+$

$m/z = 31$ likely to be caused by $[CH_3O]^+$

$m/z = 15$ likely to be from $[CH_3]^+$

The structure is CH_3COOCH_3, methyl ethanoate

(d) IR spectroscopy would give a C=O stretch in the 1720–1750 cm^{-1} region.

^{13}C NMR + DEPT spectra would identify two types of methyl carbon plus a carbonyl carbon.

^{1}H NMR spectroscopy would give two distinct single lines, one at a chemical shift close to 2 ppm, the other close to 3.5 ppm.

13.3

Yes   No   Yes
(I)   (II)   (III)

Yes   No
(IV)   (V)   Yes (VI)

13.4 $1.58 \times 10^{-5}\,mol\,L^{-1}$

13.5 $3.57 \times 10^{-4}\,mol\,L^{-1}$

13.6 (a) $(CH_3)_3N$ has no NH stretch and $CH_3CH_2NHCH_3$ does, probably about 3300–3500 cm^{-1}

(b) $CH_3CH_2CCH_3$ (with $\overset{\parallel}{O}$)

will have C=O stretch at about 1750 cm^{-1}. $CH_3CH=CHCH_2OH$ will have O—H stretch at about 3200–3400 cm^{-1}.

(c) $H_2C=CHOCH_3$ will have no peak in C=O region, $CH_3CH_2CH=O$ will.

(d) $CH_2C\equiv N$ will have $C\equiv N$ stretch at ca. 2200–2300 cm^{-1}

will have NH stretch

(e) CO_2H will have C=O stretch (carboxylic acid) and N—H stretch

CH_3 will have two intense peaks in range 1500–1570 and 1300–1370 cm^{-1}

13.7 Radiofrequency

13.8 Molecule symmetrical, so only two sets of signals

- Integrals: Consistent with 3:2 ratio for signals at 1.2 ppm and 3.5 ppm respectively.

- Shifts: Signal at 3.5 ppm is most deshielded, so will derive from H nearest to O.

- Couplings: Signal at 3.5 ppm is a quartet. Applying the 'N + 1 rule' (Pascal's triangle) it follows that coupling is to three equivalent H nuclei.

Signal at 1.2 ppm is a doublet – coupled to two equivalent H nuclei.

All data consistent with structure.

13.9 (a) (i) C_7H_7Br

(ii)

(b) (i) The UV peak is consistent with the presence of an aromatic group

(ii) The ^{1}H NMR signal at $\delta = 2.3$ ppm is consistent with a methyl group, and that at $\delta = 7.0$ is consistent with a disubstituted benzene ring

(iii) The presence of only five signals in the ^{13}C NMR spectrum suggests the presence of some symmetry in the molecule, which contains seven carbon atoms

(c) The compound is likely to be

Appendix 3 Periodic table of the elements

1	2	3	4	5	6	7	8	9	10	11	12	13	14	15	16	17	18
1 H 1.008																	2 He 4.003
3 Li 6.941	4 Be 9.012											5 B 10.811	6 C 12.011	7 N 14.007	8 O 15.999	9 F 18.998	10 Ne 20.180
11 Na 22.99	12 Mg 24.31											13 Al 26.98	14 Si 28.09	15 P 30.97	16 S 32.07	17 Cl 35.45	18 Ar 39.95
19 K 39.10	20 Ca 40.08	21 Sc 44.96	22 Ti 47.88	23 V 50.94	24 Cr 52.00	25 Mn 54.93	26 Fe 55.85	27 Co 58.93	28 Ni 58.69	29 Cu 63.55	30 Zn 65.39	31 Ga 69.72	32 Ge 72.61	33 As 74.92	34 Se 78.96	35 Br 79.90	36 Kr 83.80
37 Rb 85.47	38 Sr 87.62	39 Y 88.91	40 Zr 91.22	41 Nb 92.91	42 Mo 95.94	43 Tc (98)	44 Ru 101.07	45 Rh 102.91	46 Pd 106.42	47 Ag 107.87	48 Cd 112.41	49 In 114.82	50 Sn 118.71	51 Sb 121.75	52 Te 127.60	53 I 126.90	54 Xe 131.29
55 Cs 132.91	56 Ba 137.33	57 La 138.91	72 Hf 178.49	73 Ta 180.95	74 W 183.85	75 Re 186.21	76 Os 190.23	77 Ir 192.22	78 Pt 195.08	79 Au 196.97	80 Hg 200.59	81 Tl 204.38	82 Pb 207.2	83 Bi 208.98	84 Po (209)	85 At (210)	86 Rn (222)
87 Fr (223)	88 Ra 226.03	89 Ac 227.03															

| | | 58
Ce
140.12 | 59
Pr
140.91 | 60
Nd
144.24 | 61
Pm
(145) | 62
Sm
150.36 | 63
Eu
151.97 | 64
Gd
157.25 | 65
Tb
158.93 | 66
Dy
162.50 | 67
Ho
164.93 | 68
Er
167.26 | 69
Tm
168.93 | 70
Yb
173.04 | 71
Lu
174.97 | |
| | | 90
Th
232.04 | 91
Pa
231.04 | 92
U
238.03 | 93
Np
237.05 | 94
Pu
(244) | 95
Am
(243) | 96
Cm
(247) | 97
Bk
(247) | 98
Cf
(251) | 99
Es
(252) | 100
Fm
(257) | 101
Md
(258) | 102
No
(259) | 103
Lr
(260) | |

(For radioactive elements that do not occur in nature, the mass number of the most stable isotope is given in parentheses.)

481

Index

Note: Page numbers followed by (T), (F) or (B) indicate a table, figure or boxed material, respectively.